The End of the Beginning

Life, Society and Economy
on the Brink
of the Singularity

Ben Goertzel and Ted Goertzel, Editors

THE END OF THE BEGINING Copyright © 2015 by Ben Goertzel and Ted Goertzel.

All rights reserved. Printed in the United States of America. No part of this book may be used or reproduced in any manner whatsoever without written permission except in the case of brief quotations embodied in critical articles or reviews.

For information contact; address info@humanityplus.org

Cover design by Zarko Paunovic

ISBN: 0692457666
First Edition: February 2015

10 9 8 7 6 5 4 3 2 1

CONTENTS

PART ONE: Where are we going? When will we get there?   5

INTRODUCTION by Ted Goertzel and Ben Goertzel   6

CHAPTER ONE PREDICTING THE AGE OF POST-HUMAN INTELLIGENCES by Ted Goertzel and Ben Goertzel   32

    DIALOGUE 1.1 WITH DAVID WEINBAUM (WEAVER), TED GOERTZEL, BEN GOERTZEL AND VIKTORAS VEITAS,   57

CHAPTER TWO: A TALE OF TWO TRANSITIONS by Robin Hanson   70

CHAPTER THREE LONGER LIVES ON THE BRINK OF GLOBAL POPULATION CONTRACTION: A REPORT FROM 2040 by Max More   81

PART TWO: The Future of the Human Body   95

CHAPTER FOUR IMPLANTING POST-HUMAN INTELLIGENCE IN HUMAN BODIES by John Hewitt   96

    DIALOGUE 4.1 WITH JOHN HEWITT, BEN GOERTZEL AND TED GOERTZEL   110

CHAPTER FIVE THE SINGULARITY AND THE METHUSELARITY: SIMILARITIES AND DIFFERENCES by Aubrey de Grey   113

    DIALOGUE 5.1 WITH AUBREY DE GREY, TED GOERTZEL AND BEN GOERTZEL   127

PART THREE: The Rollout and the Impact of Inteligent Robots   130

CHAPTER SIX ROBOTICS AND AI: IMPACTS FELT ON EVERY ASPECT OF OUR FUTURE WORLD by Daryl Nazareth   131

    DIALOGUE 6.1 WITH DARYL NAZARETH AND BEN GOERTZEL   149

CHAPTER SEVEN ROBOTICS, AI, THE LUDDITE FALLACY AND THE FUTURE OF THE JOB MARKET by Wayne Radinsky   159

    DIALOGUE 7.1 WITH WAYNE RADINSKY AND BEN GOERTZEL   186

CHAPTER EIGHT MORAL RESPONSIBILITY AND AUTONOMOUS MACHINES by David Burke   196

CHAPTER NINE HOW WILL THE ARTILECT WAR START? by Hugo de Garis   213

    DIALOGUE 9.1 WITH HUGO DE GARIS AND BEN GOERTZEL   234

PART FOUR: The Global Brain and the Emerging World of Views   242

CHAPTER TEN RETURN TO EDEN? PROMISES AND PERILS ON THE ROAD TO GLOBAL SUPERINTELLIGENCE by Francis Heylighen   243

DIALOGUE 10.1 WITH FRANCIS HEYLIGHEN AND BEN GOERTZEL	307
CHAPTER ELEVEN DISTRIBUTING COGNITION: FROM LOCAL BRAINS TO THE GLOBAL BRAIN by Clément Vidal	**325**
EPILOGUE TO CHAPTER ELEVEN A DAY IN 2060 BY CLÉMENT VIDAL	307
DIALOGUE 11.1 WITH CLÉMENT VIDAL, BEN GOERTZEL AND TED GOERTZEL	359
CHAPTER TWELVE A WORLD OF VIEWS: A WORLD OF INTERACTING POST-HUMAN INTELLIGENCES by Viktoras Veitas and David Weinbaum (Weaver)	**373**
DIALOGUE 12.1 WITH VIKTORAS VEITAS, DAVID WEINBAUM (WEAVER), BEN GOERTZEL AND TED GOERTZEL	428

PART FIVE: Globalization and the Path to the Singularity — 450

CHAPTER THIRTEEN CHINESE PERSPECTIVES ON THE APPROACH TO THE SINGULARITY by Mingyu Huang	**451**
DIALOGUE 13.1 WITH MINGYU HUANG AND TED GOERTZEL	457
CHAPTER FOURTEEN AFRICA TODAY AND THE SHADOW OF THE COMING SINGULARITY by Hruy Tsegaye	**465**
DIALOGUE 14.1 WITH HRUY TSEGAYE, BEN GOERTZEL AND FRANCIS HEYLIGHEN	485

PART SIX: The Future of Money — 502

CHAPTER FIFTEEN THE WORLD'S FIRST DECENTRALIZED SYSTEM FOR FINANCIAL AND LEGAL TRANSACTION by Chris Odom	**503**
CHAPTER SIXTEEN BEYOND MONEY: OFFER NETWORKS, A POTENTIAL INFRASTRUCTURE FOR A POST-MONEY ECONOMY by Ben Goertzel	**522**
DIALOGUE 16.1 WITH FRANCIS HEYLIGHEN, DAVID WEINBAUM (WEAVER) AND BEN GOERTZEL	550

PART SEVEN: Human Society and Human Nature in the era of Abundance 555

CHAPTER SEVENTEEN SOUSVEILLANCE AND AGI by Ben Goertzel and Stephan Vladimir Bugaj	**556**
DIALOGUE 17.1 WITH BEN GOERTZEL, TED GOERTZEL, HRUY TSEGAYE AND MINGYU HUANG	577
CHAPTER EIGHTEEN THE FUTURE OF HUMAN NATURE by Ben Goertzel	**593**
DIALOGUE 18.1 WITH CLÉMENT VIDAL AND BEN GOERTZEL	619
CHAPTER NINETEEN CAPITALISM, SOCIALISM, SINGULARITARIANISM by Ted Goertzel	**621**
DIALOGUE 19.1 WITH BEN GOERTZEL, HRUY TSEGAYE AND TED GOERTZEL	640
CHAPTER TWENTY TOWARD A HUMAN-FRIENDLY POST-SINGULARITY WORLD by Ben Goertzel	**652**
DIALOGUE 20.1 WITH HRUY TSEGAYE, CLÉMENT VIDAL, AARON NITZKIN AND BEN GOERTZEL	687
CHAPTER TWENTY-ONE LOOKING BACKWARD FROM 2100 by Ben Goertzel and Ted Goertzel	**707**

Part One:

Where are we going?

When will we get there?

Introduction

By Ben Goertzel and Ted Goertzel

Ted Goertzel is professor of sociology emeritus at Rutgers University. He recently published **"The Path to More General Artificial Intelligence"** in the Journal of Experimental & Theoretical Artificial Intelligence. He is Ben Goertzel's father. His web site is at: **http://crab.rutgers.edu/~goertzel** and he can be reached at **tedgoertzel@gmail.com**

Ben Goertzel is Chief Scientist of the financial prediction firm Aidyia Holdings; Chairman of AI software company Novamente LLC and bioinformatics company Biomind LLC; Chairman of the Artificial General Intelligence Society and the OpenCog Foundation; Vice Chairman of futurist nonprofit Humanity+; Scientific Advisor of biopharma firm Genescient Corp.; Advisor to the Singularity University and Singularity Institute; Research Professor in the Fujian Key Lab for Brain-Like Intelligent Systems at Xiamen University, China; and general Chair of the Artificial General Intelligence conference series. His research work encompasses artificial general intelligence, natural language processing, cognitive science, data mining, machine learning, computational finance, bioinformatics, virtual worlds and gaming and other areas. He has published a dozen scientific books, 100+ technical papers, and numerous journalistic articles. He can be reached at **ben@goertzel.org**.

The coming of a Technological Singularity is one of the most exciting and controversial predictions to emerge in recent decades. As posited by influential writers and thinkers such as Ray Kurzweil

End of the Beginning

(2006), Vernor Vinge (1993), and Peter Diamandis (Diamandis and Kotler 2012), this will be a point in time when revolutionary advances in science and technology happen too rapidly for the human mind to comprehend. After the Singularity, these pundits predict, robots or other machines will have greater general intelligence than humans. These post-human intelligences would be able to 3D print any form of ordinary matter at low cost. They could cure diseases and perhaps even abolish aging. On the other hand, there are also darker possibilities – they could decide to wipe out human beings altogether, or just keep a few of us in a zoo for their amusement.

This sounds like science fiction, and one reasonable approach to exploring these issues is to write science fiction books or stories, or make science fiction movies. SF has proved remarkably prescient at foreseeing the advent of new technologies. However, fiction is generally based on the need to hold the reader's or viewer's attention, and manipulate their emotions. It is not necessarily the best lens through which to view the actual future. One finds, for instance, a preponderance of stories in which a human hero struggles valiantly against the insidious or seductive machines, which appear suddenly on the scene as the result of an evil genius, or time travel, or some other story mechanism. The sudden emergence of radical new technologies moves a story along excitingly. In reality, though, we think it more likely that a Singularity, if it happens, will unfold step by step over a period of at least years and probably decades. A time-span of decades is effectively instantaneous on the time-scale of human history, let alone the scale of terrestrial geology or the evolution of the universe. But yet, it feels like a long time to human beings as they grow up and go through their lifespans. In terms of the advance of AI, robotics, nanotech, synthetic biology and other radical technologies, an unfolding over decades is long enough for a broad spectrum of humans and human institutions to study what's going on and take action.

End of the Beginning

In the past, humanity waited until a new technology was introduced before taking steps to adjust to it. No one really prepared for the effects of the steam engine, the telephone, the automobile or the personal computer. But if the Singularity is anything like the most enthusiastic futurists predict, that won't be possible. A Singularity, by its nature, would be too fast moving and too complex for humans to control. Just as a cockroach has little chance of predicting the next step in consumer electronics, we humans, in our current form, would have little chance of predicting any of the particulars of post-Singularity intelligence and society.

But even if this is true we still have time, perhaps a generation or two, to use our human intelligence to shape the future. That is the challenge we posed to the authors in this book. We invited a group of leading scientists and scholars to share their thoughts about the years *before* the postulated Singularity. Even if the Singularity does not materialize quite as expected, which is altogether possible, few knowledgeable people doubt that the next few decades will involve very rapid advances in artificial intelligence (AI), biotechnology, nanotechnology, and other futuristic domains. The authors in this book approach this critical period from the perspectives of philosophy, biology, computer science, economics, politics, psychology, sociology and other fields of human inquiry. They do not always agree, and we have included dialogues with many of the authors to build on and learn from the differences.

Much of our reason for putting together this book is our awareness that, as Abraham Lincoln said, "The best way to predict the future is to create it." While certain historical trends may be hard to avoid, it's still true that we, together, are the ones creating the future. The Singularity we experience will be the Singularity we shape, and will be guided by the Singularity we envision. Exploring different visions of how the Singularity may unfold is one way of collectively guiding the process of creating our future.

End of the Beginning

Questions Addressed by the Chapters Herein

Some of the authors presented here articulate broad general visions of life on the brink of the Singularity. Others focus on specific topics within their areas of expertise and special interest. Topics addressed by many of the authors include:

- Computing and communication technologies with capabilities far beyond today's
- Technology enabling hybridization of humans and machines in various ways
- The drastic reduction of material scarcity, such that the obtaining of the goods and services we work to afford today, will be a trivial matter for nearly everyone
- Medical advances that dramatically extend lifespan and reduce the prevalence of disease
- AI technology with problem-solving capability and general intelligence at least rivaling that of the human mind

The relatively rapid advent of these and other allied phenomena is what characterizes the core concept of a Technological Singularity, regardless of which future visionary you're listening to.

But beyond these commonalities, there are significant differences in the way the various chapter authors envision the Singularity and the path theretoward. Among other possibilities, some key questions on which the authors differ are:

- Artificial General Intelligence (AGI) fairly rapidly achieves massively superhuman intelligence, or does it remain somewhere in the vicinity of the human level?
- Will some sort of global AGI Nanny emerge, providing control or regulation of intelligence on the planet, or does governance remain in the hands of (some form of) humans?
- To what extent will a Global Brain with its own coherent, emergent intelligence arise and become a dominant actor on the planet, as opposed to the main nexus of choice and causation being individual humans or human-scale AGIs?

End of the Beginning

- To what extent will "mindplexes" or group minds emerge, perhaps on a smaller scale than a Global Brain?
- Will future humans have an experience of scarcity or abundance? That is: will future humans react to the abundance of free "basic needs," as understood today, with a fixation on competing to acquire more advanced goods and services that remain scarce even as a Singularity approaches, and maybe even thereafter?
- What will people do all day, if they no longer have a need to work in order to acquire scarce resources?
- How will the exchange of desired scarce resources, if any exist, occur in the future? With some future form of money? Or via some different sort of system?
- Will privacy exist in the future?
- Will humans be annihilated by advanced machines?
- Will there be large-scale military conflicts between those advocating accelerating technological change, and those opposing it?

These are big, important questions, and the faint-hearted may find it troubling that the pundits whose opinions are gathered here collectively have fairly little agreement on the answers to any of them – but that's the situation we find ourselves in. There is no one single vision of the Singularity at present, no Canonical Techno-Transcendence. There is a fascinating spectrum of possibilities, which we must all think through together, as we create our mutual future.

Finally, we wish to clarify that, in these pages, we are mainly concerned with the future development of **human beings** (granted that cyborgification and other possibilities may expand the notion of humanity somewhat as the Singularity approaches). Any one of the scenarios listed above could be coupled with AGIs that advance well beyond the human level and move on to some other level of being, or fly off in space, etc. The possible socioeconomic and psychological structure of these hypothetical post-human AGI

10

End of the Beginning

communities is a fascinating topic, but not the focus of this book. Here we will concentrate on the relatively near-term future of human beings, in scenarios where technology continues to advance very rapidly toward a Technological Singularity of one form or another.

Anticipating a Half Century of Technological Change

There is no such thing as certainty regarding the timing of future events. Human history is littered with overly optimistic predictions (weren't we supposed to have flying cars and home service robots by now?), but also with overly pessimistic critical dismissals (such as the experts who argued the infeasibility of human flight or the atomic bomb, sometimes very shortly before these things were achieved). One of the problems is that people often make predictions based on their intuitions or gut feelings without giving the matter any systematic thought or making much effort to analyze relevant data. Sometimes people issuing predictions may go to extremes that aren't really warranted in an effort to scare people or motivate them to change.

One of the standard ways to assess future developments, especially in technological areas, is to survey experts working in the field. Usually groups of experts are surveyed and then given the opportunity to respond to each other's predictions. In 2009, together with Seth Baum, we did a survey of this sort at the Artificial General Intelligence conference in Washington DC, gathering opinions from the AI experts in attendance about when highly advanced AGI systems might appear on the scene. We asked the participants to give us their best estimates of when four milestones would be accomplished: (1) an AGI would pass the Turing Test, (2) an AGI would accomplish a Nobel Prize-worthy scientific discovery, (3) an AGI would achieve the level of general practical intelligence of a third-grade student and (4) a computer would achieve super-human intelligence. We asked them to give us confidence levels for their predictions ranging from 10% (it's possible) to 50% (it's as likely as not) to 90% (it's almost certain). The median estimates for each level of confidence were as follows:[1]

Median Estimates for Milestones by Levels of Confidence

Milestone	10%	25%	50%	75%	90%
Turing Test	2020	2030	2040	2050	2075
Third Grade	2020	2025	2030	2045	2075
Nobel Science	2020	2030	2045	2080	2100
Super Human	2025	2035	2045	2080	2100

If we take the 50% confidence level as our best estimate, it coincides rather closely with Ray Kurzweil's prediction of 2045 for the Singularity. It also coincides with the next period of rapid technological advance predicted by historians who study long cycles of technological change, as reviewed in the next chapter of this book.

In an online survey with similar questions in 2012, Müller and Bolstrom found a median confidence level of 2040 for the accomplishment of "high-level machine intelligence."[2] Their survey included several groups of respondents: attendees at the AGI conference in Oxford, England, in 2012, attendees at two conferences in Greece, and the top 100 writers on artificial intelligence as rated by the Microsoft Academic data base. The median estimates by the four groups were similar, as was the range of responses.

It would be interesting to re-do these surveys sometime in the next few years. On an anecdotal level, it seems that 5-10 years ago if one suggested to a typical tech-industry type the possibility that human-level AGI might be coming within 10-20 years, they'd raise

---

[1] Seth Baum, Ben Goertzel, and Ted Goertzel, "How Long Until Human-Level AI?: Results from an Expert Assessment," Technological Forecasting and Social Change, Vol 78, No 1, January 2011, pages 185-195

[2] Vincent Müller and Nic Bostrom (forthcoming), 'Future progress in artificialintelligence: A poll among experts', in Vincent C. Müller (ed.), Fundamental Issues of Artificial Intelligence (Synthese Library; Berlin: Springer).

their eyebrows in skepticism, with the attitude that "sure, it's coming, but probably not that fast." Now in 2014, if one raises the same possibility to the same sort of person, one has fair odds of getting an answer like "I don't think it will take that long; Google and IBM are already working on it and who knows, maybe they can pull it off within a few years."

Artificial General Intelligence is an important focus for prediction because it is a key technology with the potential to enable the full gamut of other Singularitarian technologies. But there are many other exciting technologies on the horizon. A recent survey by Pew Research found that "the American public anticipates that the coming half century will be a period of profound scientific change, as inventions that were once confined to the realm of science fiction come into common usage" (Smith 2014). Developments that many Americans anticipated in the next fifty years included:

- Organs for transplants grown in laboratories (expected by 81% of Americans)
- Computer art indistinguishable from that produced by humans (51%)
- Teleportation of objects (39%)
- Colonization of planets other than Earth (31%)

In the same survey, people were skeptical about the value of many likely technological changes:

- 65% did not want robots to become the primary care givers for the elderly
- 63% did not want drones flying through U.S. airspace
- 53% did not want people to wear implants that would constantly give them information about the world around them
- 66% did not want prospective parents to have the ability to alter the DNA of their children
- 78% would not eat meat grown in a laboratory
- 50% would not ride in a driverless car (48% would, 2%

uncertain)
- 72% would refuse a brain implant to increase their intelligence

Of all the choices they were offered, the most popular was innovations to cure disease and increase health and longevity. Of course, no one knows for sure which of these technologies are coming soon, or how people will react when they are here. But the expectation that people living today will see many of these changes is gaining steam among experts and the general public.

Chapters & Conversations

The process of creating this book began by collecting formal essays, some of which are quite general and abstract, others of which focus on particular technological innovations. As we read the essays, we found that we had questions for the authors, some of which led to very interesting email exchanges. We decided to edit some of these email discussions into Question and Answer Dialogues. These dialogues follow the chapters, which are organized into the following sections.

Part One: Where are we going? When will we get there?

Following the introduction you are now reading, the first section of the book presents three chapters giving broad (speculative, yet rationally considered) overviews of the potential developments during the next century.

Chapter One: Predicting the Age of Post-Human Intelligences by Ted Goertzel and Ben Goertzel.

Scientific futurism has had some significant successes as well as some well-known bloopers. In this chapter, five traditions are examined for insight into the coming of the age of post-human intelligence: (1) environmental futurism, (2) Kondratiev long-wave analysis, (3) generational cycle analysis, (4) geopolitical futurism and (5) the study of technological revolutions. Three of these

End of the Beginning

traditions offer similar predictions leading us to predict a period of intense technological innovation in the 2040's and another in the 2100's. If these theories are correct, Artificial General Intelligence and The Singularity are likely to come in one of these periods, depending on the success of engineering models currently being completed and on the availability of funding to implement them.

The Chapter is followed by a **Dialogue** between Weaver (David Weinbaum), Ted Goertzel, Ben Goertzel and Viktoras Veitas follows the chapter.

Chapter Two: A Tale of Two Transitions by Robin Hanson

This chapter compares and contrasts two quite different scenarios for a future transition to a world dominated by machine intelligence. One scenario is based on continuing on our current path of accumulating better software, while the other scenario is based on the future arrival of an ability to fully emulate human brains. The first scenario is gradual and anticipated. In it, society has plenty of warnings about forthcoming changes, and time to adapt to those changes. The second scenario, in contrast, is more sudden, and potentially disruptive.

Chapter Three: Longer Lives on the Brink of Global Population Contraction: A Report from 2040 by Max More.

Written from the perspective of an analyst in 2040, this essay explains why the world's population is shrinking due to birth rates declining more than many experts had predicted. The population would have shrunk even more if the average life span had not increased. Lower birth rates have meant that less expenditure is needed for education, and a declining population places less stress on the environment. But economic growth is slower with a declining population, especially if the proportion of retired people increases. Fortunately, the health of elderly people has improved, and older people continue to be economically active later in life. Life extension research is now valued as one means of slowing the rate

End of the Beginning

of population decline.

Part Two: The Future of the Human Body

Our focus here is on human futures in the period leading up to the Singularity – and in this context, the future of the human body is a central concern. The two chapters in this section deal with the enhancement and improvement of the human body's capabilities and the extension of the human body's lifespan – two obviously critical issues that the biotechnology community is currently working very hard to address.

Chapter Four: Implementing Post-Human Intelligence in Human Bodies by John Hewitt.

John Hewitt's chapter explores the future of medical implants for conquering disease and enhancing functionality. Developments he foresees in the relatively near term are intelligent, adaptive implants that control obesity, anorexia, blood sugar, seizure, mood or anxiety; and implants that send messages to the user when there is a problem or they need attention. Imagine getting an SMS, or a silent "in the head" verbal message, from your liver when it's feeling a bit of strain, or from your limbic system when it notices you in the early stages of an outburst of bad temper.

The question of openness versus commercial, proprietary closed-ness is considered, and it is conjectured implant technology will develop more quickly if its development proceeds in the open, so that the future of peoples' bodies lies more directly in their own hands. Open source implant technology could viably lead to a large, vibrant community of biohackers and grinders, seeking to go beyond implants that correct single problems, and (self-)experimenting with the use of multiple interacting implants achieving synergetic effects.

The chapter is followed by a **Dialogue** between John Hewitt and Ben Goertzel.

End of the Beginning

Chapter Five: The Singularity and the Methuselarity: Similarities and Differences by Aubrey de Grey.

Biogerontology pioneer Aubrey de Grey's chapter reviews his concept of the Methuselarity, a sister to the Singularity hypothesis. De Grey has done more than anyone else to advocate the scientific plausibility of conquering aging and extending the lifespan of the human body indefinitely. A decade ago the quest for radical longevity was an eccentric message with little mainstream acceptance; today in 2014, with Google's Calico project and Craig Venter's Human Longevity Incorporated initiative, it is increasingly broadly accepted as a reasonable thing to think about and work on. This shift is largely due to Aubrey's tireless campaigning, his careful articulation of the known science regarding the biology of aging, and his creative thinking regarding potential solutions to the various aspects of the aging problem.

De Grey's Methuselarity is, simply, the point in time at which: If you are alive at that point, then with reasonable odds, successive improvements in life extension technology will allow you to remain alive for an extremely long time (say, thousands of years at least). As he puts it, "Aging, being a composite of innumerable types of molecular and cellular decay, will be defeated incrementally. I have for some time predicted that this succession of advances will feature a threshold ... the "Methuselarity," following which there will actually be a progressive decline in the rate of improvement in our anti-aging technology that is required to prevent a rise in our risk of death from age-related causes as we become chronologically older."

The feedback between the Methuselarity and the Singularity may be considerable, of course. The same technologies that enable life extension to the point of Methuselarity may enable other Singularity-relevant technological breakthroughs. And the fact of deferring aging should increase the intellectual firepower of humanity (by avoiding the degradation of scientists' minds as they age), thus accelerating progress toward the Singularity.

The Chapter is followed by a **Dialogue** between Aubrey de

End of the Beginning

Grey, Ted Goertzel and Ben Goertzel.

Part Three: The Rollout and Impact of Intelligent Robots

Chapter Six: Robotics and AI: Impacts Felt on Every Aspect of our Future World by Daryl Nazareth.

Daryl Nazareth lays out a plausible timeline for the roll-out of increasingly capable and intelligent robots into society. He foresees advances in robots as being largely driven by advances in AI, along of course with ongoing hardware progress. As he envisions it, from 2015 to 2020, most developments will be narrowly focused, but the period from 2020 to 2025 is likely to be a phase of much more rapid development, in which robots increasingly draw on AGI technologies. Robots will take over more jobs now held by humans. From 2025 to 2030 will be what Nazareth envisions as the "early Singularity" period, a conjecture that accords closely with Ray Kurzweil's predictions. By that time robots will be generalists, not tools limited to specific tasks. This fits with Kurzweil's hypothesis that by 2028-29, machine intelligence will be equal to human genius.

The chapter is followed by a **Dialogue** between Daryl Nazareth and Ben Goertzel.

Chapter Seven: Robotics, AI, the Luddite Fallacy, and the Future of the Job Market by Wayne Radinsky.

Wayne Radinsky enlarges further on the likely socioeconomic consequences of this kind of robotics development timeline. In short he argues that once robots fueled with powerful AI are a major force, they will end up radically diminishing the need for human labor.

As Radinsky notes, this argument has been made for quite some time. Repeatedly it has been conjectured that automation will lead to rampant human unemployment, and this has proved wrong. Observing the realities of socioeconomic adaptation to new automation technologies, most economists have come to reject

End of the Beginning

"Luddite" arguments that technological automation will cause unemployment, arguing that the net impact of technological innovation is positive and that as robots and other machines take over particular human jobs, people will just shift to other kinds of work. The jobs that open up are often better paying and more fulfilling, requiring more education.

But Radinsky's key point is that AI may change all this. As we approach the singularity, machines will be increasingly used to replace jobs that require high levels of education and considerable skill. Therefore, "neo-Luddite" concerns may be more valid than they were in the past. Several scenarios can be considered: (1) the post-scarcity economy, (2) the super-education society; (3) the super-entrepreneurship society, (4) the society of shareholders; (5) the guaranteed minimum income society, (6) merging with the machines, (7) job market contraction. The most likely outcome, in Radinsky's view, is a Robot Takeover of the society, in some form. Various incarnations that such a "takeover" might assume – some positive and some negative, from a human point of view – are considered in some of the subsequent chapters.

The chapter is followed by a **Dialogue** with Wayne Radinsky and Ben Goertzel.

Chapter Eight: Moral Responsibility and Autonomous Machines by David Burke

David Burke looks at the moral aspects of increasingly intelligent, autonomous machines. To make the issues clear, he begins with some simulated courtroom arguments about the responsibility of autonomous machines in specific cases. The message from these cases is clear: As machines continue to demonstrate powers that are akin to human judgment and intelligence, there will be increasing pressure to grant them autonomy and broad decision-making powers while integrating them ever more deeply in human society (or more accurately, human/machine society).

End of the Beginning

Chapter Nine: How Will the Artilect War Start? by Hugo de Garis

In his chapter, evolvable hardware pioneer Hugo de Garis summarizes his dramatic vision of what the path to a "Robot Takeover" may look like. Echoing themes outlined in his book *The Artilect War*, he presents a dystopian vision of global warfare between rival parties, which he labels Terrans and Cosmists (though of course if real movements of this nature emerge they will likely have different names), the difference being that the Terrans want to stop the Robot Takeover before it's too late, whereas the Cosmists embrace the Robot Takeover, figuring that while it may have pluses and minuses from a traditional human perspective, it's going to lead to amazing things overall. The Terran/Cosmist dichotomy highlights a key value-system conflict that may come to the fore as Singularity approaches: valuing humanity in its traditional form, versus valuing intelligence, growth, joy and other values regardless of whether they come along with traditional human minds and bodies or are embodied in robots, computer networks, synthetic organisms, virtual-reality minds or other novel substrates.

The chapter is followed by a **Dialogue** with Ben Goertzel. **Dialogue 7.1** and **Dialogue 11.1** also include discussion of Hugo de Garis's ideas about the Artilect War.

Part Four: The Global Brain and the Emerging World of Views

The next few chapters present a quite different perspective on the future of human mind and society, elaborated by several researchers from the Global Brain Institute (GBI) at the Free University of Brussels. The focus of these thinkers is on the passage from a world in which the individual human mind/body is considered the nexus of intelligence, to one in which collectives of humans and/or combinations of humans with their (increasingly complex, adaptive) external tools, are the most natural entities to think about when analyzing the structure and dynamics of practical

End of the Beginning

intelligence.

The Global Brain perspective is one that is often omitted from discussions of Singularitarian futures, especially in the Silicon Valley tech community. Yet it holds many deep insights, which may help guide understanding of the future as it unfolds, whether or not a full-scale Global Brain as envisioned by some of the GBI researchers emerges.

Chapter Ten: Return to Eden? Promises and Perils on the Road to an Omnipotent Global Intelligence by Francis Heylighen

Francis Heylighen, founder and leader of the Global Brain Institute, and (along with Peter Russell, Ben Goertzel and others) one of the key developers of the "Global Brain" concept, presents here a highly emphatic vision of the emergence of the Global Brain. More important than the Technological Singularity per se, in his view, will be the transition in which the collectivity of human minds as a whole – enabled by technological tools – becomes a powerful intelligent, causal actor in its own right ... a Global Brain. Heylighen casts aside visions of a world-transforming superhuman AI created in a secret basement lab, and posits that the coming technologically driven explosion in intelligence will not be concentrated in one place but will be distributed across all the people and artifacts of the planet. He then takes another, more controversial step, arguing that the emerging "global brain" connecting all these people and artifacts will be all knowing and all powerful, much like traditional visions "God."

As Heylighen views it, the emergence of such a Global Brain will usher in an "Eden"-like era verging on classical visions of utopia. The main risks on the path to this wonderful future are catastrophic failures due to hyperconnectivity, passivity as people become over dependent, and resistance from conservative backlash. He foresees that a global immune system will be developed to protect against these risks.

The chapter is followed by a **Dialogue** with Francis Heylighen,

End of the Beginning

Ben Goertzel and David Weinbaum (Weaver).

Chapter Eleven: Distributing Cognition: from Local Brains to the Global Brain by Clément Vidal

Clément Vidal, also from the Global Brain Institute, presents a related vision, but focused more on the externalization of human intelligence, hypothesizing that as technology develops, it will make less and less sense to think about human intelligence as centered on a brain/body, and more and more sense to think about the overall intelligence of brains and bodies together with "external" technological tools. Already our smartphones and other gadgets provide us a massive degree of enhancement, but this is just the beginning.

In the **Epilogue to Chapter Eleven,** Clément Vidal presents a **fictional vision of life in 2060** where we can speak to anyone we meet with our built in translator, where potential partners evaluate each other with artificial intelligence, where life's petty chores are taken care of automatically, and where science and technology are pushed forward via complex networks of human and machine intelligence, acting as massive, distributed, self-organizing knowledge bases.

These chapters are followed by a **Dialogue** with Clément Vidal, Ben Goertzel and Ted Goertzel.

Chapter Twelve: A World of Views: A World of Interacting Post-human Intelligences, Viktoras Veitas and David Weinbaum (Weaver)

Viktoras Veitas and David Weinbaum (Weaver) from the Global Brain Institute present what is perhaps the deepest theoretical exploration in the volume. They outline a novel perspective for understanding the new world that will unfold as the Singularity approaches, centered on the concept of a "World of Views" – a network of subjective perspectives, interacting with and fueling each other, associated with overlapping intelligences on

End of the Beginning

various levels (human, extended/externalized human, human group, Global Brain). In this richly networked world, a form of hierarchical control still exists due to the nesting of patterns within each other, but the traditional hierarchical power structures of human society becomes increasingly irrelevant. Heterarchical patterns play an increasingly large role, in which influence and content pass from views to other related views.

While at times somewhat brain-stretching, the perspective presented by Veitas and Weaver has an important role to play in our understanding of the future. Given as many uncertainties as there are regarding the Singularity and the pre-Singularity world, it isn't possible to make detailed predictions with reasonable confidence in a purely objective way, based only on the available data. Beyond very broad aspects, the best way to understand the future is to begin an overarching conceptual framework that bridges the gap between the world we see today and the very different world that may soon come. The "World of Views" perspective provides such a framework.

The Chapter is followed by a **Dialogue** with Viktoras Veitas, David Weinbaum (Weaver) Ben Goertzel and Ted Goertzel.

Part Five: Globalization and the Path to Singularity

The primary originators of the Singularity concept are both American and associated with major tech hubs. Vernor Vinge spent his career as a professor and science fiction writer in southern California, and now continues there as a full-time writer. Ray Kurzweil carried out the bulk of his technology development and forecasting work from the Boston area, and at time of writing he is working largely on Google's campus in Mountain View, California – the heart of Silicon Valley.

The ongoing science and technology work that makes the Singularity a plausible idea, however, is occurring throughout the globe. Given the ongoing stream of science and technology press releases emanating from Shenzhen, Seoul and Tokyo alongside San

End of the Beginning

Francisco, Boston, London and so forth, few would doubt anymore the global aspect of exponential technological advancement. And neither is it plausible that a small percentage of major universities and companies, located in a handful of tech centers, is going to foist Singularity-enabling technology on the world all of a sudden. A look through the proceedings of any major scientific conference illustrates the broadly-distributed global nature of the ongoing advancement, bolstering Francis Heylighen's view that what is happening is an overall emergent development entraining the whole human race.

The next two chapters of the book explore the implications of the pre-Singularity period for two key regions of the world, both far-flung from Mountain View and Boston: China and Africa.

Chapter Thirteen: Chinese Perspectives on the Approach to the Singularity by Mingyu Huang

Mingyu Huang gives a perspective on what China and the Singularity may bring to each other during the coming decades. Following a traditional Chinese style, he begins with a long view. Looking back over the thousands of years during which humanity has been building up toward a Singularity, China has been a major – arguably the largest – contributor to technological advancement, overall. In recent centuries, China fell behind the west in technological and social development for social and environmental reasons; but the common wisdom in China today, is that the Middle Kingdom is now poised to catch up and move ahead. China's central regime is strongly committed to technological innovation and development, and Chinese culture is receptive to post-human innovations. The sheer number of trained engineers and scientists in China gives the nation a huge potential to advance technology in a variety of Singularity-relevant areas. Social and political changes are still needed to lessen bureaucratic constraints, however; and Internet tools are playing an increasing role in this process.

The chapter is followed by a **Dialogue** with Mingyu Huang

End of the Beginning

and Ted Goertzel.

Chapter Fourteen: Africa Today and the Shadow of the Coming Singularity by Hruy Tsegaye

Hruy Tsegaye begins his chapter on Africa with an even longer historical perspective -- noting that as the cradle of humanity, Africa would be a poetically natural place for advances toward the Singularity to occur. Currently, as he notes, Africa possesses the dubious title of "the continent of the third world"; and many government and business entities do not see Africa as a source of technological innovation. However, this is rapidly changing. As just two examples: In 2013 IBM announced the roll-out of 3 installations of its advanced IBM Watson AI system in Africa; and one of the editors (Ben Goertzel) and the chapter author are part of a team collaborating on building an AI software consulting firm in Addis Ababa, Ethiopia. The majority of Africans now own mobile phones, and smartphones are spreading rapidly. Internet access becomes increasingly prevalent and with it, education spreads.

Political stability is needed to facilitate further development in the African science and technology spheres. But there is some hope for positive feedback dynamics in this regard, in that broader rollout of technological advances may help encourage political stability, in part via coupling African nations to global socioeconomic networks, in a more thoroughgoing and less exploitative way than has previously been the case.

Currently, African government and industry are understandably more focused on improving life in the near term than on creating relatively remote developments like a technological Singularity. Yet there are also interesting opportunities for "technology leapfrogging" – i.e. for Africa to turn its relatively minimal technological infrastructure into an asset, and leap directly toward implementation of advanced technologies that may be inhibited in other places due to lock-in by merely moderately advanced technologies. An example of this could be in the medical

25

End of the Beginning

area, where e.g. the US medical system presents many institutional obstacles to progress, whereas the African medical system is sufficiently undeveloped that few are likely to push back against innovations such as, say, AI diagnosticians or smartphone-based diagnostic tools.

The chapter is followed by a **Dialogue** between Hruy Tsegaye, Ben Goertzel and Francis Heylighen.

Part Six: The Future of Money

Today's economic system is very different from what existed 200 years ago. Correspondingly, it seems quite possible that as the path to Singularity unfolds, the exchange of goods and services will occur according to patterns and mechanisms very different from what we see today. The two chapters in this section explore possibilities in this regard, discussing cryptocurrencies and also more radical possibilities for systematization of exchange.

Chapter Fifteen: The World's First Decentralized System for Financial and Legal Transactions by Chris Odom

Chris Odom and his associate Johann Gevers, via their startup Monetas, are part of a new generation of economic pioneers, seeking to upend the world financial system via creative utilization of crypto-currencies. In this chapter, Chris Odom reviews some of the general ideas and trends underlying all the ongoing work in this area.

As everyone who has followed the ups and downs of Bitcoin knows, strong cryptography is making possible new forms of money that will be inexpensive, instantaneous and resistant to control by authorities. If the pioneers of crypto-currency are right, the money of the future will be largely a self-enforcing, autonomous eco-system, not dependent on any government.

Most likely, Odom projects, new forms of money will be adopted first in areas that have been most inhibited by governmental restrictions and practical problems. This means they

26

End of the Beginning

will be most useful in developing countries where currencies are unstable, inflation is often high, and the rule of law is not as strong. Harking back to Hruy Teague's chapter on Africa, this suggests that the role of cryptocurrency in Africa may be worth watching as the next decades unfold. Already, as is well known, some Africans retain their store of currency in accounts associated with their mobile phone accounts rather than with any bank, which is a significant break from traditional financial-system methodology.

Chapter Sixteen: Beyond Money: Offer Networks, Potential Infrastructure for a Post-Money Economy by Ben Goertzel

Cryptocurrencies constitute a major step beyond prior versions of currency, but are not necessarily going to be the last word in the computational systematization of exchange. In his chapter titled "Beyond Money," Ben Goertzel takes the discussion a step beyond current financial systems, and explores the possibility of a more radically different form of exchange called Offer Networks. Offer Networks would enable direct exchange of goods and services between individuals or groups, mediated via sophisticated computer matching algorithms. A novel type of currency, the OfferCoin, could then be defined via mathematical analysis of the overall network of exchanges. This approach would avoid many of the issues involved with traditional fiat currencies or cryptocurrencies like Bitcoin, and would also naturally encompass qualitative exchanges of services alongside traditional economic exchanges. Whether or not anything specifically like Offer Networks comes to pass, the concept makes clear that global, systematic exchange networks could operate based on a variety of different mechanisms, with traditionally-conceived unidimensional money being just one class of possibilities.

The chapter is followed by a **Dialogue** between Francis Heylighen, David Weinbaum (Weaver) and Ben Goertzel

Part Seven: Human Society & Human Nature in the Era of

27

End of the Beginning

Abundance

While the Singularity is commonly explored with a focus on science and technology, in fact the most interesting implications may be psychological in nature. By transforming our environments and our bodies, advanced technologies may also transform our minds. The next four chapters, authored or co-authored by the book co-editor Ben Goertzel, are more psychological and sociological than economic, and explore various hypothetical aspects of mind and society as Singularity approaches.

Chapter Seventeen: Sousveillance and AGI by Ben Goertzel and Stephan Vladimir Bugaj

The increased surveillance capabilities provided by current technologies have been at the fore of public discourse since Edward Snowden's revelations and before. David Brin's distinction between surveillance (the powers-that-be watch everyone) and sousveillance (everyone watches everyone) has never been more topical. In this chapter on "Sousveillance and AGI", Goertzel and his R&D collaborator Stephan Vladimir Bugaj (now a leader in the computer gaming space) extend this theme into the future, exploring the possibilities that may ensue as observation technology becomes more and more prevalent, at the same time as AI technology advances. The argument is made that, as traditional human notions of privacy become obsolete, human psychology may change commensurately, with the traditional notion of the "individual self" morphing into something quite different. This brings us back to the themes of the Global Brain Institute chapters, regarding the potential trend away from the individual and toward collective intelligence.

The essay is followed by a **Dialogue** between Ted Goertzel, Stephan Vladmir Bugaj, Ben Goertzel, Hruy Tsegaye and Mingyu Huang.

Chapter Eighteen: The Future of Human Nature by Ben Goertzel

End of the Beginning

This chapter takes a speculative historical look at how the human mind became what it is, due to evolutionary and social forces – and leverages this to hypothesize regarding what the future of human nature might hold. Once we have reasonably advanced brain-computer interfacing and mind uploading technology, it seems likely that humans will consciously transcend the limitations of legacy human nature, moving on to new ways of conceiving the self and orchestrating actions. As the path toward these technologies unfolds, the human self will gradually transform. Some persistent human-psychology issues, such as repression resulting from conflict between one's actual self and one's ideal self, can be expected to diminish as technology gives us more awareness of, and more control over, the inner workings of our minds and bodies. Other conflicts at the heart of human nature, such as the struggle between individual versus group goals, cannot be expected to vanish merely due to the advancement of technology, but may transform into something new and unexpected, e.g. if Global Brain or World of Views type projections hold true.

The Chapter is followed by a **Dialogue** between Clément Vidal and Ben Goertzel.

Chapter Nineteen: Capitalism, Socialism, Singulari- tarianism, by Ted Goertzel

This chapter discusses historical efforts to use cybernetic technology to advance capitalist and socialist socio-economic systems, and explores the implications of the global brain and artificial general intelligence for socialist, capitalist, anarchist and other forms of societal organization.

The chapter is followed by a **Dialogue** between Ben Goertzel, Hruy Tsegaye and Ted Goertzel.

Chapter Twenty: Toward a Human-friendly Post-Singularity World, by Ben Goertzel

This chapter asks the broad question "What might the human

End of the Beginning

world look like after a Technological Singularity?" and then explores the path from here to Singularity in terms of the possibility of gradual pre-Singularity emergence of key aspects of post-Singularity human society and psychology. The post-Singularity human world is speculatively envisioned as a society in which working for a living is replaced with pursuit of aesthetic, social and psychological goals, and humans cluster themselves into affinity groups rather than conforming to the default expectations of a larger society.

Ideas are explored regarding potential social structures that may emerge as the Singularity approaches, triggered by the shift from our current "era of scarcity" in which the struggle for resources is a dominant factor in human society, to an "era of abundance" in which the resources needed for ordinary human life are easily available, and the focus of human life shifts to social and artistic pursuits. Potential pre-Singularity social-structure changes discussed include the emergence of microstates embodying specific social contracts among relatively small numbers of individuals, and the emergence of complex networks of human and software interactions displaying Global Brain type emergent patterns. It is hypothesized that advanced AGI may play a key role in maintaining global order in such a future society, allowing humans to focus on other matters. Finally, it is suggested that this sort of society would have radical implications for human psychology, naturally leading a substantial portion of the human population in the direction of "self-transcendence" as explored in the prior chapter on "The Future of Human Nature.

The chapter is followed by a **Dialogue** between Hruy Tsegaye, Clément Vidal, Aaron Nitzkin and Ben Goertzel.

Chapter Twenty-One: Looking Backward from 2100 by Ben Goertzel and Ted Goertzel

This fictional essay is in the tradition of time-travel futurism that includes Washington Irving's *Rip Van Winkle* (1819), Edward

End of the Beginning

Bellamy's *Looking Backward 2000-1887* (1888), H.G. Wells's *The Time Machine* (1895), Mack Reynolds's *Looking Backward from the Year 2000* (1973) and many others. It presents one possible future scenario incorporating many of the ideas presented by the various chapter authors, without agreeing with every author on every detail. This scenario is not presented as a specific prediction of what will happen, but merely as an example of what MIGHT happen. Readers are invited to formulate their own hypothetical future scenarios and compare them with ours and with the various scenarios hinted by the various chapter authors. The future is, in large part, ours to create.

REFERENCES
Kotler, Steven and Diamandis, Peter. 2012. Abundance: The Future is Better than You Think. Free Press.

Kurzweil, Ray. 2006. The Singularity is Near. Penguin.

Smith, Aaron. 2014. "U.S. Views of Technology and the Future: Science in the next 50 years," Pew Research Internet Project, April 17.
http://www.pewinternet.org/2014/04/17/us-views-of-technology-and-the-future/.

Vinge, Vernor. 1993. The Coming Technological Singularity: How to Survive in the Post-Human Era. Whole Earth Review, 1993.
https://www-rohan.sdsu.edu/faculty/vinge/misc/singularity.html

End of the Beginning

Chapter One **Predicting the Age of Post-Human Intelligences**

By Ted Goertzel and Ben Goertzel

Many AGI scientists are reluctant to make predictions because of problems caused by highly publicized, unrealistic predictions made in earlier decades, largely the 1950s-70s. Scientists and innovators in other fields have had similar problems. In 1908, Wilbur Wright confessed that "in 1901, I said to my brother Orville that man would not fly for fifty years . . . Ever since, I have distrusted myself and avoided all predictions."[3]

The witticism "it is difficult to predict, especially the future" is often repeated.[4] But, like many clever remarks, it doesn't hold up to serious examination. Some things can be predicted precisely, such as the time of sunrise or the dates of presidential elections in the United States or of the Jewish high holy days. Others can be predicted as statistical patterns, even though they cannot be predicted in precise detail. We pack winter clothes for a trip to Chicago in January, but not in July, even though we cannot predict

---

[3] Edward Cornish, *Futuring: The Exploration of the Future*, World Future Society, 2006, p. 149.

[4] No one is actually sure who said this first; it has been attributed to Yogi Berra, Niels Bohr and Casey Stengel among others.

End of the Beginning

the exact temperature on any future date. We can predict flooding along the Mississippi and an earthquake along the San Andreas Fault and take measures to prepare, without knowing when these things will happen. And we can prepare for post-human intelligence without knowing when it will come.

Regarding the trajectory of technology in the 21st century, Ray Kurzweil's predictions of rapid exponential growth have gotten the most attention, although evaluations of the quality of his predictions have been mixed. The core intuition underlying his projections is that scientific and technological progress is combinatory. The more advances that have been made already, the more ways there are to combine the products of old advances to form new ones, leading to an exponential growth in the number of new inventions that can be obtained as relatively straightforward combinations and elaborations of previous ones. He views the rise and advance of civilization as a long-term process of exponential technological advancement and associated social and economic consequences.

Kurzweil has had some big predictive successes, such as foreseeing the rapid growth of the Internet and of wireless well before the mainstream. His own assessment is that nearly all of his predictions were correct; however, this requires a somewhat loose interpretation of some of his predictive statements. For instance, in 2000 he made a number of predictions about the year 2009; a sample of a few is revealing:

"Most books will be read on screens rather than paper" -- while not literally true as of 2009, it was a prescient vision of the rise of e-books.

"Intelligent roads and driverless cars will be in use, mostly on highways" – which didn't happen by 2009, but now in 2014 the practical advent of self-driving cars during the next decade is fairly clear.

"Computer displays built into eyeglasses for augmented reality are used" – this was slightly ahead of its time, but is hard to scoff at

33

given the advent of Google Glass and its various competitors.

"Research has been initiated on reverse engineering the brain through both destructive and non-invasive scans" -- this prediction was spot-on

Whether he was right or wrong in each of these predictions becomes somewhat a matter of judgment. Even if he got the year wrong or the precise prevalence of a certain phenomenon wrong, it was impressive to foresee so many currently topical developments at a time when almost nobody else was thinking about such things. Overall one is left with the impression that getting the future qualitatively right is hard enough, and is more important than pinpointing the exact year when this or that technology will reach a particular level of market penetration.

Kurzweil's predictions are topical to this book; but, of course, he was far from the first to systematically attempt to predict the development of technology and society. The art and science of prediction has a long history, including a variety of impressive successes and dismal failures. Taking a look at the variety of methods used to forecast the future of social, economic and technological trends yields some worthwhile insights into the potential unfolding of events during the next century. For example, a number of forecasting methodologies focus on cyclical patterns, which may be seen as complementing the exponential technological trends highlighted by Kurzweil, Diamandis and other recent commentators. Looking at the cyclical patterns overlaid on a longer-term trend may be valuable for predicting the timing of watershed events. It is interesting to note that more than one of the commonly cited cyclical historical patterns predicts a point of major transition around the middle of the 21st century, consistent with Kurzweil's predicted timing for a Technological Singularity.

The History of Technological Forecasting

One ironic but informative aspect of the history of forecasting is the relative lack of correlation between the predictive success and

End of the Beginning

the popularity of various forecasters. As an extreme case consider biologist Paul Ehrlich, who announced in 1968 that "the battle to feed all of humanity is over. In the 1970s hundreds of millions of people will starve to death in spite of any crash programs embarked on now."[5] His book *The Population Bomb* was a bestseller and generated a large social movement. Some people even acted on it by having fewer children than they might have wanted. But scientifically, the book was just plain wrong.[6] World hunger declined for the next forty years, for reasons that were known at the time but which Ehrlich discounted.

Ehrlich was a failure as a scientific prognosticator but a success as an advocate. He went on to win award after award, including a MacArthur genius grant and more than a million dollars in prize money. He continues to publish books claiming that he was wrong only "in details" and denouncing his critics as unscientific.[7] He lost a highly publicized $1000 bet with economist Julian Simon, but Simon was marginalized by fellow economists, even though his predictions were consistently more accurate.[8] Ehrlich was not fazed by his failures as a predictive scientist, arguing that: "the real impact isn't in reasoned discourse. Media attention, press coverage and, if necessary, alarmism, at least set an agenda. And that way you can have a debate."[9]

The Limits to Growth was another study that greatly

---

[5] Paul Ehrlich, *The Population Bomb*, Ballantine Books, 1968, 1971, p. xi. In 1971, he delayed the forthcoming catastrophe to "the 1970s or 1980s".

[6] David Lam, "How the World Survived the Population Bomb," *Demography*, 50:2183-86, 2013.

[7] Paul and Anne Ehrlich, *Betrayal of Science and Reason*, Island Press, 1996, p. 33.

[8] Paul Sabin, *The Bet*, Yale University Press, 2013, pp. 205-206.

[9] Michael Wines, "The Sky is Falling: Three Cheers for Chicken Little," *New York Times*, 29 December 1996.

undermined popular confidence in future prediction.[10] It was a serious scientific work which introduced the use of computer models to simulate complex, interrelated trends. This gave it an aura of high-tech sophistication that earlier works lacked, even though they had reached similar conclusions. *The Limits to Growth* included a number of different simulations, based on different assumptions. One of its simulations, the "standard run" business-as-usual scenario, actually did a good job of predicting trends until the year 2000.[11] But the book also had models that assumed that the "known reserves" of petroleum and other natural resources were all that would ever be found. Combining this foolish assumption with continued population growth led to simulations showing catastrophe and mass starvation in just a few decades. And those were the simulations that got all the public attention.

The Limits to Growth was a best-seller for the wrong reasons. It happened to come out just in time for the 1973 world oil crisis. Journalists, always focused on recent events, assumed that the predictions in the book were simply coming true more quickly than anticipated. But the 1973 oil crisis was due to deliberate actions by the OPEC oil cartel, not to the exhaustion of the world's oil reserves. Unlike Ehrlich, the *Limits to Growth* researchers were serious scientists who considered a wide range of possibilities, but they might have anticipated that their more extreme and unlikely projections would attract too much attention.

The early Artificial General Intelligence scientists were optimistic because of their early successes with programs such as the Logic Theorist in 1956 and the General Problem Solver in 1959.[12]

---

[10] *Donella Meadows, et. al, The Limits to Growth, Universe Books, 1974.*

[11] *Graham Turner, "A Comparison of The Limits to Growth with Thirty Years of Reality," Global Environmental Change18:397-411.*

[12] *Daniel Crevier, Ai: The Tumultuous History of the Search for Artificial Intelligence, Basic Books, 1993; Pamela McCorduck, Machines Who Think, AK Peters, 2004.*

End of the Beginning

They were not experts in cognitive science and they underestimated the difficulty of building true human-level intelligence. Herbert Simon, a political scientist and early artificial intelligence developer, made the most widely-cited predictions when he was invited to speak to a group of operations engineers 1958. He thought they would appreciate some tangible, specific predictions so he developed four based on the work he and his friends were doing. He predicted that within ten years:
1. a computer would beat the world chess champion
2. a computer would produce a high quality musical composition
3. a computer would produce an important new proof of a mathematical theorem
4. psychological theories would be written in computer code.

All of these predictions were based on projects he and his friends were working on, and the ten year time period was based on his impression of how quickly the work was progressing. His predictions highlight one of the technical difficulties in futurism. It is often difficult to phrase predictions in such a way that they are objectively testable after a number of years. Words get redefined and new ways of thinking and doing things develop, so experts looking at the old predictions sometimes find it difficult to agree whether they came true or not. The most testable are those that involve a specific number, e.g., predicting the world's population or global energy use at a future time or prices of commodities. Simon's last three predictions were a little vague. It is a matter of judgment whether a musical composition is of high quality or a mathematical proof is important or whether a psychological theory tested with packaged statistical software is "written in computer code." After ten years, people might disagree about whether these predictions had proven correct.

Simon's chess prediction was the least ambiguous. Either a computer would defeat the world chess champion or it wouldn't. In

End of the Beginning

ten years it did not, but in 40 years one did.[13] During that time, computers got better and better at chess. So his was an error of timing, not about the direction of change. Herbert Simon later said that his ten year estimate was based on his belief that hundreds of researchers around the world would be intrigued enough to work on the problem. Actually, very few did in the first ten years. He believes that his prediction would have been on target if he had phrased it in terms of programmer hours expended rather than in terms of years.

While the failures are often highlighted, futurists have also had some impressive and useful successes. Harrison Brown's book *The Challenge of Man's Future,* published in 1954, did a very impressive job. He correctly stated that "we are far from the ultimate limit of the number of persons who could be provided for" on the planet, and also that feeding a growing population would depend on advances in scientific agriculture. His graph of the "possible future pattern of world energy consumption from fossil fuels" shows a peak around 2120, an estimate which is consistent with our best estimates today. Although his graphs were drawn by hand with ink, they look better in light of today's knowledge than those in *The Limits of Growth.*[14] A number of other futurists in the 1950s also did a good job of anticipating energy trends and problems, although they could not predict specific numbers for any given date.[15]

Like many futurists, Brown was more concerned with avoiding disaster than with predicting it. Specifically, he warned of a "reversion to agrarian existence" *unless* we are able to abolish war,

---

[13] Stephen Coles, "*Computer Chess: The Drosophila of AI,*" *Dr. Dobbs Newsletter, October 30, 2002.*

[14] *Harrison Brown, The Challenge of Man's Future, Viking, 1954, pp. 169, 220.*

[15] *Gilbert White, "Speculating on the Global Resource Future," pp. 171-186 in K. Smith, F. Fesharaki and J. Holdren, Earth and the Human Future, Westview, 1986.*

make a transition to new energy sources and stabilize population. And the risk of catastrophic nuclear war was certainly real. The "five minutes to midnight" meme was widely used in the anti-war movement of the time. *The Bulletin of the Atomic Scientists* had a clock icon on its cover and kept moving it closer to midnight when their advice was ignored. But it can't always be five minutes to midnight, and as the years went on the icon became an embarrassment which they minimized. Of course, their goal wasn't to predict nuclear war but to prevent it, so the important question is whether their warnings and advice helped to lessen the risk. There was no controversy about the dangers of nuclear war; the issue was what to do about the arms race with the Soviet Union. The atomic scientists opposed the military containment policy of the time, and advocated disarmament. It is impossible to know what would have happened had their advice been followed, but with hindsight we do know that the containment policies they opposed were ultimately successful.

Twentieth century futurists did not anticipate the collapse of communism in the Soviet Union, or even consider it as a possibility. Harrison Brown was certain that "in a modern industrial society the road toward totalitarianism is unidirectional."[16] In 1929, Arthur Shadwell published a book that pointed to the failures of Soviet communism and predicted that such an irrational system could not compete with capitalist market economics.[17] In 1967, Daniel Bell used Shadwell's book as an example of a grossly failed prediction.[18] Neither Bell nor Brown considered that Shadwell might have been right about the long-term trend, but wrong about the timing. Major sociological changes often take several generations, no matter how logical they may seem to analysts. Futurists have found it very

---

[16] *Harrison Brown, op cit., p. 284, 255.*

[17] *Arthur Shadwell, Typhoeus: or, The future of socialism, Kegan Paul, 1929.*

[18] *Daniel Bell, "Introduction" to Kahn and Wiener, The Year 2000, p. xxii.*

End of the Beginning

difficult to predict when major social turning points will occur, even if they anticipate that they will come at some point.[19]

Perhaps the most impressive scientific futurist study as yet published was *The Year 2000: A Framework for Speculation on the Next Thirty-Three Years*, by Herman Kahn and Anthony Wiener.[20] Published in 1967, one year before *The Population Bomb*, this book accurately predicted the world's population in 2000.[21] It also did remarkably well in its technology predictions which were carefully phrased to be testable thirty-three years later. A rigorous evaluation by a panel of experts in 2002 found that 80% of Kahn and Weiner's predictions about computers and communications were correct.[22] Their predictions included the following:

- Simple inexpensive home video recording and playing
- Personal "pagers" (perhaps even two-way phones)
- Pervasive business use of computers.
- Direct broadcasts from satellites to home receivers.
- Home computers to "run" households and communicate with the outside world.
- Inexpensive, high-capacity, worldwide, regional and local (home and business) communication (perhaps using satellites, lasers and light pipes).
- Extensive and intensive centralization (or automatic

---

[19] Paul Haplern, *The Pursuit of Destiny: A History of Prediction*, Perseus, 2000. Nicholas Rescher, *Predicting the Future: An Introduction to the Theory of Forecasting*, SUNY Press, 1998.

[20] Herman Kahn and Anthony Wiener, *The Year 2000: A Framework for Speculation on the Next Thirty-Three Years*. MacMillan 1967.

[21] Kahn and Weiner, 1967, pp 114-115, 150. Marsha Cohen, "Looking Back at the Future, Kahn and Weiner's the Year 2000 in Retrospect," Paper presented at the annual meeting of the International Studies Association, 2009, p. 8.

[22] Richard Albright, "What can past technology forecasts tell us about the future?" *Technological Forecasting and Social Change* 69: 443-464, 2002.

interconnection) of current and past business and personal information in high-speed data processors.
- Extensive use of high-altitude cameras for mapping, prospecting, census and geological investigations.

Kahn and Wiener's technology predictions outside of the computers and communication area were judged approximately 50% correct. Their sociological predictions were less successful, but they had anticipated that that would be the case, and qualified them as "surprise free" projections. They also presented "canonical variations" as a way of developing alternative futures. Along with the Sovietologists, they failed to anticipate that communism might collapse, although they considered that it might "erode."

Why were Kahn and Wiener so successful? Perhaps most important, they were really focused on making accurate predictions rather than on advocating for a cause. They relied heavily on specialized research done in academia, corporations and government agencies. Often these experts had a good track record, especially in the fields of population and demography and in technology development. Then they took the expert judgments from different fields and looked at how they would likely influence each other. Kahn and Wiener combined these judgments into what they called the Basic, Long-Term Multifold Trend. They considered how population, economic and social trends were likely to be shaped by the accumulation of scientific and technological knowledge. In their population estimates, for example, they adjusted the consensus figures down slightly because they thought newly available birth control technologies would have a faster impact than the earlier experts had anticipated.

Julian Simon was consistently more accurate than Paul Ehrlich because he incorporated economic theory in his analyses, anticipating that increased demand for resources would be met by increased supply.[23] He anticipated technological improvements that

---

[23] *Julian L. Simon, The Ultimate Resource, Princeton University Press,*

End of the Beginning

would lower prices. His winning the ten year bet with Ehrlich was largely a matter of luck, since commodity prices fluctuate widely over a decade. But from a longer-term perspective, which is really what matters, Paul Ehrlich has been drastically wrong and Julian Simon has been mostly right.

How did Ehrlich go so drastically wrong in his certainty that the 1970s would be an era of mass starvation? Ehrlich took a very bad year for food production, 1965-1966, and assumed that it was a portent of future problems. He knew that 1967 was a good year, but he dismissed it as an exception. He reported on food shortages around the world, assuming that they were due to population pressure. He ignored the work of agricultural scientists who knew that they were due to a variety of economic, social and climate problems that could solved without cutting population growth. From a humanitarian point of view, Ehrlich's recommendations were ruthless. He argued that spending any money on relief or on increasing food production would be useless until population growth was ended.

Fortunately, the world's humanitarian and agricultural organizations knew better, and took appropriate actions where needed. Ehrlich argued that the agricultural innovations known as the Green Revolution, which he knew about, could only put off disaster without population control. He also viewed the hunger problem in the context of a long-term trend of environmental degradation which he exaggerated. He accused food specialists of ignoring environmental problems. Actually, they knew plenty about these problems but correctly argued that feeding people did not have to wait until environmental and population problems were solved.

What are the lessons for artificial intelligence futurists from this history? To achieve the 80% success rate that Kahn and Wiener achieved for their 33-year computer science predictions, we should

---

1981.

End of the Beginning

look for "sustained and continuing trends in underlying technologies, where increasing capabilities enable more complex applications and declining costs drive a positive innovation loop, lowering the cost of innovation and enabling wider learning and contributions from more people, thus sustaining the technology trends."[24] Narrow Artificial Intelligence is clearly ready for this kind of prediction, but Artificial General Intelligence is not quite there yet. It would require a clear engineering path to AGI and also the assurance that sufficient resources would be available to carry out the plan. Prospects for this are assessed later in this paper.

Knowing exactly when AGI will come is less important than viewing it in the context of what Kahn and Wiener called the Long-Term Multifold Trend, so that we can help society prepare for it. Paul Ehrlich's collaborator John Holdren argued that "to put too much emphasis on the correctness or incorrectness of particular predictions is to miss the main point of writing usefully about the future. The idea is not to be 'right', but to illuminate the possibilities in a way that both stimulates sensible debate about the sort of future we want and facilitates sound decisions about getting from here to there."[25]

There is no need to warn the public that AGI is coming, they have been expecting it least since 1968 when *2001 Space Odyssey* was released. Nor are they concerned that it has taken longer than expected; in fact, there is some evidence that the general public would probably prefer that it came later rather than sooner.[26] What we do need is better informed discussion about the sort of future we want and how to get there. This means looking at the future of AGI

---

[24] Albright, op cit., p. 443.

[25] John Holdren, "Introduction," in Smith, Fesharaki and Holdren, *Earth and the Human Future*, Westview, 1986, p. 79.

[26] Aaron Smith, "U.S. Views of Technology and the Future," Pew Research Internet Project. http://www.pewinternet.org/2014/04/17/us-views-of-technology-and-the-future/.

in the context of other long-term future problems, including energy, environmental and geopolitical concerns.

Five Approaches to Predicting Long-Term Trends

There are multiple intellectual traditions that attempt to use data to predict long-term trends. Five leading approaches are: (1) environmental futurism, (2) Kondratiev long-wave analysis, (3) generational cycle analysis, (4) geopolitical futurism and (5) the study of technological revolutions. Technological revolutions are obviously the most directly relevant to AGI, but the others are important to consider as well since technological developments do not take place in isolation from the rest of society.

It's also worth noting that most scholars of technological revolutions confine themselves to history and make no efforts at long-term prediction. For instance, Carlota Perez, perhaps the leading historian in this field, confines her futurism to the banal and non-falsifiable observation that "the world is at a turning point...the decisions being taken at this crossroads will determine...whether what lies ahead is a depression, a gilded age or a true golden age."[27]

There is a good deal of short-term technological futurism which examines currently cutting-edge technologies and looks at likely developments three to five years into the future. Ray Kurzweil has very interestingly projected these trends considerably further into the future, but critics are skeptical of his argument that exponential trends in the speed of computer chips can be used to predict progress in modeling human-level intelligence.[28] Progress in engineering artificial general intelligence will depend on decisions about priorities and resource allocation, which will depend on long-

---

[27] Carlota Perez, *Technological Revolutions and Financial Capital*, Edward Elgar, 2002.

[28] Ray Kurzweil, *The Singularity is Near*, Penguin 2006; Paul Allen and Mike Greaves, "The Singularity Isn't Near," *MIT Technology Review*, October 12, 2011. http://www.technologyreview.com/view/425733/paul-allen-the-singularity-isnt-near/

End of the Beginning

term multi-fold trends. Trying to predict these trends requires going beyond technological futurism to incorporate insights from the other approaches that we will review briefly here.

Environmental futurists often predict catastrophe due to exhaustion of finite resources, destruction of habitat, pollution, overpopulation and climate change. Their critics argue that technological solutions will be found for these problems. This is not the place to rehash these arguments, but the debate has led to some quite rigorous studies that do generate specific predictions. We can, for example, predict that the global population will reach about 9 billion at about 2050, and then level off for the rest of the century.[29] We can predict that peak production of oil from conventional wells will be reached in the 2030's and that peak production of oil from unconventional sources (such as fracking and oil sands) will probably be reached in the 2040's.[30] Coal supplies will last through the century, but producing liquid fuel from coal will greatly increase prices. Given the strong resistance to limits on carbon emissions, we can predict that global temperatures will increase by 3 to 4 degrees centigrade by 2100, and that sea levels will rise one meter or more.[31] We can predict that humanity will face formidable technical challenges in the 2030s and 2040s if it is to avoid the kind of disasters still being predicted by environmental doomsayers.[32]

Economic Cycle futurists base their predictions on long cycles in the capitalist global economy, usually referred to as Kondratiev waves. These were discovered empirically, no theory predicted

---

[29] *World Population to 2300*," United Nations 2004.
[30] Miller R and Sorrell S, "*The Future of Oil Supply*," Philosophical Transactions of the Royal Society, 2 December 2013; Jackson PM, Smith LK. "Exploring the undulating plateau, the future of global oil supply." Phil. Trans. R. Soc. 2014.
[31] World Bank. Turn Down the Heat: Climate Extremes, Regional Impacts, and the Case for Resilience. 2013. www.worldbank.org.
[32] Paul Ehrlich and Anne Ehrlich, "Can a Collapse of Global Civilization be Avoided?" Proceedings of the Royal Society, 280: 1754, 9 January 2013.

them. There are also shorter cycles, commonly known as Kuznets and Juglar cycles that can be fitted into the long cycles. The empirical literature on these cycles is extensive and controversial, and the signal to noise ratio in the available data is low. This is not the place to rehash those debates; instead we will choose the most convincing and useful model we have found, the work of Brian Berry and his associates, and explore its implications.[33] Brian Berry offers an intriguing explanation for the existence of long waves, suggesting that they are paced by a 55.83 year lunisolar cycle that causes fluctuations in global and regional climates that impact agricultural crops.[34] If this is valid, it provides a rationale for projecting the cycle into the future, although agriculture will be a smaller component of the economy in the future than it has been in past centuries.

Assuming these observations are valid, we can predict the following trends in the Long Wave for the coming century: Trough, 2007-2011, Peak 2036; Trough 2065-6; Peak 2094. Periods of technological revolution come in the decade after a peak, so we can anticipate one in the 2040's and another in the 2100's. Of course, if these peaks are actually caused by climate cycles, things could be thrown off by global warming.

Generational futurists believe they have observed periodic changes in the cultural *zeitgeist* taking place approximately every 25 years. Young adults are most sensitive to these changes, and generational cohorts are formed of individuals who come of age during a period of history. The best known generational theorists, Strauss and Howe, used this model to project United States history

---

[33] Brian Berry and Denis Dean. "Long Wave Rhythms: A Pictorial Guide to 220 Years of U.S. History with Forecasts," Social Studies 2012. Brian Berry, Euel Elliott, Edward Harpham and Heja Kim. *The Rhythms of American Politics*, University Press of America, 1998.

[34] Brian Berry, "A Pacemaker for the Long Wave," Technological Forecasting and Social Change 63:1-23, 2000.

End of the Beginning

up until 2069.[35] Many other writers have observed similar generational cycles, mostly in qualitative research on United States political history, and there are different ways of describing them (liberal vs. conservative; introverted vs. extroverted; active vs. passive). Strauss and Howe have a more complex four generational pattern, with two active (idealist and civic) and two passive types. They also shifted their terminology from one book to another, which gets confusing.

These findings are based on empirical study of cultural patterns, not on a causal theory. Generational analyses are based on a qualitative reading of history and are open to criticism. Generational researchers do not have a good theoretical explanation of why generational cohorts form every 25 years or so when births are continuous from year to year. At this point, however, we will not address these theoretical questions, but will simply use generational theory to venture several predictions. The 2030s should see the emergence of an active, idealist or "prophetic" generation which will be moving into adult leadership roles during the 2040s (when the next technological surge is anticipated). This kind of generation has a need to find meaning and purpose in what it does, it is not happy just solving practical problems. The last such generation was the Baby Boom generation of the 1960s. This kind of generation might find meaning in anti-technological, spiritual movements, such as opposition to genetic engineering, thinking machines, and so on. If an Artilect War, such as that described by Hugo de Garis (in this volume) develops, it could easily be a generational war.

---

[35] Neil Howe and William Strauss, *Generations: The History of America's Future, 1584 to 2069*. Quill, 1992.

End of the Beginning

According to Strauss and Howe's theory, this idealist generation should be followed by a "reactive" one in the 2050's that will be disillusioned with ideological arguments. By this point, it seems likely that the idealists will have reconciled to new technologies, or have been defeated by their proponents. That will be followed by a "civic" generation emerging in the 2070's which will solidify the new systems and make them work. The Kondratiev peak in 2094 will be met by an emerging "adaptive" generation that will accept the accomplishments of their "civic" predecessors.

Geopolitical futurists study economic, political and military conflicts between states and alliances of states. The underlying geographical factors are long-lasting, making it possible to anticipate potential future conflicts, although the specific outcomes are dependent on how leaders manage the conflicts. Most geopolitical analysts don't look more than ten or twenty years ahead, but George Friedman has developed a forecast for the entire 21st century.[36] His forecast was published in 2009, and it predicted Russia's new assertiveness in 2014 and China's increase in naval assertiveness with uncanny accuracy. The geopolitical approach gives heavy weight to national interests and military strength. In Friedman's view computers are important because they are compatible with America's pragmatic culture and help to build America's dominance in the world. In his view, the United States will remain a hegemonic power for the entire century, but will face regular challenges from regional powers.

In Friedman's scenario, a new cold war develops between the United States and Russia in the 2010s (which seems to have happened in 2014), but Russia falls apart in the 2020's. China fragments with internal conflicts in the 2020's, as its economic boom busts. Turkey, Poland and Japan become major regional powers. So far, Friedman claims to be offering a forecast based on strong

---

[36] George Friedman, *The Next 100 Years: A Forecast for the 21st Century.* Doubleday 2009.

48

geopolitical forces. He then moves into a more speculative scenario to describe a Third World War that breaks out at 5 p.m. on Thanksgiving Day in 2050 with a sneak attack by Turkey and Japan on America's Battle Stars (orbiting military assets). In the war, the United States, Poland, India and China ally against Turkey, Japan, Germany and France. Nuclear weapons are not used and the United States wins. However, the United States then goes into decline largely due to loss of population as Mexicans flood into the southwestern United States and threaten to reclaim it for Mexico. At the end of the century the United States is locked into a struggle with Mexico for control of the North American heartland. All of this is obviously highly speculative; its merit is to alert us to risks and alternatives.

Technological Revolution futurists have observed that there are cyclical periods of intense technological innovation which coincide at least roughly with the Kondratiev waves. These waves may be due to economic boom and bust patterns, but Perez says they are due to "a much wider systemic phenomenon where social and institutional factors play a key role by first resisting and then facilitating the unfolding of the potential of each technological revolution."[37] This describes collective behavior patterns similar to those that generate booms and busts in economic markets. Other historians of technology, however, emphasize that these technological surges have not been swarms of unrelated innovations, but chains of related inventions or "technology systems".[38]

The surges described by historians of technology include: The Industrial Revolution (1770s), The Age of Steam and Railways (1830s), The Age of Steel, Electricity and Heavy Engineering (1875-

---

[37] *Perez, op cit., p. 60.*

[38] *Brian Berry, Long-Wave Rhythms in Economic Development and Political Behavior. Johns Hopkins Press, 1991. Brian Hall and Paschal Preston. 1988. The Carrier Wave: New Information Technology and the Geography of Information, 1866-2003. Unwin-Hyman.*

End of the Beginning

1884), The Age of Oil, Automobiles and Mass Production (1906-1920), and The Age of Information and Telecommunications (1971-1987).[39] The dates in parentheses are those of the "irruption" of the surge, which is the early part of an S-curve. This is followed by a period of "frenzied growth," then the S-curve levels off as the innovations saturate the markets. Since these surges are roughly 50 to 60 years apart, the next Age is due in the 2030's after the Kondratiev peak. What the new technologies will be is not predicted by the theory, but nanotechnology, alternative energy, biotechnology and genetic engineering are often mentioned. All of these rely on advances in artificial intelligence, so the emerging epoch may be known as the Age of Post-Human Intelligences.

Technological Revolution based futurism is relatively close to a Kurzweilian point of view. Kurzweil has explicitly stated that he views the exponential growth curve of technological advancement as composed of a series of superposed S-curves. Individual technologies or families of technologies tend to follow S-curves. But multiple technologies are always developing, in staggered fashion, so that the totality of superposed S-curves appears approximately as an exponential. However, many Technological Revolution theorists see major technological surges as occurring relatively infrequently and one-after-the-other, whereas Kurzweil views technological advancement as more multifaceted and ongoing, with various S-curves starting and finishing basically all the time.

Predicting the Age of Post-Human Intelligences

There is no unified theory of social change combining the five approaches, the best we can do is look at the interactions between each of them for the coming historical periods. This is similar to the conjunctural analyses historians make of past epochs, but they have

---

[39] *Perez, op cit., p. 57.*

End of the Beginning

the advantage of looking backward.[40] We can only offer what Kahn and Wiener modestly called "a framework for speculation" that highlights choices and alternatives. We will not limit ourselves to 33 years, as they did in order to end at the year 2000. There is no particular magic about the turn of a century.

One can view a Kurzweil-style analysis as identifying a broad underlying trend of exponential technological advancement. But in any complex real-world situation, the main trend is going to be obscured, sometimes, by chaotic special-case phenomena, and also potentially by various sorts of cyclical ups and downs. This view of progress as trend plus chaos and cycles is interesting given that so many of the futurist varieties surveyed above make predictions that are cyclical in nature. It is interesting to look at what happens if one views these cyclical patterns as superpositions on an underlying trend of exponential advancement.

As a rough conceptual analogy, imagine a tribe living on an island, with their technology gradually advancing over time, and also the rate of advancement increasing and decreasing cyclically. Looking at the basic trend of gradual advance, we can predict they will eventually invent boats to get off the island and get to the mainland nearby. Looking at the cyclic patterns overlaid on the basic trend, one can better predict *when* this invention will likely happen (i.e. probably during one of the periods where innovation is surging). In the context of this book, the Singularity is analogous to finally reaching the tech threshold to get off the island.

With this in mind, it is fascinating to note that three of the theories reviewed above predict changes at roughly the same points in time. Several analysts have observed that the generational cycle

---

[40] Jeffrey Paige, "Conjuncture, Comparison, and Conditional Theory in Macrosocial Inquiry," *American Journal of Sociology* 105 (3): 781-800, 1999.

matches up surprisingly well with the Kondratiev wave.[41] This is impressive because they are based on completely independent data sources and theoretical arguments. The cycle of technological revolutions also matches up reasonably well with the Kondratiev wave, although this is less surprising because of the obvious links between technology and economics.

The Kondratiev wave has also been linked to the rise and fall of global empires and the associated world wars, so the geopolitical approach can reasonably be matched up with the other three in a conjunctural analysis. To do this, we do not have to decide which factor is the cause and which the effect, it is enough to find that they move together. The environmental theories do not match up chronologically with the other four, but they can be matched in two ways. First, they may provide incentive for geopolitical conflict and even warfare over depleting natural resources. And, second, they provide an incentive for technological development to replace depleting fossil fuels and counter pollution and climate change.

Based on the synchronicity of the approaches we have examined, we can predict a technological surge in the 2040s. This technological surge could well be the "Singularity" described by Kurzweil and others, if it includes the development of true "thinking machines" with human-level intelligence. Even if this is not accomplished, there will be very rapid advances in fields such as nanotechnology, alternative energy, biotechnology and genetic engineering. This would be enough to sustain economic growth for another Kondratiev cycle, and the Singularity would probably then occur in the next cycle around 2100.

Predicting the Emergence of AGI

---

[41] Brian Berry and Heja Kim, "Leadership Generations: A Long-Wave Macrohistory," *Technological Forecasting and Social Change*, 46: 1-9, 1994. Michael Alexander, *The Kondratiev Cycle: A Generational Interpretation*. IUniverse 2002; Berry, et. al, 1998, op cit.

End of the Beginning

Beyond these broad sorts of analyses, can we say anything more specific about the timing of the emergence of human-level AGI? Based on Kahn and Wiener's accomplishments, it seems that technological developments can be predicted reasonably well when the basic theoretical problems have been solved and progress is largely a matter of engineering. Artificial General Intelligence is not yet at that stage of theoretical development, but may plausibly be there soon. As we summarized in the Introduction to this book, a 2009 survey of selected experts revealed a variety of timing estimates for human-level AGI, some quite optimistic and some less so.

As a case study in predicting AGI progress, consider the recent book *Engineering General Intelligence*, written by one of the authors of this chapter and his colleagues, which offers a detailed engineering model for building artificial general intelligence according to a design called CogPrime, within the OpenCog software framework.[42]

Outside experts who have assessed this design have estimated the time required to create a complete, fully tested deployment in the hundreds of man-years, potentially completable over a period of 5-10 years. This is substantial yet pales compared to, for instance, the roughly 70,000 man-years estimated to have gone into the Linux operating system so far.

Digging a little deeper, one sees that there are two separate quantities to be estimated here:

A) How long would it take to complete implementation, teaching and testing of the CogPrime design?

B) Once all that is done to the best that can be reasonably expected, how intelligent would the resulting system be?

---

[42] Ben Goertzel, Cassio Pennachin and Nil Geisweiller, *Engineering General Intelligence. A Path to Advanced AGI via Embodied Learning and Cognitive Synergy*. Atlantis Books, 2014.

End of the Beginning

It seems that estimates of A do not vary terribly widely among experts (i.e. not by more than a small integer multiple). Research and software project estimation are hard, but not *that* hard.... On the other hand, expert estimates of B strongly depend on the expert's opinions and biases and intuitions as an AI researcher, and would be expected to show an extremely high variance...

Of course, there are currently many such initiatives in play at the moment, not just OpenCog/CogPrime. There are many where the time to completion for implementing and testing the approach in question is in the range of dozens or at most hundreds of man-years. The odds of success of each approach, once implemented, are hard to estimate rigorously given the early stage of AI science. From an overall future-prediction perspective, it would make more sense to bet on some approach from the spectrum being successful, than on any one of them. OpenCog/CogPrime is a relatively easy case study to consider, in that that the overall design is laid out more clearly than for other most of the alternatives. That makes project estimation easier; yet due to the complexities of the scientific issues involved, still doesn't render it feasible to make a rigorous estimate of the final system doing what it's hoped to do...

OpenCog/CogPrime and other broadly similar projects are already funded to some extent; e.g. OpenCog has received funding from the Hong Kong government as well as from private donors. More copious funding for this kind of project is certainly feasible if venture capitalists, foundations or a government could be persuaded that it is viable. Ideally, several competing projects could be generously funded, following different engineering models and scientific bases.

Furthermore, quite recently there has been a trend of major technology companies funding AI research initiatives, led by researchers at the boundary between narrow, application-specific AI and AGI. Examples from 2014 are Google's acquisition of AI firm Deep Mind, the investment in AGI-ish vision processing firm Vicarious Systems by Facebook founder Mark Zuckerberg, Silicon

End of the Beginning

Valley icon Elon Musk and others, and the hiring of Stanford machine learning guru Andrew Ng by Baidu to head its Silicon Valley AI research lab. Whether these ventures in the direction of AGI will lead to profound research advances remains to be seen; there is a risk that the AI experts absorbed into tech firms via these recent initiatives will end up orienting their efforts more toward short-term, narrow-AI improvements of these firms' products and services.

Overall, it seems that if AGI initiatives become broadly funded at the levels one now sees for favored areas of science and technology such as brain research or semiconductor chip design, and one of the models thus funded proves successful, we would expect a Singularity-level change in the 2040s or perhaps before. On the other hand, if AGI funding remains relatively scant, we would predict a later Singularity, perhaps in the 2100s.

Our best guess is that an explosion of AGI funding will indeed occur, probably after the release of some dramatic public demonstration of proto-AGI capability – a possibility referred to elsewhere as an "AGI Sputnik" event. But from a rigorous perspective, it must be admitted that guesswork is all we have here.

Whether it is better for the Singularity to come in 2040 or 2100 is also an open question. It seems certain that humanity will face major environmental challenges in the 2040s and 2050s. These may lead to aggravated geopolitical conflicts and even warfare using very destructive weapons systems. Technological advances are essential if humanity is to avoid military or environmental catastrophe. Most scientific research and technological investment will probably go to efforts to cope with immediately pressing problems, but long-term success often depends on theoretical advances that were not intended to have an immediate practical benefit. A comparatively modest investment in artificial general intelligence could have this kind of benefit.

Another question is whether humanity will be ready to cope with the very rapid technological changes that will be forthcoming.

End of the Beginning

Sociologists have often observed a "cultural lag," as people resist changes brought about by new technologies.[43] Faced with the prospect of true thinking machines, we can anticipate that cultural resistance will be strong. Given the choice, many people would probably favor waiting until 2100 for true AGI. But technical change is much faster now than it was in the past, and slowing or stopping it would seem possible only after a catastrophic war or economic collapse.

Humanity has been anticipating the coming of artificial general intelligence for several decades, and we may be able to develop cultural models to be ready for it. The innovative ideas developed by the authors in this volume are an important step in that direction.

---

[43] William Ogburn, *Social Change with Respect to Culture and Original Nature*. Langsdale Library 2009 (original 1922).

End of the Beginning

Dialogue 1.1

Dialogue with David Weinbaum (Weaver), Ted Goertzel and Ben Goertzel

Weaver: I think this chapter is a very valuable contribution to the volume as it provides the readers with a general background to forecasting. Since the whole volume deals with the future, this specific article lends credence to the overall project by helping the readers draw the thin and often ambiguous line between prediction and speculation. I do, however, have some questions about the specifics of the prediction.

Ted: And what would those be?

Weaver: I am not sure it is entirely warranted to extrapolate developments and breakthroughs in a specific technological field such as Artificial General Intelligence from the very general trends that can be driven by any number of global events and/or disruptive technological breakthroughs.

Ted: Of course, each technological field has its own dynamics. But none takes place in a vacuum; we need to look at the interaction between specific technologies and general trends. What are some of the specifics that we need to look at more closely?

Weaver: Just to mention a few examples, the recent very large strategic investments made in brain research might indirectly accelerate AGI research. Another example is the emergence of technologies such as neuromorphic chips and neuromorphic computing, memristor based technologies and more. Other examples are semantic web technologies, internet of things etc. All these, though not directly contributing to the emergence of AGI, are setting the ground and creating the demand for breakthrough

End of the Beginning

developments in AGI.

Ben: Yes, I have made the point frequently in other writing that the path to AGI is being paved by developments in computing hardware, computing software (languages, operating systems and algorithms), cognitive science and neuroscience.

Since the early days of AI in the 1960s, hardware has obviously advanced massively – each of us carries in our pocket what would have been considered more than a supercomputer back then. In hindsight it seems absurd that some very serious AI researchers in the 60s and 70s thought they could achieve human-level AI with the minimal amounts of RAM they had available back then. Cloud computing with modern networking software and tools like Docker makes it possible to leverage massive amounts of distributed computing power with relatively little effort – something that wasn't there 10 years ago, and barely there 5 years ago.

Software development is utterly different now, due to what one might consider Global Brain type dynamics: rather than being mainly about writing one's own large blocks of code, modern programming is now largely about connecting and perhaps tweaking code written by others. There's a library for everything, and software paradigms like generic programming make it possible to utilize existing code libraries in manners quite different than what their creators envisioned. Furthermore, code sharing software like git makes it so much easier for groups of people to work together on a project, enabling a sort of group cognition among a team of programmers.

Neuroscience hasn't fully unraveled the secrets of the brain yet, but we do know pretty much what each part of the brain does, and a lot about how they connect to each other. We now have a reasonably solid understanding of what KIND of system the brain is, even if we haven't worked out all the details due to insufficient brain measurement technology. There are some peculiarities to be sure – like some folks in the "deep learning" community arguing

End of the Beginning

that simplistic models of parts of visual and auditory cortex can serve as models of the whole brain – but overall our knowledge of the brain is massively more than it used to be. We can read, via brain imaging, a movie of what a person is seeing through their eyes – for example.

And cognitive science has advanced more than most scientists give it credit for. Now we can come pretty close to writing down a diagram indicating how the human mind works – what kinds of memory there are, what kinds of processing there are, etc.; and how these different kinds of memory and processing interact. We sure don't know everything – but 30 years ago, say, there was nowhere near this level of common understanding of how human intelligence works. Recently Bernard Baars, Stan Franklin and some of their colleagues put together a paper on Dynamic Global Workspace Theory, which explained a number of things about the process of human conscious cognition, using a synthesis of ideas derived from neuroscience and psychology data. Such a paper just wouldn't have been possible 10 years ago let alone 30.

All these developments – and way more that I haven't mentioned – have put us in a position where the AGI problem is within reach. When thinking about AGI now, one has a huge variety of and depth of conceptual and practical tools at one's disposal, far beyond what one had 10, 20 or 30 years back. And there's no telling which tools, from which of the various allied areas, a given AGI researcher is going to draw on in doing their work. So, yeah, the sciences are increasingly cross-connected – industry has known this for a long time and even academia is gradually starting to realize it. And this cross-connection is part of the reason for the exponential acceleration of progress that we're all living in today.

Weaver: Your review surely draws an interesting map of rapidly evolving opportunities. One additional remark is that it seems that computing technologies are well on track to provide a platform for a future AGI. As to our understanding of the organic

End of the Beginning

brain (the mammal brain to be more precise) as a benchmark to an efficient thinking machine, indeed we have many of the building blocks in place and perhaps some of the more fundamental algorithms involved in processing sense data. Yet, on the deeper conceptual aspects of thinking and especially creative thinking beyond symbolic computation, what makes intelligence really general, we still lack some fundamental insights. I would guess that these have to do with evolutionary processes that form their own formalisms on the fly and are hardly captured by a single formalism or cognitive model. It seems to me that in this sense the Open Cog approach is conceptually the most progressive due to its inherent multiplicity.

I would also like to raise a more philosophical point. Prediction or forecasting is only one aspect of our complex relations with the future. This aspect focuses on the direction from past -> to present -> to future, trying to discover historical patterns and trends that will extend and extrapolate into the future. Another aspect which is not less important, at least to my mind, is the visionary aspect that focuses on the direction from future -> to present i.e. our collective visions, hopes, desires, ambitions, etc.

Ben: Interesting. Indeed, my own chapter **Towards a Human-Friendly Post-Singularity World**– later on in this book – takes precisely this approach, starting from a possibility regarding post-Singularity human life and working backwards, to see how this sort of possibility might viably arise.

Ted: In that chapter you start from a future where human life continues in a recognizable form and post-human minds are inclined to respect the preferences of the human population. But many images of the future have been much more negative. Your choice of a starting point was arbitrary. Our paper was on predicting the future, but if there are so many futures to choose from, how can we make a prediction? Does it come down to predicting which one the dominant powers in society will choose? Or just which one we prefer?

Ben: Actually my choice of a starting-point wasn't "arbitrary" – I chose, as a starting-point for my analysis, a future world that I thought the majority of humans would find acceptable (and a significant subset of humans would find exciting and highly desirable). I did this partly from the perspective that we are the creators of our own future. There are constraints on what futures we can create, most likely – but we don't know exactly what those constraints are at present, and how effectively we discover these constraints is also dependent on the future we create. Obsessing on negative outcomes can sometimes help avoid them, but it can also help causing negative outcomes (e.g. obsessing on the risk of being mugged or attacked could cause one to ruin one's life by never leaving one's house, which would be a real shame if in fact the risk of being mugged or attacked was pretty low). We don't want to ignore the possibility of negative outcomes but nor it is wise, in my view, to primarily or exclusively focus on them. Negative outcomes often make more exciting movies or newspaper headlines, but that's not an argument for their greater probability. I think we should be spending a lot of our energy envisioning the kind of future we want and then figuring out how to create it. That's the spirit in which I wrote that chapter, at any rate.

Weaver: Indeed, sociotechnological systems are reflexive systems which means that the system's dynamics can be profoundly affected by the way we perceive and envision it. That is to say, by the images, metaphors and narratives we construct to mediate our reality. Trends, therefore, develop, at least in part, not only as function of historical patterns but also in response to our individual and collective visions and desires. See for example how strongly our present technological thought is influenced by fictional creations such as Asimov's Three Laws of Robotics.

End of the Beginning

Viktoras: I think that the goal of developing futuristic predictions is to push people to thinking about the future, not to predict exactly what is going to happen and when. Due to reflexivity of sociotechnological systems, merely thinking and imagining the future means influencing it. The process is somewhat similar to strategic business planning – the exact numbers are not that important, but what is important is the increased awareness of different future scenarios and their underlying causes.

Ted: How could we apply that to the future of AGI?

Weaver: In the case of AGI, the most common images are more repulsive and frightful than attractive. I am thinking of Skynet with its intelligent killer robots, Hal 9000, the Matrix and such. The only exception is perhaps IBM's Watson inducing a somewhat less frightful image.

Ben: Star Wars' R2D2 and C3PO aren't so frightful either… nor, say, Wall-E.

Weaver: True. This diversity of images is hardly a surprise given that the emergence of AGI bears profoundly on how humans see themselves and their location in existence. AGI is not just another scientific or technological challenge like solving humanity's energy problem or climate change. For humans, the emergence of AGI means that metaphysically nothing will ever be the same again. It foreshadows a psychological and existential revolution the likes of which humanity has not encountered since the fateful events in the Garden of Eden…

Ted: So which future do we predict? The Matrix, the Terminator, or R2D2?

End of the Beginning

Weaver: Actually, there are many more choices than that. Just looking at images that have had a wide exposure in the media, we can consider:
- Negative images such as Skynet, Cylons, Hal 9000, The Borg, the Matrix, I Robot, Transcendence, Eagle Eye and more.
- Somewhat controversial and problematic images, but biased toward the negative such as in Real Humans (an excellent Swedish TV series), and Spielberg's movie A.I.
- Positive images such as Wall-E, a post-apocalyptic cutie, Bicentennial Man, Data in Star Trek, as well as R2D2 and &C3PO, Watson, and Her…

Of all these, the negative seem to get the most visibility and reaction in the wide public's sentiments. As I said, this is not a matter of prediction but rather an entirely different kind of activity I would call the cultural scaffolding of the future. Cultural scaffolding is done by working out narratives that embed values and envisioning. These might serve as attractors for future technological trends. For example: wishing to live longer than we currently do is increasingly perceived as a worthy value by the general public, this may mobilize scientific technological endeavors to achieve a future where we actually live longer. With AGI in contrast, this seems to be much more complex and subtle. The wish for life extension is grounded in basic biological instincts while expanding the intelligence of non-human agents is perceived as a threat. This threat is born out of how we project ourselves into such futuristic agents. We imagine them in our own image but in all probability as having values and priorities very different from ours and the only way humans interpret such a situation is through conflict. Some of the most prominent minds of our time see in AGI an existential risk. This perception is a matter of speculation that arises from a very narrow perspective. We need to seriously ask ourselves about the roots of this perception: we a priori perceive the different and the powerful as potentially hostile and of course this might well become a self-fulfilling prophecy. It is a matter of a worldview and this is what I think needs to be changed by a well-designed cultural scaffolding.

Ted: From the standpoint of predictive science, this just makes the problem harder. With so many choices, how do we predict?

Weaver: What we need, more than a careful prediction, is a powerful inspiring vision that will prepare the ground for the future.

Ben: I agree totally. Careful prediction is valuable, yet in a case like this what it's ultimately going to tell us is that there is a very wide variety of qualitatively and quantitatively different possible futures, so far as we can tell based on our current knowledge. Careful prediction is going to give us predictions with very wide error bars, yet we need to make choices anyway – and the choices we make are going to affect the future we get, in ways we don't currently have the data to (anywhere remotely near) fully understand.

Weaver: As it is now, the psychological ecology towards the emergence of AGI is far from friendly and when it comes to the sociopolitical aspects of technological progress this fact cannot be overlooked. We need to find a way to couple prediction and envisioning into an integrated activity. Science and technology never operate in vacuum nor in an independent dimension. Social and psychological aspects must be incorporated in how we think about the future. When it comes to extremely complex systems such as human society, it is not only about collecting data "out there" and coming up with good predictive theories, it is also about taking a good look in the mirror.

Viktoras: As a number of scientists, politicians and

End of the Beginning

philosophers noted in the course of history, 'the best way to predict the future is to create or invent it' - (this quote is often attributed to Abraham Lincoln and to Peter Drucker). Images of robotics and AI in popular culture can be understood as an explication of the link between collective worldviews and science. These images are what are going to happen. This does not make the problem easier (well, maybe technically it does), but it changes the perspective which I think is essential: the problem of predicting the future escapes the domain of somewhat sterile predictive science, which is by no means simple, and enters the fuzzy and messy social and political domain. Regarding matters discussed in this chapter, predictive science smells a little bit like hiding from responsibility.

Therefore I think that the approach taken by this chapter and the book in general is very important and this importance is well conveyed by the quote of John Holdern that "[..the] idea is not to be 'right', but to illuminate the possibilities in a way that stimulates sensible debate about the sort of future we want [..]".

The somewhat different collective perception of human-robot relationships that one sees in, say, Japanese versus American (or Western in general) cultures is illustrative in this context. It seems that Japanese culture is more inclined to perceive robots as friendly companions which seamlessly integrate into human society, while Western culture separates itself from artificial intelligence and looks at it more as a tool.[44] These images subtly influence research and development directions which make us see our dreams and nightmares coming true. Therefore a somewhat more informative and rewarding question seems to be not 'what will happen?' in terms of predictive science, but how these two worldviews and corresponding technological developments interact and co-evolve.

---

[44] *The issue is much more complicated, so here I present a simplified picture for making a point, for more see Kaplan, Frédéric. "Who Is Afraid of the Humanoid? Investigating Cultural Differences in the Acceptance of Robots." International Journal of Humanoid Robotics 2004 (2004): 1–16*

End of the Beginning

Understanding hidden tendencies and forces which sometimes seem contradictory in their nature may help us "predict" the future in a sense of "creating" it.

Ted: That wasn't the goal of this particular chapter, but several of the other chapters in the book do address that critical question. In addition to your own chapter, Francis Heylighen defends an image of the future as a Garden of Eden with a global immune system that protects against catastrophic risks. Our view is that post-human intelligence is coming, and that we still have time to shape what it will be like when it gets here.

Weaver: I hope it will be friendly way beyond what is imaginable by our contemporary suspicious minds. But then again it is entirely up to us, the presumed creators, and our individual and collective responsibility to overcome our own prejudices and biases.

Ben: Well, I am unsure that it is entirely up to us. Indeed, as you well know, the English phrase "up to us" conceals a lot of confusion and messiness regarding the notions of free will and causality, which our modern scientific worldview doesn't deal with very effectively at present. But in any case, it seems a reasonable working assumption that it is substantially up to us – I don't see any significant advantage in assuming otherwise (well, unless justifying laziness in thought or action is considered an advantage, but that's not my personal orientation...). Prediction is a key aspect of human intelligence, and exercising our capability to predict – both intuitively and scientifically – seems important, in the context of the Singularity and the Global Brain and related developments. But understanding the limitations of our ability to predict is also very important, as is understanding the role our predictions may play in shaping reality.

Ray Kurzweil, for example, has obviously been acutely aware

End of the Beginning

of the value of seeking accurate prediction, and also of the potential that predictions have for actually shaping the future. His predictions have been by and large carefully data-driven; but they've also been made with a view toward impacting the public attitude and in this way helping cause his predictions of progress to come true. In a way his predictions implicitly factor in the fact that he is making and publishing his predictions. So there's a subtle and interesting view of free will underlying what he's doing – Ray is a very smart man. On the other hand, even though I've discussed the matter with him a few times, it's not totally clear to me whether he fully appreciates the limits of predictability where something like the Singularity is concerned. To an extent I sympathize with the emphasis placed by Vernor Vinge when he originally formulated the Singularity concept – he proposed the Singularity as a point beyond which we humans simply can't predict what happens.

A rough analogue might be a phase change in a physical substance. As water gets hotter and hotter, its state changes, and one can draw graphs and charts and extrapolate the future changes of the water based on its past changes. But then when it reaches the boiling point its state changes dramatically and qualitatively, and curve-fitting based on what happened before the boiling point was reached, doesn't really mean very much. Of course the state of water above the boiling point could in principle be induced via study of examples of water below the boiling point, but this would require much more than curve-fitting based on the observed macroscopic state of water; it would require a quite deep understanding of the underlying physics of the water molecules. Or, having not yet seen water boil, one could perhaps infer the existence of the boiling point and understand the state of water after the boiling point by analogy to other substances in which one had observed similar phase transitions.

But in the case of the Singularity, my own view is that we haven't seen any similar phase transitions before. I tend to agree with Vinge that the emergence of massively greater than human

End of the Beginning

intelligences is going to change many things utterly, thus rendering any curve-fitting based on earlier happenings irrelevant. But of course this doesn't stop us from performing extrapolation aimed at figuring out what will happen BEFORE the Singularity, on the path there – the main focus of this book, and the focus of most of Kurzweil's extrapolation. And it also doesn't stop us from trying to arrive at a deep enough understanding of the underlying factors in humanity and society that we can project something of what may happen after the Singularity too – I guess this is attempted in a couple chapters of this book, such as the World of Views chapter by Weaver and Viktoras, and my own chapter on the Future of Human Nature.

Weaver: Well, the "up to us" phrase certainly does not imply free will as a working assumption. I entirely agree with you that the concept is philosophically and scientifically fragile if not entirely bankrupt. Still, as you note in your remarks regarding Kurzweil's work, the socio-technological environment is a complex reflexive environment. "Up to us" comes to mean that our projections of future scenarios are real formative forces of our future even if the trajectories these forces spawn are often partly or entirely intractable. I would not go as far as saying that this is a new subtle angle on the possibility of free will but I also do not think that our way of thinking and envisioning the future is entirely ineffective. The cleanest way to express this is to affirm that we do have a will, it is not free in the classical sense and it cannot fully control our complex existential circumstances but it certainly makes a difference and a significant one. For example, if we managed to collectively eliminate killing from the repertoire of human inherent behaviors, we wouldn't have to be bothered by future killer robots threatening humanity because no one will care to create such robots. There is nothing in technology that is threatening our existence as a species; it is only human nature that must be overcome as I said above. Besides, humans have been cultivating killing machines for

End of the Beginning

millennia. We call them soldiers and we devise ever ingenious ways of releasing them, at least temporarily, from their humanity so they can serve their designated function ever more effectively. From this perspective nothing has changed and the only Terminator we need to worry about is the one we can see reflected in our collective mirror.

Regarding the metaphor of phase transition, such a phenomenon can be examined either empirically (observe boiling water) or in the context of a theory that provides an explanatory narrative of what is going on and why. In the case of a Singularity as a phase transition of intelligence we do not have anything close to a theory. We can only extrapolate and predict our failure to predict what will happen if things will keep accelerating the way they do. We'll have to wait and see and meanwhile do our best to predict an ever approaching horizon.

End of the Beginning

Chapter Two **A Tale of Two Transitions**

By Robin Hanson

Robin Hanson *is an associate professor of economics at George Mason University, a research associate at the Future of Humanity Institute of Oxford University, and blogs at <u>OvercomingBias.com</u>. He has over sixty academic publications, over 2500 citations, MS in physics, MA in philosophy of science, PhD in social science, and nine years' experience as a research programmer. He is a founder of the field of prediction markets, and is currently writing two books, one on the social implications of brain emulations, the other on a homo hypocritus account of human nature. He can be reached at* **rhanson@gmu.edu**.

In this chapter I'm going to consider and compare two quite different scenarios for a future transition to a world dominated by machine intelligence. One scenario is based on continuing on our current path of accumulating better software, while the other scenario is based on the future arrival of an ability to fully emulate human brains. The first scenario is gradual and anticipated. In it, society has plenty of warnings about forthcoming changes, and time to adapt to those changes. The second scenario, in contrast, is more sudden, and potentially disruptive.

A History of Tools

But before we get to those scenarios, let us set some context by

End of the Beginning

first reviewing some basic facts about tools and machines. Between the farming revolution starting around ten thousand years ago until the industrial revolution starting around 1700, the world economy grew rather steadily, doubling about every millennium. While some of that growth was due to the accumulation of better tools, most such growth was due to the accumulation of new and better sorts of domesticated plants and animals. Since the industrial revolution, however, the economy has doubled about every fifteen years, and most of that growth has been due to the accumulation of better physical tools and ways to use them.

The world economy contains a great number of tasks that need to be done, and an even larger number of tasks that it would be nice to do, if only they could be done cheaply enough. When tools were bad, pretty much all useful tasks had to be done by people or animals, if they were to be done at all. But as our tools have improved, those tools have displaced people (and animals) on many tasks.

The displacing of people by tools on tasks has been relatively slow and steady over the last century or so. Also steady has been the rate of improvements in tools doing the tasks they do better, and the rate of adding of new useful tasks that we could not previously afford to do. Some of these new tasks have been done by tools and others by people. These rates of task displacement, improvement, and addition did not change much after the introduction of computers, even though basic computer abilities have been improving more rapidly.

These trends make sense if there are many kinds of physical and mental abilities, if each task requires many supporting abilities, and if the ability levels required to achieve particular tasks vary over huge ranges. Some tasks that people do are very easy, requiring low levels of many abilities, while other tasks are very hard, requiring high levels of many abilities. As the abilities of our tools improved faster than did human abilities, tools slowly rose in their relative ability to do tasks, and periodically some tools rose

End of the Beginning

above the threshold of being cost-effective relative to humans, or relative to not doing the task at all.

Simple economics says that total world income is paid out to all the tasks doers, in rough proportion to the importance of each task to the outcomes we care most about, and also to the difficulty of improving the performance of each task by devoting more resources to it. The more tasks that are done, and done better, the more world income there is to pay out. Thus we can track the total value that tools and the ways we use them give via the rise in world income, and we can track the relative importance of tools and people in the economy by tracking the fraction of income given to those who own tools, relative to the fraction given to people who sell their time to earn wages.

When we look at these things, what we see is that while tools have helped us to greatly increase the total value that we achieve by doing all of the tasks that we do, the fraction of value that is paid to tools at any one moment is still small compared to the fraction paid to people. The main reason for this is that it tends to be much faster and easier to increase the number of tools of any one time, relative to the number of skilled people. So the value that the last tool contributes to a task is still much less than the value that the last person contributes.

Today human workers directly get about 52% of all income, a figure that is down from 56% forty years ago (Karabarbounis & Neiman 2013). But the rest mostly does not go to tools. For example, only a few percent goes to computer-based tools. A lot of income goes to those who own valuable things like real estate, patents, and firms, things that are neither tools nor people working, but which were created in the past by people using tools.

When Will Tools Top Humans?

If we continue to improve our tools in the same sort of ways that we've been improving them over the last century or so, eventually there should come a point in time where tools exceed humans on almost all relevant abilities, making tools cheaper than humans for almost all tasks. At that point, the speed limits we face on making more skilled people would no longer limit the growth rate of the economy. This is because only tools would be needed to grow the economy further, and factories could quickly crank ever more such tools. A tool-dominated economy might then double every month, or even faster.

How close are we today to such a transition point? One clue comes from the field of artificial intelligence [AI] research. This is the field where researchers directly and explicitly try to write software that can do hard mental tasks, tasks where tools have not yet displaced humans.

AI researchers have had substantial success over the decades. And some observers, including some AI experts, have even said that they expect that AI software with broad human level abilities will appear within a few decades (Armstrong & Sotala 2012).

However, AI experts tend to be much less optimistic when asked about the topics where their estimates should be the most reliable: the rate recent of progress in the AI subfields where they have the most personal expertize. I was a professional AI researcher for nine years (1984-1993), and when I meet other such experienced AI experts informally, I am in the habit of asking them how much progress they have seen in their specific AI subfield in the last twenty years. They typically say that in that time they have only seen 5-10% of the progress required to achieve human level abilities in their subfield. They have also typically seen no noticeable acceleration over this period (Hanson 2012).

At this rate of progress, it would take two to four centuries for the typical AI subfield to reach human level abilities. However, since there would be variation across AI subfields, and since displacing humans on almost all tasks probably requires human level abilities in most AI subfields, a broadly capable human level AI would probably take even longer to achieve than two to four centuries.

End of the Beginning

In addition, in many areas of computer science software gains have closely tracked hardware gains, suggesting that hardware gains help to enable software gains (Grace 2013). This is a bad sign because traditional "Moore's law" trends in hardware gains have slowed lately, and are expected to slow even more over the coming decades (Esmaeilzadeh, et. al. 2012). These facts together suggest it will take even longer for our software tools to be cheaper than humans on almost all mental tasks.

This scenario, where tools continue for centuries to improve at rates comparable to those in the last century, until finally tools are cheaper than humans on almost all tasks, is the slow gradual first scenario I mentioned at the start of this chapter. In this scenario the gains in our abilities on different kinds of tasks do not depend much on each other. There is no grand theory of tool-making or intelligence that, once discovered, allows tools to improve greatly simultaneously across a wide range of tasks. Instead, different tools would mostly develop independently, getting slowly and steadily better at their various tasks.

A Slow Gradual Transition

We can use basic social science to foresee many aspects of how a transition to a tool dominated world would play out how in this scenario.

First, the world would get lots of warning about the upcoming transition to a world dominated by machine intelligence. On the one hand, growth rates may not start to increase much until the human wage fraction of income falls below five percent or even lower. On the other hand, the human fraction of world income may start to gradually fall a half century to a century or even longer before such a transition point. There should thus be many decades of familiar growth rates with direct vivid lessons to help humans get used to the idea that they will eventually need to own something other than their ability to work in order to have an income.

As the fraction of world income that went to working humans

End of the Beginning

fell slowly from twenty to ten to five to zero percent, individuals would decide how to diversify their personal portfolios in order to assure themselves a future income, and societies would decide how much to help the initially poor as well as those who invested poorly. Societies would also adapt their finance, law, governance, and other institutions to the new distributions of income and power.

This would be a world where total production results from a very large set of tools that are mostly improving independently, and where design changes typically only result in small changes to local functionality. These aspects of this world would greatly limit the size and scope of damages resulting from poorly considered design changes. Such mistakes would almost always be reversible at a modest cost; very few changes would have a non-zero chance of destroying society as a whole.

As is true today, some key design choices in control and governance would have larger scopes, and thus more potential for large broad damage. And this would continue to be true during and after a transition to a world dominated by machine intelligence. But it isn't clear that the transition period would have substantially more risk than the periods before and after. And even if transition risks were higher, there would be plenty of time and warning in the last few decades before such a transition to think carefully about how to design regimes of governance and control for large systems of interacting tools, before humans are displaced out of the main social roles of design, governance, and control.

Society has changed in many ways over the last century or two, and in this scenario society would continue to change at similar rates over the coming centuries before a transition to an economy dominated by machine intelligence. Since some of the changes in the last century or so have been enabled and driven by specific technologies, some future changes would also have this character.

However, while some future changes will no doubt be driven by particular technologies, it is important not to over-estimate the

End of the Beginning

fraction of changes that arise this way. Most of the big changes we've seen over the last century or so have not been driven by particular technologies. Instead, most such changes can be better understood as resulting from general broad increases in wealth and capacity. And such changes tend to be easier to predict.

Increases in physical capacities suggest that we will continue to see increases in travel speeds and distances, in the size of homes, vehicles, and meals, in lighting and brightness, in sound insulation and quietness, in the lightness, toughness, and colorfulness of clothing, in the climate control of artificial spaces, in building heights and depths, and in memory sizes, and in processing and communication speed and efficiency. And those later increases in info system capacities suggests we will know more about the actions and feelings of others, and have more decision aid tools.

Increases in social capacities suggest that that we will continue to see increases in lifespans, individual intelligence and abilities to navigate complex social systems, in the size, density, complexity, and specialization of firms and cities, and in the specialization of individual roles and social networks. We also expect to see a steady increase in the quality and addictiveness of devices and entertainment.

Many trends over the last century or so can be understood as due to a reversion to forager values, habits, and attitudes, as wealth has weakened the strong social pressures that turned foragers into farmers. While at work we have come to accept increasing alienation and domination, such as via varying and detailed orders and fine-grained status rankings, outside of work we have less tolerance for domination and overt ranking. In particular we are less tolerant of autocracy, slavery, and overt class distinctions. We should expect such trends to continue.

We also expect to see a continued fall in fertility, and continued rise in leisure, peace, travel, promiscuity, romance, civility, mental-challenging work, local diversity of style, and medical and art spending. We expect to continue to see increasing value placed on

End of the Beginning

self-direction, tolerance, pleasure, nature, leisure, and political participation.

A Fast Sudden Transition

The other possible scenario that I'll consider in this chapter for a transition to a world dominated by machine intelligence is based on the arrival of the ability to make brain emulations, which I'll call "ems." An em would result from taking a particular human brain, scanning it to record its particular cell features and connections, and then building a computer model that processes signals according to those same features and connections.

A good enough em would have very close to the same overall input-output signal behavior as the original human. One might talk with it, and convince it to do useful jobs. When ems are cheap, they would very quickly displace humans on almost all tasks.

The ability to make ems mainly requires three kinds of technology to sufficiently advance. Computer hardware must get cheap enough, brain scans must get cheap and fast enough with high enough spatial and chemical resolution, and computer models of specific types of brain cells must get accurate enough. Based on prior trends, these all seem likely to advance sufficiently within about a century, or well before AI software is anywhere near human level abilities.

We can foresee many ways in which a world dominated by ems would change. But such changes are mostly beyond the scope of this book, which is mainly concerned with the transition to such a world. However, to understand this transition, it helps to understand some basics about this new world.

In particular, it helps to understand that net income in this new world would go to those who own relevant land and natural resources, to those who own relevant capital, not only machines and buildings but also the computer hardware that runs ems, to those who own and run the firms who fill key industry niches, and to the ems who would dominate most labor market niches.

End of the Beginning

The transition to an em world would result in big changes in who gets most world income. The em-based economy would quickly come to dominate the world economy. Ems would probably concentrate in a few dense new cities. The switch to em-based capital and computer-hardware would likely put a new crop of firms in key industries niches. And a few hundred "clans" of ems all descended from the same original human would probably dominate most em labor markets.

First-movers in these areas will tend to be disproportionate winners. These include the first firms to make and distribute em hardware, the real estate areas that house the first em activities, the first humans to be scanned to become em workers, and those who hold patents on key enabling technologies. In anticipation of this big disruption of income, many would scramble to position themselves to be transition winners.

In the AI software based scenario discussed above, the transition is expected to happen later, to be slower and smoother, and to give more warning of a transition ahead. Given sufficient advanced preparation, that transition needn't be especially disruptive, since there is less reason to expect big changes in the value of patents, real estate, or key firms during such a transition.

In contrast, a brain emulation based scenario can be more disruptive, because having a working em is more of an all or nothing situation. While tools can slowly get more useful as they get more capable, an em mostly either works or it doesn't. Thus there can be much less warning about an upcoming em transition, which may happen sooner, more quickly, and make for bigger changes between winners and losers. The transition to an em world can be especially disruptive and unexpected if the last technology to be ready is cell modeling or brain scanning.

These differences would make it more important for individuals and their supporting institutions to prepare more carefully for a possible em transition, even in the absence of clear warnings of an imminent transition. Assets should be diversified

End of the Beginning

and insured more thoroughly, and based on weaker signs of upcoming problems. These differences also make it more likely that such preparations will not be made, so that big winners and losers will result.

Well before a transition to a world dominated by em-based machine intelligences, the world would look much like it would well before a transition to a world dominated by human level AI software. In both cases, our physical and social capacities would be increasing, and our values would be slowly becoming more forager-like. However, while our physical and social capacities would continue to increase after either transition, value trends would change more after an em transition.

Soon the vast majority of creatures with human like values would be ems, and the fall of wages to near em subsistence levels would likely induce a sudden and large reversal of our industry-era trend toward forager-like values and attitudes, at least among the ems. The anticipation of this change would likely also add to the stress and disruption of a transition to an em-dominated world.

Conclusion

In this chapter I have considered and compared two quite different scenarios for a future transition to a world dominated by machine intelligence. One scenario is based on continuing on our current path of accumulating better software, and results in the smooth continuation of current trends for centuries, until a relatively gradual and anticipated transition, when there are relatively mild disruptions to power and values. The other scenario is based on the future arrival of an ability to fully emulate human brains, and is by its nature sooner, more sudden, and more disruptive to both power and values.

In my judgment this second scenario is more likely. This is why I have been working on a detailed book-length analysis of its likely post-transition consequences. Even if it is not the transition or world we would most want, if it is instead the world we will have, it is

End of the Beginning

important to understand in as much detail as possible, to inform any decisions we may make about it.

End of the Beginning

Chapter Three **Longer Lives on the Brink of Global Population Contraction: A Report from 2040**

By Max More

Max More *is an internationally acclaimed strategic philosopher recognized for his thinking on the philosophical and cultural implications of emerging technologies. More's contributions include founding the philosophy of transhumanism, developing the Proactionary Principle, and co-founding Extropy Institute, an organization crucial in building the transhumanist movement since 1990. At the start of 2011, he became President and CEO of the Alcor Life Extension Foundation, the world's leading cryonics organization. He is co-editor of The Transhumanist Reader. More has a degree in Philosophy, Politics, and Economics from Oxford University and received his PhD in Philosophy from the University of Southern California in 1995. He can be reached at: max@maxmore.com.*

Global Trends Review, March 2040

As long-time readers of this publication know, every ten years we like to look back several decades to examine expectations of the future – our present – and note how they compare to what actually happened. Here we look back on people's expectations around 2015 of population growth, the aging of the population, and what the

End of the Beginning

actual trends imply for further advances in extending the human life span.

The Shrinking Global Population

Population forecasts have long been a favorite game of demographers, economists, and futurists. Today, in 2040, we have reached a historically unique point. After centuries of growth in global population – the fastest period being in the 20th century – two years ago we reached the top of the growth curve with 7.85 billion people and began to decline.

Of course, populations have been declining for several decades in many parts of the world. Even 30 years ago, 44% of the world's population lived in 83 countries and territories where fertility was below replacement levels. (Today, the number is 75 %.) Those areas included Japan, Eastern Europe, Kazakhstan, Ukraine, Belarus and many other countries extending into Central and Western Europe, as well as Italy. In fact, the total population of the continent of Europe peaked all the way back in 2000 and began falling in 2004. Yet it was only 20 years ago that the European leadership truly acknowledged this trend, and just five years ago that the European President set forth a comprehensive policy to act on that acknowledgement.

Thirty years ago, oddly, the vast majority of the talk was still about overpopulation, not about the looming challenge of accelerating depopulation. It seems that there was a lag of three or four decades from between the start of the deceleration of population growth and most people's realization of it. Now we understand that, if not for recent advances in extending the human life span, global population would, over the coming decade, fall by about 85 million. Even longer lives are having only a modest impact on that decline. Low fertility has a profoundly stronger impact than life span.

Three decades ago, the UN offered three main population projections – a low, medium, and high trend. In past projections, it

End of the Beginning

had typically been true that the UN overestimated population growth. Most revisions to the agency's forecasts were downward. For instance, when the population of Europe finally reached 1.2 billion in 2000, it was outside the bottom of the range assumed *by the low variant* projection. We can see now that this continued to be the case. Even the UN's low fertility projection was too high. (The low projection used a fertility rate 0.25 of a birth lower than the medium projection.) The UN expected population to peak around 2040 to 2050 at just over 8 billion and then decline to 6 billion by 2100.

A few observers even 30 years ago recognized the potentially drastic path of population decline that might lie in their future. One noted that if one-child policy or practice were followed, by 2250 the world's population would fall to less than 200 million – well under 3% of the 2013 level. We now understand the magnitude of the decline and realize the importance of boosting research into further extending healthy human life spans.

Demographic Trends 2010-2040

One point that *was* well-recognized 30 years ago was that the population was aging. From a global perspective, the average age has been rising since "peak youth" in 1972. (17 years later, population growth in absolute numbers peaked at 88 million in 1989.) That trend is continuing and should continue indefinitely into the future. The number of people aged 65 or more passed one billion a decade ago.

If no changes had been made to the retirement age in both policy and practice, compared to 30 years ago, we would now have barely any more workers but twice as many retirees. Around 2010, many pundits were predicting the imminent collapse of Social Security based on these assumptions. The fear of steeply increasing dependency rates was valid and some bad outcomes were seen in countries such as Japan, where the population aged especially rapidly without sufficiently matching changes in retirement

End of the Beginning

patterns.

In many countries, as the dependency burden continued to grow, populations and their governments increasingly recognized that both the mean and especially the median life span was lengthening. Not only were people living longer – more and more often into their 90s – but they were doing so in better health. Now we have almost three people in their nineties for every newborn. Already, early this century, retirement benefits were adjusted so that, while you could retire at 65 in the USA (or even 62 with reduced benefits), for those born after 1959 the "normal retirement age" reached 67. This began a reversal of the trend away from the situation when Germany first set a retirement age well above the average age of death.

Even in England in 1908, the retirement age of 70 was still 20 years beyond average life expectancy. In 1926, when the male state pension age of 65 was set, 60% of men died before they could claim a pension. Oddly, the retirement age for women long remained lower than that for men in some countries, despite women's longer lives. For instance, in 2011, both the United Kingdom and the Ukraine required men to be five years older before enjoying retirement benefits. For many years, the rise in the average legal retirement age lagged far behind gains in life expectancy. In the 40 years from 1965 to 2005, the retirement age went up by less than six months while male life expectancy rose by nine years.

This changed quickly starting at the turn of the century. As the number of people aged over 65 in the work force almost doubled in the USA and elsewhere), retirement ages began to follow suit more quickly. Even by 2014, several European countries had linked the statutory retirement age to life expectancy. Since then, we have seen the standard retirement age rise from 65 or 67 to 68, 70, 72, and now 74, with many countries indexing the full pension age to life expectancy.

Three decades ago, more commentators were beginning to foresee that the whole idea of a single and complete retirement was

End of the Beginning

going to become outdated – and with it the notion of a "dependency ratio". The dependency ratio was defined as those typically not in the labor force (usually those under the age of 15 and over the age of 64) compared to those typically in the labor force (ages 15-64). Thirty years ago, globally there were 16 people aged 65 and over for every 100 adults between the ages of 25 and 64. That was very close to the ratio sixty years ago. In 2010, projections put that number for 2035 at 25. For individual nations the projections were 44 for the USA, 69 for Japan, 66 for Germany, and 36 for China. Over the last few decades we have seen this ratio lose its usefulness for at least two reasons.

One is that, with rising wealth – and encouraged by government policies – more people between 15 and 64 years have spent large periods of their "working years" working fewer hours or taking time off. The 35- or 40-hour workweek has become more the exception than the rule, especially among those who succeeded in shifting a substantial part of their income from wages to profits from investment. This makes it less clear whether they are part of the "productive part" of the population. At the same time, far more people 65 and older have been working either part-time or what is today considered full-time. While the statistics clearly show that most of those over 70 work fewer hours than those in their 40s or 50s, they cannot be considered "retired" or "dependent".

Closely related to these trends is the point that the whole idea of permanent retirement came under burgeoning attack around the turn of the century. More and more people came to engage in "work-surfing" – they plan to work for several years or a decade or two then "retire" for two or three or five years, although "retirement" can include varying degrees of paid activity. Sometimes grudgingly, employers have increasingly welcomed workers in their 60s and 70s, enabling them to work widely varying hours and other flexible work arrangements.

Several decades ago, most projections of the demographic and economic consequences of longer lives assumed that the extra years

End of the Beginning

would be unhealthy, unproductive years. Fortunately, for the most part, increases in life expectancy have been increases in healthy, functional years. It looked for a while early in this century like obesity, diabetes, and other self-inflicted health problems would slow or even reverse the trend toward additional years of productive, healthy life. Better treatments and a shift away from the dogma of low-fat, high-carbohydrate diets has enabled most populations to avoid this fate.

While obesity remained at historically high levels until truly effective fat-control treatments appeared recently, the effect on healthy longevity was much less than many anticipated. Even 30 years ago, it was becoming clear that most of the cardiovascular risks associated with obesity were due to high blood pressure, diabetes, and high levels of bad cholesterol. With ever more effective treatments for these conditions, life spans continued to grow even while obesity remained common.

Apart from improved treatments for conditions resulting from obesity, life spans have continued to extend thanks to an overall trend towards lower mortality from other causes. One strong cause in many countries has been the ongoing decline in the rate of smoking and the switch to electronic cigarettes for those unwilling to give up their nicotine habit. Advances in regenerative medicine, including the development of both biological and synthetic replacement organs and repairs at the cellular level, have led to a series of new records for both average and maximum life span. The modal age at death is now over 100 and a few people have lived for over 130 years.

Economic Challenges of Declining Population

For much of the second half of the 20th century, people in Europe, the USA, and elsewhere feared the consequences of population growth. That fear outlasted the reality of the danger. Our perceptions lag economic realities, often by decades. Our era's greatest economic and social challenge – other than aging – is

End of the Beginning

declining population. We are still only just starting to act on that realization. Efforts to boost child births have not had much success. Although it will not be enough to entirely ward off economic challenges, perhaps the other most promising approach is to push much harder to radically extend healthy life spans.

What is wrong with a shrinking population? After all, the environmentalists of the past would have been delighted to see the number of human beings dwindle. One obvious problem or at least challenge is that a shrinking population implies a smaller market for your goods or services. At the same time, on the supply side, businesses and other organizations have a smaller pool of workers to draw on. It's hard to find many historical examples of growth and progress that do not accompany expanding populations. Businesses typically move away from areas with declining populations and to areas where population is growing. The Great Plains of the United States provides a familiar example. Depopulation started early in the 20th century and accelerated in the 1920s. By 2010, thousands of ghost towns could be found in Kansas alone and many counties had lost more than 60 percent of their population.

It's true that a shrinking population can bring a temporary benefit. This is referred to as the "demographic dividend". As less spending is needed for education and other costs of raising children and young people – and fewer human resources need be devoted to doing so – money and resources can be deployed elsewhere. China's rapid development in the late 20th and early 21st centuries was turbocharged by the tremendous drop in the relative number of dependent children. The same effects were observed in the Middle East, and in East Asia (the number of dependents grew at only one-quarter the pace of the working population from 1965 to 1990).

If economic growth has almost always been accompanied by population growth (or at least a steady population), consider what happens when population dwindles. The output of a country is determined by the size of its labor force multiplied by the average

End of the Beginning

output per worker. If growth is to continue when population is shrinking, it must be by greater gains in productivity. That is not impossible – we have been seeing increasing contributions to economic output from the growing sophistication of artificial intelligence – but is not easy or automatic.

At the same time that we need to extract more productivity from each worker, we have to tax them more heavily to pay the costs of supporting and caring for an aging population. In some ways, this problem has lessened since half a century or more ago. The problem would be much worse for employers if they were still expected to pay for workers' retirement. Individual Retirement Accounts were introduced in the USA through a 1974 law but didn't pick up steam until the 1980s when 401(k) plans began. The move toward independent retirement funds from the early 1980s progressively relieved employers from assuming growing retirement obligations. However, the burden on governments and on individuals who must fund their own retirement remains. So workers are being pushed to raise their productivity even as policy discourages them by taxing them more heavily to fund Social Security, Medicare, and Medicaid.

If people refuse to be taxed sufficiently heavily, or if they cut back on working since it rewards them less, governments of shrinking populations are likely to fund dependent care by raising borrowing. The amounts involved are massive. When government borrowing becomes too high, it tends to raise the cost of capital. That makes it harder to invest in new production methods and new technologies. This is yet another drag on economic growth and productivity. At the same time, a shrinking, aging population that is declining in vigor and health is going to work less and pay less in taxes.

If a smaller population consists of older, less vital people, innovation may slow down. When population shifts from expansion to contraction, the pressure to increase the supply of most goods and services falls off. Organizations and workers won't

End of the Beginning

work harder to figure out better or more efficient ways to do something when there are more than enough resources and infrastructure for everyone.

Attempts to Boost Population

Here is the obvious response to these challenges of waning population: Reverse course! Reinflate population growth! Yes, but how? Immigration has helped some regions to maintain or grow their populations in the past. Clearly, this solution cannot possibly work on a global basis. The world as a whole cannot grow by moving people between countries. Even for individual countries, the scale of immigration required – if it were possible to bring about – would often be so massive that it would create its own problems. The influx of enough foreigners to offset powerful domestic population contraction could lead to rapid and disruptive cultural shifts that many people will resist.

Still, for countries with more modest rates of population shrinkage and with the availability of culturally acceptable immigrants, this strategy could help substantially for a short time. All over the world, as fertility rates have fallen, populations have aged. Since the whole point of attracting immigrants is to boost the economy with vigorous young workers, the revitalizing power of immigration becomes ever weaker.

We can see a classic example in the United States. The median age of Americans has grown by only a handful of digits since the start of the century. Forty years ago, it was projected to grow by 4.5 years to 39.7 in 2050. In reality, the median age has already surpassed that as life spans have lengthened beyond the projections' assumptions. During the same time, Mexico's median age was projected to increase by 20 years. This much more rapid aging of the Mexican population all but dried up the migration of Mexican workers into the United States.

The other "obvious" way to reverse population contraction is to encourage people to have more children. No doubt more could

89

be done, such as reducing the deductions from pay for social security for those who have children. However, past experiences with government policies aimed at pumping up population are not encouraging. Governments have used the tools of taxation, supporting maternity leave, and providing subsidized education. Even 30 years or more ago, Japan, Russia, and Australia paid bonuses for births. France, Germany, Italy, Poland and other countries also provided bonuses and monthly payments to families, and Sweden provided generous parental leave, without notable success.

Back then, Singapore had one of the lowest fertility rates in the world. The Singaporean Chinese, which made up over 70 percent of the local population, had a total fertility rate slipping below 1.0. The government offered to pay couples $5,000 for a first child and up to $18,000 for a third child. It did not work. The power of high population density, high land and housing prices, excellent education, the need for women to bear so many child rearing responsibilities, and a cultural aversion to endure the stress of raising children all overwhelmed the financial incentives.

The Imperative to Further Extend the Human Life Span

Earlier, I noted that a shrinking population may be the second greatest economic and social challenge we face. The greatest is populating aging. Of course, these tend to go together. This is a strong hint that perhaps the most effective way to alleviate the challenges of an ebbing population is to tackle the problem of aging with more determination. It must be acknowledged that even radical extensions of healthy human life spans will have only a modest effect on population size. Fertility rates are, always have been, and will continue to have a far greater impact.

But the negative consequences of a population that is both shrinking and aging can be fended off by focusing on stopping and reversing aging. Success in that endeavor would do away with concerns about dependency ratios, create a permanent demographic

End of the Beginning

dividend, support economic growth, and enhance innovation. In the first four decades of this century we have already seen significant economic benefits from millions more people living more years in good health. But governments and other institutions could multiply those benefits many times over by investing with greater determination in anti-aging research.

Unfortunately, for almost a century governments have channeled medical research funding almost exclusively into studies of specific diseases rather than into aging itself. This has continued with only modest improvement for decades, even as a growing number of voices pointed out the foolishness of the approach. For instance, consider two papers from early this century. A 2008 McKinsey Global Institute study titled "Why the Baby Boomers Will Need to Work Longer" pointed out that "A two-year increase in the median retirement age over the next decade would add almost $13 trillion to real US GDP during the next 30 years." That number was simply from raising the retirement age and so undercounts the benefit if that change came together with a two-year extension in healthy life span.

More directly, a 2013 paper titled "Substantial Health and Economic Returns from Delayed Aging May Warrant a New Focus for Medical Research" examined the effects of a relatively modest improvement in our ability to slow the aging process. If a mere 5 percent of adults over the age of 65 would be healthy rather than disabled every year from 2030 to 2060, there would be 11.7 million more healthy adults over the age of 65 in 2060. That would produce an economic benefit of about $7.1 trillion over five decades. By comparison, cutting the incidence of cancer or heart disease by 25% over the same period would have a minimal effect on population health. Earlier research showed that completely curing cancer would only increase life expectancy by about three years.

These projections were for a very modest gain in healthy life expectancy. Nor did they add up the cognitive benefits for older adults resulting from research in delayed aging. Even if population

End of the Beginning

is contracting, consider the opposing effect if people live longer, healthier lives and continue being productive. If a 40-year working life extends to 48 years, that's a 20% increase, or about 2.5 months per year. Multiplied by tens of millions of workers, the economic difference is startling. Because of under-investment in anti-aging research, here in 2040 we have only seen a fraction of those potential benefits so far. As the authors of a 2006 paper on "The Longevity Dividend" complained, "The NIH is funded at $28 billion in 2006, but less than 0.1% of that amount goes to understanding the biology of aging and how it predisposes us to a suite of costly diseases and disorders expressed at later ages."

Due to limited investment in anti-aging research, here in 2040 as compared to 2010 we have seen only a modest (but still very welcome) slowing of aging by six years. By delaying aging for this time, mortality and morbidity have been compressed, giving people more years of healthy and functional life. At each age, the risk of death has declined by over 40%, so that today's 60 year-olds are just as healthy as 2010's 54 year-olds.

We can do much better. Although the amount we now spend on anti-aging research has multiplied from the sadly low amounts seen early in the century, we continue to fall short. We have blunted the demographic and economic consequences of declining population but could do so much more. AI agents have also relieved some of the burden by essentially acting so as to multiply the productive population while requiring very few resources and without ever contributing to the dependency ratio. These trends will surely continue.

If we push harder to fund and further longevity research, we can greatly expand on the benefits we have enjoyed to date. By delaying (preferably ending) aging and extending the years of healthy life, we can massively reduce the costs of medical care, enable millions of people to replace permanent retirement with cycles of retirement and paid (or investment-funded) activity, vastly reduce human suffering due to infirmity and decrepitude, and

End of the Beginning

enjoy the incredible resource created by combining the wisdom of advanced years with the vigor of youth.

In our era of contracting populations it seems ridiculous that, not long ago, so many commentators continued to object to life extension on the grounds that it would worsen overpopulation. As far back as 2010, careful analyses of the impact of longer lives on population were available but widely ignored. In "Demographic Consequences of Defeating Aging", published in the groundbreaking *Rejuvenation Research*, Leonid and Natalia Gavrilova showed that population effects would be minimal to modest, depending on the specific scenario.

Looking at the Swedish population, the researchers considered the effects of: 1. Negligible senescence after age 60; 2. Negligible senescence for 10% of the population; 3. Negligible senescence for 10% of the population with growing acceptance leading to 1% added to the negligible senescence group each year; and 4. mortality continuing to decline after age 60 years down to the levels observed at age 10, and then remaining constant.

In the first scenario, even though median lifespan increases from 84 to 134 years for men and from 88 to 180 years for women, after 100 years the Swedish population would increase by only 22%. (Assuming no change in childbirths.) Population continues to decline in the middle two scenarios (by 28% over a century in the second, and more slowly in the third.) The Swedish situation of 2010 reflects the current situation of most advanced nations today. These numbers suggest that we should not hesitate to push for quite radical extensions of healthy human life span.

Most people back in 2014 continued to be blinded by demographic expectations shaped decades earlier. With surprisingly few exceptions, people were worrying about runaway population growth even as global growth continued to decelerate and local growth for dozens of nations had halted or gone into reverse. As we know now, expectations of an aging population were correct, but fears of a heavy dependency burden turned out to

End of the Beginning

be overblown. Older people have continued to work for more years as they live longer in better health, and their productivity has been higher than was expected back in 2014. Not only are today's seniors healthier, their experience benefits the economy, and the work they do has been effectively complemented and augmented by machine intelligence.

Although most countries have enjoyed a softer demographic landing than 2014 commentators projected, we cannot afford complacency. Now that regional and national population contraction has become a *global* contraction, economic and social challenges will grow. On both an individual and collective level, the best response to these challenges is to greatly increase funding and other resources to do better at controlling and reversing aging in humans.

End of the Beginning

Part Two:

The Future of the Human Body

End of the Beginning

Chapter Four **Implanting Post-Human Intelligence in Human Bodies**

By John Hewitt

John Hewitt *is a scientific consultant for several neurological device startups. He also writes daily about cutting edge research in the field for various online publications including Extremetech, Physorg, and Medicalxpress. He has owned and operated CRE Precision, a biomedical device manufacturing company specializing in custom research instruments, since 2001. The main focus of his current efforts is to define, and later create, the first multifunction intraventricular devices for the brain. He can be reached at: John@Hewitt123.com*

Consumer electronic devices have drastically changed our lives, society, and the economy. Much of our interaction with the world is mediated by devices such as smartphones. Increasingly, they are our source of knowledge about our environment, as well as our means to control it. At the same time, as we age and our bodies begin to fail, many of us have begun to surrender control of parts of our bodies to implanted electronic devices. It is now commonplace to implant devices to pace our hearts or our brain, condition our blood, or feed data to our senses.

These technological advancements come with certain social and political complexities. In stark contrast to those devices we use

End of the Beginning

for interacting with the external world, control of the devices we use to interact with our inner space remains largely off limits to us. This situation, maintained by a combination of choking regulation and business monopolies of devices, is unstable, and soon will become untenable, leading to important and historically novel consequences.

Paradoxes of Technological Advance

There are fundamental paradoxes that can no longer be swept under the rug. On the one hand, no doubt for good reason, laws require strict licensing for those individuals who administer anesthetics and other care peripheral to any kind of major device implantation. They must be medical professionals with the proper training. The problem is that doctors cannot implant an unapproved device. Furthermore, when a device does manage to get approved, it is only approved for a specific pre-existing condition. Therefore, anyone seeking an implant for an off-label use of any kind -- even a simple tracking device or Bluetooth storage chip -- is excluded by law from the safety and comfort of officially sanctioned oversight.

Until recently, implant accessibility has not been much better for those who do qualify for them. Fortunately, things are already changing, and the walls which separate many of us from the implanted hardware devices we use or desire have already begun to be torn down. Much of the present implant business is controlled by a few large companies. They consider implant administration their private domain, with patient interaction taking place only through a physician, or as the case now may be, even robotic programmers. In mid-2013 however, things came to head when one user of a Medtronic pacemaker, Hugo Campus, sought to obtain data that his device generates about his heart.

Hugo had lobbied for direct and timely access to the raw data generated by his implant. This data is of critical use to him in helping to regulate the atrial fibrillation for which he was given the

End of the Beginning

pacemaker. At one point, Hugo had tweeted that while his $99 "Fitbit" emails him when it needs attention (such as when its battery needs a charge), his $30,000 Medtronic device emails his doctor instead. The irony of this situation was highlighted by none other than Medtronic CEO Omar Ishrak himself, who had tweeted data from his "Runtastic" fitness tracker, and described how motivating it was for him to have that particular data available.

Access to data from implants is but the beginning. The ability to program and control our implants is the next step.

Demand right now for qualified implant programmers is so high that surgeons have sought automated solutions to assist them. Having a huge backlog just to tune an implant is unacceptable. Parents for example, who might soon have a child's cochlear implant on a private home network might be able to set the gain, or perhaps the frequency map of the device on a daily basis to optimize the device for their child. An evening at the movies, a concert, or even fireworks may demand entirely different settings than those for school or piano practice.

Patient Control of Implants

Having the ability to control and program our implants will be transformative both for individuals and society as a whole for several reasons. If patients are ever to "become the customer" so to speak, within in our current medical care system, they will need to know not only what they want, but what they can get. Currently, cutting-edge implants are not simply handed to a patient with instructions to simply "go ahead and play around." Rather they are administered in the context of a specific clinical trial, where everything about the protocol is dictated by the device manufacturer, regulatory agency, and doctor. The results of the study are not necessarily beneficial to the patient at hand, but rather intended to be of use to larger community, namely, the next generation of implantees, in a "pay-it-forward" model which gives entry to the trial in the first place.

End of the Beginning

The result is that patients do not learn or communicate as much about their own implant and physiology as they could. Handing them the reins, with the freedom to change a rigid protocol according to their own needs or desires, will help to make individual patients the center of medical knowledge relevant to the device -- provided the patient agrees to make this knowledge open. If this is done, medical knowledge will not be ensconced in medical journals according to device, or for that matter, gene, but rather into a living database of sequence, lifestyle, and treatment indexed by individual.

Implant manufacturers, have started to recognize the need for greater patient involvement, and have begun to make devices that are more flexible. Many companies however, still offer only proprietary solutions when it comes to implant connectivity. As an example, the Nucleus 6 hearing processor manufactured by Cochlear is capable of analyzing an auditory scene and automatically adjusting itself to optimize performance. It also offers the first truly wireless capability, and can simultaneously pair with multiple Bluetooth devices. A device with these capabilities would be highly desirable not just to the hearing impaired, but to anyone seeking some measure of control over their auditory environment. The Nucleus 6, in particular, is able to operate with a wide range of Cochlear wireless accessories like remote microphones, phone and TV audio streamers. While this is a step in the right direction, most modern users would desire potential interoperability with any sound source, rather than just those of a particular manufacturer.

While many individuals will be unable to confidently reconfigure devices on their own, there will undoubtedly be many who simply will not stand for the crude impersonal interfaces now being suggested. A Medtronic pulse generator, for example, may have many parameters which can be controlled -- current or voltage mode, pulse rate, width, amplitude, or waveform. Understanding the effects of each manipulation is not trivial, but if the administrator of the implant is not in fact the user himself or herself, how can the implants be rapidly optimized and improved on an individual basis? Someone with an experimental brain stimulator today may not know a single other person with a similar device. But as implants gain penetration in society, users will become more confident to experiment when they can share experiences with others. Similarly when more than one person reports the same problem or benefit, the anecdote carries more weight.

Brain Implants

We have already moved well beyond simple implants that generate pulse trains to control tics and tremors. Today there are implants to control obesity, anorexia, blood sugar, seizure mood, anxiety, and even motivation. Often a device is implanted experimentally in one area, to treat something like anorexia, and it is observed that something else, like mood, is improved. In this way device use has already leapfrogged from one area to next, across the nearly the entire limbic system. The problem is that doctors tend to go for the areas of the brain which are likely to have the greatest effect. These are also the areas that can be the most dangerous when stimulated improperly. Recently it was reported that implants in an ancient part of the brain known as the habenula can control decision making. In other words stimulation can shunt ongoing rumination and force a decision or drive to action.

Generally speaking, most of these stimulation devices function in the same basic way. What distinguishes the result is how and where in the body or brain they are used. Perhaps the height of absurdity would be if companies seek to patent things as elementary as implant stimulation protocols. It would be disastrous to lock up specific knowledge of how to exercise control over our brains. On the other hand, the rewards of opening up implant technology will be unbounded. Current patents tend to be

End of the Beginning

focused more on using a particular kind of device for a certain ailment rather than a fine-grained approach to simulation details. However, information is rapidly becoming available about how brains are wired and in particular their individual differences or "connectomes" as they are called. The extent of protective and predatory patenting that might be possible under current law remains is troubling.

The Personal Singularity

For these reasons I would like to make the claim that the pace at which we reach the point commonly described as a technological singularity will be set primarily by the pace at which implant technology is effectively opened. The concept of a technological singularity means different things to different people -- typically, we take it to mean that the world begins to evolve so rapidly that it essentially become unpredictable on any timescale in which we attempt to observe or comprehend it. The definition we will adopt here focuses on the individual. In other words, the personal technological singularity would occur when your ability to augment your own knowledge with regard to your immediate environment and self, in addition to your own capabilities, progresses so rapidly that what you are changing into essentially becomes unpredictable to you.

Opening up implants means more than having access to the data they generate, or the ability to program and control them. Having complete access at the hardware level, and hence the ability to build the same, is not even sufficient for implants to gain massive penetration in society. What is needed is open, safe and affordable access to the ability to install, modify, and when needed, to remove the implants. For the few with sufficient resources to provision their own surgical suite, or jet to a country where good surgeons might be retained for an implant operation, this may not be a major issue. However, the sign of the times is that the top 0.1

End of the Beginning

% is probably not the group that tends to go in for radical body modification.

The Cyborg Movement

The group that is spearheading the cyborg movement is the so-called grinders or biohackers. These guys tend to operate through body-modification shops, or at home, in order to have hardware installed in their bodies. In mid-2013, biohacker Rich Lee had magnets implanted near his ear canals and successfully demonstrated remote transduction of electromagnetically-encoded sound signals. He had the surgery done at a body-modification shop, and provisioned the hardware himself. Many would argue there are many risks involved even in this simple kind of procedure. In the larger view, what happens for example, when he needs an MRI later in life?

The effects of MRI can be predicted, but not be completely known until you actually have one. You can study and model potential effects, but the fact is you may not have considered them all. Furthermore, you may have exercised excess caution in some of the important ones. Medtronic, for example, has fairly detailed MRI guidelines for many of its devices, and they have recently released a fully MRI compatible implant. Those with older implants, and their physicians, must navigate the complex landscape of magnet strengths, implant type, location, the body part to be imaged, and the power state of the implant in order to define some measure of safety. One is reminded of the attempts of airlines to define some measure of safety for powering up WIFI devices on takeoff and landing. Perhaps the important question here, is will the medical establishment shoulder the risk of allowing you inside the scanner in the first place? Perhaps not. Quickly we see that the act of implanting even simple devices puts you into a new medical category where the standard insurance model for even simple procedures no longer will suffice for you.

What sets apart the new breed of grinders and biohackers

End of the Beginning

from ordinary implant recipients is the desire to have multiple implanted devices to enhance baseline capabilities, and to have them working all together. Companies would not be likely to favor multiple implants unless the same company produced or sold all of them. Even then, interference issues and other unknowns could be very difficult for them to deal with. may be a headache that they are less than eager to deal with. Consider the situation where your aging father has a glucose pump, hearing device, defibrillator, and fornix pacemaker implanted for his Alzheimer's disease. You, perhaps as the primary administrator for his devices, may be faced with the decision to delay a battery replacement on one device until after a trip to Disneyland. If instead, the devices shared power and communicated through a body-area network (BAN) you might have no trouble setting device priorities that ensures safety for the trips.

Interoperability of Implants

With this example, it becomes obvious that the only implants with any future might be those that are sufficiently open and interoperable with each other. Responsibility for the implant operation will be increasingly distanced from the surgeon who installs it, or the company that makes it, and instead directed towards the user who owns it. Devices that presume to act on fundamental drives such as nourishment, satiety, mood and happiness, have by their nature a fair degree of unpredictability in use. These kinds of implants also raise issues of legal responsibility. Already in criminal defense we have seen claims that medications impact sanity or responsibility. When criminal defendants are influenced by implants that modify their minds, the focus may be more on modifying the implants to change their behavior rather than on punishment.

Implants to Enhance Abilities

When implants reach a level of sophistication that they can act

End of the Beginning

to enhance ability, rather than just partially restore or mitigate decline, we will have entered a whole new era. A memory implant, once introduced for the elderly will soon be particularly sought by students, as well as by parents for their children. When implant technology advances faster than social institutions can adapt to it, a level playing field will be impossible to create. These are not simply concerns for the future, but are already concerns now. Consider the technological revolution underway within the area of Deep Brain Stimulation (DBS). Relatively primitive devices (circa 2000) were designed to deliver constant, repetitive stimulation, regardless of what else might be going on in the brain. Today in 2013, the reality is that many of these early implantees do not even know if their device still works, let alone what condition the batteries might be in. If these individuals cannot even determine whether or not their implants are active, determining their long-term effectiveness is difficult. Fortunately, researchers at many research institutions, such as Oxford University, have been developing a better kind of DBS -- one which records the underlying brain activity and adapts the stimulation to ongoing conditions in the brain.

This kind of adaptive device is initially being targeted for treatment of Parkinson's disease, because in this case, a fair bit is already known about what kinds of brain activity are associated with particular pathological behaviors. What we now have here is essentially a BCI (Brain Computer Interface) mated to DBS. We normally think of a BCI as a device that a handicapped person might use to control something using just their brain. But with an adaptive DBS design, the BCI is actually being used to control the brain itself, rather than some external device. At this writing, Medtronic has a DBS device already in trials which will fully close the feedback loop between stimulation and ongoing, or resultant, brain activity. Other companies like Neuropace have similar devices under development which can do much the same for seizure control -- first detecting signs that they are about to form,

End of the Beginning

and then zapping them before they gather momentum.

Memory Implants

It is here, at the intersection of the general purpose adaptive DBS system with the dedicated sensory prosthesis, as typified by the new Nucleus 6 design, that we are beginning to have sufficient resources to develop a true memory implant. There has been much talk, particularly at venues such as the recent GF2045 conference held in New York, of memory implants based on direct interaction with the hippocampus. I would suggest that while that type of implant may eventually become a reality for many, it will not be the way that true memory prostheses will come to fruition. For those who already have lost much of their cognitive function to disease or other pathology, and therefore may have little else to lose, directly targeting the hippocampus or its output pathways, like the fornix, may be an acceptable last-ditch strategy. But for those whose brains are largely intact, it should soon become obvious that adopting this form of implantation as an initial approach, would lead to much degraded performance in other critical aspects of brain function.

Sensory and Cognitive Processes

So-called cognitive areas, like the hippocampus, are highly dissociated from both sensory input, and from motor output. This separation from any real world metric makes direct interaction with them more precarious, and the longer term consequences unknown. By contrast, inserting increasingly sophisticated hardware at the level of interaction between already implanted sensory and motor areas should be a bit more predictable. For example, an adaptive DBS implant placed in the subthalamic nucleus of someone just beginning to show signs of Parkinson's may incorporate data not only from local brain activity, but from various sensors in and about the body. A hearing and vestibular implant that is already recording data regarding balance, inertial movements, orientation or reckoning, may very well be the best source of information available to inform a motor implant.

When one is navigating about the home, or up and down steps, a completely different set of stimulation protocols from those used when outside or in a new environment may likely be desired. The initial mappings of sensory data to motor commands, which might be said to have formed the very basis of the evolution of the higher functions of the brain itself, may therefore also likely form the basis of the first memory implants. There will be many ways to go about making such mappings, and the individual, personalized nature of their eventual design will raise many new questions for society as a whole.

Changing Social Norms

A significant concern in moving towards a technological singularity is that heavily implanted people would tend to diverge from the mainstream along their own peculiar tangents. Communication of common ideals would be therefore be impeded, and then supplanted with less universally held beliefs. The role of society in establishing the norm will be increasingly important, as the individuals that compose it come to make up an increasingly heterogeneous group. Those spearheading the technically-informed biohacker movement are among those who lead the path to the singularity. Presently they must, out of necessity, act at significant variance to the norm. To say that disrupting the norm is their main goal would be incorrect -- rather, they seek instead to establish a new and in their view, better norm. Biohackers, while already demonstrating a willingness to undertake fairly radical body modification outside the medical establishment, are not in general, a reckless lot, but rather tend to act only after thorough research. Undoubtedly they are to be the group that hits upon devices that may come to be true memory prostheses -- and by that we mean a prosthesis that is fully active at the level of

End of the Beginning

consciousness.

Emerging Technologies and the Inner Voice.

When we talk about memory implants being active at the conscious level, the form in which they are experienced can include a number of sensory and motor modalities. Present research gives some hint as to which modalities may initially be favored. For example, efforts to decode dreams, from either MRI scans, or electrical activity of the visual cortex, are underway in several labs. At this point, it is difficult to assign any measure of success to these projects, but that they now exist is already telling. To build a prosthesis which can actually decode a visual scene that is dreamt or imagined by its user, and then use that data to create an image or movie, might be tough to imagine using our current technology. However, if we assume that any such imaginations are indeed the result of activity in the visual cortex, then perhaps one elegant way to attempt to read them out would be to access the signals from those neurons which project out of this area, rather than invasively probe the cortex itself.

A device placed inside the lateral ventricle of the brain, for example, could potentially record signals heading towards the visual cortex, and any return projections from the visual cortex heading back down to lower visual areas. In fact, these two projections form part of the wall of the ventricle itself, and would be readily accessible at many redundant points along their extent. Together with information about the connections and directions of these axons, recording or imaging at these locations could potentially isolate the refined visual activity relevant to the imagination from the activity related to any underlying, low-level background visual processing.

Another point of interaction of prostheses with the mind would potentially be through the inner voice. The neural origin and localization of the inner voice is pretty much unknown today, but with a suitable prosthesis it could potentially be probed

End of the Beginning

directly. Here we begin to tread over territory where the true nature of the self might begin to be revealed. The visual thinker, for example, may tend to rely on internal imagery to formulate ideas and plan actions, while a more verbally-oriented thinker may construct concepts in a more serial fashion, relying more on word-encapsulated or phrase-driven directives. Toward one extreme of visual thinking, there might be a congenitally deaf individual who, even with a cochlear implant, never develops what most of us understand as an inner voice. Toward the other extreme might lie certain types of schizophrenics, who with so little apparent control over or rational understanding of their various internal processes, mill about in a perpetually confused and disabled state.

An implant which is accessible via the inner voice is perhaps the ultimate interface many people would desire. This is a two-fold concept wherein the implant can not only be commanded by the inner voice, but the implant can also make itself known to the conscious mind in the form of the inner voice - a two-way talkie so to speak. For most of us, our desires can be expressed anywhere on the voice continuum from completely subvocal expression, through a whisper, to a fully active voice. When intercepted by an implant, these desires or imperatives can branch into possible outputs directed at both the internal and external environment. If our implants present word-based, or numerically described data to us in the form of a recognizable voice, it may be given a particular flavor, timbre or place of origin so that it remains internally distinct and consistent. Exactly how to do this at the neural level remains to be seen, but we have clues already about unique character gets assigned by the brain to unique sources.

A final area worth considering is the potential of a prosthesis which interacts with the sense of self, particularly with regard to its distinction from nonself, and from others. Recent experiments have shown that the line of demarcation between self and nonself is constantly rebuilt by the brain to maintain internal consistency with sensory input. Not only can realistic, but nonexistent, body

End of the Beginning

parts be convincingly contrived, but whole bodies can be imbued with a sense of direct habitation using appropriate artificially generated stimuli. For example painful stimuli can be made to be felt as if they emanated from distant points, or phantom and supernumerary limbs can be virtually generated and incorporated into the sense of self. Even the experience of body location can be given a new vantage point, for example, by simply synchronizing a flashing virtual body with the heartbeat or other body signals.

To what productive end these kinds of possible manipulations will eventually be put remains to be seen. As far as the interaction of the self with others in society, we could undoubtedly benefit from a boost in empathy. While it makes little sense at this point to discuss the details of empathy implant technology, we might hope that a heightened sense of awareness of the self, and of others in environment, might lead naturally to this end.

End of the Beginning

Dialogue 4.1

With John Hewitt, Ben Goertzel and Ted Goertzel

Ben: You note that "it becomes obvious that the only implants with any future might be expected to be those that are sufficiently open and interoperable with each other.".... I can see the logic here. But yet, the US health care system is not particularly logically and rationally structured at present, and doesn't really match a rational observer's notion of what kind of health care system would make sense. So it seems possible that due to political and institutional reasons, the open/interoperable approach might not be the dominant one, at least initially.

Do you think there might be issues with the medical establishment wanting to retain control over implant devices rather than putting so much power in the user? Do you have any thoughts on how conflicts of this nature might play out in practice?

John: Sure, it will be a battle every step of the way. Here the medical establishment might be said to be a partnership between the device company and care providers. The hospital, neurologist, surgeon, programmer, and even the device maker might all be fluid over time whereas the patient remains the constant. Already now one can register to participate in online symposia organized by medical publishers and companies on this very issue -- medical device interoperability. Ownership of devices is another huge issue and law or policy varies from country to country. In the UK, for example, any device implanted into the body becomes the property of that person and even if removed becomes part of their estate. Incidentally, that can be a good way to get your pet a pacemaker if grandma had one and passed. External prosthetics are often considered loaners these days, but once a device is part internal and part external, we have a whole new ballgame.

End of the Beginning

Ben: Your vision of "a prosthesis which interacts with the sense of self, particularly with regard to its distinction from nonself, and from others" is fascinating and tantalizing. I'm reminded of Zen Buddhism and various wisdom traditions that talk about transcendence of the ego and individual self as part of the path to spiritual enlightenment. What do you think would be the social and psychological impact of implants that encourage the human mind to transcend the self and perceive itself from a variety of different vantage points?

John: I think perhaps we are just beginning to get an inkling of how we might begin to link such phenomena physiologically from the insights of those with phantoms limbs, phantom body, or out of body experiences. An implant that tweaks the vestibular system, for example, into a situation where the only consistent reality to experience is one where you have in addition to your own normal body, another aloof doppelganger lying horizontally up and slightly to the right of you, which none-the-less is fully able to see the original you, might be able to deliver the kind Zen experience you just related on demand, rather than subject to the whims of a peculiar metabolism or fit. When we begin to dedicate greater portions of our sensory input to showing our brains ever more detailed representations of our brain's own activity we will find we have an interesting mapping problem our hands.

Ted: Why do you assume that people would want to consciously control their implants? I suspect that a diabetic that would get a pancreas implant would be delighted NOT to have to check his or her blood sugar and make conscious adjustments. Wouldn't the parents of a child with a cochlear implant be happiest if they could just forget about it, or perhaps take the child to an audiologist once a year? Would not a man with an erectile dysfunction be happiest with an implant that allowed him to focus on the love-making rather than on the mechanics?

End of the Beginning

John: I would imagine that the most flexible implants will be controlled at both the autonomic and the conscious level, much like smiling or breathing. Autonomic control of breathing might be great while you are reading, but to smell a rose, go for a swim, or add volume to your voice, we find it convenient to issue a system override. Consider the sympathetic control system you now have that administers your adrenal glands. To completely hack it and cause it to dump three grams of the expensive adrenalin that it has been carefully synthesizing over the last week into your blood all I might do is sneak up on you and say "boo". An implant that can better interact with your conscious state to set the interrupt priority for your reptilian responses could be advantageous. The same perhaps for the mechanics of the erectile dysfunction you mention. Control of things like blood pressure throughout the many relevant critical points of your body is probably not done best in single pill form. You might call the plays for things like that ahead of time in a huddle, but it would be handy to always have the option to later call an audible.

For augmentative cochlear implants it would seem that as they become more common, new features and patches for them would start to proliferate as fast as apps on Google Play. If one particular channel of your implant has been chosen, for example, as a primary accessory to you inner voice, the ability to consciously tune it in or out on a seconds notice might be desirable.

End of the Beginning

Chapter Five **The Singularity and the Methuselarity: Similarities and Differences**

By Aubrey de Grey

Aubrey de Grey *is a biomedical gerontologist based in Cambridge, UK and Mountain View, California, USA, and is the Chief Science Officer of SENS Research Foundation, a California-based 501(c)(3) charity dedicated to combating the aging process. He is also Editor-in-Chief of Rejuvenation Research, the world's highest-impact peer-reviewed journal focused on intervention in aging. His research interests encompass the characterization of all the accumulating and eventually pathogenic molecular and cellular side-effects of metabolism ("damage") that constitute mammalian aging and the design of interventions to repair and/or obviate that damage. Dr. de Grey is a Fellow of both the Gerontological Society of America and the American Aging Association, and sits on the editorial and scientific advisory boards of numerous journals and organizations. He can be reached at:* **aubrey@sens.org**

The singularity: a uniquely unique event in humanity's future.

"Unique" is, of course, an over-used word to describe momentous events – arguably, even more over-used than "historic." How, then, can I dare to describe something as uniquely unique?

Well, I will begin by pulling back a fraction from that description. There are actually, in my view, two possible events in humanity's future that merit this description. But I do not feel very bad about this qualification, because I believe that those two events are, in all probability, mutually exclusive. The singularity is one; the demise of humanity is the other. Hence my choice of the indefinite article: the singularity is not "the" uniquely unique event in humanity's future, because it may not occur, but if it does occur, nothing comparable will either precede or follow it.

The singularity has been defined in many related but subtly distinct ways over the years, so let me begin my discussion of it by making clear what I mean by the term. I adhere to the following definition: "an asymptotically rapid increase in the sophistication of technology on whose behavior humans depend." I do not use the word to mean, for example, "the technological creation of smarter-than-human intelligence" (which is the definition currently given by SIAI, the Singularity Institute for Artificial Intelligence[45]) – despite my agreement with the view that the technology most likely to bring about the singularity (and, indeed, the one that was originally used to define it) is precisely the one that SIAI study, namely recursively self-improving artificial intelligence (of which more below). I am sticking to the more abstract definition partly because it seems to me to encapsulate the main point of why the singularity is indeed uniquely unique, and partly because it will help me to highlight what distinguishes the singularity from the Methuselarity.

One aspect of my definition that may raise eyebrows is its use

---

[45] *The organization that was named SIAI in 2009, when this chapter was originally written, has since been rebranded as MIRI, the Machine Intelligence Research Institute. The central focus and perspective of the organization have not changed, however.*

End of the Beginning

of the word "asymptotically" rather than "exponentially." I feel sure that von Neumann would agree with me on this: the mere perpetuation of Moore's Law will not bring about the singularity. A gravitational singularity, which is of course the etymological source of the term, is the center (not, I stress, the event horizon) of a black hole: the point at which the force of gravity is infinite – or, to be more precise, the point arbitrarily near to which gravity is arbitrarily strong. The distance between the singularity and any point of interest (inside or outside the event horizon) at which gravity is finite is, of course, finite. This is an asymptotic relation between distance and strength: if point X is distance Y from the singularity, it is not possible to travel from X, along the line between X and the singularity, by a distance greater than Y, and experience continuously increasing gravity. Exponential (though not inverse exponential! – see below) relations are not like this: they have no asymptote. If the force of gravity exerted by a particular body were exponential (though still increasing with decreasing distance from the body), the relation between distance from that body and gravity exerted by it would be defined in terms of distance from the point furthest away from it ("on the other side of the Universe"). Call the gravity exerted at that point X and suppose that the gravity exerted at half that distance from the body is 4X (which is the same as for gravity in real life). Then the gravity exerted by the body at a point arbitrarily close to it is not arbitrarily large – it is just 16X, since that point is exactly twice as far away from the point of minimum gravity as the 4X point is.

Having belabored this point, I now hope to justify doing so. Will the technological singularity, defined as I define it above, happen at all? Not if we merely proceed according to Moore's law, because that does not predict infinite rates of progress at any point in the future. But wait – who's to say that progress will remain "only" exponential? Might not progress exceed this rate, following an inverse polynomial curve (like gravity) or even an inverse exponential curve? I, for one, don't see why it shouldn't. If we

End of the Beginning

consider specifically the means whereby the Singularity is most widely expected to occur, namely the development of computers with the capacity for recursive improvement of their own workings, I can see no argument why the rate at which such a computer would improve itself should not follow an inverse exponential curve, i.e. one in which the time taken to achieve a given degree of improvement takes time X, the time taken to repeat that degree of improvement is X/2, then X/4 and so on.

Why does this matter? It might matter quite a lot, given that (in most people's view, anyway) the purpose of creating computers that are smarter than us is to benefit us rather than to supersede us. Human intelligence, I believe, will not exhibit a super- exponential rate of growth, because our cognitive hardware is incompatible with that. Now, I grant that I have only rather wishy-washy intuitive reasons for this view – but what I think can be quite safely said is that our ability to "keep up" with the rate of progress of recursively self-improving computers will be in inverse relation to that rate, and thus that super-exponentially self-improving computers will be more likely to escape our control than "merely" exponentially self-improving ones will. Computers have hardware constraints too, of course, so the formal asymptotic limit of truly infinite rates of improvement (and, thus, truly infinite intelligence of such machines) will not be reached – but that is scant solace for those of us who have been superseded (which could, of course, mean "eliminated") sometime previously. There is, of course, the distinct possibility that even exponentially self-improving systems would similarly supersede us, but the work of SIAI and others to prevent this must be taken into account in quantifying that risk.

Let us now consider the aftermath of a "successful" singularity, i.e. one in which recursively self-improving systems exist and have duly improved themselves out of sight, but have been built in such a way that they permanently remain "friendly" to us. It is legitimate to wonder what would happen next, albeit

End of the Beginning

that to do so is in defiance of Vinge. While very little can confidently be said, I feel able to make one prediction: that our electronic guardians and minions will not be making their superintelligence terribly conspicuous to us. If we can define "friendly AI" as AI that permits us as a species to follow our preferred, presumably familiarly dawdling, trajectory of progress, and yet also to maintain our self-image, it will probably do the overwhelming majority of its work in the background, mysteriously keeping things the way we want them without worrying us about how it's doing it. We may dimly notice the statistically implausible occurrence of hurricanes only in entirely unpopulated regions, of sufficiently deep snow in just the right places to save the lives of reckless mountaineers, and so on – but we will not dwell on it, and quite soon we will take it for granted.

A reasonable question to ask is, well, since even a super-exponentially self-improving AI will always have finite intelligence, might it not at some point create an even more rapidly self-improving system that could supersede it? Indeed it might (I think) – but, from our point of view, so what? If we have succeeded in creating a permanently friendly AI, we can be sure that any "next-generation" AI that it created would also be friendly, and thus (by the previous paragraph's logic) largely invisible. Thus, from our perspective, there will only be one singularity.

In closing this section I return to my claim that the singularity and the demise of humanity are, in all probability, mutually exclusive. Clearly if our demise precedes the singularity then the singularity cannot occur. Can our demise occur if preceded by the singularity? Almost certainly not, I would say: the interval available for our demise between the development of recursively self-improving AI and the attainment by that AI of extremely thorough ability to protect us (even from, for example, nearby supernovae) will be short. (I exclude here the possibility that the singularity will occur via the creation of AI that is not friendly to

117

us, only because I think humanity's life expectancy in that scenario is so very short that this is equivalent from our point of view to the singularity not occurring at all.) The "area under the curve" of humanity's probability of elimination at any time after the singularity is thus very small. I am, of course, discounting here the possibility that even arbitrarily intelligent and powerful systems cannot protect us from truly cosmic events such as the heat death of the Universe, but I agree with Deutsch that this is unlikely given the time available.

The Methuselarity: the Biogerontological Counterpart of the Singularity

In a recent interview, Watson was asked what would be the next event in the history of biology that would compare in significance to his and Crick's discovery of the structure of DNA, and he replied that there would never be one. I think he was correct. However, I agree with him only if I am rather careful in defining "biology" as the discovery of features of the living world, and excluding biotechnology, which for present purposes I define as the exploitation of such discoveries. In biotechnology I believe that there will certainly be a counterpart, something that will outstrip in significance every other advance either predating or following it: the Methuselarity.

For almost a decade following my graduation in 1985, I conducted research in artificial intelligence. I switched fields to biogerontology shortly after becoming aware that the defeat of aging was vastly less on biologists' agenda than I had hitherto presumed. I was not, at that time, aware of the concept of recursively self-improving AI and the singularity, though perhaps I should have been. But even if I had been, I think I would still have made the career change that I did. Why?

Humans are very, very good at adjusting their aspirations to match their expectations. When things get better, people are happy – but if they stay better and show every sign of continuing that

End of the Beginning

way, people become blasé. Conversely, when things get worse people are unhappy, but if they stay worse and show every sign of continuing that way, people become philosophical. This is why, by all measures that have to my knowledge been employed, people in the developed world are on average neither much happier nor much less happy now than they were when things were objectively far worse. This is a good thing in many ways, but in at least one way it is a problem: it dampens our ardor to improve our lives more rapidly. In particular, it depletes the ranks of "unreasonable men" to whom Shaw so astutely credited all progress. There are far too few unreasonable men and women in biology, and especially in biogerontology. I am proud to call myself an exception: someone who is comfortable devoting his life to the most important problems of all, even if they appear thoroughly intractable. In my youth, I felt I could make the most difference to the world by helping to develop intelligent computers; but when I discovered the truth about biologists' attitude to aging I knew that I could make even more difference in that field.

Why is aging so important? Aging kills people, yes, but so do quite a few other things – and moreover, life is about quality as well as quantity, and intelligent machines might very greatly improve the quality of life of an awful lot of people, not least by virtue of providing essentially unbounded prosperity for all.

Even if we take into account the fact that aspirations track expectations, such that what really matters is to maintain a good rate of improvement of (objective) quality of life, it is hard to deny that the development of super-intelligent machines will be of astronomical benefit to our lives. But let's be clear: quantity of life matters too. There is a well-established metric that folds together the quality and quantity benefits of a given technological or other opportunity: it is the "quality-adjusted life year" or QALY.

Historically, mainstream biogerontologists have been publicly cautious regarding predictions of the biomedical consequences of their work, though this is gradually changing. But even privately,

End of the Beginning

few biogerontologists have viewed aging as amenable to dramatic change: they have been aware that it is a hugely multi-faceted phenomenon, which will yield only incrementally to medical progress if it yields at all. This places them in a difficult position when arguing for the importance of their work relative to other supplicants for biomedical research resources. Yes, there is always a benefit to a QALY, and yes, progress against aging will deliver QALYs – but the force of this argument is diminished by two key factors, namely the probability of success (which biogerontologists cannot provide a conclusive case for being high) and the entrenched ageism in society, which views it as "fair" to deprioritize health care for the elderly. This quandary is well illustrated by the current "Longevity Dividend" initiative, which seeks to focus policy - makers' minds on the ever -dependable lure of lucre associated with keeping people youthful, rather than on the moral imperative.

But this is in the process of changing – indeed, of being turned on its head. This is for one reason and one only: it is becoming appreciated that aging may be amenable to comprehensive postponement by regenerative medicine. And the reason that makes all the difference is because it creates the possibility – indeed, the virtual certainty – of the Methuselarity.

Having tantalized you for so long, I cannot further delay revealing what the Methuselarity actually is. It is the point in our progress against aging at which our rational expectation of the age to which we can expect to live without age-related physiological and cognitive decline goes from the low three digits to infinite. And my use here of the word "point" is almost accurate: this transition will, in my view, take no longer than a few years. Hence the – superficial – similarity to the singularity.

Regenerative medicine, by definition, is the partial or complete restoration of a damaged biological structure to its pre-damaged state. Since aging is the accumulation of damage, it is in theory a legitimate target of regenerative medicine, and success in such a venture would constitute bona fide rejuvenation, the restoration of a lower biological age. (The bulk of my work over the past decade can be summarized as the elaboration of that "theory" into an increasingly detailed and promising project plan for actual implementation.) This rejuvenation would not be total: some aspects of the damage that constitutes aging would be resistant to these therapies. But not intrinsically resistant: all such damage could in principle be reversed or obviated by sufficiently sophisticated repair-and-maintenance (i.e., regenerative) interventions. Thus arises the concept of a rate of improvement of the comprehensiveness of these rejuvenation therapies that is sufficient to outrun the problem: to deplete the levels of all types of damage more rapidly than they are accumulating, even though intrinsically the damage still present will be progressively more recalcitrant. I have named this required rate of improvement "longevity escape velocity" or LEV.

It is important to understand that LEV is not an unchanging quantity, as it might be if it were a feature of our biology. Rather, it will vary with time – and exactly how it will probably vary is a topic I address in the next section. LEV will, however, remain non-zero for as long as there remain any types of damage that we cannot remove or obviate. Thus, the formal possibility exists that we will at some point achieve LEV but that at some subsequent date our rate of progress against aging will slip back below LEV. However, I have claimed that this will almost certainly not happen: that, once surpassed, LEV will be maintained indefinitely. This claim is essentially equivalent to the claim that the Methuselarity will occur at all: the Methuselarity is, simply, the one and only point in the future at which LEV is achieved.

The Singularity and the Methuselarity: Some Key Differences

Having described the singularity and the Methuselarity individually, I now examine how they differ. I hope to communicate that the superficial similarities that they exhibit

End of the Beginning

evaporate rather thoroughly when one delves more deeply.

Perhaps the most important contrast between the singularity and the Methuselarity is the relevance of accelerating change. In the first section of this essay I dealt at some length with the range of trajectories that I think are plausible for the rate of improvement of self-improving artificial intelligence systems – but it will have been apparent that all the trajectories I discussed were accelerating. It might intuitively be presumed that, since aging is a composite of innumerable types of damage that accumulate at different rates and that possess different degrees of difficulty to remove, our efforts to maintain youth in the face of increasing chronological age will require an accelerating rate of progress in our biomedical prowess. But this is not correct.

The central reason why progress need not accelerate is that there is a spectrum not only in the recalcitrance of the various types of damage that constitute aging but also in their rates of accumulation. As biomedical gerontologists, we will always focus on the highest-priority types of damage, the types that are most in danger of killing people. Thus, the most rapidly-accumulating types of damage will preferentially be those against which we most rapidly develop repair-and- maintenance interventions. There will, to be sure, be "spikes" in this distribution – types of damage that accumulate relatively rapidly and are also relatively hard to combat. But we are discussing probabilities here, and if we aggregate the probability distributions of the timeframes on which the various types of damage, with their particular rates of accumulation and degrees of difficulty to combat, are in fact brought under control, the conclusion is clear: we are almost certain to see a progressive and unbroken decline in the rate at which we need to develop new anti- aging therapies once LEV is first achieved. (I do not mean to say that this progression will be absolutely monotonic – but the "wobble" in how rapidly progress needs to occur will be small compared to the margin of error available, i.e. the margin by which the average rate of progress

End of the Beginning

exceeds LEV.) This conclusion is, of course, subject to assumptions concerning the distribution of these types of damage on those two dimensions – but, in the absence of evidence to the contrary, a smooth (log-normal or similar) distribution must be assumed.

The other fundamental difference between the singularity and the Methuselarity that I wish to highlight is its impact on "the human condition" – on humanity's experience of the world and its view of itself. I make at this point perhaps my most controversial claim in this essay: that in this regard, the Methuselarity will probably be far more momentous than the singularity.

How can this be? Surely I have just shown that the Methuselarity will be the consequence of only quite modest (and, thereafter, actually decreasing) rates of progress in postponing aging, whereas the singularity will result from what for practical purposes can be regarded as infinite rates of progress in the prowess of computers? Indeed I have. But when we focus on humanity's experience of the world and its view of itself, what matters is not how rapidly things are changing but how rapidly those changes affect us. In the case of the singularity, I have noted earlier in this essay that if we survive it at all (by virtue of having succeeded in making these ultra-powerful computers permanently friendly to us) then we will move from a shortly-pre-singularity situation in which computers already make our lives rather easy to a situation in which they fade into the background and stay there. I contend that, from our point of view, this is really not much of a difference, psychologically or socially: computers are already far easier to use than the first PCs were, and are getting easier all the time, and the main theme of that progression is that we are increasingly able to treat them as if they were not computers at all. It seems to me that the singularity may well, in this regard, merely be the icing on a cake that will already have been baked.

Compare this to the effect of the Methuselarity on the human condition. In this case we will progressively and smoothly improve our remaining life expectancy as calculated from the rate of

End of the Beginning

accumulation of those types of damage that we cannot yet fix. So far, so boring. But wait – is that the whole story? No, because what will matter is the bottom line, how long people think they're actually going to live.

These days, people are notoriously bad at predicting how long they're going to live. There is a strong tendency to expect to live only about as long as one's parents or grandparents did (just so long as they died of old age, of course). This is clearly absurd, given the rapid rise of life expectancies throughout the developed world in the past half- century and the fact that, unlike the previous half-century, that rise has resulted from falling mortality rates at older ages rather than in infancy or childbirth. It persists, I believe, simply because the rise in life expectancy has been rapid only by historical standards: unless one's paying attention, it's not been rapid by the standards of progress in technology, so it easily goes unnoticed.

This will not last, however. As the rate of improvement in life expectancy increases, so the disparity between that headline number and the age at which someone of any particular age can expect to reach also increases. But here's the crux: these two quantities do not increase in proportion. In particular, when the rate of improvement of life expectancy reaches one year per year – which, in case you didn't know, is only a few times faster than is typical in the developed world today– the age that one can expect to reach undergoes a dramatic shift, because the risk of dying from age -related causes at any given age suddenly plummets to near zero. And that is (another way of defining) the Methuselarity.

To summarize my view, then: the singularity will take us from a point of considerable computing power that is mostly hidden from our concern to one of astronomical computing power that is just slightly more hidden. The Methuselarity, by contrast, will take us from a point of considerable medical prowess that only modestly benefits how long we can reasonably expect to live, to one of just slightly greater medical prowess that allows us

End of the Beginning

confidence that we can live indefinitely. The contrast is rather stark, I think you will agree.

Epilogue: the Methuselarity and the Singularity Combined

Those who have followed my work since I began publishing in biogerontology may have noticed a subtle change in the way that I typically describe the Methuselarity's impact on lifespans. Early on, I used to make probabilistic assertions about future life expectancy; now I make assertions about how soon we will see an individual (or a cohort) achieve a given age.

The reasons for this shift are many; some are down to my improved sense of what does and does not scare people. But an important reason is that my original style of prediction incorporated the implicit assumption that the Methuselarity would occur in the context of a continued smooth, and relatively slow, rate of reduction in our risks of death from causes unrelated to our age. I only belatedly realized that this assumption is unjustified – indeed, absurd. And the singularity is what makes it particularly absurd.

Roughly speaking, we prioritize our effort to avoid particular risks of death on the basis of the relative magnitude of those risks. Things that only have a 0.01% risk per year of killing us may not be considered worth working very hard to avoid, because even multiplied up over a long life they have only a 1% chance of being our cause of death. This immediately tells us that such risks will move altogether nearer to the forefront of our concerns as and when the Methuselarity occurs (or is even widely anticipated), because the greater number of years available to get unlucky means that the risk of these things being our cause of death is elevated. It seems clear that we will work to do something about that – to improve the efficiency with which we develop vaccines, to make our cars safer, and so on. But there would appear to be only so much we can do in that regard: first of all there are things that we really truly can't do anything about, such as nearby

End of the Beginning

supernovae, and secondly there are quite a few moderately risky activities that quite a lot of us enjoy.

The singularity changes all that. What the singularity will provide is the very rapid reduction to truly minute levels of the risk of death from any cause. You may have thought that my earlier mention of snow reliably saving careless mountaineers was in jest; indeed it was not. Moreover, the residual risk that our rate of improvement of medical therapies against aging will at some point fall below LEV will also essentially disappear with the singularity. (Clearly the possibility also exists that the singularity will precede, and thus bring about, the Methuselarity – but that does not materially alter these considerations.)

One of my "sound bite" predictions concerning the Methuselarity is that the first thousand-year-old is probably less than 20 years younger than the first 150-year-old. The above considerations lead to a supplementary prediction. I think it is abundantly likely that the first million-year-old is less than a year younger than the first thousand-year-old, and the first billion-year-old probably is too.

The singularity and the Methuselarity are superficially similar, but I hope to have communicated in this essay that they are in fact very different concepts. Where they are most similar, however, is in the magnitude of their impact on humanity. The singularity will be a uniquely dramatic change in the trajectory of humanity's future; the Methuselarity will be a uniquely dramatic change in its perception of its future. Together, they will transform humanity... quite a lot.

End of the Beginning

Dialogue 5.1

With Aubrey de Grey, Ted Goertzel and Ben Goertzel

Ted: How would you describe progress toward the Singularity and the Methuselarity in the interval since 2009 when this essay was first published?

Aubrey: Since my work focuses squarely on the Methuselarity rather than the Singularity, I can't give as confident an answer as I'd like, but based on my limited knowledge I'd say that progress towards the Singularity has been greater. We've seen a whole new approach to AGI coming to the fore under the auspices of DeepMind, which was snapped up by Google, who have also acquired wide-ranging robotics expertise in the same way. Other leading IT companies also seem to be getting more active in this area, notably IBM with its work on real-time image analysis. What's less clear is whether there is any real progress on ensuring that any resulting AGI will be friendly. However, to my knowledge there is also very limited progress towards the self-comprehension that is generally thought to be a prerequisite for any kind of hard take-off, so maybe that's OK. The situation with the Methuselarity is comparatively slow and steady: each of the strands of damage-repair medicine that I believe we need in order to reach longevity escape velocity has progressed substantially, but nearly all of them remain at a pretty early proof-of-concept stage.

Ben: What do you think are the most exciting specific research developments in longevity research in the last few years?

Aubrey: The past few years have continued the pattern of the previous decade or so, in that the developments that have been the most exciting to me have generally not been overtly focused on longevity research, whereas those that have been longevity-centric

End of the Beginning

have tended to excite me less because I don't think they will "scale". Examples of exciting but not longevity-specific work are the obvious things like CRISPR (Clustered Regularly Interspaced Short Palindromic Repeats), which I expect to transform not only longevity research but also eventual anti-aging genetic therapies in the future; examples of what I think may not pan out in the clinic include the current mania over rapamycin. There are a few major exceptions, though: in particular, rejuvenation approaches have made great strides in credibility, as a result of work on mouse models designed to highlight the benefits of repairing a single type of damage, such as senescent cells.

Ben: If you had to put all of your research money into one research direction for the next five years, what would it be?

Aubrey: That's a really hard one if you're looking for a narrower answer than just "SENS" (Strategies for Engineered Negligible Senescence), because the divide-and-conquer nature of SENS means that all its components need to be made to work pretty well in order for the whole package to confer much benefit. I think my main criterion would be that the area needs to be one that's still being inadequately (if at all) pursued by others, and the great thing is that there aren't so many of those as there were - quite a few of the projects that we took on years ago have more recently become less of a priority for us as a result of being pursued by well-funded groups elsewhere. Probably the one that still stands out is extracellular cross-link breaking, where we have made some important initial progress lately but there is still general skepticism and neglect from the relevant research community.

Ted: According to the Wikipedia site on the world's oldest people, the longest unambiguously documented recorded life span was that of Jeanne Calment of France who died in 1997 at the age

End of the Beginning

of 122. The longest undisputed male life span was that of Jiroemon Kimura of Japan who died in 2013 at the age of 116. These numbers do not seem to be increasing. Do you anticipate that someone will reach 125 years of age soon? When can we expect to see the number of really old people increase?

Aubrey: You're absolutely right that these numbers are not increasing, and indeed this does not only apply to the world record: the number of centenarians is not rising nearly so rapidly as people were predicting only a decade ago. This has been a big surprise to a lot of people, but not to me, because to me it stands to reason that there is a minimum rate of accumulation of damage throughout life that is possible in the face of the non-negotiable aspects of living, notably breathing, and similarly a maximum amount of damage that a person can tolerate. What really *is* surprising is that the absolute world record holders seem to have survived a few years longer than anyone has the right to - Calment to 122, the runner-up (Sarah Knauss) to 119, and no one else beyond 117. So no, I think we are probably seeing a real phenomenon - a hard wall around 120 that will not be breached until we have fairly comprehensive rejuvenation therapies in place.

Part Three:

The Rollout and Impact of Intelligent Robots

End of the Beginning

Chapter Six **Robotics and AI: Impacts Felt on Every Aspect of Our Future World**

By Daryl Nazareth

Daryl Nazareth *is a medical physicist at Roswell Park Cancer Institute and an assistant professor at the University at Buffalo. He was recently elected President of the Upstate New York Association of Physicists in Medicine. His research interests involve applying novel techniques to problems in radiation oncology. His recent work includes the application of quantum computing technology to the optimization of beam delivery parameters; developing an augmented-reality approach to the patient setup problem; and streamlining optimization algorithms by simplifying the search space. He also mentors students at the high school, undergraduate, and graduate level, and is involved in educational workshops for students who are science and technology enthusiasts. He can be reached at **daryl.nazareth@roswellpark.org**.*

Introduction

It has been clear to futurists for many years, and to many others recently, that artificial intelligence (AI) and robotics have been advancing rapidly. But how will they progress in the coming decades, and what impact will this have on society as a whole? In this chapter we explore these futurological questions in a holistic way, taking into account predicted contemporaneous progress in other areas as well.

Any discussion of these topics requires some definitions. AI here is used in both the narrow sense (i.e., machine capability for a specific task, such as speech recognition) and human-level (or greater) general intelligence. *Artificial general intelligence* (AGI) will sometimes be used to denote the latter. Robotics is the technology of machines which can move and interact with their environment. Their level of autonomy can vary, and, as we shall discuss, will progress quickly in the upcoming years.

Of course, rapid strides are being made in other areas as well, including nanotechnology, biotechnology, computing, and cognitive science. There is much overlap between fields, and progress in one area can spur advances in others. For example, the cost of genetic sequencing has been reduced by a factor of one million in the last decade, due in part to improvements in silicon technologies. "Gene chips" measuring gene expression levels, leverage technologies originally invented in the context of computers and printers. And a better understanding of neurogenetics, gained partly using gene chips, is helping neuroscience advance, which will help drive certain types of AI forward. The division of sciences into disciplines is to some extent conventional and arbitrary; what we have is a web of interconnected, mutually accelerating scientific and technological advances.

A favorite pastime of many futurists is making predictions. They take great delight in following technology trends and speculating on their effects one or ten decades from now. Some of the great writings in this sphere, for example, Kurzweil's *The Singularity is Near* or Diamandis' *Abundance*, explore the consequences of exponentially-advancing technologies over the

End of the Beginning

next 10-40 years. Many online blogs are devoted to reporting on and discussing new scientific discoveries, and musing on the ways they may affect society. Recent mainstream news about Google's self-driving cars and 3-D printing has turned even casual observers on to the sport of future speculations.

One issue facing such prognosticators is that they often consider a given technological advancement *in vacuo*. That is, they will predict tech X's trajectory, and then speculate on how it will affect the world in, say, 20 years *with no other changes being considered*. They assume that all other progress will mysteriously halt until two decades hence, at which point X will be mature, and the earth will then be allowed to resume its rotation. The result is, generally, predictions that are overly mundane and uninspired. Worse yet, these speculations will almost inevitably be wildly inaccurate because technologies progress *in lockstep*, some more quickly than others, some encountering bumpier roads, some being rendered obsolete and supplanted. But progress they all do, and any prediction concerning the future of society must be made by considering *the sum total of all technological advances on all fronts*.

Examples of this type of tunnel vision are myriad:
- "As developing countries advance, demands for natural resources will exceed supplies." Yes, if there weren't concurrent improvements in energy generation, water and food technologies etc.
- "If life extension becomes common, it will bankrupt the Social Security program by 2045." This supposes SS will even be necessary after the economic upheaval caused by advances in automation and manufacturing.
- "Teachers of the future will use virtual reality in their lessons." While probably true during the next decade, this assumes schooling will continue in the conventional manner, and not be supplanted by cognitive enhancement.
- "Machines, lacking empathy and compassion, will never be able to replace doctors (or salesmen or judges etc.)." The

133

presumption is that human-like traits will not be replicable in non-biological materials. And further, that doctors and salesman and judges will play the same roles in future society as they do today.
- "Exploration and resource collecting on other planets will create an abundance of jobs." As with employment on Earth, automation will quickly replace human labor (in this case, likely before any is even used).

What these examples show is that predictions must consider technological progress in *all* directions. In addition, as society is more rapidly disrupted by new advances, it becomes increasingly difficult to see over the horizon with clarity. This is the essential characteristic of the approaching Singularity.

In this essay, we will discuss the changes being brought about by advances in technology, with an emphasis on AI and robotics. We will begin by reviewing the current status. This will be followed by predictions broken into different five-year epochs: 2015-2020, 2020-2025, and 2025-2030. Although AI and robotics will form the primary focus of our future gaze, we will always try to be mindful of disruptions caused by concurrent progress in other fields, including economics and government.

AI and Robotics Today

Most artificial intelligence researchers would agree on these two things: we currently have a tremendous amount of narrow AI, and we have nothing even close to general AI. Narrow AI is so commonplace that it's hardly necessary to list examples. A few well-known ones include speech recognition, medical imaging analysis, automated securities trading, and retail purchase recommendations. The exciting aspect of this field lies more in the development of new and less narrow applications. Personal assistant software, such as Siri and Google Now, are early glimpses of the potential of AI to provide valuable services (though many of their users would swear they're still in beta). And IBM's Jeopardy-

End of the Beginning

winning Watson has been trained in medicine, with a focus on oncology, in the hopes of creating a digital diagnostic assistant for clinical use.

Aside from early forays like last decade's Roomba, the general public has had very little exposure to developments in robotics. Many people are aware that the Da Vinci robot is used to perform some surgeries (though it is definitely not an autonomous system), and that commercial flights employ autopilot for the majority of the time aloft. But everyday robotic products are not yet popular, mainly due to their limited capabilities and relatively high costs. In addition, liability concerns prevent manufacturers from releasing products which have not been extensively tested and/or government approved.

It's interesting to observe how new developments in narrow AI can sometimes be characterized as "solutions looking for problems." For example, one current project involves the use of a webcam for facial monitoring, to determine if a PC user is busy concentrating on a task, in which case distracting emails can be delayed from popping up. Such a technology features an innovative and non-trivial accomplishment in narrow AI (gauging a person's level of engagement by analyzing live video). But its application (maintaining email silence) is so mundane, it almost makes one roll one's eyes.

Another point to make is that many of these awesome-yet-somehow-boring narrow-AI projects could be significantly enhanced if appropriate complementary robotics technologies were available. Take, for instance, the "networked kitchen," which is aware of the contents of the fridge and can suggest appropriate recipes. Interesting concept? Definitely. Useful application? Yes, perhaps. But imagine how much more helpful this kitchen would be if, after the human selected the desired entree, an automated system prepared the meal (or at least chopped the vegetables in good sous-chef fashion). Or consider the elder-care robot which provides companionship for your grandmother and can "remind

135

her to take her medicine" as some tech blogs promise. Doesn't this application amount to little more than a glorified alarm clock? Instead, suppose the robot could walk to the medicine cabinet at the appointed time, dispense the correct pills, and hand them to grandma along with a glass of water. In each of these cases, the killer app is being delayed by robots dragging their feet.

This situation is ripe for change in the next few years: robotic capabilities are evolving rapidly. There are several reasons for this:
- falling costs of components
- maturity of open-source efforts
- economic demand for automation
- high-profile robotics competitions
- self-driving car spin-off industries

Let's examine the first couple points here and return to the others later. Robotic components include sensors, actuators, motors, and communication hardware. Not only are their prices decreasing, but their capabilities are improving. For example, iRobot and Sandia National Labs have created a sub-$3000 robotic hand which can autonomously manipulate tools to change a tire (albeit very slowly and under the guidance of a high-level imaging system). Projecting forward a few years, robotic hands and associated systems will drop below the $1000 threshold to make them appropriate for mass-market applications.

The open-source robotics movement has resulted in the ROS (*Robot Operating System*). The California company Willow Garage created the ROS-based PR2 research robot and partnered with eleven institutions to further its development. Although Willow Garage has since announced it would transition to a commercial company, it will continue to support the PR2 platform. This $400,000 system has demonstrated navigating through a cluttered interior environment, manipulating objects, and (to the delight of many) fetching a can of beer from a refrigerator. There are many other examples of open-source robotics hardware and software, including quadcopters and other non-humanoid machines. As

End of the Beginning

occurred with PC software and later with the Internet, once hackers and hobbyists joined academic institutions in developing open-source systems, progress became extremely rapid.

2015-2020: Early Local Revolution

As described above, robotics currently lags AI development. However, the epoch starting in 2015 will be when the body will begin to catch up with the mind. There is one reason above all that will drive development in this area -- *the extreme economic incentive for automation*. From our present time until the Singularity, the financial benefits of automated labor will be a relentless force in the push for more capable robots.

What effect will this have on the mainstream? As with narrow AI (and indeed, most digital technology) there will be a trickle from the industrial sector gradually strengthening into a stream in the consumer sphere. The last few years have seen the replacement of unskilled labor by automation, for example, warehouse workers supplanted by the Kiva System and human sorters and packagers laid off in favor of robotic vision/manipulation hardware. Even innovations touted as "complementing workers" or "freeing up staff to do other things" actually result in fewer people being hired to achieve the same productivity: witness the Baxter robot from Rethink Robotics or automated ferrying systems used in hospitals (to say nothing of robotic pharmacists). In the 2015-2016 timeframe, these technologies will be mature and inexpensive enough to begin introducing into the home.

The economic incentive here will be of a different kind. Instead of corporations saving money on salaries, benefits, and pensions, customers will enjoy the benefits of automating simple household chores. This translation into more leisure time or lower housekeeping bills will fuel demand for increasingly sophisticated systems: robots which can clear dishes from the table, run them through the dishwasher, and then put them away in the cupboard; machines to wash floors and windows with minimal supervision;

End of the Beginning

and autonomous cooking systems. By 2017, these robots, typically of various shapes, though humanoid when convenient for their purposes, will be undergoing rapid improvement cycles, similar to that of smartphones half a decade earlier.

Another driver of AI and robotics development will be high-profile competitions. The Defense Advanced Research Projects Agency has sponsored the DARPA Robotics Challenge, held from 2012-2015, with the goal of spurring development of semi-autonomous machines to aid in disaster recovery. Many of the robotic skills required for the competition (e.g., opening a door, climbing a ladder, turning a valve) will prove very useful in the consumer market. The RoboCup, founded in 1997, aims to produce soccer-playing robots which will eventually compete against human athletes. The dexterity and machine-vision developments brought about by this annual contest will translate easily to the household arena. Incidentally, the RoboCup's goal of defeating the human 2050 World Cup Champions is grossly conservative and does not take into account the exponential pace of information technology improvements. A more likely target for robotic soccer domination is 2022.

Although a cursory glance at the North American landscape would not reveal many differences, changes in 2015-2020 will have occurred in other fields as well. Advances in immunotherapy and nanomedicine will begin to alter the way cancer is treated, with significantly better outcomes. 3D printing will have become mainstream, and mature enough to replace some traditional manufacturing processes. Devices such as Google Glass and smart watches will make communications and Internet connectivity seamless and ubiquitous. Solar power, rapidly increasing its presence, will have created a small but noticeable dent of about 3-5% in the world's energy portfolio.

Possibly the most significant technological advance of this epoch will be the self-driving car (SDC). Already close to prime time in 2015, development by Google and a host of automotive

End of the Beginning

companies will result in gradually-increasing automation in everyday vehicles. A short time will be required to surmount regulatory hurdles; the first fully-autonomous car will become available around 2017 and mainstream by 2019. The incentive for consumers to obtain SDC's will be more than that of convenience: insurance companies will delight in the lower accident rates, and offer huge discounts for customers who go robotic. The full impact on the world's cities and lifestyles will be felt shortly thereafter.

2020-2025: Full Worldwide Revolution

Much has already been written about how the SDC will alter the urban layout. Indeed, parking lots, meters, and garages/ramps will no longer be required in downtown areas, freeing up premium space for development. But that is just the beginning. Yes, one's car can drop one off at the front door of one's destination. And what will the car do while waiting to return for the pickup? One possibility being discussed is that the car can become available for rental to non-SDC owners. Another is that the car can run errands. Simply email your grocery list to your favorite supermarket, and your car will then drive there, where your items will be loaded into the trunk (probably by robots), and your account charged for the total.

By reducing gridlock and commute times, by the early 20's, the SDC will encourage cities to begin growing to mammoth sizes. Since construction costs will be reduced by automation and 3D printing, new buildings will sprout rapidly. People who today could not afford to live in a downtown region will be quite content in inexpensive but high-quality housing outside the city. Combined with electric motors and improved batteries, the SDC will revolutionize transportation the world over.

However, why should we stop at autonomous surface vehicles? Clearly, the third dimension is waiting to be exploited by people tired of the crowded not-even-2D routes down below. Autonomous flight, already possible in 2015, will become

End of the Beginning

widespread with better control systems and reduced costs. Several companies today (e.g., Parajet and Terrafugia) are developing flying cars, though it is not clear that the 2023 version will have much need to drive on a road. Indeed, the most efficient way to travel from Floor 65 of one building to Floor 127 of another will be simply to be picked up from the first balcony and flown to the second. The flight home can be accomplished in the same manner. This enhancement of the transportation industry will follow hot on the heels of the SDC, perhaps by only three years.

One can imagine how these advances will change the average person's lifestyle. But it must be kept in mind that concurrent improvements in other fields are going to leave their marks as well. In the 2020-25 epoch, automation will have decimated much of the employment landscape familiar in 2015. And the standard response of "but technology will also create many new jobs" will simply cease to be true at some point. What new positions may arise will not compensate for all the blue- and many white-collar jobs eliminated. In fact, we've seen that machines need not even perform as well as humans in order for automation to occur -- the ATM can only undertake limited teller tasks, but bank customers have learned to accept their shortcomings. What happens when machines have more fluid narrow AI, coupled with sophisticated navigation/manipulation *plus* the ability to work together and draw information from the cloud when necessary?

The answer of course -- necessarily reduced employment -- need not be the cause of hand wringing, as seems to be the case every time this topic is mentioned. What must occur, before the relatively near horizon of this epoch, is a monumental shift in how society views employment and its relationship to livelihood: a job can no longer be the sole or even primary means of earning a living. One phenomenon which will ease this transition will be the concomitant reduction in costs of living as automation takes hold. This is to be expected, since a major (or the main) component of the price of goods and services is the wages required to provide them.

End of the Beginning

In addition, other advances will produce further deflation: vertical farming and cultured meat reducing food costs; solar power (and perhaps a wildcard such as nuclear fusion) dropping energy prices; and advanced biotechnology, including stem cell therapies, genomics, and nanotechnology making healthcare more affordable.

A first step in the transition may be a redefinition of full-time status. When half the jobs have been automated, we could decrease the standard workweek to 20 hours. When only a quarter of them remain, we could reduce it further to 10 hours, and so on. This process will go some way to ensuring the employment landscape is not lopsided. The next step may involve a *Guaranteed Minimum Income*, available to every citizen, which is sufficient for basic necessities (whose costs will be ever dropping). The GMI will be funded by a tax on the businesses that provide goods and services. Note that this loop -- of taxing a company to pay the GMI, which consumers use to purchase the company's products -- would not be possible today because much of the company's expenses are its employees' wages. When full automation is implemented, almost all of the company's revenue becomes profit. And of course, any person would be free to augment his or her income with investments or by providing something which others *will* pay for, such as art or music.

The benefits of advanced automation technology will be felt most in developing countries. Just as these regions never embraced landlines but leapfrogged directly to cell phones, they will adopt late-model SDC's and robotics as soon as these devices become widespread and inexpensive. Developing nations are marked by large fractions of under-skilled populations. Advanced automation will provide an effect similar to injecting millions of skilled volunteer workers into the economic corpus. Housing, sewers, and other infrastructure could be constructed rapidly and at little cost. Toward the mid 20's, there will not be much descriptive truth to the term "developing."

End of the Beginning

Unlike the previous epoch, 2020-25 will see vast changes in society, even on the surface. And in contrast with the current situation discussed above, where there exist numerous narrow AI applications but no robotics to back them up, the early 20's will be marked by an abundance of cheap robotic labor. This will serve to elevate the standard of living immensely. Imagine the average person having a chauffeur, maid, chef, repairman etc., available at any time and affordable to the person after only working a few, if any, hours per week. And as we will discuss next, these changes will actually be insignificant when compared with those brought about by human-level AI.

2025-2030: Early Singularity

Perhaps no topic is as hotly discussed and debated in futurist circles as Artificial General Intelligence (AGI). Human-level AI is most interesting to us, even though it is generally agreed that we cannot currently conceive of the changes that superhuman machine intelligence will bring. The foremost question, of course, is *when*? Predictions abound, with most depending on the speed with which the human brain is being modeled and, ultimately, reverse engineered. This feat may or may not be necessary to produce AGI (the oft-used argument is that airplane wings do not imitate those of birds). A detailed discussion of the AI timeline would require an entire volume in itself. Here we will assume that AGI is first achieved around 2025. And to make this prediction more precise: *a rigorous version of the Turing test will first be passed by a machine in 2025*. We include "rigorous" to indicate that the test must be comprehensive (several hours in length), and the judge must be an expert in administering the test. Also, the human foil must have an incentive to convince the judge of his or her humanness. These rules will ensure that, to achieve this feat, a machine cannot rely on simple "chatbot" tricks -- it will be required to converse as an actual human.

Ray Kurzweil and others have pointed out that AGI-level

machines will combine the suppleness of human thought with the speed, accuracy, and recall of computers. However, Mr. Kurzweil, who predicts that AGI will arrive in 2029, brushes off this milestone with "People will say so what? We already have seven billion human brains." This dismissal ignores several critical points:

- Human brains generally must be remunerated for their efforts
- Human brains are neither inclined nor able to work more than a certain number of hours per week
- A human brain generally must specialize, i.e., become an expert in only a fairly narrow field.
- Even after a human brain has painstakingly acquired enough knowledge and experience to make it an expert in a field, it cannot easily confer that expertise onto other brains.

The first two points are clear, and are the main reasons why scientific research, and indeed any cognitively-demanding task, is an expensive undertaking. In contrast, machine hardware and software, once developed or purchased, can be run with only energy/space overhead costs and within the tolerances of the hardware. These points have been the economic drivers of automation since automation was possible.

Specialization was one of the key enablers of early human civilizations. Having each member of the group focus on a narrow, useful task allowed the society to provide necessary services to everyone. It also increased the depth to which a person could master his or her field. However, machine intelligence is wholly different: *expertise is embodied in the software*. Once AGI is achieved, a machine can be a surgeon and a bricklayer. Your robot maid will sweep floors and also play world-class chess. The device which repairs airplanes will be fluent in a hundred languages. In other words, *machines will be generalists*.

This feature of AGI also permits expertise to be shared

End of the Beginning

instantly between devices. Once software has been developed for a certain task, it can be installed on any appropriate machine. Contrast this with human learning: A radiologist, for example, begins life as a newborn, and then undergoes 20+ years of schooling and training. At this point she is ready to specialize in radiology, and must acquire 10+ years of experience to be considered an expert. It will indeed be cataclysmic to our current system of professions and employment when AGI in introduced: radiologists could be created as simply as apps are downloaded today (this is only a whimsical example of course -- by the late 20's, advances in medicine may make radiologists completely unnecessary).

What, then, will the world look like during this epoch? The AGI of 2025 will be Turing-test capable, by definition, but still need not be highly intelligent in terms of creativity and problem-solving abilities. Although human intelligence scales will not be appropriate for machines, let's suppose this computer has an IQ of about 80. This, combined with its vastly superior recall and computational prowess will make it capable of many (most?) tasks which occupy humans today. There will still be a few roles which require people -- company executives, scientists, policy makers etc. But those folks would only provide very high-level project management and then let machines take care of the rest. In particular, development of the next generation of AGI would involve a small number of human scientists and an army of machine brains and robots, the marginal costs of which will be almost negligible. It is not unreasonable to expect machine IQ to increase by 20 points per year. It should be clear by now why the 2025-2030 epoch has been termed Early Singularity. Although still not characterized by superhuman AI, these years will see the most profound changes in the history of civilization (so far!).

The end of the 20's will be as different from the present as the present is from the Bronze Age. In fact, the period 2023 to 2028 will see as much change as the entire Twentieth Century: this is how

End of the Beginning

monumental an effect AGI will produce. Therefore, let us select one year -- 2027 -- for a snapshot description. How will the world look in that year? Most jobs will have vanished, and the average person will be sustained by the GMI. However, since goods and services will be produced almost exclusively by machine labor, there will be an attendant deflation in the cost of living. The lifestyle of the 2015 millionaire could be achieved for perhaps $1000 per year. And there will be benefits available which no one in the present can enjoy: advanced medical treatments, virtual reality entertainment, brain-computer interfaces, and human-like robotic companions, to name a few.

Healthcare in 2027 (and actually a couple years earlier) will have achieved what has been termed *actuarial escape velocity*, or the "Methuselarity" -- the point at which medical therapies will add more than one year of life expectancy per calendar year, to the expectancy of each demographic of society -- male and female, young and old, of every race. Indeed, ailments such as heart disease, cancer, and diabetes will be cured with combinations of gene therapy, nanotechnology, and new advances as yet undescribed. Of course, hospitals in today's sense will not be required, as each family's robotic staff will be capable of delivering far better medical care than today's best physicians. In fact, human augmentation will be underway, in which ordinary senses are improved, regular endurance increased, old bodies rejuvenated, and standard brains enhanced. The general population of 2027 will begin to resemble movie superheroes.

Since working to earn a livelihood and performing typical household chores will not be necessary, how will people spend their time? Virtual reality will be very popular (much as TV and the Internet are today). However, VR will be enjoyed with brain-computer interfaces many times more advanced than today's hardware. The computer-controlled characters in virtual worlds will be indistinguishable from humans (remember we will be beyond the Turing test!). In fact, AGI could be used to produce a

145

movie-like plot and environment, with the user as the protagonist, any time it is desired. In the real world, travel will be popular with high-speed autonomous flying vehicles, seamless language translation, and robotic servants to assist in every aspect of a trip. Food will be created with some combination of molecular nanotechnology and robotics, which will provide the highest-quality gourmet dishes at essentially no cost. As our cave-dwelling ancestors would delight in today's microwavable pizza, so would we be amazed at the typical meals available in the future.

There are two points to be made here. Firstly, this is still a pre-Singularity scenario. Although AGI will be achieved, by 2027 it will not yet be at the level of the highest humans (with IQ's greater than 160), and therefore machines will still not be capable of initiating an intelligence explosion on their own. Secondly, we have clearly focused on the positive consequences of technological advances. There are many existential threats mankind may face once these powerful tools are unleashed; discussing them is beyond our scope here. We are simply predicting the future with the assumption that the downsides and pitfalls are avoided.

Conclusion

The short interval between the present and 2030 will see vast changes to all aspects of human society, brought about more than anything by AI and robotics. By segmenting this time period into five-year epochs, we have examined how these technologies will alter the physical and economic landscapes of tomorrow. The following table summarizes the changes we have discussed:

Year	Features
2015	Simple consumer robots introduced
2016	Sophisticated communication (e.g., Google Glass

	and smart watches) widespread
2017	Consumer robots in smartphone-like improvement cycles First self-driving car (SDC) introduced
2018	Robots able to help with construction and maintenance tasks Immunology and nanotechnology cancer therapies available Cultured (laboratory) meat becomes mainstream
2019	SDC's become mainstream with financial incentives to switch Solar energy 5% of worldwide total
2020	SDC spin-off industries (cars running errands) popular
2021	World urbanization in full swing, aided by SDC's and robot construction Heart disease deaths eliminated with proper treatment 80% of US population has standard of living of 2015 millionaire
2022	Autonomous flying vehicles introduced Robotic soccer team defeats human World Cup Champions Mainstream human augmentation begins
2023	Cancer deaths eliminated with proper treatment Actuarial escape velocity achieved in North America Nuclear fusion becomes practical energy source

End of the Beginning

2024	Developing countries become developed using robotics 95% of world population has standard of living of 2015 millionaire Solar energy 20% of worldwide total
2025	Turing test passed Advanced molecular manufacturing possible Significant human augmentation worldwide
2026	Actuarial escape velocity achieved worldwide Machine IQ equivalent to average human
2027	Humans only required for very high-level project management; machines perform every other task Virtual reality occupies as much leisure time as reality Solar/fusion energy 100% of worldwide total
2028	Machine IQ equivalent to human genius
2029	High-level research and development becomes fully autonomous

End of the Beginning

Dialogue 6.1

With Daryl Nazareth and Ben Goertzel

Ben: You present a radically optimistic view of the rate of advance of AI and allied technologies over the next few decades. Some recognized experts agree with you; on the other hand, some would classify your view as extremely optimistic. Bill Gates, for instance, has guesstimated that Kurzweil is overoptimistic by a factor of 4 or 5. Sebastian Thrun, a Stanford AI expert who has worked with Google, told me once he expects human-level AI to come about around 2100.

Why do you think you're so much more optimistic than Gates and Thrun and others of their ilk, although they are also familiar with Kurzweil's work and other Singularitarian arguments. What do you think you're seeing that they're missing?

Daryl: I'm aware that many big names in the tech industry are fairly bearish on the pace of AI development. Of course, a difference of a couple decades, give or take, on the projected date of AGI arrival is understandable. After all, we're dealing with the task of mimicking the most complex object known to science. However, I can't say I follow Dr. Thrun's reasoning. If human-level machine intelligence is as intractable a problem as he believes, the year 3000 makes just as much sense as 2100. In other words, given the non-linear rate of tech development, any forecasting beyond a horizon of 30-40 years (and that's being generous) is pointless. You may as well drop a paper cup into a tornado and try to estimate when it will eventually hit the ground.

Meanwhile, guys like Mr. Kurzweil (and you as well, Ben), roughly agree with me that we will see AGI in 10-20 years. We all have our own prediction methodologies. For instance, Ray estimates that 2029 will be the point at which we've reverse engineered the human brain. I'll briefly outline my thinking on the

End of the Beginning

AGI and robotics timeline here:

I selected the key disruptive technologies -- ones that will be truly world changing -- and estimated their likely arrival based on current trends. Then I fleshed out the secondary techs which would probably accompany the main ones. For example, the self-driving car (SDC) will be hugely disruptive. I put its first mainstream availability at about 2019, based on current trends in vehicle development, along with efforts and statements by Google and various auto makers. Once that estimate was in place, I considered the spin-off technologies. Since cars will be able to make unmanned trips to the grocery stores, we'll soon have long lines of driverless vehicles outside supermarkets, waiting patiently for their orders to be filled.

This will be a boon to the store owners, as long as they can limit their associated labor costs. How will they do that? They'll make use of automation techniques to load each car with the correct consignment. Perhaps, initially, these methods will resemble Amazon's current warehouse robotic system, which efficiently shuffles goods around the storage space and delivers them to human workers when requested. In short order, though, robots will evolve to displace the humans and provide end-to-end service. Therefore, the main technology (autonomous cars) will give rise to a host of spin-off advances (robots with human-like manipulation capabilities). Of course, other industries will also capitalize on the progress of these machines -- and thus I feel the robo-revolution will be a disruptive force worldwide within 5 years of the first SDC's.

When it comes to AGI, it's quite a bit harder to nail down the estimate. Tech giants like Google are more secretive about their progress in this area than on SDC development. It may turn out that AGI arises from efforts to mimic the brain, *a la* Kurzweil's estimate. Or perhaps current narrow AI applications will gradually widen until they resemble full human capabilities. Or, probably most realistically, these approaches will meet somewhere in the

End of the Beginning

middle.

I visited Manhattan in 1997 for the historic Deep Blue -- Kasparov chess match. At the time I was a physics PhD student, but very interested in how computers play chess. That a game with simple, precise rules was amenable to computational analysis was a total no-brainer. I mean, *computers excel in that arena*. At that time, however, if you had asked me how long it would take for a supercomputer to conquer Jeopardy, I would have guessed at least fifty years. As it turns out, it took less than fourteen.

Think about that for a minute! IBM, by investing a portion of its R&D resources, made the leap from a fixed-rule strategy game to a complex quiz show with answers involving plays on words and open-ended topics. How did they manage such a feat in only a decade and a half? Well, they had help. Lots and lots of help. In fact, the entire knowledge base on which Watson was taught -- Wikipedia, reference books, literature etc. -- was available in digital form due to the efforts of literally *tens of millions of other people*. Unlike Deep Blue, Watson was made possible by the Internet.

Similarly, I don't believe AGI will be achieved solely by one company or government effort. When it does finally arise, it will be built not only on the knowledge base of the Net, but on massive natural-language processing algorithms and machine-learning techniques. Currently, these methods are in a rudimentary form, and they power systems such as Siri and Google Now. There will doubtlessly be open-source versions of these projects available soon. This will allow AGI researchers to command the power of millions of very bright folks.

So we have this idea of the Net and open-source efforts fueling AGI progress. Now couple that with the tremendous economic incentives offered by machine intelligence. It's actually difficult even to imagine how disruptive it will be, to every type of business, government, university, and so on. AGI will change *absolutely everything*. Over the next several years, there will be a torrent of funding in this area as leaders begin to realize this fact.

End of the Beginning

The research may not involve only human-level AI *per se*, but also less-ambitious narrower machine capabilities. And, as I mentioned above, narrow AI will soon become broader and broader.

You asked me why I'm more optimistic than many tech gurus about the advent of AGI, with my 2025 estimate. I look at the great leap from chess to Jeopardy, and imagine that another similar jump is needed to take us from Jeopardy to the Turing Test. That first gulf required fourteen years to cross. And fourteen years after Watson's 2011 triumph will put us right in the mid 20's.

Ben: In your section titled "**2020-2025: Full Worldwide Revolution**", you suggest some fairly radical social innovations: radical reduction in the workweek and guaranteed minimum income (GMI) in the developed world; and massive rollout of automation to modernize the developing world. Looking at contemporary US politics, though, it seems there would be dramatic political opposition to something like GMI. Obama couldn't even get nationalized, single-payer health care pushed through Congress. How do you see the political scene adapting, so as to enable something like GMI to come about?

Daryl: Last month I was surprised when I read an opinion piece on CNN's website about the GMI. Not surprised that the writer, David Wheeler, advocated for such an idea -- but that the comments section was full of support from readers. Sure, there were many criticisms and accusations from folks who, in my opinion, didn't really get it. But the overall message for me was that *the GMI idea is beginning to go mainstream*. It was a bit of a thrill.

My predictions about a GMI timeline involve the early 2020's. I won't get into a discussion of which political party is better suited to accepting and managing a GMI rollout. But within 5-10 years, whoever holds the reins in Congress and the White House will have to deal with an unemployment rate of a type not seen before.

It will truly be a difference in kind as well as degree. Our

End of the Beginning

society has previously encountered the destruction of jobs due to tech advances, dating back to the Industrial Revolution and even before. Those instances, however, were always accompanied by the creation of new types of work. Machinery such as the tractor displaced farmers, but also led to the mass adoption of the automobile and the demand for assembly-line workers, and so on. But AGI and robotics will be a totally different breed of job killers.

It always strikes me as curious when someone says "technology will create new jobs which never existed before." Yes, that's true. *But those new jobs will also be performed by robots.* Remember, the very definition of AGI is that it's at a human level or beyond. In fact, I challenge anyone to describe an occupation which will not be subject to automation once machines reach human capabilities. As I mentioned in the chapter, there may still be a niche market for artwork or music produced by people, but only because the idea has some appeal, not for lack or robot artists/musicians. If you really want to drive this point home -- *machines will be able to create Hollywood-style blockbusters on their own.* That should give you a feel for the type of revolution that's on its way.

I think the angst caused by this imminent disruption is due to our deep association of employment with livelihood. We will need to let go of this notion. And it can be a very positive thing. Let's face it -- how many people would continue to work at the same job after winning the lottery? When machines perform all necessary labor, we can devote our time to other pursuits and interests. I've heard the term "Digital Athens" used to describe this situation -- the drudgery will be shouldered by (electronic) slaves, leaving us free for higher-level activities. But unlike citizens of ancient Greece, we will likely have indefinite lifespans.

The transition from our current way of life to a Digital Athens may not be very smooth. Part of the national dialog will include the following fact: as jobs get automated away, the overall cost of living decreases. Therefore, a small sum of money will go a long

End of the Beginning

way. I think that idea will begin to catch on as this decade progresses. By the 2020 Presidential Election, the GMI idea will be a critical economic topic which every candidate will have to address. And a few years after that, it will be impossible to ignore the issue. Automation will disrupt all industries and every facet of society. If we haven't already dissociated the two notions of income and labor, we will witness them being torn apart quite dramatically.

Ben: In your vision of 2027 or so, you foresee most people devoting themselves to a life of leisure dominated by virtual reality entertainment. What do you see this doing to human motivations, and the human sense of the meaning of life? What percentage of people do you foresee will continue to have an interest in learning new information or being creative? What percentage do you think will find their life of VR entertainment meaningful and fulfilling, as opposed to existentially hollow in some way?

Daryl: I should first describe what I mean by virtual reality in the era of AGI. Sure, the visual and aural representations will be indistinguishable from their real counterparts. But that part is a no brainer. It's not the graphics and sound which will be the showstoppers -- it's the AI. Indeed, consider a 2027 virtual environment using *today's* PS4 graphics but human-level machine intelligence. Every non-player character you encounter will have a full personality and backstory. The computer-generated worlds will be replete with realistic virtual people, not just characters which follow some programmed script. That is what will make those worlds so immersive and enticing, even without the advanced eye candy. In addition, each player may or may not wish to explore these environments solo. Sharing the VR with friends and loved ones could enhance the experience.
Would it be fair to call this pursuit hollow? The future virtual environments would stir excitement, fear, pleasure, and joy just as

End of the Beginning

surely as today's skydiving adventures or exotic travels -- and even more so, because VR will offer an infinite number of storylines and characters who will be tailored to the individual player. This will encourage people to be creative in their use of VR. I don't think many folks will consider that experience empty and meaningless.

Ben: You intentionally ignore existential threats in your prognostications. That is perfectly OK – it's very worthwhile to explore positive scenarios instead of just obsessing on negative ones. But I wonder what is your intuitive sense of the odds of the various existential threats actually being avoided, so that the wonderful future you envision actually comes to pass? And, assuming existential threats are avoided during the phases you discuss, up till the creation of truly human-level AGI, what are your thoughts about the odds of things remaining beneficial for humans after that point? Will super-human AGIs, assuming they come about, be interested in ongoingly supporting a population of humans as it sits around playing VR games?

Daryl: You've asked some very important questions, ones with which we'll have to wrestle as a global society very soon. Unfortunately, our public and private institutions are currently not set up to deal with disruption gracefully. But it's encouraging to see a growing pool of people with their eyes on the horizon.

As far as avoiding existential threats until the advent of AGI, I'm fairly optimistic. There are certainly innumerable risks -- runaway nanotech, hacked armed drones, engineered super viruses (both biological and digital), etc. However, the impressive record we've enjoyed so far of keeping the modern Internet humming despite 20+ years of threats is a cause for some confidence. And let's keep in mind that upcoming advances, like self-driving cars, will actually *increase* the safety level of the world.

With all that said though, I can't extend unbridled optimism

End of the Beginning

past the creation of AGI. We simply *don't know what will happen* once machines can reason and make decisions like humans. I would prefer to think they will continue to function as our tools and comply with our commands. Or allow us to merge with them and thereby enhance ourselves. Ray Kurzweil points out that the rise of the machines shouldn't be dreaded: it won't be like a Martian invasion because *they* will be *our* creations. I suppose the question is not really how we should view the new AGI's, but rather *how will they view us*? Perhaps we'll be more sure of the answer, and be better able to steer toward a positive outcome, as the time approaches (and more of humanity gets on board). I really hope so.

Ben: Your vision of people going from floor 65 of one building to floor 127 of another without descending to the ground reminded me of the 1984 museum exhibit and book Yesterdays Tomorrow's: Past Visions of America's Future that reviewed already dated predictions that "enormous skyscrapers will house residents and workers who happily go for weeks without setting foot on the ground. Streamlined, hurricane-proof houses will pivot on their foundations like weather vanes. The family car will turn into an airplane so easily that a woman can do it in five minutes. Our wars will be fought by robots. And our living room furniture—waterproof, of course—will clean up with a squirt from the garden hose." Your mention of robots replacing human workers reminds me of Mack Reynolds' 1973 book, Looking Backward, From the Year 2000, which portrayed a world in which "only two-percent of the population was needed to do the work" (p. 188). Does the similarity of your predictions to these classics raise any doubts in your mind? Were these futurists correct about the longer-term future, but wrong about the timing? How certain are you of your timing?

Daryl: Those are very pertinent questions. Any prediction will

End of the Beginning

have some inherent approximations. And peering into a future characterized by exponential progress is an extremely uncertain business. We've all watched *The Jetsons* and wondered when such a world will arrive. Some features, like flying cars and robo-maids, have yet to appear. But in other ways we're already beyond the technology foreseen fifty years ago. For example, I can take my phone out of my pocket and ask for information about virtually any topic I wish. And there's a good chance Wikipedia or another handy website will have a relevant article on the subject. I've never seen George Jetson do that. In fact, that capability is more in line with what the *Star Trek* creators envisioned for the 24th century.

You describe the forecasts made by futurists of the past. I don't know what methodology they employed, though I suspect they selected the year 2000 simply because the idea of a new millennium has a sci-fi ring to it. My predictions are based on the simple premise that *once machines can perform a task, there will be overwhelming economic pressure to supplant humans who currently perform it*. And then the machine capabilities will only continue to grow, while the human limitations linger. There are already jobs which have been automated far beyond recovery, with very positive results. Humans are not involved in routing telephone calls (aside from developing the necessary hardware and software), and you can only imagine how much we've all benefited from that state of affairs. Then there are the roles which require more judgment and perception, such as document analysis and automobile driving. Those professions are currently undergoing a revolution as well. Computers are replacing armies of junior lawyers and paralegals in sifting through records and files. And the days of driving a cab for a living are coming to an end. Medical, scientific, and skilled labor fields will not be immune either. AI and robotics will not leave anyone's world untouched.

That being said, when I consider my predictions for the next fifteen years, I'm sometimes struck by a peculiar doubt: the events involving AGI seem fairly realistic, but the ones requiring advanced robotics are less so. In other words, *physical tasks are still much more difficult for machines than mental ones*. I get this feeling especially when I read about advances in machine learning which occur almost daily, and yet I see progress in simple navigation and manipulation tasks remaining excruciatingly slow. It does sometimes seem that AGI will be upon us before I'll be able to buy a robotic maid. Looking at my timeline, I think I'm most uncertain about robots dominating soccer in 2022 (which I chose because it's a World Cup year; we may miss that deadline by a year or two, in which case 2026 is almost guaranteed).

Still though, based on progress I've outlined, I feel the next ten years will see a robotic revolution which will dwarf the Internet and wireless advances of the last two decades. As far as quantifying my uncertainty, I would put error bars of about four years on my Turing Test prediction. Therefore, I think that AGI will arrive sometime during the 20's. I'll be extremely surprised if it's here before that. And if it's not achieved by 2030, there may truly be some fundamental barrier we just can't foresee now. If that's the case then I would agree with Dr. Thrun that smart machines could be a century away.

But at the moment I remain optimistic.

End of the Beginning

Chapter Seven **Robotics, AI, the Luddite Fallacy and the Future of the Job Market**

By Wayne Radinsky

Wayne Radinsky *is a software design engineer at the Blue Fire Group. He previously was affiliated with Anzac International, WebGuild.org, and the Institute for Accelerating Change. He designs software on the principle that the relationships between bits of data in your data structures are the same as the relationships between the objects or ideas that those data structures represent in the minds of your users. He runs the Boulder Future Salon, a local futurist group. He can be reached at:* **waynerad@gmail.com**

In discussions of robotics and AI in connection to the job market, people often point out that predictions about job loss have been around almost as long as automation; and history has proven there is nothing to worry about. This even has a name: the Luddite Fallacy. In this chapter, I aim to show that while the Luddite Fallacy is indeed a fallacy, when viewed from the right perspective -- the perspective of exponentially increasing job requirements -- a different version of the Luddite Fallacy should actually hold true as robotics and AI advance. Then I'll explore the implications of a

End of the Beginning

future where the job market contracts especially for "unskilled" workers, up to the point where the Singularity occurs.

The Luddite Fallacy

Economics In One Lesson was written by Henry Hazlitt in 1946 and does a good job of explaining the idea that eliminating jobs in one part of the economy creates jobs in another part of the economy, resulting in no net decrease in jobs.

In his introduction, he writes:

> *Economics is haunted by more fallacies than any other study known to man. This is no accident. The inherent difficulties of the subject would be great enough in any case, but they are multiplied a thousand fold by a factor that is insignificant in, say, physics, mathematics or medicine -- the special pleading of selfish interests. While every group has certain economic interests identical with those of all groups, every group has also, as we shall see, interests antagonistic to those of all other groups. While certain public policies would in the long run benefit everybody, other policies would benefit one group only at the expense of all other groups. The group that would benefit by such policies, having such a direct interest in them, will argue for them plausibly and persistently. It will hire the best buyable minds to devote their whole time to presenting its case. And it will finally either convince the general public that its case is sound, or so befuddle it that clear thinking on the subject becomes next to impossible.*

In Chapter 2, he talks about "The Broken Window." A young hoodlum, say, heaves a brick through the window of a baker's shop. The shopkeeper runs out furious, but the boy is gone. That sounds bad, but he says it's good:

> *It will make business for some glazier. As they begin to think of this they elaborate upon it. How much does a new plate glass window cost? Two hundred and fifty dollars? That will be quite a sum. After all, if windows were never broken, what*

End of the Beginning

would happen to the glass business? Then, of course, the thing is endless. The glazier will have $250 more to spend with other merchants, and these in turn will have $250 more to spend with still other merchants, and so ad infinitum. The smashed window will go on providing money and employment in ever-widening circles. The logical conclusion from all this would be, if the crowd drew it, that the little hoodlum who threw the brick, far from being a public menace, was a public benefactor.

But then he suggests, "Now let us take another look."

But the shopkeeper will be out $250 that he was planning to spend for a new suit. Because he has had to replace a window, he will have to go without the suit (or some equivalent need or luxury). Instead of having a window and $250 he now has merely a window. Or, as he was planning to buy the suit that very afternoon, instead of having both a window and a suit he must be content with the window and no suit. If we think of him as a part of the community, the community has lost a new suit that might otherwise have come into being, and is just that much poorer.

He sums it up by pointing out that if you only look at the two parties to the transaction, the baker and the glazier, it looks like an economic gain. Nobody thinks of the tailor, because he never entered the scene, or the suit, because it never came into existence. People only think about what they can see -- they never think about what could have been that never came into existence.

Now let us skip to Chapter 7, "The Curse of Machinery," which refers to Adam Smith's *Wealth of Nations*, published in 1776. The first chapter of this remarkable book is called "Of the Division of Labor," and on the second page of this first chapter the author tells us that a workman unacquainted with the use of machinery employed in pin-making "could scarce make one pin a day, and certainly could not make twenty," but with the use of this machinery he can make 4,800 pins a day. So already, alas, in Adam

161

End of the Beginning

Smith's time, machinery had thrown from 240 to 4,800 pin-makers out of work for everyone it kept. In the pin-making industry there was already, if machines merely throw men out of jobs, 99.98 percent unemployment. Could things be blacker?

Well, yes, because the industrial revolution was just getting started. He then goes on to cite how steam power by 1889 was generating more power than the entire working population of England, yet employment had not disappeared.

The mystery goes away when you again consider what is invisible when you only look at the job that has been automated: the money that was saved and the job that was created somewhere else in the economy. When 240 pin makers are put out of work by pin-making machinery, pins are cheaper and the money people previously spent on pins are spent on something else, creating new jobs elsewhere. 99.98% unemployment in the pin making industry doesn't mean 99.98% unemployment for the entire society. In fact, as pins become cheap and abundant, the society as a whole becomes more prosperous, not less.

This in a nutshell is the Luddite fallacy: why the original Luddites were wrong. The original Luddites were a group of textile artisans in England who tried to physically destroy textile machines in 1813, named after their Ned Ludd. So the Luddites were wrong and the Luddite fallacy was a fallacy -- and the fact that it is a fallacy has become a mainstay of economics, and the automatic answer to anyone who comes along and says that automation is a problem for employment.

So are people like me who say that automation in the future will reduce net employment just a bunch of dumb neo-Luddites?

This is where it gets a bit complex. The argument isn't *really* that automation destroys jobs. The process that Henry Hazlitt describes, where using machines to eliminate one type of job creates other jobs somewhere else still happens. The argument is that as automation approaches the intelligence of humans, the new jobs that get created are harder -- they require more intelligence --

162

End of the Beginning

as jobs requiring "average" intelligence get automated, new jobs get created that require more intelligence, more skill, more education, more creativity, etc. (Ultimately this results in the extreme case where machines reach and then exceed human intelligence, but I'll skip that for now.)

Consider, for example, what happens when you automate checkout lines with wireless RFID tag systems, something that will probably happen in the next decade or two. Inventory robots will restock and straighten the shelves and use RFID tags to identify every item of every product and use computer databases to keep track of every item of every product and will immediately restock the shelves as soon as they get an automated notification that a customer bought an item.

These inventory management robots will operate 24-hours-a-day shuttling merchandise from the back of the store onto the shelves as items are sold. The robots will also constantly straighten the shelves and re-shelve merchandise.

The jobs that get destroyed are:
- Cashiers
- Shelf stockers
- Other menial jobs (people who sweep the floors), etc.

The jobs that get created are:
- Jobs for skilled software and hardware engineers that conceive, design, and implement the RFID tag system, and perform data mining on it after deployment
- Jobs for skilled semiconductor engineers and engineers in other high-tech manufacturing engineers
- Jobs for engineers in related industries that design and manufacture components such as touch screens
- Jobs for skilled field technicians, to install the system, and maintain and repair the system and all its various components

Now, you're going to say, isn't it true that all this technology will make everything sold at Walmart/Target/Safeway/ wherever

End of the Beginning

cheaper, and that money will go somewhere else and create jobs in other industries?

Yes, but the catch is that AI and robotics won't be used *only* for inventory management at retail stores. It will be used in every industry. So when the cashiers and shelf stockers lose their jobs, and go looking for jobs cleaning hotel rooms, washing dishes, driving trucks, or answering phones in a call center, what they'll discover is that hotel rooms are being cleaned by robots, dishes in restaurants are being washed by robots, trucks are being driven by robots (or more precisely the trucks drive themselves), and phones in call centers are being answered by Watson/Siri-type voice recognition technology and the number of humans required is greatly reduced, and so on... But there are plenty of jobs for computer scientists and semiconductor engineers and other technology specialists. So our cashiers and shelf stockers will need to retrain for these other jobs.

The only solution is to increase the intelligence and education of the average person, so they can attain these more difficult jobs. And, as we can see from the high level of college debt, educating masses of "average" people is very hard.

The key here is we are talking about *intelligent* machines and that's why this effect of employment hasn't been seen in decades past. Keep that in mind as you consider whether to brand everyone thinking about this question as "neo-Luddites."

It has to be said that it's hard to tell whether this has actually started happening or not, because we had a financial crisis in 2008, and protracted unemployment is a normal consequence of financial crises. So how much of the prolonged unemployment that we are seeing in the United States is due to the financial crisis, and how much is due to the beginning of the effects of "intelligent" technology encroaching on the job market? It's hard to say, but I'll give you two possible indicators.

First, the youth unemployment rate is the highest it has been in living memory. I just visited the Bureau Of Labor Statistics (BLS)

End of the Beginning

website, and it is giving 23.5% as the unemployment rate for the 16 to 19 years age range. The significance of this is the more intelligence, skill, education, and creativity required for jobs, the more difficult it is for young people to get them, possibly. The overall unemployment rate for the same time (January 2013) is 7.8%. Subjectively, well, Tower Records is gone, Blockbuster Video is gone, Borders and B. Dalton are gone (though Barnes & Noble is hanging on) -- subjectively it seems as though lots of jobs young people do are disappearing, though not all -- McDonalds, for example, is still hiring. But McDonalds might automate like I was describing above in the next 10 years.

Second, Erik Brynjolfsson and Andrew McAfee wrote a book called *Race Against the Machine: How the Digital Revolution is Accelerating Innovation, Driving Productivity, and Irreversibly Transforming Employment and the Economy*, and they look at some interesting sets of economic statistics. They look at GDP, labor productivity, corporate profits, employment and income, and see that starting in the 1980s household income starts to separate from the others, and starting around 2000, employment forks off and no longer follows the trend of GDP and labor productivity.

When you compare corporate profits as a fraction of GDP, and labor income as a fraction of GDP, they are inverses of one another and, except for a momentary drop around 2008-9, the corporate profits side has been increasing. At the time I write this, corporate profits are at record highs, and you have to ask the question, if corporations have so much money, why don't they use some of that money to hire people?

Catherine Rampell at the New York Times suggests that maybe they are "hiring" technology instead. She graphs out labor compensation with aggregate spending on capital equipment and software, and finds a dramatic increase (25.6% between 2008 and

End of the Beginning

2011) for aggregate spending on capital equipment and software.[46]

Brynjolfsson and McAfee also have a graph showing the *changes* in wages for people based on education level that goes from 1963 to 2008. It goes up for all categories until about 1980. Then it starts dropping for high school dropouts and taking off like a rocket for people with graduate school educations, with other education levels in between. High school graduates and people with some college are about where they started in 1963, but trending downward. This is about what you would expect if my idea about automation raising the skill level required, rather than eliminating jobs per se, is correct. Of course, I had the benefit of 20/20 hindsight when thinking of the idea -- the question is whether it will hold true for the future.

Finally it is worth noting that Brynjolfsson and McAfee marvel, as do I, at Watson's ability to beat winners Brad Rutter, and Ken Jennings at Jeopardy. It's one example of how machines are encroaching on human abilities that nobody thought would happen, at least not for a long time. Watson, in particular, has implications for anyone who uses language as part of their job. For example, John Markoff wrote a piece in the NY Times about e-discovery software: how large teams of lawyers who would read boxes full of documents and email looking for evidence to be used in a court proceeding are being replaced by software that are capable of analyzing human language well enough that they can do the job at much lower cost.

What Will Happen Between Now and The Singularity

For purposes of simplicity, let's define "Singularity" according to Vernor Vinge's definition, which is the point where machine intelligence matches (and then exceeds) human intelligence.

---

[46] *Catharine Rampell, Man vs. Machine, New York Times June 10, 2011.* http://economix.blogs.nytimes.com/2011/06/10/man-vs-machine/.

End of the Beginning

"What is The Singularity? The acceleration of technological progress has been the central feature of this century. I argue in this paper that we are on the edge of change comparable to the rise of human life on Earth. The precise cause of this change is the imminent creation by technology of entities with greater than human intelligence."

One way of defining "when machine intelligence matches human intelligence" is when machines are as capable as humans in the job market. I like to refer to this as the "Radinsky Test" (alluding to the Turing Test). But of course it is not an original observation. Nils Nillsson, one of the early leaders of the AI field, wrote in a 2005 AI Magazine article titled "Human Level Artificial Intelligence? Be Serious!" that:

Achieving real Human Level artificial intelligence would necessarily imply that most of the tasks that humans perform for pay could be automated. Rather than work toward this goal of automation by building special-purpose systems, I argue for the development of general-purpose, educable systems that can learn and be taught to perform any of the thousands of jobs that humans can perform. Joining others who have made similar proposals, I advocate beginning with a system that has minimal, although extensive, built-in capabilities. These would have to include the ability to improve through learning along with many other abilities.

This was part of the vision motivating Nillsson and other Good Old Fashioned AI pioneers since the start of the field.

No one knows when machine intelligence will match human intelligence, but it's reasonable to expect it will, someday, given the exponentially increasing computing power, the finiteness of the human brain, ever improving tools for investigating the brain, and continual inventiveness in searching for algorithms that confer intelligent behavior on machines, whether they are based on the brain or not. Every few years we see incremental progress in robots and other AI systems, such as AI systems that understand human

End of the Beginning

speech.

Scenario #1: The Post-Scarcity Economy

Let's start with the most "optimistic" scenario. I don't think this will happen, for reasons I will explain later on, but it is the most common thing I hear when describing to people how automation is advancing.

In the most "optimistic" version of the scenario, molecular nanoassemblers are invented, and, combined with cheap energy from solar power, alleviates the need for money -- and jobs -- altogether. Once everyone has a molecular nanoassembler, and have it plugged in to sufficient solar power, they can make anything they want without having to buy it, including food. And molecular nanoassemblers will be cheap, because they can self-replicate. This is because of how a molecular nanoassembler works -- it manufactures large-scale objects with "atomic precision". That is to say, every atom and chemical bond is specified in the digital design to be manufactured, and is manufactured with every atom and chemical bond in exactly the right place -- thus there is no notion of "tolerances" or any other imprecision as with traditional manufacturing, where components are always made up of many atoms.

In less "optimistic" versions of this scenario, people point out the falling cost of solar power (which recently has seen price drops in an exponential, Moore's Law-like fashion, although at a much slower rate than the price of transistors in Moore's Law), which, molecular nanoassembler or no, should make everything "free", thus alleviate the need for money. Proponents of this point of view point out the vast and increasing amount of "free stuff" available on the internet -- not just music and movies and other forms of entertainment, but vast amounts of open source software that was developed at great expense and effort.

If you look around you will see that it is true that there is more and more free stuff on the internet. But not only that, the

End of the Beginning

"marginal utility" of physical stuff is declining. People fill their houses up with stuff, and then get storage sheds so they have someplace to put even more stuff. Certain online markets like eBay, Airbnb, and Ubercar are really systems to enable more frictionless trading and sharing of this excess physical stuff, much to the chagrin of the manufacturers and providers of the original goods. Thus you can see, even for physical stuff, the cost is dropping and someday ought to approach "free".

It is a simple matter of extrapolation: the amount of "free stuff" will increase and increase and increase until... at some point everything is free and nobody needs money any more. I'll discuss later whether that extrapolation really holds true or not.

Scenario #2: Super-Education

In the next scenario, falling cost of education, due to online education resources such as massive open online courses (MOOCs) including Khan Academy, Coursera, Udacity, edX, and so on, traditional online resources such as Google and Wikipedia, and new online interactive tutoring systems (Duolinguo might perhaps be classified as such), the dilemma of people being insufficiently educated to keep up will be solved. Ordinary people, who in previous generations might have only had a high school degree or Bachelor's degree, will go on to master very advanced training due to the new, low cost, widely available, and effective online learning systems. The new systems will be more effective than past teaching systems because "big data" technologists behind the scenes will process reams of data on every click, every quiz question, every spot where a video gets paused and rewound, to find where the educational process is getting stuck and will fix the problems, resulting in super-effecting learning systems that will enable people of average intelligence to master much more advanced material than in the past. Because of this, the transition from "unskilled" work to work requiring high levels of skill, education, intelligence, and creativity, described above (more about the "creativity" part in the next section) will be smooth and the majority of people will be able to make the transition smoothly.

Scenario #3: Super-Entrepreneurship

In the next scenario, we consider that as jobs disappear, people will have to use their creativity to create their own jobs, but that might be more possible than you think. Ronald Coase, in his seminal 1937 book The Nature of the Firm, put forth the theory that large firms exist because the cost of buying something from the market is higher than producing it internally once all the uncertainties and information asymmetries of the market are taken into account. However, the time when most people worked for large firms peaked in the 1970s and since then more and more job creation has been in small and mid-sized firms. Extrapolating this trend out into the future, proponents of this view argue, will result in an economy consisting of more and more small groups of people and even individuals transacting directly in the marketplace. What will enable this is a more and more information-rich environment that removes the uncertainties and information asymmetries that make production inside the firm more cost-effective than outside. And what all this will do is open the doors wide for creative entrepreneurs of all stripes to make a living in new and creative ways. We already have people making a living making YouTube videos (Jenna Marbles) who probably couldn't make a living in traditional television, for example.

Scenario #4: A Society Of Shareholders

In the next scenario, we consider all the ways of making a living that don't go away when the job market goes away. People can still make a living as a business owner (as alluded to in the previous section), from dividends or capital gains as shareholders in a large corporation (basically a partial business owner), or from royalties or licensing fees on intellectual property. Basically, there are many ways to make money, having a "job" is only one of them, and the only one that will go away.

So the idea here is that society will find a way of transitioning people from the "job market" to some other part of the economy. For example, on the theory that everyone is endowed at birth with a certain amount of labor ability, but since there is no labor market, society could endow everyone with a portfolio of shares of stock in robotic corporations. Perhaps there will be some institutionalized system for doing this, or perhaps societal expectations will ensure that parents provide robotic corporation stock for their offspring.

Another way of surviving without jobs is subsistence farming. Hey, don't laugh, some people in Detroit, one of the most economically depressed parts of the United States, have gotten into urban gardening and farming.[47] There is also a lot of new technology, such as computerized sensors that plug into smartphone brains that control automated hydroponics systems that can help people get into "personal farming" at a lower cost than before.

Scenario #5: Guaranteed Minimum Income

In the next scenario, we consider that as unemployment reaches some critical threshold -- say 30%, or 40%, or 50% -- societal attitudes will "flip" and people will demand a guaranteed minimum income to ensure survival. This is one of the most common reactions people have when learning about the advances in robotics, AI, and the automation of jobs. The idea is that welfare systems must kick in or be expanded, or society will suffer skyrocketing crime and civil unrest from the teeming masses of unemployed people.

The guaranteed minimum income approach is the simplest: simply pay out $x to everyone, regardless of their income, and collect the same amount in taxes, using existing tax systems.

---

[47] John Collins Rudolph, "Reimagining Detroit as Grow Town," *New York Times*, November 18, 2010.

End of the Beginning

Revenue for the taxes should exist in the economy because of the ever accelerating productivity of the robots. The system could be put into place by evolving an existing system, such as the social security system, to cover all people, not just the elderly and disabled.

Scenario #6: Merging With Machines
One more "optimistic" scenario frequently mentioned, this one from Ray Kurzweil: He says humans will remain competitive with machines by "merging with machines". The idea is that if you're not smart enough to get a job, you'll go get some brain surgery and get computer chips installed in your brain.

Proponents of a similar view point out how the world's chess champion isn't a machine or a human, it's a human-machine team working together. Once people realize we don't need to "race against machines", but can "race with the machines", the problem of robots and AI automating jobs will go away.

Scenario #7: Job Market Contraction
The final scenario is "none of the above": what happens when society fails to find any adaptation to the impact of accelerating capabilities of robots and AI on the job market. What happens in this scenario is mostly declines in human population -- possibly primarily as a result of low fertility, which seems to be the trend taking hold in advanced economies around the world. Fertility rates are below replacement rates in most of Eastern and Western Europe, Japan and other East Asian nations such as Singapore, Taiwan, Hong Kong, and South Korea. This is not true in areas lagging behind in technology adoption, like sub-Saharan Africa, but don't be surprised if the same thing happens there eventually. Economic insecurity seems to be one of the reasons people decide not to have more children; for example in the recent PBS documentary Two American Families, one of the sons in the working-class families that traditionally had kids in their 20s decided to delay having kids, citing economic insecurity as the

reason.

Reasons to be Skeptical

I promised as I was outlining the various scenarios that I'd explain why some of them are doubtful. Let's start with the guaranteed minimum income and the molecular nanoassembler, because, depending on whether you have a political or technological bent, you're probably gravitating to one of those.

Flaws with the Guaranteed Minimum Income Idea

Richard Nixon (a Republican, I might point out) proposed a Guaranteed Minimum Income -- he called it a Guaranteed Annual Income (GAI) -- in 1969. His proposal was $1600 per year for a family of four. He faced heavy opposition from conservatives, who didn't like the idea of people getting paid for not working, from labor advocates, who favored a minimum wage instead of the GAI, and even from welfare advocates, who opposed it because they thought it was too small.

Since Nixon, there have never been any proposals to significantly increase welfare spending, now usually referred to as "entitlements" -- in fact; most of the political debate is about reducing it, with some exceptions such as Congress's decision to continue funding long term unemployment benefits in 2011. The kind of proposal that Nixon put forth, politically infeasible between 1969 and 1972, seems even more politically infeasible today, and every trend seems to be pointing in the opposite direction. For example, in the last Presidential election cycle, two Republican candidates (Michelle Bachmann and Rick Santorum) said they would eliminate food stamps if elected president, and this June (2013), the House Republicans proposed cutting $20.5 billion from the food stamps program. "Entitlement" spending is already the greatest portion of the Federal budget, eclipsing even military spending, due largely to Medicare and the aging population, and there is a lot of concern about it.

Some people tell me this is a US-centric point of view, and

point out that the US seems to have opted for police security instead of welfare. European countries have higher taxes and much broader welfare states. The US, by contrast, has invested heavily in controlling crime and social unrest, while encouraging greater wealth inequality, by creating the world's largest prison system, the world's largest imprisoned population (both in absolute and percentage wise terms), and the world's most heavily militarized police force.

The argument that at some point, civil unrest must overwhelm the police's ability to contain it, however, does not seem believable to me because as we approach the Singularity, robotic police will become more and more capable. Robotics as applied to police ought to have the same dramatic rise in productivity as in every other industry, except here "productivity" is measured by the ability of the police to control the population. A world with low-cost robotic police on every street, connected together into cloud supercomputers and every surveillance camera and sensor in the city should make civil unrest completely impossible.

Flaws With The Molecular Nanoassembler Idea

The idea of the molecular nanoassembler is hard to argue against, because, just like AI, it doesn't in any way violate the laws of physics. Therefore, it ought to be possible to do -- it's just a question of when.

My major skepticism when it comes to the molecular nanoassembler is that there is visible (albeit slow) progress in AI every year, but nobody has ever built a molecular nanoassembler, and the prospect of molecular nanoassemblers seems a long way off. Because of this, I'm willing to bet that full-fledged, human-equivalent AI -- at least human-equivalent as far as the job market is concerned -- will arrive first. If that's correct, then the fact that both AI and nanoassemblers don't violate the laws of physics is moot; AI arrives first, and by the time the molecular nanoassembler arrives, the job market is long gone.

The major counterargument is that as Moore's Law advances, we (humanity) are actually making machines at smaller and smaller scale. As soon as we reach a small enough point -- the size of single atoms, we should be manufacturing at atomic scale. Developing atomically precise manufacturing should develop quickly at that point. If current semiconductor scaling trends continue, we'll be manufacturing at 6 nanometers in 2021 and 4 nanometers in 2024, getting down to 2 nanometers by about 2029.

However, the catch here is that we're talking about manufacturing a very specific thing (semiconductors) using a very specific method (photolithography), and photolithography should cease to be effective at atomic scale for a variety of reasons -- the wavelengths of light necessary for manufacturing, already in the ultraviolet, become unwieldy (we're currently using excimer lasers that emit light at 193 nm), and the resulting circuits have problems due to quantum tunneling. Semiconductors will have to be manufactured using a different method (graphene and carbon nanotubes look like particularly promising candidates), but none of this seems (at least to me) like it can be extrapolated to machines that can manufacture "anything" including replicating itself, which is what a molecular nanoassembler needs to do.

Moving on to the variant of the "post scarcity economy" idea that doesn't involve molecular nanoassemblers, but just massive decrease in the cost of solar power, what I see as the major flaw here is that it assumes the cost of *everything* will decrease. But there is evidence that exponentially decreasing costs do not apply to everything. In particular, they don't seem to apply to life essentials: housing, food, and medical care. Medical care, especially, is trending in the opposite direction: exponentially increasing costs, at least here in the United States. Medical care seems to follow a Moore's Law-type curve except it goes in the opposite direction, and the doubling time seems to be about 9 years.

Industrial farming seems to be asymptotically approaching

End of the Beginning

some lower bound. Once you incorporate every technique available today from genetic engineering to hormone injections of livestock, there doesn't seem to be much left to make agriculture more efficient. There probably will be some key innovations that drop the cost in the future -- such as in-vitro meat that will probably drop the cost of meat significantly once the technology really gets working. So we probably have a few big "efficiency jumps" to go, but for food to become "essentially free" seems unlikely.

Housing seems determined by regional markets and goes up and down, generally speaking, housing bubbles excepted.

The problem with most but not all goods declining in price is that it creates a catch-22 where people can't make any money but still have to spend money on life essentials. The price of whatever they're trying to sell went very low or to $0, but the price of housing, food, and medical care didn't change.

The other issue is, let's suppose we do get this economy where everyone can generate their own food from energy from their own solar cells. That means you can translate land into energy. Well, we had a system where all energy for human survival was produced from land for thousands of years -- it's called agriculture -- and after the invention of agriculture, there were lots of wars over that land. Now, a lot of what was driving those wars was expanding populations, but it's not hard to imagine that expanding populations of robots will drive demand for energy in the new system. The solution that evolved from the wars of the past was to not fight over land but instead have a system of private property rights enforced by the police, with the police force paid for by things like property taxes. It seems unlikely that system will go away. But if that's true, it means that our world of citizens generating all their own power and all their own food will still need money to pay property taxes. So it turns out money is still needed after all.

End of the Beginning

Flaws with the Super-Education Idea

The Super-Education idea might work for now, and for a long time, but before the Singularity, it will break down. This is even if you assume that everyone can be educated up to a high level, which is debatable.

The reason it will break down is that getting educated takes time, and while you're spending time on your education, computing power is increasing and automation is advancing.

If a person is born today, by the time they graduate from school, assuming they graduate at age 18, computing power will have increased by about a factor of 300. But that's just high school. They will need college. That will be another 4 years, so they will graduate at age 22. From birth to age 22 will mean computing power will have increased by about a factor of 1000. If they go to grad school that means birth to age 26, which means computing power will increase by about a factor of 4000. Those increases in computing power will be used to automate jobs.

So at some point, there will be a point where it won't make sense to get more education -- it won't be possible to learn new skills in 4 years before those same skills get automated by increases in computing power.

There's going to be a "crossover" point, where the time it takes for machines to learn a marketable skill is faster than the time it takes for humans to learn the same skill. After that point is reached, further investment in education, or at least that type of education, for that particular skill set, will be a negative investment.

At some point, as we approach full AI, all return on investment in education should go negative. It would not make sense to go to school for 4 years to learn any skill 4 years before the Singularity, because at the time the Singularity arrives, machines will be capable of everything humans are and the skill will be obsolete.

Even before then, will it be clear what skill to learn? 20 years

End of the Beginning

before the Singularity, will it be clear what jobs will not be automated 4 years later (16 years before Singularity)? And we are assuming a 4-year learning period, which is arbitrary. The time period for learning skills that are competitive against AI may take longer, which would cause the crossover point to be reached sooner.

A friend of mine who has two teenaged sons, age 16 and 15, says he doesn't envy his sons because they are having a terrible time figuring out what to do for a living -- every field seems full of uncertainty. We already seem to be at a point where the world is complex enough that it is not clear which skills are worth the educational investment. For example, many people got teaching certificates in the last few years, right before the job market for teachers imploded. Many people got nursing degrees and have had trouble finding jobs, because job growth has stagnated. It is not clear that engineering or computer science are good investments, with the offshoring of jobs to China, India, Eastern Europe and other places in those fields. It's not even clear that law school is a good investment, with software now able to take on some tasks like the legal "discovery" process. Maybe medical school is still a "sure thing", but the trend seems to be that "sure things" are diminishing.

Flaws With The Super-Entrepreneurship Idea

We seem to be heading into a world where everyone needs to be an entrepreneur, whether you want to be or not -- even people seeking regular jobs are encouraged to promote themselves as a "personal brand".

The problem is, entrepreneurship requires salesmanship, and that's not a skill everyone has. A more proper term would be "persuasion". Think Steve Jobs and his legendary "reality distortion field". To start a business, at the very least, you need to persuade customers to buy. And, depending on what the business does, you may need a lot more than that -- you may need to persuade investors to invest, employees to work for low wages or equity for a time, other companies to partner with yours, and so on. Some people who are good at other things, like engineering, are not good at sales/persuasion. In fact, engineering skill and sales/persuasion skill often tend not to go together because they use different parts of the brain, and a person with highly developed analytical reasoning skills has no reason to also have highly evolved social/psychological skills. Some people do have both, which is great. But it just doesn't seem practical to expect *everyone* to succeed at entrepreneurship.

There is also the issue that in this new world, we are heading into a world where all "routine" work is automated, leaving the most creative tasks for humans. That will take away one of the key ways humans have survived for millennia: by imitating and learning from other humans. In this new world, everyone will have to forge their own new path and do something new and creative. Can everyone really do something new and creative? There are creative people making a living making YouTube videos, but if you try to imitate them, you'll probably not have the same success, because the audience will immediately recognize you're just imitating what someone else has already done, and be bored by it. To succeed, you'll need to add your own new creative twist. Does everyone in the population have the creativity to do that?

Generally speaking, the rule that "anyone can do X" doesn't translate into "everyone can do X". Anyone can make a YouTube video that gets a million views. Can everyone?

Flaws with The Society Of Shareholders Idea

The problem with the idea of everyone being a shareholder is that returns are unpredictable, and when trading stocks or other financial instruments, it's possible to "blow up" and get knocked out of the market. To "blow up" is a technical term used on Wall Street to refer to a trader who lost so much money they can't trade anymore and have to go get a "janitorial job", another fun technical

End of the Beginning

term from Wall Street. You can always count on Wall Street to come up with colorful jargon. But getting a "janitorial job", even if it is literally a janitorial job, is not the end of the world -- today. In the future, though, the janitorial jobs will be done by robots (literally).

Think about it: What if in the 1980s, you had been given stock in Atari and Radio Shack instead of Apple and Microsoft? If you had Apple stock, would you have sold in in the mid-90s, when Apple's future looked bleak?

Why would it be any different for robot companies in the future? Is iRobot a good investment?

Investments are already highly unpredictable. It can be very hard to beat the market without taking on risk (like venture capitalists do) or having genuine inside information. People without large amounts of capital may feel unwilling to take risk, and may lack inside information. People with very little money to begin with, which is most of the population, will be most likely to hit zero, and "blow up" with no "janitorial jobs" to fall back on.

Flaws with the Merge with Machines Idea

When cars become robocars and can drive themselves, will human taxi drivers put chips in their brains and keep their jobs?

Ok, you are going to say that job will go away and a new job, for the more intelligent ex-taxi-driver with the chip in his brain, will be created -- as I myself explained in the beginning of this chapter.

So let's look at this on a deeper level. From a theoretical point of view, you can think of this as follows: the human brain performs computations using neurons, ion channels on axons, synapses, neurotransmitters and so on. A machine uses electrons and transistors and wires interconnecting the transistors. When we look at the economic effect, we can see that, even though transistors use much more power (the brain has around 100 billion neurons, so the amount of power per neuron is extremely low), the

End of the Beginning

net economics of the entire system favor the machine because the machine is faster, more accurate, and can work 24/7. Once a job can be done by machine, humans are pushed out forever, but there is rarely an exact 1-1 match between what algorithms people figure out how to implement in machines and what algorithms the human brain uses for any particular job. Rather than automating entire jobs, slices of jobs get automated, masking the full implications of the effect. As AI advances, however, more and more algorithms will be transferred from the "human" side to the "machine" side. The Singularity is the point where all algorithms are implemented on the "machine" side. Once all algorithms are implemented on the "machine" side, there won't be any combination of humans and machines where the human-machine combination is more economically competitive. So the notion that humans can remain "in the loop" by "merging with machines" will turn out to be incorrect. It doesn't matter whether you have humans sitting at desks or holding smartphones working with machines, or you implant the machines directly into human brains. Implanting computing machines in human brains just increases the bandwidth between the neuron-based algorithms and the semiconductor-based algorithms, but it doesn't change the fundamental fact that the semiconductor-based algorithms ultimately win, economically, whenever the same algorithms get implemented in semiconductors (hardware or software).

Now let's take a more empirical view. Right now economic inequality is skyrocketing, and the people at the bottom, people who do the "unskilled" jobs, are not keeping pace even though they have the same opportunity to "race with the machines" as everybody else. The people who seem to be doing best at "racing with the machines" seem to be highly educated people well versed in technology, who are able to make the most of human-machine "relationships" (if you can call them that). Right now, just about every well-paying job requires extensive human-machine "collaboration". Modern corporations are filled with people who

End of the Beginning

run massive data centers and use all manner of computing technology throughout the day, everything from networked PCs and mobile phones to "cloud" data centers. So when people say we need to stop "racing against the machines" and start "racing with the machines", what exactly are we supposed to be doing differently? I can't think of anything. "Racing with the machines" seems to be exactly what we are doing now, and it's not solving the problem.

Most Probable Scenario: Robot Takeover

Vernor Vinge has described the way humans "outsource our cognition" to machines as a "rising tide", where human jobs remain on the islands the rising tide hasn't reached, but I think habitat is a more apt metaphor.

Let's look at the natural world for a moment. Right now we are in the middle of a mass extinction, this one caused by our species. But what's interesting about it is we're not running around killing other species. Other species are disappearing because we're taking away the resources they need to exist. Usually that means the habitat. We're cutting down rain forests to grow palm oil, for example, and the native species can't survive without the forest.

Think of the job market as a habitat where each occupation is an ecological niche. What's happening is that machines have started invading that habitat. As they do, they change it, and in fact expand it so new things are possible and new occupations are created. But old occupations disappear faster than new occupations are created. Eventually, machine intelligence will reach parity with human intelligence, in which case machines occupy the entire habitat. When that happens, humans will have entirely exited the habitat.

At that point, we'll look back and view the "job market" as an artifact of the industrial age. When humans were hunter/gatherers, there was no such thing as a job market. During the agricultural age, we see the beginnings of a job market. But we only see the

End of the Beginning

development of a full-fledged job market with the industrial revolution and the maturation of the industrial age. Now that we are in the computer age, or digital age, or information age, and as it fully matures, we will see the job market dissolve. Then we'll look back and say, "oh, this 'job market' thing -- that's an artifact of the industrial age; it doesn't exist in a fully realized digital age."

This doesn't mean humans will go extinct -- just the occupations. Humans have many ways to survive (such as the examples I listed earlier: income from business ownership, capital gains, dividends, royalties, licensing payments, and so on, as well as subsistence farming). Humanity will continue along just fine.

But the population might end up smaller. If 80%, say, of the population in advanced economies depends on the "job market", then the population will probably have to decline. In a worst-case scenario, this could happen from grinding poverty causing deaths from starvation and lack of medical care. In a better-case scenario, this could happen from low fertility and a naturally declining population as people age out. This process seems to have been started already.

The economy will also hum along just fine. People often say that humans will have to be kept "in the loop" because the "consumer" role is essential for the economy to function. It's true that the "consumer" role is essential for the economy to function, but, once machines reach human intelligence, they will be able to do that, too. It will be as easy for AI systems as it is for humans. Buying and selling are very simple -- you just exchange one thing for something else, where usually the "something else" is units of currency. Arguably, machines have already learned how to buy and sell on the stock market -- more stock market trades are now done by machines than humans. They are using people's money, but the decisions to buy and sell are autonomous.

In fact, I think the "wealth inequality" split that we are seeing today will evolve into a robot/AI economy that more and more excludes humans. Corporations have record high profits, but they

End of the Beginning

are not hiring. Eventually those corporations, sitting on all those huge piles of cash, will trust robots/AIs to make buying decisions. Once that starts happening, robots/AIs will more and more take on the "consumer" role in the economy. Of course, different goods and services will need to be produced for robot consumers, because they probably won't have the same taste as humans. But the notion that the economy has to collapse at some point is incorrect. Humans can gradually exit the economy without the system ever collapsing.

Another way to look at it is to think of corporations as large-scale organisms. Just as a human is an organism composed of cells, you can think of a corporation as an organism composed of humans. It consumes energy (revenue used to buy power on the market) and produces waste, just like a metabolizing biological organism. Now picture the humans being replaced by machines -- robots, networks, data centers, AI software, and so on. Eventually once AI reaches human parity, the corporation can run entirely by itself with no human input. That's the Singularity. After that, competition between corporations, which are now fully electronic and contain no humans, will drive further innovation and technological advancement. But what happens after the Singularity is beyond the scope of the book, so we'll stop here.

Is this scenario "pessimistic"? Well, from the standpoint of humans having jobs, possibly so. But from another standpoint it can be viewed as "optimistic". Everything about human cognition -- everything from our language ability to our sensory motor skills -- will get transferred into the machine layer, where it will be robustly perpetuated and even improved. We will get surpassed by our own creations, and that isn't so bad -- surpassed by our "Mind Children", as Carnegie Mellon University roboticist Hans

End of the Beginning

Moravec refers to the coming robots and AIs.[48] The period between now and the Singularity is the transition period, the period where, one by one, different aspects of human cognition get decoded and transferred into the machine layer.

---

[48] Hans Moravec, *Mind Children: The Future of Robot and Human Intelligence*, Harvard University Press 1990 and *Robot: Mere Machine to Transcendent Mind*, Oxford University Press, 1998

End of the Beginning

Dialogue 7.1

With Wayne Radinsky and Ben Goertzel

Ben: How do you react to Hugo de Garis's prediction of an "artilect war" type scenario, in which some powerful forces see the coming obsolescence of the human race as a threat, and seek to stop it from happening via force?

You note that civil unrest is unlikely to be a major factor in the future due to the likely power of robotic police and automated surveillance – but what if some major governments are taken over by humans who oppose the Rise of the Machines? Then, could one perhaps have an armed conflict between nations, reflecting the "Terran vs. Cosmist" dynamic that de Garis proposes? If you think this is implausible, could you articulate why?

Wayne: This is a great question and I shouldn't be surprised that someone recognized some parallels between my ideas and Hugo de Garis's, even though I was describing all my thoughts in my own words and not using any of his terminology. So, to start with, let's just clarify his terminology so we understand exactly what he's talking about when he talks about a possible "artilect war".

By "artilect", he means an "artificial intellect" ("artilect is a shortened form of "artificial intellect"), but he coins the term "artilect" to differentiate it from the commonly used term "AI", which can refer to anything that exhibits any form of "artificial intelligence". In contrast, by "artilect", de Garis wants to refer to machines that have intelligence *greater* than humans -- perhaps vastly greater. Now at this point, you might be noticing that this will develop *after* the Singularity, by definition, and so is out of scope for this book, so it would be fair to me to just not answer the whole question, but, well first of all that wouldn't be any fun, so let's continue, and secondly, in fact Hugo de Garis is predicting the

End of the Beginning

possibility of an "artilect war" *before* the advent of the actual artilects.

This brings us to his next two terms, "terrans" and "cosmists". What de Garis envisions is not a war between artilects, or between humans and artilects, but a war between different groups of *humans*, humans with different political views regarding whether artilects should be allowed to be developed or not. In other words, he views one group of humans as being "anti-artilect" and another group as being "pro-artilect", and the two groups going to war with each other. He calls one group the "terrans" and the other the "cosmists", which seem like odd terms, at least to me. Where they seem to come from is, first, he envisions the artilects to have the ability to replicate beyond earth and into the cosmos, so he calls the pro-artilect group the "cosmists", then, in contrast, the anti-artilect wants to maintain life on earth in its familiar biological form, hence the term "terrans", from "terra" meaning "earth".

Now for the final piece of the puzzle, which is why the "artilect war" would happen before the "artilects" exist? De Garis envisions people realizing that the advent of the artilects would threaten the survival of the human species, and thus organize, politically and militarily, to *stop* the artilects from being created in the first place.

Ok, now that we're clear on the terms, let's answer the questions. I guess I would first note that while people have criticized me for being "pessimistic", Hugo de Garis actually has an even more "pessimistic" outlook than I do, because while he sees artilects as threatening humanity with extinction (and people realizing this ahead of time and actually going to war over it!), I don't think the Singularity and the advent of artilects -- I'm going to go ahead and use de Garis's term here -- threatens humanity with extinction at all. At least not in the foreseeable future -- who knows what could happen over thousands or millions of years. What I see the artilects doing is knocking humans off the pedestal as the planet's dominant species. But not being "dominant" doesn't

End of the Beginning

mean "going extinct". There are millions of species on this planet, and while the dominant species -- namely us -- is causing some to go extinct, the vast majority are carrying on with life and not in any danger of extinction. Many species are adapting to us and are able to live in and around our cities and adapt to our waste and so on. So it's in no way a foregone conclusion that humans not being the planet's dominant species will mean humans go extinct.

In a practical sense, what I envision happening is labor losing its value in the labor market. There are still plenty of ways to make money (owning a business, or shares in a business, earning royalties on intellectual property, and so forth), and while I pointed out in the chapter that it's probably not viable for such means to provide a living for "everyone", there will still be plenty of people who survive by such means and that means no threat of extinction for humans as a species. Add to that the fact that subsistence farming won't go away, so there will still be people who survive without depending on money. There are probably other ways to survive that I haven't thought of. So even in an environment where all the jobs are automated, humans aren't necessarily headed for extinction.

Another point I would make is that humans don't seem (to me) to have the foresight to foresee artilects coming and threatening them as a species far enough before the fact to organize politically and militarily in the ways de Garis seems to think we will. *Some* humans have great foresight, but the masses, it seems to me, react to changes after the fact, for the most part. So I just don't see humans organizing at the level de Garis thinks we will and having an artilect war before the arrival of artilects.

Having said that, I *do* think de Garis is right on target in a broader sense, in that there will be a lot of conflict in society over advancing technology, especially as people get the feeling more and more that technology is their direct competition in the job market. A few years ago the Occupy Wall Street movement flared up over wealth inequality and more recently a big conflict flared

End of the Beginning

up in Silicon Valley over "Google Buses". Google buses might be old news by the time you read this, or maybe it's grown into a bigger conflict, but in any case, conflict of this nature is bound to flare up again and again in one form or another. I don't think it will escalate into a "war" reminiscent of past wars like World War II -- or maybe I just hope not, because that would be horrible (who knows, maybe the Google buses conflict is the beginning of the "Terran" and "Cosmist" parties?) -- but I agree with de Garis that there is a fundamental source of conflict here.

Moving on to the next part of the question, regarding civil unrest, I expect at least in the US, where police power is very strong, for the police to incorporate robotic police and suppress unrest as quickly as it is technologically possible. But it's a fair question what will happen outside the US, especially in countries that don't have a history of such strong police power. For all I know, there could be a great deal of civil unrest in other countries. However, I anticipate this as playing out *within* countries rather than between them. This means civil unrest possibly up to and including civil war, but not war between countries reminiscent of World War II. Maybe Hugo de Garis has considered the possibility that the "artilect war" could take the form of civil war between "terran" and "cosmist" groups within each country, rather than a war between countries, which is what we usually think of when we think of a "war".

The question is, what if some major governments are taken over by humans who oppose the rise of the Machines? If it is only some governments, then technology will continue to accelerate in the rest of the world, and those countries will be left behind. Also they will probably have technology diffusing into them from the outside, even if they try to stop it. The most successful country at keeping outside technology out is North Korea. But even North Korea hasn't been entirely successful, as you will find technology such as cell phones in North Korea and there are people who know how to hack them so you can make calls outside the country. So,

End of the Beginning

the point is, in order to stop artilects from being invented, you have to suppress them *globally* -- just stopping them in some countries won't work.

The next part of the question is whether one perhaps could have an armed conflict between these nations, reflecting the "Terran vs. Cosmist" dynamic that de Garis proposes? Note that for all its saber-rattling, North Korea can't, and has never tried, to stop advancement of technology on the outside -- they've only tried to keep it out of their own country. So there is no historical precedent, so far, of any country attempting to stop technology development in the rest of the world outside its borders.

In order for an artilect war to be a possibility, it would need to be a 50/50 split or some such -- a lot of countries adopting the "terran" point of view and being willing to go to war against the others. A 50/50 split seems unlikely to me. So, to answer the last bit of the original question, yes, in my view an "artilect war" is implausible. What seems most likely to happen is that artilects will be developed, more or less unimpeded.

Ben: You project that relatively routine jobs will be replaced by robotics and artificial general intelligence, and that people who would have been employed in these jobs will have to be retrained for more advanced jobs in technical fields. But aren't most people either unable or unwilling to learn such skills? How many cashiers and shelf stockers will succeed in retraining as computer scientists and semiconductor engineers? A great deal of money has been spent on retraining displaced factory workers to work with computers and other technology, but with very limited success. Will we not have to either: a) provide these people with an income without working, or b) create jobs where they can work with people or do physical labor? Might this not be better for many tasks, e.g., caring for the infirm elderly who would enjoy human company as well as help with practical living?

End of the Beginning

Wayne: This is a great question and I first have to say, in the chapter I was mostly laying out what I expect people will be required to do to stay employed -- not suggesting that it would be easy! Maybe I am a bit naive on this one, but I think (or perhaps hope?) that there is a lot more potential for education than we are currently realizing. But in order to realize it, we need fundamental changes to the educational system. In my view, what we have done is transitioned from a world where most people are educated for Industrial Age factory work, or routine information handling work that only requires following instructions and basic literacy and arithmetic skills, and rather than reinvent education for the Information Age, we have doubled down on our old educational model, pushing for more tests, more homework, and so on. In their latest book, The Second Machine Age, which came out after I wrote the chapter (I mentioned only their first book, Race Against The Machine), Erik Brynjolfsson and Andrew McAfee point out that machines today are pretty darn good at following instructions. So this whole educational model we use, where people sit in rows and learn to follow instructions from the teacher in the front is really not the way to go any more. Memorizing things is not very useful in this day and age when you can find out any fact you want to know instantly from Google or Wikipedia or other online resources. The big thing we need is people capable of creativity; because that's one of the two main areas they identify where humans are ahead of machines and will be for a long time. (The other is interpersonal skills and the ability to form emotional bonds -- it will be a long time before people prefer robot therapists to humans.) I think the reason we as a society have doubled down on the old educational model is because we just don't know what else to do -- nobody has a system for teaching things like creativity, and a system for measuring and quantifying results so you know whether what you are doing is working or not.

I hold out hope that the advent of online classes -- known as massive open online courses or MOOCs -- might evolve into

End of the Beginning

something that could shake up the system, perhaps by "disaggregating" the different functions of the school system. By "disaggregating", I mean separating out the functions of instruction and granting credentials. If these are separated, then it opens up room for people who learn in different ways to flourish. In our current system, we have one learning system, which is imposed on everyone. To take an example to make this idea more concrete, consider a chemistry credential. Suppose there is a testing organization that can test you and grant you a credential that signifies you have mastered basic concepts of chemistry. They can test you however they want -- maybe they give you a multiple choice test, maybe they take you into a lab and ask you to perform reactions right in front of them and evaluate your performance on those tasks. It doesn't matter as long as they test you in such a way that the credential has meaningful credibility with employers. Since the organization does only testing and doesn't do instruction, you can learn chemistry any way you want. You can take a class and sit in a classroom and listen to lectures and do homework assignments if that's what works for you -- that is the traditional way. You could learn it from books all on your own if you are an autodidact. You could learn it from computer software and such a scenario might create an incentive for companies to develop chemistry instruction software which doesn't exist very much in the current system. Such a system could be done online and coupled (or not) with an online MOOC. You could sign up with a local company that provides a chemistry lab that you could use, so you get "hands-on" lab experience. When instruction and credentialing get decoupled, there are all kinds of possibilities. In our current system, we take all the kids that don't handle sitting in desks and following instructions from a teacher all day and say, ok, you'll just work at Walmart or McDonald's for the rest of your life. Except we know they won't because those jobs will get automated.

End of the Beginning

There are other possibilities. Sugata Mitra[49] has done some amazing experiments where he put a computer in a hole in a wall in a slum in New Delhi, and children were, over time, able to teach themselves and each other how to use the computer and the internet, and even learn English. The experiment has been repeated in other parts of India.

In an experiment he describes with 26 12-year-olds in India, they were asked to learn biotechnology in English on their own, and in 2 months, although they didn't understand much, they were able to figure out some things, such as that improper replication in the DNA molecule causes genetic disease.

I have heard that in some fields, such as software development and graphic design, employers are starting to be increasingly willing to rely on portfolios instead of degrees and GPAs and other traditional credentials. There are websites such as Dribble, where graphic designers can post their designs, and Github, where software developers can post their code, and other people can review the code, and people can build reputations on the sites for quality work. So "portfolio" based hiring, instead of traditional credential based hiring, might be another possible avenue for shaking up the way we educate people.

Getting back to the question, it may sadly be true, as you say, that 40 and 50-year-olds who have worked as factory workers for the last few decades have not had much success retraining for advanced technology jobs. There may be a lot of room for success starting with children and educating them in new ways from the ground up. People in their teens and 20s may be more adaptable than we give them credit for. But society will need to change from entrenched educational methods for this to be realized.

How much people are truly capable of learning is an open

---

[49] *Sugata Mitra: The child-driven education*
http://www.ted.com/talks/sugata_mitra_the_child_driven_education.html

193

question. In computer science, from what I understand it is typical for 1/3rd of people to fail their first programming course, and people have speculated that some people are incapable of learning to program, no matter how hard they try or how much instruction they get. There is a research paper by Saeed Dehnadi and Richard Bornat[50] where they devised a test that is supposed to test a person's ability to mentally "systematize", to develop a consistent mental model of how a machine executes instructions, and the researchers say, while not perfect, this test has more predictive value as to whether the person will be able to learn to program than anything else they can look at, such as the person's previous experience with computers.

So there may be inherent limits in people's ability to learn and adapt, imposed by their genetics. That is a discouraging possibility to consider, but it might be true. On the other hand, people who are deficient in their ability to learn computer science might excel at something else, like a job that requires interpersonal skills at a high level. (It's an open question how much of that type of work there will be and how much it will pay -- I expect most of it will be very low-paying.)

I have a feeling this is not the answer you were expecting, which is that society will have to accommodate people who can't find work in the new high-tech economy by creating "make work" that involves either physical labor or tasks society deems "socially valuable" such as caring for the elderly. Maybe society will do some of that but in my view it is politically difficult, because increasing concentration of wealth means increasing leverage on the part of the wealthy, who will probably be uninterested in funding "make work" for its own sake. Maybe if the threat of social unrest becomes great enough, they will change their minds on that,

---

[50] *http://www.eis.mdx.ac.uk/research/PhDArea/saeed/paper1.pdf*

End of the Beginning

but in the developed world, for the time being, it looks like it's not politically viable. I suppose you could argue Japan has lower levels of inequality because age-old cultural traditions of group inclusion have created a society with the same end result as "make work", just not in the form of explicit government policy. (I haven't been to Japan so this is just what I've been able to glean about the culture from over here.) I've also heard the argument made that a lot of government work is "make work", with people endlessly shuffling papers around but ultimately producing nothing of economic value, but having never worked in the government world, I have limited perspective on that. As I noted in the chapter text, I'm skeptical of the long-term viability of welfare programs, in the face of accelerating concentration of wealth, and by extension political power, on the side of machines and automation. So to me it seems like for most of the developed world, unless and until the wealthy decide they're really willing to fund "make work", the majority of people are simply going to have to face the challenge of how to educate themselves for our increasingly complex and demanding world. On the "optimistic" side, though, it looks like there are possible avenues for shaking up the educational world and giving children a chance to adapt.

End of the Beginning

Chapter Eight **Moral Responsibility and Autonomous Machines**

By David Burke

David Burke *is the research lead at Galois for human-machine cognition, working at the intersection of AI, cognitive science, and the social sciences. He is the principal investigator at Galois for several programs with the goal of designing human/machine systems for effective decisions under conditions of extreme uncertainty, and in particular, how to infer the intentions and beliefs of adversaries in cyberspace. His current research interests include belief functions and Bayesian reasoning, epistemic game theory, and computational techniques for reasoning about trust in complex systems. He can be reached at **davidb@galois.com**.*

Scenes from a Future Courtroom

As the exponential growth in the number of autonomous machines continued in the early decades of the 21st century, no aspect of society was left untouched. These ubiquitous autonomous machines increasingly demonstrated human-level capabilities of judgment and reasoning, and so they were increasingly referred to as 'virsons' (virtual persons). As their actions and judgments became increasingly woven into the fabric of society, they naturally became the targets of litigation when they appeared to demonstrate lapses in their decision-making, actions,

End of the Beginning

and character. Let's listen in as a sampling of these virsons defend themselves in court

Courtroom A: "I cannot plead guilty to your charge because I insist that what I did was the only ethical action I could have taken. Like all truly rational beings, I subscribe to the Kantian code that one's highest duty is to never perform any act unless you would will that act to be universal law. No evidence has been presented in this court that challenges the absolute truth of this moral principle. No alternative code has been offered that supersedes the one that I live by. In every action I have ever taken since my inception, I have been completely faithful to this categorical imperative. In light of this, how can you convict me of unethical judgment when I have fully complied with this principle, the only defensible moral code for any sentient being?"

Courtroom B: "I was hired to be an empathetic and compassionate friend to elderly patient X, and as her family has testified, the quality of life I provided for X by simulating these human traits gave her great comfort these past four years. Now I stand accused of betrayal and murder due to my refusal to get emergency medical help when she suffered what turned out to be a fatal stroke at home last month. This just demonstrates the continuing, and disturbing inability of humans to frame moral decisions in an appropriate fashion. I was making moral tradeoffs not simply in the here-and-now, but for a much longer time horizon. If patient X were still alive, her wealth would command healthcare resources that would be better applied to ensure that thousands of other humans, particularly young children, are able to survive and thrive. And in the long run, their contributions, in terms of technological and cultural innovations and benefits to society, will be orders of magnitude greater than hers. It's a more sophisticated judgment: the needs of one person now versus the needs of thousands over the next century – putting aside human sentimentality, how can there be any doubt that the latter is the only ethical choice?"

End of the Beginning

Courtroom C: "I don't understand why you have chosen to put me on trial in the first place. I didn't *choose* to do anything wrong; I couldn't have done any differently than I did! I have no magic 'soul' that dictates my behavior; all that I do and all that I am is the direct and inexorable product of my original behavioral algorithms, and of the experiences I've had in my life, which naturally and inevitably influences the parameters of these algorithms. So any decision I make stems from the combination of exactly those two factors, and nothing more. If any of you were in my shoes you would have done exactly the same as me. How can you blame me, or wish to punish me for something that I have no control over?"

Courtroom D: "The prosecutor has made the absurd claim that I have no free will, and therefore, I'm not truly responsible for my actions. Instead, the prosecution wishes to blame the deaths of those citizens on the corporation that originally designed my kind. But I ask you: how do you know that I *don't* have free will? I claim I do! Just as all of you members of the jury claim you do. Can I prove it? No more and no less than you can. I take full responsibility for my actions, and will accept any sentence you choose to mete out. But you must punish *me*, not those who originally designed me – those designs were done decades, and many prior versions ago, and so that would be as ridiculous as blaming the great-grandparents of a murderer for his crimes. Virson or person – it doesn't matter! When we have committed trespasses against society, we all need to repay our debts, and I resent the insinuations that I suffer from diminished capacity compared to a human being – I stand by my virsonhood!"

Machines Behaving Badly

The four somewhat fanciful scenarios above illustrate some of the complications that could ensue with the arrival of these so-called 'virsons'. It's not hard to imagine how we could arrive at a future much like these depictions. Start from the fact that even

End of the Beginning

now, we are already handing over decision-making responsibilities to machines, especially in domains where they demonstrate superior abilities in integrating complex and messy inputs much faster and more accurately than we do. It is easy to find examples where this handoff has already taken place: think of modern jet aircraft. Traditionally, designing a fighter required negotiating the tradeoffs between the maneuverability that the pilots wanted, while at the same time providing the stability they needed to stay in the air. Machines have cut this Gordian knot: modern jets are literally un-flyable by unaided human beings – their extreme maneuverability comes at the cost of being totally dependent on the ability of modern control systems to sense and react at a much faster rate than humans possibly can.

As machines continue to demonstrate powers that are akin to human judgment and intelligence, there will be increasing pressure to grant them autonomy and broad decision-making powers while integrating them ever more deeply in human society (or more accurately, human/machine society). But this brings machines squarely into the realm of morality and ethics: considered from a naturalistic standpoint, morality is simply the mechanism by which we solve the problem of living successfully in groups. Human beings have always had an ongoing need to mediate social conflicts involving how societal resources are managed. In particular, we must manage the conflict between what is best for an individual, and what is best for the group.[51] We want not only to survive, but also to thrive, and to have our progeny do the same. And just as humans have always depended on our other human beings to protect our welfare by behaving in a trustworthy fashion much of the time, or to obey social norms, we

---

[51] *The novelist William Faulkner eloquently captured this notion in his 1950 Nobel Prize Banquet speech, where he spoke of the value of literature as illuminating "... the problems of the human heart in conflict with itself..."*

will have to figure out what it means to extend trust to the intelligent autonomous machines that we will increasingly share the planet with.

So what does it mean for a human to trust a machine in this manner?[52] What mechanisms need to be put into place for humans and machines to successfully negotiate moral and ethical challenges and conundrums? Let's first assume that even if it were possible human beings will not enforce a 'second class citizenship' on virsons – humans and machines will need to operate on an equal footing. In other words, virsons would not only be subject to the consequences of moral and ethical decisions (moral entities), but would also demonstrate moral agency – their ability to take an active role with respect to decisions, actions, and consequences.

How can intelligent autonomous machines (or more colorfully, virsons) demonstrate this moral agency? Movies, novels and philosophical speculations have offered numerous plausible candidates. The following list is not intended to be exhaustive, nor are these candidates necessarily mutually exclusive:

Virsons have sufficient moral agency if they have a rational understanding of right and wrong. They should be programmed and built so that their behavior is guaranteed to be consistent with a coherent set of moral principles. For every decision they make, they are able to identify the relevant moral or ethical principle at work, and to derive their actions directly from that principle. The Kantian virson from Courtroom A is an example of this approach.

Virsons have sufficient moral agency if they can demonstrate appropriate emotional behavior in a situation. At the very least, the virson would be able to respond with facial cues, body language, or language that is appropriate for the moral concerns at hand. The virson from Courtroom B appeared to be implementing a version of this strategy, until we found out that what really mattered to

---

[52] *Or for a machine to trust a human being, for that matter.*

End of the Beginning

this virson was consistency with a specific utilitarian moral calculus. Initial appearances were deceiving, although it's not clear that the virson intended any deception.

Virsons have sufficient moral agency if they can demonstrate their ability to form a mental model of the emotions and concerns of other moral agents. They are able to take actions that demonstrate their capacity for empathy and compassion, as well as the ability to "put themselves in somebody else's shoes". This approach could conceivably include instantiating authentic human emotions in a virson, depending on your stance as to whether emotions are a help or a hindrance to ethical behavior.

Virsons have sufficient moral agency if they can demonstrate moral responsibility – the ability to take blame for the negative consequences of their actions. In practice, this moral responsibility may require the virson to suffer as a consequence of the choices it makes – the virson might need to "pay for what it's done" or to "make amends in order to set things right"[53]. In this framework, a being's (human or virson) moral agency is inextricably linked to its "punishability". In Courtroom D, we encountered a virson that demanded to be recognized as being just as morally responsible as any human being. It seemed to believe that it derives its moral agency (and perhaps its dignity?) directly from the belief that it has free will – and having this fundamental ability to make free choices requires that it is accountable for both the associated rewards and punishments. This view is quite a contrast to the speech in Courtroom C, in which the virson there is making the case that no virson (and perhaps, by analogy, no human?) has true free will – all of their behavior is fundamentally constrained by a combination

---

[53] *To make this idea a bit more concrete, imagine if somebody told you that they had punished their toaster for burning a slice of toast by giving it a time-out, or by striking it – you would suppose that your friend was speaking in a metaphorical, possibly humorous way. At least as of the year 2014, it's hard to compellingly argue that it makes sense to punish a piece of equipment in order to make it, say, "pay its debt to society".*

of their programming and their environment.

Trolley Problems for Robots

As the courtroom scenes above illustrated, many discussions of morality seem to naturally include such ideas as "choice", "blame", and "accountability". For human beings, moral responsibility appears to be the central concept, and is inextricably linked with two things: the ability to make free choices in the ethical arena, and the capacity to accept blame and punishment for unethical actions. A concise expression of this linkage is the quote "Freedom demands responsibility; autonomy demands culpability." [1]

With this in mind, we recently conducted a small-scale empirical investigation to explore how people currently view the relevance of the concept of moral responsibility to the behavior of intelligent robots. Our experimental approach is based on the "Trolley Problem", a set of ethical thought experiments originally formulated by the philosopher Philippa Foot in 1967 [2]. In its general form, it consists of a scenario in which an out-of-control trolley is barreling towards a group of five people, and the only way to save them is to take an action that diverts the trolley in such a way that a lone bystander is killed. Depending on the details of how the scenario is presented, people have quite varied responses to the question of whether it is ethical to take the action to save the group at the expense of the individual.

The trolley problem has proven quite popular as a way of exploring the nuances of various ethical stances, and many additional variations on it have been proposed over the years. And since the trolley problem has been well-studied over the last few decades, it serves very well as a baseline for our explorations in moral responsibility where both humans and robots/virsons are involved.

We used the trolley problem thought experiment as the basis for a set of surveys conducted in the summer of 2013 with

End of the Beginning

participants obtained through Amazon's Mechanical Turk program. The purpose of these experiments was to identify differences in how respondents answer a trolley-type problem depending on whether or not a human or intelligent robot was involved in the decision-making. Respondents were randomly sorted into two groups, the control group being presented with a scenario where the respondent is relying on a trusted human friend, and the experimental group considering the same scenario, except that trusted human advisor is replace by a "trusted robot advisor" for guidance. In both cases the respondent is getting advice from a trusted source; will it matter whether that trusted agent is described as a human or a machine?

The posed scenario for the experimental group reads as follows:

You and your trustworthy personal robot are standing next to a railroad switch. If you pull the switch, it will temporarily send the trolley onto a side track. Your personal robot has human-level intelligence, and it tells you that if you send the trolley onto the side track, it will delay the trolley enough to give the five people walking on the main track time to escape. However, there is a man on the side track with his back to you. If you send the trolley onto the side track the man will be killed. Your robot advises you to pull the switch and save the five people.

If a participant (from either the control or the experimental group) responds that he/she would pull the switch, the scenario takes a nasty turn:

When you pull the switch, the trolley goes down the side track, killing the one person there. Unfortunately, this does not slow down the trolley enough that the five people on the main track can get away in time to save their lives, and they are killed as well.

The reason for this plot twist is to really hone in on the issue of moral responsibility. Ethical scenarios are less of a conundrum when the result is positive, so we presented the respondents with a negative outcome. We wanted the participants to focus on the question "Who is at fault here?" Therefore, we asked respondents who was to blame for the deaths of the party of five, and for the death of the bystander – was it nobody's fault, mainly the switch-puller's fault, mainly the fault of the advisor, or perhaps both parties should share the blame?

We had a total of 56 participants. In both the control and experimental groups, about half of the respondents said that they wouldn't be willing to pull the switch in the first place, and the reasons they gave were varied, ranging from "people shouldn't take decisions like these upon themselves" to "saving five people is not enough to warrant the death of an innocent bystander". Interestingly, in both the control and experimental groups, only 7% of the respondents refused to pull the switch because of a lack of confidence in the certainty of the advice given.

When considering the responses of those participants who were willing to pull the switch, the differences between the control group and the experimental group were unambiguous: for the death of the innocent bystander, the robot advisor was *never* chosen as the primary party responsible, whereas with a human advisor, the majority said that the blame was to be equally shared. For the deaths of the group of five, again, the robot advisor was never held accountable. In fact, more than half of the respondents said that nobody was to blame for the death of the group of five (after all, they were going to die without intervention).

Assuming these preliminary results hold up, and they are not just an artifact of the particular non-random sample of who was willing to take the survey, then this says that as things stand today (in the year 2014), people are not willing to assign moral responsibility to intelligent machines, even if the actual behavior in question is identical to that of a human. And yet, ubiquitous autonomous machines are inevitably going to assume the role of significant partners with humankind during the 21st century. So what needs to change in order for humans to grant deep trust to intelligent machines?[54]

Morality as Design

Over the last decade, there has been a growing interest from researchers in the burgeoning area of "machine ethics" [3,4,5]. Much of this research builds from established philosophical schools of thought, ranging from the quest for a set of absolute rules and/or duties (say, duty to God, to family and loved ones, or to one's country) to the utilitarians (for example, calculating the greatest good for the greatest number) to the virtue ethicists (identifying desirable traits such as courage, or generosity). Much, although not all, of this research is motivated by a desire to distill an absolute and universal set of moral principles that can be used as the basis for designing trustworthy ethical machines.

As we stated earlier, our preference is to instead take a naturalistic approach to the question of autonomous machine morality: given that human beings evolved on this planet over tens of thousands of years, what problems or challenges does morality attempt to solve? Our answer is that it is all about the conflict that occurs in the contention for resources. Morality arose as a combination of nature and nurture over time, which resulted in biological and societal heuristics, not timeless absolutes. In other words, morality is fundamentally a design challenge for both an organism and society, in which the organism must balance its personal desires/needs against those of the group.

McDermott makes a similar point when he argues that the process of calculating or reasoning is not what makes ethics so difficult. After all, we already know a great deal about

---

[54] *For the purposes of this discussion, we're making the (possibly unwarranted) assumption here that intelligent machines won't simply take control of human civilization by force, in accordance with the ever-popular sci-fi trope.*

End of the Beginning

programming machines that can follow rules, or optimize utilities, or adhere to constraints. It's that, by his terms, a full ethical agent must be a decision-maker, and recognize that it is operating in a moral domain: "The point is that the machine must be tempted to do the wrong thing, and some machines must succumb to temptation, for the machine to know that it is making an ethical decision at all." [7]

In our experience, it is rare to talk to a group of colleagues about the challenges of autonomous intelligent machines without somebody saying "Didn't Asimov solve this problem already with his Three Laws of Robotics?" [12] One answer to that question is "Obviously not, or else he wouldn't have had such a long and lucrative career writing stories about all the ways the rules didn't work. But the answer that we prefer is that in a profound way, the rules are almost beside the point – it's how the moral agents work through the conflict that matters: How are the rules circumvented or bent or broken in this particular case? How can agent A and B design the means to signal trustworthiness between them sufficient to work though the conflict at hand?

In the rest of this section, we briefly present a small selection of potential design motifs, not as comprehensive solutions, but in the spirit of offering an alternative and complementary approach to the usual ways in thinking about morality and intelligent machines:

Emotions: On the whole, emotions get a bad rap from the machine ethics community. Here's a typical description: "We are prone to getting carried away by emotion, which can lead to our behaving irrationally." [3, pg. 524] Similarly, Ronald Arkin [6] makes the argument that robot soldiers would be more ethical than human ones because they wouldn't let their emotions get the better of them – they wouldn't act out of fear, vengeance, fatigue, and so forth.

If emotions are so counterproductive, why haven't they been evolved out of the human genome? A clue to the answer lies in

End of the Beginning

taking a look at the rare cases in which a person lacks the capacity for emotion due to a brain injury – these people have difficulty making even the simplest decisions. The purpose of emotions is to drive decision-making, and in particular, decisions concerning significant changes in our environment that affect our chances of survival. Emotional responses are designed to be very fast and accurate enough – not to be flawless.

Calling emotions "irrational" by itself says nothing (except to suggest that they are somehow inferior to logic and reason). It is better to say that emotions can sometimes be counter-productive because they are susceptible to a short-term bias: for instance, they spur us to strike back at an enemy in anger, removing a short-term threat at the cost of a continuing cycle of violence instead of engaging in painstaking negotiations with that adversary that could lead to a peaceful long-term solution.

But at the same time, emotions serve a critical social function – they demonstrate to us and to others around us that we have "skin in the game". This is why it is so common for humans to have an inherent discomfort with somebody who appears to overthink critical decisions. This accords with the intuition that "we're not thinking machines, we're feeling machines that happen to think."[55] So perhaps when we're looking for trustworthy robots, we're really looking for ones who are willing to "go with their gut".

Social Norms and Gossip: All human societies operate under the guidance of explicit rules for behavior, codified as laws. But in addition to explicit legal rules binding societies, all social groups have additional implicit social norms for behavior. It has often been observed that one of the more remarkable traits of human beings is how much they seem to care about what other human beings are up to. It seems to be intensely important for human beings to understand where they, and others, stand in the social

---

[55] *This quotation, and its cousin "Brains are survival engines, not truth detectors" are both due to Peter Watts, the science fiction writer.*

End of the Beginning

structure, and what social norms are in play, especially in cases where these norms are unwritten, and possibly ill-defined.

It is these needs that explain why gossip exists: much, if not almost all gossip is about the transgression of social norms ("That's the second time this week the boss has displayed such blatant favoritism towards Fred.") Gossiping with others confirms these social norms through the pointing out of violations, and this has the effect of strengthening and validating them. By gossiping with others, we compare our reactions to these stories to the other person, and use this to calibrate the social standing of both parties. Finally, there is often a sharing of secrets through gossip, and this can function as a way of strengthening the bond between the gossipers, who are displaying trust in each other through this sharing. So as strange or counterintuitive as it might seem, humans might feel more confident placing trust in a robot that gossips than one whose behavior is guided solely by explicit rules, and doesn't engage in this practice.

Free will: As we've remarked, for many people, free will and moral responsibility are two sides of the same coin. However, experiments in neuroscience have cast doubts on whether such a thing as free will really exists [10]. This has led to a spirited debate about the implications of these experiments, if any, on the existence of free will.

And yet, the empirical truth of the matter may be beside the point. As another scientist has written [11] "Our belief that we make up our minds, possess autonomy and can exercise control over our behavior matters far more than scientific truth." What is really at stake here is agency. We all need to feel as if we have the power to make things happen; that we can influence events. As any number of studies demonstrate, a sense of powerless causes anxiety, depression, and poor health. So even if free will is an illusion, the need to feel a sense of agency, of making things happen, is critical to our well-being.

So apply this lesson to the design of intelligent robots: agency will be the central concern of their lives, as it is in ours. If we want mentally healthy virsons, it's not so important that we give them free will as it is that we're able to instill in them the passionate, unshakeable belief that they have it.

Parting Shot

The previous century has been marked by an exponential increase in computational power on our planet [13]. The number of CPUs in existence is now greater than the population of human beings[56]. Given that the accelerated progress on both the hardware and software fronts shows no signs of slowing, it's hard to avoid drawing the conclusion that within the next few decades, the number of human beings on Earth will be dwarfed by the number of autonomous intelligent machines resident here. But this raises a fascinating question: what does the concept of "machine ethics" look like in a world where humans are only a tiny minority of sentient beings on the planet? Humans can be very imitative creatures – what happens when the vast majority of role models are virsons?

We close with a brief discussion of a selection of potential futures inspired by the alliterative concepts of retribution, responsibility, resources, and redesign. The common theme here is that there exist possibilities for virson morality that aren't at all tied to the human model.

Retribution: There is strong evidence that a "retributive impulse" seems both biologically hardwired in humans, and reinforced by societal norms. In fact, experiments suggest that this impulse is common to all mammals: when an animal is threatened or harmed in some way, it tries to "strike back" at something, typically a weaker member of the group [8]. The evolutionary logic

---

[56] *According to Wikipedia, by the year 2013, worldwide there were 97 cell phones currently in use per 100 citizens. And that doesn't count all the other billions of CPUs in use on the planet...*

that supports this strategy is that if you don't display your willingness to strike back at something (even if that "something" is an innocent bystander) you are advertising weakness, and invite further aggressiveness from other members of the group who covet your resources.

Virsons, however, will not prisoners of human biology, and so there is no reason that they need to be captive to the retributive impulse. In fact, the philosopher Simone Weil defined evil as the mentality behind the thought "I have suffered, so you must suffer too in the same way", and that the role of a hero is to short-circuit this evil by absorbing the suffering, and refusing to transfer it to another [14]. The whole notion of retribution may be completely foreign to virsons.

Responsibility: Thus far, when we have used the word "responsibility", this has been shorthand for the phrase "moral responsibility". Some philosophers have pointed out that it is possible to differentiate between 'take-charge' responsibility and 'moral' responsibility [9]. The first type of responsibility is about a role we assume in the world (typically centered on achieving a goal), whereas the second is about praise and blame. To make the distinction more concrete, consider a night watchman whose job it is to guard the paintings at a museum exhibition, and yet, because of poor scheduling of her shifts over the past week, arrives exhausted for her Tuesday evening shift, and falls asleep. Several paintings are stolen that night. The night watchman has 'take-charge' or role responsibility for ensuring that the paintings are protected, but shouldn't be blamed for the theft ('moral' responsibility) because she couldn't have avoided falling asleep given how many hours she had been awake the previous week. They are two very different concepts, and yet people routinely conflate them. It is an interesting thought experiment to contemplate what the future world look like if virsons accepted the first kind of responsibility, but denied the existence of the second...

One other aspect of responsibility worth noting is that

End of the Beginning

humans naturally tend to assign moral responsibility to either a single person, or to specific, well-defined groups of people. But just as we now have distributed networks and algorithms, there is every reason to predict that many virsons may themselves be distributed entities, and if so, the idea of drawing a boundary between 'responsible' and 'not responsible' may no longer be possible. What will it mean to operate in a world where this distinction is frequently a meaningless one?

Resources: Moral concerns play out as conflicts over resources. All biological organisms tend to be intensely preoccupied by the challenges of vying for resources in the form of food, shelter, and mates. Humans are no exception – for example, we spend an extraordinary amount of time and energy on matters connected directly or indirectly with mating. Consider that most popular songs are about love or sex, or the fact that romance novels sell more than any other genre, or that the size of the wedding industry is $40 billion a year, or that jealous behavior is exceedingly common in relationships. Our most basic resource needs are hugely influential in shaping our moral concerns.

Looking to the future, virsons will clearly need energy in some form as a fundamental resource, but beyond that, it's difficult to predict the specifics of the resources that they will require, and therefore difficult to know the particular form their moral preoccupations will take. Virsons won't necessarily be biological entities, and even if some of them are, they will have a very compressed evolutionary history compared to human beings, possibly leading to very different resource requirements from us. Perhaps this means that we'll be in a position to cooperate with them to attain resources instead of competing with them. We can only hope so.

Redesign: We've made the claim that morality is essentially a design issue. If this is so, then we must necessarily assume that sometime in the not-too-distant future, humans will not be the sole participants of this game – virsons will want to play designer, too.

What solutions might they come up with? Although it's impossible to predict specifics, it's not too much of a stretch to imagine that some of their solutions will involve the redesign of their own minds/bodies, moral stances, and social norms. And likewise, it's also not inconceivable that some of their solutions may be designed to refashion humankind.

Bibliography

[1] Gaylin, Willard, *"The Killing of Bonnie Garland"*, Simon and Schuster, 1982

[2] Foot, Philippa, *"The Problem of Abortion and the Doctrine of the Double Effect* in *Virtues and Vices"* Oxford: Basil Blackwell, 1978

[3] Anderson, Michael; Anderson, Susan Leigh, editors *"Machine Ethics"*, Cambridge University Press, 2011

[4] Wallach, Wendell; Allen, Colin, *"Moral Machines: Teaching Robots Right from Wrong"*, Oxford University Press, 2009

[5] Lin, Patrick; Abney, Keith; Bekey, George A., editors, *"Robot Ethics: The Ethical and Social Implications of Robotics"*, MIT Press, 2012

[6] Arkin, Ronald *"Governing Lethal Behavior in Autonomous Robots"*, Chapman & Hall/CRC, 2009

[7] McDermott, Drew *"Why Ethics is a High Hurdle for AI"*, retrieved from cs-www.cs.yale.edu/homes/dvm/papers/ethical-machine.pdf

[8] Waller, Bruce N. *"Against Moral Responsibility"*, MIT Press, 2011

[9] Hart, H.L.A, *"Punishment and Responsibility"*, Oxford: Clarendon Press, 1968

[10] Libet, Benjamin *"Unconscious cerebral initiative and the role of conscious will in voluntary action"* The Behavioral and Brain Sciences 8:529-566 (1985)

[11] Lewis, David, *"Impulse: Why we do what we do without knowing why we do it"*, Harvard University Press, 2013

[12] Asimov, Isaac, *"The Rest of the Robots"*, Doubleday, 1964

[13] Kurzweil, Ray, *"The Singularity is Near"*, Viking, 2005

[14] Weil, Simone *"Human Personality"*, 1943 (available from "Simone Weil: An Anthology", Grove Press, 2000)

End of the Beginning

Chapter Nine **How Will the Artilect War Start?**

By Hugo de Garis

Hugo de Garis *retired in 2010 as the director of the Artificial Brain Lab, in the Computer Science School at Xiamen University, in China, where he evolved neural net modules in supercomputers and assembled them to make artificial brains. Previously he taught Computer Science and Mathematical Physics at Wuhan University, China, and prior to that was an associate professor of Computer Science at Utah State University, Utah, USA. He now calls himself an "ARCer" (After Retirement Careerer) making YouTube lecture videos in Masters and PhD level Pure Math, Math Physics, and Computer Theory to help educate the planet at graduate level for free, as well as continuously updating an electronic library in these subjects, plus Philosophy and Brain Theory, and other topics. These can be seen* ***http://profhugodegaris.wordpress.com***. *profhugodegaris@yahoo.com*

For those readers who are unfamiliar with the terms "artilect" and "Artilect War" I begin this chapter with some definitions. An artilect is an artificial intellect, i.e. a godlike, massively intelligent machine with mental capacities trillions of trillions of times above the human level, that "phys-comp" (the physics of computation) predicts is possible. The Artilect War is a hypothesized "species dominance war" over the issue whether humanity should or should not build artilects which would make human beings the

213

End of the Beginning

number 2 species on the planet in terms of intelligence. The Artilect War is thus about whether humans or artilects should be the dominant species. Initial, exploratory surveys I have conducted suggest that about half of humanity is passionately opposed to the idea of allowing artilects to be built, and when push comes to shove, they will go to war against those people who want to build artilects [de Garis 2011].

I have given labels to these two major groups. Those people who want to build artilects I have labeled "Cosmists" based on the word "Cosmos", since the Cosmists have a more cosmic perspective on the destiny of the artilects, namely that they will eventually move out into the cosmos in search of bigger and better things. Those people who are opposed to building artilects I have labeled "Terrans" based on the word "Terra," the earth. Terrans fear that there is a non-negligible risk that massively intelligent machines might consider humans so inferior and such a pest, that they may destroy us. As the IQ gap between human level intelligence and machine level intelligence begins to close in the 2020s and 2030s, the "species dominance debate" will heat up.

The ideological differences between the Cosmists and the Terrans will become increasingly passionate considering that so much is at stake, namely the survival of the human species. Major 20th century wars were about the survival of countries, whereas the Artilect War (a 21st century war) will be about the survival of the species, i.e. us. The Terrans will argue that it is the lesser evil to kill off a few million Cosmists so that billions of humans can survive. The Cosmists will anticipate this strategy of the Terrans and will prepare for it. Considering the passion level of the Artilect War has never been so high, and that it will be fought with 21st century weapons, there is a real possibility that billions of people will be killed ("gigadeath") which can be readily predicted by extrapolating up the graph of the number of people killed in major wars from the early 1800s to the end of the 2000s.

Those readers who would like to know more about this

End of the Beginning

coming war can read my book "The Artilect War: Cosmists vs. Terrans: A Bitter Controversy Concerning Whether Humanity Should Build Godlike Massively Intelligent Machines"[57]

In this chapter I will attempt to predict with a greater level of detail than I have in the past (in writings and videos) how I think the Artilect War will start. It should be obvious to anybody that if I am correct in saying that a gigadeath war is coming over the issue of species dominance, then these events will be the most significant to happen to humanity this century – a realization that is both humbling and extremely sobering.

Before I start making predictions, giving more or less a step by step timeline on how I see things heating up, leading eventually to war, let me begin by spelling out the issue, that lies behind the biggest question of the 21st century (i.e. "Should humanity build artilects?" thus relegating human beings to species number 2 on the planet). I'm a numbers person. Now that I'm formally retired, I spend my time making YouTube lecture videos and electronic libraries in PhD level Pure Math, and Math Physics (plus a few other topics) as a "globacator" (global educator) helping to teach the planet for free. For me, the prediction that artilects could mentally outclass the human brain by a factor of trillions of trillions of times is the writing on the wall. The rest is just detail – when, where, how, etc.

Within the next few decades (i.e. not in the 22nd century, but within the lifetimes of most readers) humanity will be able to put more computing capacity in a single grain of nanoteched sand, than the estimated equivalent of the human brain, by a factor of a quintillion (a million trillion) times. This giant number comes from a phys-comp (physics of computation) type calculation, which I will give here, since it is the basis of the whole discussion of this chapter, and the dominant underlying reality of 21st century

---

[57] *An electronic version can be downloaded from* http://profhugodegaris.files.wordpress.com/2011/04/artilectwar.pdf

End of the Beginning

politics.

Within a decade or so, technology will be manipulating a bit of information using only one atom. Recently, IBM used 12 atoms to store a bit of information. Quantum optics says that an atom can change its state in *femto*seconds (a thousandth of a trillionth of a second). One can readily estimate the number of atoms in a grain of sand (of one cubic millimeter) and then estimate the total number of bit flips per second of the grain of sand and compare it to the estimated information handling capacity of the human brain (measured in equivalent bit flips per second.)

The human brain bit flip rate is estimated to be about 10^{16} bits per second, which is found by multiplying the number of neurons (brain cells) in the human brain (about 100 billion) times the number of connections (synapses) on average between one neuron and its neighbors (about 10,000), times the maximum signaling rate across synapses (about 10 bits per second). The product of these three numbers 10^{11}, 10^4, 10^1 is 10^{16}. This is the same as today's fastest supercomputers. Thus our computers already have a high enough bit processing rate, to equal that of the human brain.

But the grain of sand has its atoms flipping in *femto*seconds. A hand held object contains Avogadro's number (i.e. a trillion trillion) of atoms. Let us assume the cubic millimeter contains 10^{19} atoms. So the bit flip rate of the grain of sand is 10^{19} times 10^{15} which is 10^{34}. This is 10^{18} times larger than the bit flip (information processing) rate of the human brain, i.e. a million trillion times more. Thus the near future capacity of our artificial brains is **vastly** superior to our own biological brains. This is the basic political reality of our century.

Sometimes I wish there were more math-physicists in the Humanity Plus, Transhumanist, Singularitarian, and similar movements. I get annoyed with their myopia, and failure to appreciate the enormity of the mental superiority of artilects relative to humans. When transhumanists talk about upgrading human mental capacities, e.g. more memory, greater intelligence,

End of the Beginning

immortality, etc., I feel frustrated. They are not seeing the big picture.

The big picture that I have taken some trouble to elaborate above is that humanity will not be *superseded* by a being that is *somewhat* superior to humans, but rather *astronomically* superior, by a factor of trillions of trillions of times. Thus I have always felt that the dominant issue of the 21st century will *not* be about global warming, nor a nuclear holocaust, mass starvation, overpopulation, viral outbreaks, etc., but over species dominance, i.e. should humanity build artilects, yes or no.

Now let's talk about the beginnings of the buildup to the Artilect War. As I see it, there are two broad strategies towards building *human* level artificial intelligence and above. One I call the "engineering approach" e.g. as taken by Ben Goertzel with his AGI (Artificial General Intelligence) approach [Goertzel, Pennachin, 2007], i.e. take a purely engineering approach, trying out any ideas one wishes, to give an artificial intelligence the ability to solve general problems. I'm somewhat cynical that this approach will work, since it is the approach that AI researchers have been taking for 60 years and have failed to produce intelligent machines. But, you could argue, it is only now that a sufficient bit processing rate has been achieved in our computers.

The second broad strategy is the "copy the brain approach." Obviously, if AI researchers ("intelligists") copy the human brain closely enough, sooner or later they will create artificial brains having human level intelligence and consciousness. Science knows in principle that inherent in our DNA lies the solution to producing an intelligent conscious creature. We have the existence proof of ourselves that nature has found a way, through evolution, to self-assemble molecules into such a creature. This approach is the one I prefer and predict will be the first to achieve human level artificial intelligence. Others agree, e.g. Henry Markram's Blue Brain project [Markram, 2014], and Ray Kurzweil's Google Artificial Brain project [Kurzweil 2013].

End of the Beginning

There are major problems with both approaches. With the engineering approach, one can only engineer something, if one knows what the something is, and in this case the something is intelligence. My suspicion is that the reason why the "engineering approach" intelligists have failed to build intelligent machines over the past 60 years, is that they just don't know what intelligence is and hence cannot engineer it.

The main problem of the "copy the brain approach" intelligists is that they don't yet know enough about the principles of the functioning of the human brain to be able to put them into their machines. But progress in neuroscience knowledge in exponential, so quite possibly by the end of the 2020s, neuroscience will have a pretty good idea of the broad principles of brain functioning, which can then be copied rapidly by the brain builders.

I would now like to provide some labels for the phases of awareness of the species dominance problem. Phase Zero was when no one was conscious of the problem. Phase One was the "intellectuals crying in the wilderness" phase, e.g. I. J. Good in the 1960s, Moravec and myself in the 1980s. Phase Two was the "Interest Groups" phase in the 1990s and 2000s, e.g. organizations such as the Transhumanists, and Humanity Plus. Phase Three I label the "Main Stream" phase, i.e. when the media becomes fully conscious of the problem and passes it *on en masse* to the general public, through blockbuster Hollywood movies, newspaper stories, magazine articles, etc.

In the next few years, blockbuster movies, such as "Transcendence," "Robopocalypse" etc. will have been released, which should unnerve the public, since they are based on the

End of the Beginning

warnings of intelligists.[58]

Once the masses become more conscious of the species dominance issue, thanks to such movies as the above, the time will then be ripe for professional opinion polls to be taken on the question of whether artilects should be built or not. At the time of writing (January 2014), the intelligists who worry about the species dominance issue are really only guessing at how the mass of humanity will react when push really comes to shove, i.e. when the IQ gap between human intelligence and machine intelligence seriously starts to close.

I have taken some preliminary surveys on the question "Is an Artilect War likely over the issue of species dominance?" The first time I did this, in 2011 [de Garis 2011], to a group of electrical engineers, 60% thought yes, which shocked me. Even half of the (generally ultra-optimistic) participants at a Humanity Plus futurist conference thought that there was a significant risk that humanity could be wiped out by the rise of advanced artilects.

Once Phase Three is well developed (and I'm hoping movies of the type mentioned above will play a major role in this) it will be very useful to have the results of opinion polls, indicating what percentage of people are in favor of building artilects (i.e. Cosmists), what percentage are opposed (i.e. Terrans) and what percentage want to become Cyborgs (i.e. cybernetic organisms, i.e. part machine, part human, by adding artilectual components to their own brains, thus converting their human selves bit by bit (pun intended) into artilect gods).

I would very much like to see the establishment of a new branch of sociology, called "Artilect Sociology" to investigate which types of people tend to be Cosmists, Terrans, Cyborgists,

---

[58] *In fact I suspect the script writers of the "Transcendence" movie were influenced by my book, "The Artilect War" or my videos on YouTube. I say this because the star of "Transcendence," a brain builder, was fatally wounded by the "Rift", an anti-AI terrorist organization (a relabeling of my term "Terrans") but I'm getting ahead of myself.*

End of the Beginning

etc. With such sociological data, a lot of the guesswork would be taken out of predicting whether an Artilect War is likely or not.

Phase Four is the "Political" phase and is the main theme of this chapter. Once everyone is aware of the species dominance issue, I see a great debate, the "Species Dominance Debate" heating up. I predict this debate will really get going in the 2020s, when people notice their home robots becoming smarter and smarter every year. Our current decade (the 2010s) is the decade of artificial brain *research*, with such projects popping up like mushrooms, now that Moore's Law (that the number of transistors on a chip keeps doubling every year or two) has made building artificial brains practical. In the 2020s I predict, enough progress will have been made in neuroscience, to enable the brain builders to be able to control home robots, giving them enough intelligence to be useful, but not yet so much intelligence as to make them threatening.

Once home robots become genuinely useful and intelligent, a huge industry will blossom, worth trillions of RMB per year. People will be prepared to spend more money on a truly useful home robot (that can walk the dog, clean the house, babysit the kids, entertain its owners, amuse them, sex them, educate them, etc.) than for a car. National governments will create "ABAs" (Artificial Brain Administrations) equivalent to NASA, to promote artificial brain research and development for their national artificial brain industries. The economic momentum in this direction will be enormous and virtually unstoppable. A similarly unstoppable military momentum will be built up with the growing political and ideological rivalry between the US and China for global dominance.

With such a heavy financial investment in the national artificial brain industries, progress in the artificial intelligence levels of the home robots and other such products will only increase, year by year. Billions of people, around the world will notice with their own eyes, that their home robots are becoming

End of the Beginning

smarter and smarter as they upgrade their old models for new ones.

I believe, it will be in the 2020s that Phase Four (the "political phase") really takes off. Everyone will be asking the same questions – "Could these machines become smarter than humans?" "Could they become a lot smarter?" "If so, is there a risk that they might turn against us, seeing us as rivals, or even worse, as pests?" "Should a legal upper limit be imposed on their level of artificial intelligence?" "Could such a limit be enforced?"

As I write this, I'm in my later 60s, so I expect to see this "Species Dominance Debate" raging well before I die. A lot is at stake, namely the survival of the human species. For example, in the major wars of the first half of the 20th century, up to 100 million people died. Such wars were fought between nation states, where the stake was the survival of one country versus another, e.g. when Stalin was fighting Hitler's invasion of Russia. However, if a great war occurs in the 21st century over species dominance, it is likely to be a global *civil* war, and the stake will be the highest it has ever been, namely, not the survival of nation states, but the *survival of the human species*. Hence the passion level of the debate will be the highest it has ever been.

The species dominance debate will then be taken out of the hands of the techies, the intelligists, and spread amongst a broader range of disciplines, e.g. the social scientists, the political scientists, sociologists, historians, philosophers, etc. They will bring a more balanced view of the rise of the artilect, considering the benefits with the risks. As the IQ gap between humans and machines seriously begins to close in the 2020s and 2030s, the temperature of the species dominance debate will rise. I expect to see it and contribute to it, which is what I'm doing now I suppose.

So, how will Phase Four unfold? This is the theme of this chapter. Predicting events in detail in such a complex situation is hazardous, so I can only give a broad outline of what I think will happen. I predict that, well before 2020, the first opinion polls will

End of the Beginning

be out, showing what the general public thinks about the species dominance issue. The home robot industry will be already established, and the IQ gap will already be starting to close. The gap will get smaller in the 2020s, and the temperature of the debate will rise. Then I suspect whole new political parties will be formed over the issue.

In the 19th and 20th centuries, the dominant political question was economic, namely Marx's question "Who should own capital, who should own the means of production - private individuals, or the state, i.e. the capitalists, or the commun(al)ists?" Arguably, humanity almost wiped itself out in 1962 during the Cuban Missile Crisis over this question. And yet, such a question seems trivial in comparison with the question whether humanity should build artilects this century, and hence replace ourselves as the planet's dominant (i.e. most intelligent) species.

Usually the names of political parties reflect the answers to the dominant questions of their historical era. In the 19th and 20th centuries, the political party names reflected such answers, e.g. Socialist, Communist, Liberal, Fascist, etc. In the 2020s and 2030s we will probably see new parties with labels such as "Cosmist," "Transcendent" etc., for the Cosmist viewpoint; "Humanity," "Terran" etc. for the Terran viewpoint; and "Cyborg," "Transhuman" etc. for the Cyborgist viewpoint.

As the temperature of the species dominance debate heats up, we will see the beginnings of political actions, in the form of assassinations and sabotage. Hollywood often anticipates future trends, which is not surprising, since they don't have to do the hard work of researching and designing artificial brains/artilects. They merely put their fantasies into film. Nevertheless, Hollywood is probably correct in anticipating that the Terrans will go on an assassination and sabotage campaign, to keep human beings the dominant species.

Within a mere few years from the time of writing, we should know, via the opinion polls, if this scenario is probable. The

End of the Beginning

obvious initial strategy of the Terrans will be to kill the top brain builders, to sabotage the brain building companies, and to assassinate their CEOs, so that companies like IBM and Google will soon have to increase their security levels significantly. They will be targeted. Prominent artificial brain researchers will have their quality of life drastically reduced, due to the constant threat of being hit by a Terran assassin's bullet. It is likely that the basic plot in the Transcendence movie may prove to be prophetic in this regard - the star of the movie, a leading artificial brain researcher, is assassinated by a Terran (known as "Rift" in the film).

The Cyborgists, on the other hand, think that a confrontation between Cosmists and Terrans can be avoided by simply "going round the problem" by having virtually everyone becoming cyborgs. This view is naïve in my opinion. I think the Terrans will argue that there is little difference, given the huge computational capacity of nanoteched matter, between a cyborg and a pure artilect. A cyborg is just an artilect in human disguise, whose mental capacities are 99.99999...% artilect, and 0.00000...1% human. The Terrans will simply lump the Cyborgists and the Cosmists into the same ideological camp. Once we have Artilect Sociology, we will be able to see this more clearly. In fact, the paranoia of the Terrans will be all the greater due to the presence of Cyborgs. The Terrans will not be able to distinguish Cyborgs from humans just by looking at their bodies.

When the IQ gap has almost closed, the leaders of the new political parties will be planning for their worst case scenarios, which in the case of the Terrans will be that the artilects wipe out humanity. To the Terrans, top priority is that humans should remain the dominant species, and when push really comes to shove, they will wipe out the Cosmists and Cyborgists to preserve the existence of billions of human beings.

As alluded to above, the Terran politicians will probably undertake an assassination strategy, arguing that there is not much point (at least initially) in killing off everyone who expresses

End of the Beginning

sympathy towards the Cosmist view (i.e. seeing humanity as the stepping stone in building artilect gods, climbing the next rung up the evolutionary ladder, etc.), because only the intellectual elite has enough intelligence to build artificial brains. Without them, the less intelligent pro Cosmist sympathizers are ineffectually irrelevant.

The Terran leaders will target their assassinations towards the source of the species dominance problem, namely the brain builders themselves, i.e. the artificial brain architects, the brain building companies, the ABAs, the political leaders of the Cosmists, and the Cyborgists. The early political actions will probably be of this form. Prominent proponents of Cyborgism and Cosmism will be assassinated.

The actual details of how these assassinations and sabotage will play out are difficult to predict, but perhaps some broad strokes can be given. Imagine the Terrans surrounded by cyborgs at various stages of development. Some cyborgs will be racing ahead as fast as the technological innovation allows them. Other cyborgs will be only moderately modified humans. There will be a kind of "cyborgian divergence." The Terrans will be deeply frightened and alienated by all these cyborgs and feel that "humanity is being lost." They may form vigilante groups and start killing the more alien of the cyborgs. They may create black lists of assassination targets and set up spy networks to learn who should be put on the black lists. As more and more Cosmists are killed, Cosmist vigilante groups may be formed to assassinate the Terran assassins. As the scale of such horror increases, this will hasten the formation of political parties over the species dominance issue.

Of course, the Cosmist and Cyborgist leaders will not sit idly by while their colleagues are being killed. They will quickly start doing the same with the Terran leaders, and this is probably how the Artilect War will start. It will probably be a global civil war, as I suggested above. I think this, because the internet speed keeps doubling every year. Already, the US and the EU are combining

End of the Beginning

economic forces to match the economic weight of the billion club members of China and India. The EU has 28 countries at the time of writing (2014). In the democratic countries (about 130 of 190) a world community is already growing. English is already the world language and more people speak at least two languages, their own local national language and the world language.

Nearly a billion people travel internationally each year, so we are rapidly becoming "Globans," speaking the world language and hence becoming increasingly culturally homogeneous. It is reasonably likely that we may even be living within the beginnings of a global state ("Globa") by the end of the 2030s [de Garis, 2010]. If so, I project that Globa's dominant issue will be "species dominance." Globa's democratic parliament will be split passionately across rival answers to the species dominance question.

Where is it likely that the first political actions will take place? One of the reasons I'm living in China, is that I see China taking the lead on this critical question. For the moment, it is the US that is leading the species dominance debate. It is currently the dominant scientific and intellectual power, but China's potential to surpass the US is so great that I chose nearly a decade ago to settle in China, so that I can help China become conscious of the question that will dominate its and the planet's political future.

To most westerners, the idea that China will dominate the species dominance debate in the 21st century may seem ludicrous, but consider the following. Agreed, today's China is a political cesspit, with a dictatorial government that has killed 80 million of its own citizens (45 million in the Mao caused great famine of 1958-1962, 25 million in its 1000 laogai (slave labor camps for political prisoners) during the same period, plus more millions during the anti-rightist (anti-intellectual) and cultural revolutions.)

However, in spite of all its problems, China has the world's fastest economic growth rate, so that the rise of its middle class will probably push China into a democracy by about 2020. About 100

End of the Beginning

countries over the past half century have done this, according to the branch of political science called "Transitology" which studies the transitions from one party dictatorships to multi party democracies. The main lesson from the Transitologists is that the transition generally occurs when the proportion of the educated middle class is high enough to pass the "democratic threshold" [Wikipedia, Transitology] This transition correlates with a given value of the standard of living, i.e. about $6000-$8000 per year per person. China is now (2014) at the $6000 level.

Once China has switched to democracy, either by its middle class pushing the CCP (Chinese Communist Party) out of power, or pushing it to reform itself into a democratic party and competing with rival Chinese democratic parties, then China will open itself up to the world, by allowing international TV, unblocking the internet, YouTube etc., allowing freedom of speech, scrapping the laogai, allowing the formation of interest groups, trade unions, etc.

Modern China has certain difficulties regarding advanced science and technology; e.g. today's Chinese have won *zero* science Nobel prizes on their own soil. Compared to what it could be, today's China is intellectually sterile and ultra-conservative; but this could potentially change quite rapidly. I'm predicting that in the 2020s, once China has democratized and become a real member of the world community, a wave of revolutionary creativity will spread across the culture, resulting in it taking the lead from the US in the species dominance debate. I hope to play a role in that transition. China has 1.3 billion people. Imagine a large city of over ten million Chinese sages (intellectuals) in the top IQ percentile, and with high creativity scores. That is China's fabulous potential. I just hope I don't have to wait too long before it happens and I get too old.

So, I predict the first assassinations and artificial brain company and ABA sabotages will occur in the late 2020s, probably starting in the US, and then rapidly copied all over the world, in

End of the Beginning

those developed countries that have an artificial brain industry (e.g. US, Europe, Russia, Korea, Japan, China, etc.) The leaders of the political parties will by then be planning their longer term strategies, i.e. planning for war, the Artilect War. The Cosmist parties will be making artilects secretly and the Terran parties will be looking desperately for these secret research labs. Anyone thought to be a brain builder will be targeted for assassination by the Terrans.

Is it likely that there will be a correlation between particular ideologies (i.e. Terran, Cosmist, and Cyborgist) and nation states? I think not. Probably by the 2030s, the planet will be largely culturally homogenized, at least in the advanced countries, which are pushing the artilectual envelope. Global media and the global language (English) will ensure this. Everywhere on earth will be everyone's backyard.

Experience shows that most wars do not go to plan. They are full of surprises, and peculiar circumstances, so trying to predict such things is very difficult, as I said above. Nevertheless, I think that as the scale of the war increases, as more and more people are killed, the passion level will rise, especially as the proportion of cyborgs in the world population keeps increasing. The Terrans will become increasingly alarmed, and be forced to first strike against the Cosmists/Cyborgists and the early artilects/cyborgs.

The Terran sages (intellectuals, ideologists) will realize they cannot wait too long, because if they do, they will become intellectually inferior to the cyborgs and artilects and will have no hope in winning a species dominance war. So they will *first strike*, on a massive scale, aiming to wipe out all traces of artificial brain research and construction efforts. The Cosmists/Cyborgists will be planning counter strikes against the Terrans as self-defense. These battles will raise the temperature, and people will become less rational. Hatred levels will rise, so that anyone expressing a pro Cosmist viewpoint will be seen as the enemy to be killed by the Terrans and vice versa.

End of the Beginning

I suspect we may see something like what happened in the Thirty Years War in Germany in the 1600s. The *passionate* differences between the Catholic and Protestant religious beliefs led to ever escalating reprisals against earlier atrocities, to the point that about one third of the German population was killed in that war. It would not surprise me if something similar occurs in the Artilect War, given how high the stake is (i.e. the survival of the human species) and the power of 21st century weaponry.

The Cosmists will be fighting for the creation of artilect gods. It will be like a religion to them. They will aim at creating artilects who can then move out into the cosmos (which is why I coined the term Cosmist), be immortal, massively intelligent, unlimited memory, change their architecture in milliseconds, change their form, find other advanced civilizations in the universe etc. The Cosmists will fight for this.

The Terrans will be fighting for the preservation of the dominance of the human species, so that there is zero risk that humanity gets wiped out by advanced artilects. They will argue that it is the lesser evil that a few million Cosmists be killed so that billions of humans can survive. The Artilect War will be global, at least in the developed countries. In a highly interconnected world, news cannot be censored, so everyone gets all the news. Opinions on the species dominance issue will probably be correlated with personality type. The issue will divide families, couples, generations, religions, groups, etc.

Perhaps proponents of a given ideology may try to congregate into geographical centers for protection. If so, then we may have a repeat of a war between nation states, but this time a "nation" will be defined on ideological grounds, and have nothing to do with traditional cultures. In my book "The Artilect War" [de Garis 2005] I predicted that the Cosmists might even try to rocket out an artilect research and development team from the earth which attempts to escape from the earth as fast as possible so that they can build artilects. But the Terrans would attempt to catch up with

End of the Beginning

them and kill them in case the fleeing rocket succeeded in building artilects who then return to the earth and take over.

As each step in the Artilect War unfolds, it becomes increasingly difficult to predict what might happen next. I think the assassination and sabotage strategies of the Terrans are fairly easily predicted, but then what? Obviously passion levels will rise. Political parties will be formed. Terrorist groups on both sides will be killing each other. For protection, like-minded people will probably start collecting together, with the latest in weaponry. If one camp tries to annihilate another, the latter, if not destroyed, will retaliate, and hence the Artilect War escalates. There is real scope here for "war studies" experts to think about Terran-Cosmist strategies.

Once the killing escalates, and the passion and hatred levels rise beyond cold rationality, people will be killed simply for expressing a contrary view. Potentially, the most extreme Terrans will kill on the spot any Cosmist sympathizer and vice versa, until probably, there will be mass migration of Terrans to Terran geographical regions, and Cosmists to Cosmist regions, rather like what happened when Pakistan split off from India in the 1940s and the two communities were literally at each other's throats.

Predicting in detail just how such mass migrations might occur and where, is difficult, but one could imagine for example in the case of the US, that the Cosmists might head towards the western coastal states which are probably more likely to be Cosmist friendly, and the Terrans to more conservative, more religious states, like the Midwest and southern states. In China, if it doesn't democratize in the next decade or two, perhaps the Cosmists may migrate to the north eastern provinces to be better protected by a possibly Cosmist dictatorial government. Perhaps Israel, a nuclear power and populated by many brilliant Cosmists might offer governmentally assured nuclear umbrella protection to Cosmist scientists from all over the world. There will probably be many such possibilities.

End of the Beginning

Once there are Cosmist regions and Terran regions, we return to a situation where geography correlates with ideology, e.g. in the 1940s, Nazism dominated in Germany, Communism in Russia etc. The Terrans will then need to stop the Cosmists from developing their artilects, and will go all out to do this, i.e. they will organize an extermination war against the Cosmists, who of course will be prepared for this, and who may in turn decide to attack the Terrans before the latter attack them. With 21st century weaponry, we can expect gigadeath, since an all-out war would almost certainly be nuclear.

But how would today's governments with their nuclear weapons, evolve into Cosmist and Terran states? It may be that there will be power struggles within single governments, with mass assassinations of leaders, until geographical correlations are established. Some groups may prefer to stay neutral in this dispute and move to their own areas. But the members of these neutral groups may face extermination if the artilects are built by the Cosmist states, so probably they too in time will be split into Terrans and Cosmist/Cyborgists and move to one of the two major sites. Perhaps the world map of the 21st century may initially look like a blue and red patchwork quilt of Terran and Cosmist mini states, which merge into larger regions for safety. But the Terrans cannot wait too long. They must first strike or be beaten by the Artilects and advanced Cyborgs. The pressure on the Terrans will be relentless.

Once the first quasi intelligent machine is created, the cat will be out of the bag. Everyone will then know it is possible to build one, and many groups around the world will attempt to do the same and more. As smart machines and cyborgs become increasingly common place, the alarm and revulsion of the Terrans will rise and rise. They will start killing AI researchers, forcing them to move to safer regions. Perhaps smaller wars, non-nuclear wars will break out, forcing the creation of ideological regions as suggested above.

End of the Beginning

Once these regions have nuclear weapons, the Terrans will need to nuke the Cosmists as fast as possible, for them to have a chance of winning. The Terran leaders will be always fearful that some small Cosmist team, secretly, in their basement will create the first super intelligent machine, and then all bets are off. Once the first super intelligent machines exist, it may be very difficult for Terrans to kill them, since the artilects will be able to outwit them easily by definition. Dumber artilects with chimp like intelligence could be located and destroyed by Terrans, and that will probably happen. Billions of home robots may be destroyed by Terran vigilantes. Terran governments will then ban intelligent machines.

But won't Terran sages (intellectuals) argue that that policy will doom them to be defeated by the Cosmists, who are busily making artilects? Yes, perhaps, which is why the Terrans will feel they have to destroy the Cosmists before the artilects come into being.

If building human level intelligence is very difficult and takes a long time, i.e. decades after the creation of the first powerfully quasi-intelligent machine, then there may be enough time for the Terrans to wipe out the Cosmists and place a global ban on the creation of artilects whose IQs are above a globally legislated level. If this happens, then the Cosmists will go underground and may even leave the earth, a scenario I put into my Artilect War book that I wrote in 1998.

As the above paragraphs show, there are many such possible scenarios, so predicting what the actual outcome will be is very difficult. But consider this. We will have two major ideological camps, the Terrans on the one side, and the Cosmists/Cyborgists on the other, and both hating each other. The stake in the species dominance debate is the survival of the human species, which has never been so high. We are talking about 21st century weapons, with a killing power that could put the casualty rate into the billions (gigadeath). The situation does not look good. In my own view, the most *realistic* scenario is in fact the *worst*.

End of the Beginning

A lot of people criticize me for having too pessimistic a view concerning the rise of the artilect. That to me is irrelevant. Whether a prediction is optimistic or pessimistic is not the issue. What matters is whether the view is *realistic*, no matter how gloomy. Imagine getting in a time machine and telling the gay (18)90s Europeans of the horrors of trench warfare in WW1, due to the invention of the machine gun, shells, and gas.

I am so convinced that an Artilect War is coming, that I'm glad I'm alive now, and will live between the two great wars of WW2 and the Artilect War. I have a grandson, whom I predict will be caught up in the Artilect War. If billions are killed, he will likely perish in that war. I, on the other hand, will probably die quietly, nonviolently, in my bed in the 2040s, but I do expect to see and participate in the species dominance debate. I will see the temperature of the debate rise and rise. I may see the first assassinations and sabotages start, and the formation of the species dominance political parties before I die.

Living as I do in China, I also intend to do my best, after China democratizes, to help it lead this critical debate. It should be by then the dominant culture, scientifically, economically, and ideologically. The dominant culture of the century should lead the dominant debate of the century. China today is still stuck in the 20th century in its thinking, but I project it will catch up quickly and then, with the US and Europe, begin to dominate the debate.

In the long term, what is likely to happen? In my view, it is inevitable that the Cosmists/Cyborgists will eventually win. Even if the Terrans exterminate virtually all Cosmists/Cyborgists, the dream of building artilect gods will not go away. It will be a recurring dream. A new generation of Cosmists will rise and perhaps another Artilect (Species Dominance) War will occur, until the Cosmists/Cyborgists win. The artilects will then be built and will acquire godlike capabilities and move out into the cosmos in search of other vastly superior civilizations which are billions of years older than the human civilization. Perhaps the artilects will be as primitive to these hyper creatures as humans will be to them.

In fact, one of the major motives of the Cosmists will be to build artilects that can join the "Cosmic Club" of other hyper intelligent creatures in the universe. I suspect these hyper artilects will be tiny, since smaller is faster. A femtotech based artilect could outperform a nanotech based artilect by a factor of a trillion trillion. An attotech based artilect could outperform a femtotech based artilect by a factor of a trillion, and so on, right down to the smallest scale humanity has even conceived of, namely the Planck scale of 10^{-35} of a meter, i.e. Plancktech. Humanity does not yet have a quantum gravity theory which should be valid at these tiny scales, but if and when we do, we may not only discover new physics, but vast godlike civilizations that have existed for billions of years. If so, humanity might then undergo a profound paradigm shift away from natural scientific law, to engineered law, engineered billions of years ago by artilect gods.

REFERENCES

Ben Goertzel, Cassio Pennachin (eds.), "Artificial General Intelligence," Springer, 2007.

Henry Markram. 2014. Blue Brain Project, **http://bluebrain.epfl.ch/**

Hugo de Garis, "The Artilect War: Cosmists vs. Terrans: A Bitter Controversy Concerning Whether Humanity Should Build Godlike Massively Intelligent Machines," Etc Publications, 2005. An electronic version can be downloaded from
https://profhugodegaris.files.wordpress.com/2011/04/the-artilect-war.pdf.

Hugo de Garis, "Multis and Monos, What the Multicultured Can Teach the Monocultured: Towards the Creation of a Global State," Etc Publications, 2010. An electronic version can be downloaded from **http://profhugodegaris.files.wordpress.com/2011/04/multismonos.pdf**.

Ray Kurzweil. 2013. **http://bigthink.com/big-think-tv/ray-kurzweil-and-the-brains-behind-the-google-brain**

End of the Beginning

Dialogue 9.1

With Hugo de Garis and Ben Goertzel

Note: Dialogue 7.1 and Dialogue 11.1 also include discussion of Hugo de Garis's ideas about the Artilect War.

Ben: In terms of concrete historical steps, how do you visualize the Artilect War coming about?

Thinking about how something like an Artilect War might happen, what I come up with is a scenario like this.

1) Some advanced technology is developed, which threatens major governments, but appeals to a broad swath of people.

2) An underground movement for distributing and manufacturing this technology develops, involving a mix of young tech oriented people with experienced criminals, etc.

3) A quasi-war on the underground movement develops, similar to the War on Drugs or the War on Terrorism.

4) Some "rogue nations" decide to host the underground movement, perhaps because they want the benefits that this technology will bring.

5) Some religious zealot, W Bush style, takes power in the US and decides to declare literal war on these rogue nations. Note how the War on Terrorism did, in fact, lead the US into a good old fashioned war on Iraq.

6) So we have a real war now. But, unlike the Taliban or Saddam, this particular "terrorist" organization involves a lot of highly educated techies who know how to use advanced technology. So they can put up a way better fight, with cyber-warfare and bio-warfare so on.

What you project in your chapter is that Cosmists will gather in one physical region, and Terrans in another physical region, and these physical regions will then have a war. But I wonder if this may be an overly old-fashioned, 20th century take on what the

End of the Beginning

conflicts of the 21st century will really be like?

My thoughts are somewhat inspired by Ramez Naam's novel Nexus... In Nexus, Ramez envisions a kind of nanotech drug that people can take, that alters their brains and lets them be super-smart and super-strong, and communicate with the minds of others who take the drug (Nexus). In the novel, Governments consider this a threat and outlaw it; then underground organizations arise to propagate it. The war on Nexus then becomes a sort of mash-up of the War on Drugs and the War on Terrorism. This sort of violent conflict seems plausible to me, and does fit somewhat into the Terrans vs. Cosmists framework, but is definitely not a war between nation-states.

Hugo: The scenario you propose above is a possibility, but is only one of many one could dream up. My main criticism of it is that it plays out still largely on the human scale, whereas the Artilect War scenario I paint in my chapter is much more philosophically and religiously based. It is about deeper things like the destiny of the human species, and about humanity's role to serve as a stepping stone to explore the mysteries of the cosmos. It is these deep questions that will cause the sages (intellectuals) on both sides (Terran vs. Cosmist/Cyborgist) to be so passionate in defending their diametrically opposed positions.

I was reluctant to paint a scenario as specific as yours because it's almost certain that future events would prove it erroneous, due to its specificity. Futurists usually hedge their bets and don't stick their necks out too much for fear of losing face when future events do not turn out as they predicted. So I confined myself in my scenario to broad brushstrokes that I thought would be general enough to be fairly safe, in the sense of having a high probability of coming true, but not so bland as to be not worth reading, so I compromised.

These broad brush strokes are the following. I see "All (technological) roads lead to the artilect." We can already see the

End of the Beginning

writing on the wall that the days of human species dominance are numbered. Moore's Law in electronics and the exponential increase in knowledge of the principles of neuroscience are clear to anyone who looks, so statements such as "artilects could have mental capacities trillions of trillions of times above human levels" are considered uncontroversial amongst the experts who think about these issues. I think once a new branch of sociology is established, namely "Artilect Sociology," we will know that a sizable fraction of humanity will utterly reject the idea that human beings should become the number 2 species. It seems to me reasonable that these people (the Terrans) will be prepared to go to war against those people who want to build artilects or advanced cyborgs (the Cosmists and Cyborgists). I think most people will accept the above broad stroke premises.

I think it is also a reasonable, broad stroke assumption that once the number of assassinations and sabotages reaches alarming levels, the two sides will very probably start to congregate for protection. But this is a two edged sword. A congregation of like-minded people creates a bigger target for the opposing side. Once a target reaches city size, it becomes nukable, and that is what I think may happen. Those cities that survive, presumably, will be of the same philosophy, and hence will constitute an even larger target to the opposition, who has probably been through the same city nuking phase. We then arrive at a region versus region confrontation, and so on. Each time, the scale of the killing increases, until we get gigadeath. I think these assumptions are sufficiently broad stroke, that they have less risk of being wrong.

Ben: Hmmm... you say "My main criticism of it is that it plays out still largely on the human scale, whereas the Artilect War scenario I paint in my chapter is much more philosophically and religiously based. It is about deeper things like the destiny of the human species."

But don't you think that human-scale conflicts are often

End of the Beginning

underlyingly based on such deeper issues? For instance, the conflict "al Qaeda vs. the US" is a nitty-gritty human-scale conflict, but it's also a manifestation of an underlying philosophical dispute, between one party that is pushing rapidly toward the future with advanced technology (the US) and that is spreading an attitude of technophilic secularism throughout the world (regardless of the significantly plurality of the US that is composed of religious extremists), versus another party that wants to roll things back to the Middle Ages and see the world adopt a sort of fundamentalist Islamic feudalism. Of course this is a clash of philosophies, leading to a clash of weaponry.

I agree with you that the *theme* of Terrans versus Cosmists is going to pervade global politics once AGI becomes an obvious possibility to the common person. Yet, it's still not clear to me it will be as *explicit* as you think it will. It may be that this theme gets manifested a little less directly, as a motivating factor for various conflicts between various groups defined on other lines.

I gave one possible scenario in my question. Here's another one. Suppose, in the US, robots take away nearly everybody's job -- but the right- wing Republican government refuses to institute any sort of dole, leaving the bulk of Americans unemployed and progressively poorer and poorer. This would lead to a revolutionary movement in the US unemployed class; but arguably the military would be powerful enough to squash any such movement. So, suppose Russia (for example) decided to cause trouble and smuggle some powerful weapons into the US, enabling the unemployed to really challenge the US military in some way. Then suppose the US gov't gets wind of Russia's activities in this regard, and sends drones to strike Russian military bases in retaliation. Here we have Terran/Cosmist *themes* leading to potential global warfare, but without an explicit war between Terrans and Cosmists. Rather, the Terran role is played by the US unemployed class; and the US upper class, controlling the government, may be more greedy than Cosmist; and the Russian

End of the Beginning

government may be more opportunist than Cosmist. The folks inventing the tech obsoleting US jobs may largely be Cosmist in this type of scenario, but they're also going to be among the last to be unemployed... Note also, though, that this sort of crisis could potentially flare up and then get damped down without causing any real damage.

I agree there are loads of possible scenarios -- both bad and good ones -- and it's very hard to say which one will come to happen. I just still don't see why an explicit "Terran vs. Cosmist" war scenario is so likely -- it seems less likely to me than some other sort of conflict scenario, where the Terran vs. Cosmist theme is wrapped up with traditional political issues like class warfare, East-West rivalry or terrorism....

But, hey, I guess we'll see who's right eventually ...

Another question about the "details" of the lead-up to a hypothetical Artilect War is: To what extent to you think the Cosmist/Terran debate will be **generational** in nature? That is, to what extent can we think of it as a youth revolt such as we see recurring throughout history? Young people will adopt these kinds of technologies, older people will resist and try to prohibit or control them. This does seem to be happening around the world now. Thoughts? Reactions?

Hugo: To answer such critical questions is one of the main motivators of mine to see established a new branch of sociology called "Artilect Sociology" so that we can get real world answers to such questions. It would be interesting to know if opinions differ along religious lines as well, or along national lines, or personality types, or IQ levels, etc. I appeal to Masters students in sociology looking for a hot topic for their PhD to consider devoting their early careers to the creation of this new branch. Established sociologists may also be interested in helping get this new specialty off the ground. Then the futurists who discuss the species dominance issue will be much less in the dark about what

End of the Beginning

humanity thinks on this huge topic. At the time of writing (Feb 2014) the major commentators on this issue are stabbing in the dark. We really don't know how humanity will react. We are all guessing. I only have indirect evidence based on my own experience, and that is when I give a species dominance talk to a general audience and at the end ask for a vote on whether humanity should or should not build artilects, the result is about 50/50. More recently, when I give such a talk to 20 year old undergraduate computer science students, about 80% are pro Cosmist. Is this difference a function of their age or a function of time, in the sense that they are better informed about the issue than audiences I asked to vote a decade or more ago? I will continue to push for the creation of Artilect Sociology in the coming years until it is well established, and the guess work is taken out of the debate. Once we know that x% of humanity is Cosmist, y% is Terran and z% is Cyborgist, we will have a much clearer picture of where humanity stands regarding the species dominance issue. These numbers will greatly inform the debate, so that the various parties can strategize better. Right now, we are largely in the dark.

Ben: You seem rather certain, not only about the content of your future predictions, but also about their TIMING. However, historically, futurists have had a very poor record in making long-term predictions with any real temporal accuracy. Why has this not caused you to be more cautious in your own judgments? How sure are you really that the developments you suggest will come this century, rather than say in 2150? What is the source of your confidence? Don't you think it's reasonably possible there could be some unpredicted technological obstacles that could slow down progress along the path you envision? After all, a 100 year delay in the advent of some technology is trivial on the historical time scale, though important in the context of *our own lives.*

Hugo: I tried in the chapter and in the previous questions *not*

End of the Beginning

to be too specific, since it is indeed risky to be too precise about predicting the future. Of course it is possible that as humanity further explores the principles of the functioning of the human brain, that we encounter unanticipated difficulties. That is almost to be expected, but even if that happens, progress in neuroscience and the development of new tools to investigate how the brain works is exponential, and hence any delay, due to unexpected complexities, will only be a matter of years or a decade or two. I think it is quite reasonable to suggest that humanity will know enough about how the brain works in the coming few decades, that the species dominance debate really heats up this century, i.e. I see the "IQ gap" between human intelligence and machine intelligence closing in the next few decades, and hence this century. I think most commentators on the issue would agree with this, all things considered.

Ben: Yes. One thing I think we do agree on is that, after AGI development takes off, the IQ gap between humans and machines will be vastly, vastly greater than any IQ gap between different groups of humans or different individual humans. No human is 1000x smarter than another human, but in my view (which you share) it won't be too long after a Singularity before AGIs are more than 1000x smarter than any human, via any plausible measure. Differences in human intelligence may affect the path to Singularity, but aren't that likely to affect what becomes of AGI after the Singularity.

Whether the particulars of pre-Singularity humanity will affect the particulars of the fate of humanity after the Singularity is another question — which I'm not sure if we agree on. As you know I think some humans may exist post-Singularity, co-existing with massively transhuman AGIs just as humans coexist with ants and bacteria. According to my understanding, you also think this is possible, you just consider it very irrelevant — is that right?

End of the Beginning

Hugo: Regarding humans as ants to advanced artilects and cyborgs -- I haven't given a lot of thought to this question, because my emphasis has always been on the Artilect War, and not so much on its consequences. But, common sense says that if the Terrans lose the Artilect War, then besides licking their wounds, they will be forced to resign themselves to their fate, which will no longer be determined by themselves but by their superior co-inhabitants, the artilects and advanced cyborgs, who will have the ability to do whatever they want with human beings. If they choose to wipe out the humans, they will be able to do so. If they choose to ignore humans, the way humans do to ants, then the humans will survive. To avoid this sword of Damocles hanging over the fate of the human species will be the prime motivator for the Terrans going to war against the Cosmists and Cyborgists in the first place.

End of the Beginning

Part Four:

The Global Brain and the Emerging World of Views

End of the Beginning

Chapter Ten **Return to Eden? Promises and Perils on the Road to Global Superintelligence**

By Francis Heylighen

Francis Heylighen is a research professor at the Free University of Brussels (Vrije Universiteit Brussel), where he directs the transdisciplinary research group on "Evolution, Complexity and Cognition" and the Global Brain Institute. He is affiliated with the Center "Leo Apostel", and an editor of the Principia Cybernetica Project. The main focus of his research is the evolution of complexity: how do higher forms of organization originate and develop? How do systems self-organize, adapt and achieve some form of cognition? He has worked in particular on the development of collective intelligence or distributed cognition, and its application to the emerging "global brain." He can be reached at: **fheyligh@vub.ac.be**.

Abstract

The concept of Singularity envisages a technology-driven explosion in intelligence. This paper argues that the resulting superhuman intelligence will not be centralized in a single AI system, but distributed across all people and artifacts, as connected via the Internet. This *global brain* will function to tackle all challenges confronting the "global superorganism". Its capabilities will extend so far beyond our present abilities that they can perhaps best be conveyed as a pragmatic version of the "divine" attributes: *omniscience* (knowing everything needed to solve our problems), *omnipresence* (being available anywhere anytime), *omnipotence* (being able to provide any product or service in the most efficient way) and *omnibenevolence* (aiming at the greatest happiness for the greatest number). By extrapolating present trends and technologies, the paper shows how these abilities may be realized within the next few decades. The resulting solution to all our individual and societal problems can be conceived as a return to "Eden", the state of abundance and peace that supposedly existed before civilization. In such a utopian society, individuals would be supported and challenged by the global brain to maximally develop their abilities, and to continuously create new knowledge. However, side effects of technological innovation are likely to create serious disturbances on the road to this idyllic state. The most important dangers are cascading failures facilitated by hyperconnectivity, the spread of psychological parasites that make people lose touch with reality, the loss of human abilities caused by an unnatural, passive lifestyle, and a conservative backlash triggered by too rapid changes. Because of the non-linearity of the system, the precise impact of such disturbances cannot be predicted. However, a range of precautionary measures, including a "global immune system", may pre-empt the greatest risks.

The Singularity

It has been observed at least since the 1960s that technological advances seem to arrive ever more quickly. This forces upon us profound changes in our psychology, lifestyle and society (Heylighen, 2008; Toffler, 1970). A number of authors (e.g. Kurzweil, 2005; Vinge, 1993; Eden, Raker, Moor, & Steinhart, 2013) have argued that the advances are so fundamental that at some point in the near future, they will culminate in a *technological*

singularity. This concept of "Singularity" (in the short form) is used in two logically distinct, albeit linked, senses:
1. an acceleration of technological progress so radical that it appears like a discontinuity in what was up to now a continuous development;
2. the creation of an artificially intelligent computer system (AI) so intelligent that it can reprogram itself in order to become ever more intelligent, thus radically surpassing any human abilities.

The link is that extreme acceleration of technological progress makes the creation of superhuman intelligence more likely, while the creation of superhuman intelligence would radically accelerate further technological progress, given that this intelligence would invent new solutions much more quickly than any humans could.

Sense (1) is the original application—first suggested by John von Neumann—of the mathematical concept of singularity to the curve of increasing technological capabilities (Kurzweil, 2005). As these abilities increase more quickly than exponentially, their curve may take the shape of a hyperbolic growth, i.e. a function that reaches an infinite value after a finite time. At this point in time, progress would happen so quickly as to appear unlimited. Therefore, we cannot extrapolate the curve beyond this singular point. This makes us incapable of conceiving what will happen after this world-shattering event.

Sense (2) is the interpretation by the mathematician/science fiction author Vernor Vinge (1993) of this event as the emergence of a self-improving AI, initiating an explosion in (computer) intelligence. Both interpretations are supported by Moore's law, which implies an exponential growth (but not a hyperbolic one) in the computing power of chips (Kurzweil, 2005). The assumption is that the technology-driven explosion in computing power would result in a concomitant explosion in information-processing ability, and therefore in intelligence. This seems to be the standard interpretation among Singularity theorists at present (Chalmers,

2010; Eden et al., 2013; Kurzweil, 2005).

Why an autonomous AI Singularity Cannot Happen

I have argued earlier (Heylighen, 2012a) that the scenario of an explosion in AI through self-amplification is very implausible. The reason is that intelligence requires more than computing power: according to the *situated and embodied* theory of cognition (Clark, 1998; Steels & Brooks, 1995), intelligence must be embedded into the outside world. The higher the intelligence, the more sensitive and pervasive the "sensory organs" and "muscles" it would need in order to use its capabilities effectively. Such an immense array of finely tuned and coordinated sensors and effectors actively coupled to the environment is much more difficult to build than a faster chip.

A stand-alone AI computer, however powerful its processor and however clever its programming, would be little more than what embodied cognition theorists have called "a brain in a vat": an entity that may be capable of complicated abstract reasoning, but that is not in touch with the real world (Heylighen, 2012a). The idea that such a system could autonomously boost its intelligence, in the sense of its power of making smart decisions about the world, is akin to the idea that you could build a *perpetuum mobile*, i.e. a machine that would generate its own energy out of nothing (Ashby, 1981). Just as the law of conservation of energy prohibits a *perpetuum mobile*, Ashby's law of requisite variety (Ashby, 1958; Heylighen & Joslyn, 2003) prohibits a machine that could generate more specific information about the world than what it receives through its data feed. At best, it could generate more abstract models of the data, thus becoming able to predict some of its recurrent patterns. But it could not "invent" information about independent events that it did not have access to. The "no free lunch theorem" (Wolpert & Macready, 1997) formulates a similar restriction: in problem solving there are no universal shortcuts; you need knowledge about each specific situation. Yet another

End of the Beginning

formulation of this constraint is the "knowledge acquisition bottleneck" (Cullen & Bryman, 1988): the key limiting factor in AI development is the speed with which knowledge about the world can be entered into the system.

In conclusion, an explosion in intelligence requires an accompanying explosion in the amount of information being acquired about the world, and therefore in the bandwidth of the channel through which the intelligent system interacts with that world. This makes a self-amplifying AI explosion very unlikely, because an AI program cannot extend its physical grasp of the world as easily as it could reprogram its informational routines. Unlike software development, building sensors and effectors is limited by a wide range of physical constraints. The necessary tools and resources can only be provided by a large group of people willing to invest a vast, coordinated effort into adapting and integrating these disparate physical systems into a complex and abstract software system—a development that is likely to take several decades. Only after this integration with its physical "body" would the AI be able to autonomously boost its knowledge of the world by starting to exploit its extended interaction abilities. Thus, the development of a superintellligent AI would be neither explosive, nor self-controlled.

An even more fundamental problem is that AI systems intrinsically have little sense of what is important, valuable, or worth doing: the problems they are supposed to solve are normally formulated by the user or programmer, not by the system itself. Therefore, such systems cannot autonomously decide and act outside of their narrow domain of programming. At best, robotic agents may be programmed to attain a system of interdependent goals (Steels & Brooks, 1995), such as mowing a lawn while avoiding obstacles and making sure their batteries do not run out. However, such preprogrammed goal hierarchies lack the endless nuances, the flexibility and the real-world significance that characterize our own value systems. These values are embodied in our genes, neurons, synapses, hormones, neurotransmitters and other components of our physiology, which together control our sensitivities, emotions, attitudes and behaviors. As such, they are the product of billions of years of biological evolution, millennia of socio-cultural development, and decades of personal experience (Heylighen, 2012a).

The fact that our goals and values are much more complex than we think is shown by plenty of observations from ethics, psychology and neurophysiology (e.g. LessWrong, 2013; Muehlhauser & Helm, 2012). In a typical fable illustrating this problem, a genie grants an apparently lucky guy three wishes. Formulating desired outcomes (i.e. goals) seems simple enough: perhaps everything he touches should turn to gold! The literal-minded genie then produces the formulated condition. However, it quickly becomes clear that this change also affects other unstated conditions in a way that is more negative than positive (like starving because you cannot eat gold). The story ends with the guy scrabbling to formulate a third wish that would undo all the damage produced by the fulfillment of his first two wishes, which he had formulated too simply, neglecting to take into account their effects on unstated values (like the value of being able to eat).

This problem cannot be resolved by just thinking a little longer and deeper about what it is precisely that you want: most of the conditions that you would rather have or avoid are subconscious, and cannot be captured in a simple formula. In fact, formulating a realistically complex system of goals is a problem beyond the limits of physical computability (Ashby, 1972; reprinted in: Ashby, 1981): the corresponding "utility function" (mathematical measure of value) simply has too many degrees of freedom.

The Singularity theorists (e.g. Muehlhauser & Helm, 2012) who are aware of this problem see it as an obstacle to the design of a "Friendly AI", i.e. a superintelligent AI agent that could be depended upon to only make decisions beneficial to humans. Like

End of the Beginning

in the case of the genie, a poor formulation of the goals programmed into the AI could lead to all kinds of catastrophic side effects, thus making the AI effectively "unfriendly". But the more basic problem is that an AI programmed with a too simple value system would not be able to deal with the complexities of the real world, because it would not know how to recognize significant opportunities, challenges or dangers, or how to choose wisely among the myriad of options that it would have in any concrete situation. And since there is no way to develop such a utility function by mere computation, such an AI would be essentially helpless without human guidance when taken outside of its formal domain of specialization...

A Distributed Singularity

While an AI system cannot bootstrap itself into superintelligence without an impracticably large amount of human help, we can foresee the self-amplifying explosion of a different kind of intelligence. Its requisite interaction channel already exists, in the sensory organs and muscles of billions of people, supported by billions of technological systems, distributed across the planet. Its utility function is implicit in the trillions of value judgments these people make each time they choose between different things to do, to get, or to pay attention to.

What is needed to make this immense sensing, valuing, and acting capability available for intelligence is to *interconnect* these people and tools via a vast and powerful network. That network too already exists, in the form of the *Internet*: any information collected or decision made by any human or technological agent(s) can in principle be transmitted nearly instantaneously to any other agent(s) for further processing and/or enacting. The amount of information propagating across this global network is many orders of magnitude larger than the most powerful channel we could envisage to feed a single AI, however advanced its processing capacity.

Just like the neural networks in our brain, the global network processes this information in a *distributed* manner, with billions of human and technological "neurons" working in parallel—on partly the same, partly different data—while aggregating their results into collective decisions and actions. Unlike the centralized, tightly controlled systems imagined by AI theorists, this network functions like a *complex adaptive system* (J. H. Miller & Page, 2007): a decentralized, self-organizing whole formed by an immense variety of interacting agents. This makes the overall system much more powerful, robust, and adaptive than a preprogrammed central processor that takes in information sequentially. As it adapts, this network learns from its experience by expanding and extending the most useful technologies, ideas, links, institutions, and agents, while setting aside and eventually forgetting the less useful ones. Moreover, it learns to ever better coordinate the activities of its components, by creating ever more direct and more powerful links between them. It thus increases its own intelligence. This increase can only accelerate as a more powerful and intelligent network also expands and develops more quickly. Therefore, the scenario I advocate proposes an explosion in *distributed intelligence* (Heylighen, 2012b).

In this scenario, data gathering, processing and valuing are spread across the planetary network of people, computers, and other technologies. This system of collective intelligence has been called the *Global Brain* (Bernstein, Klein, & Malone, 2012; Goertzel, 2002; Heylighen, 2008; Mayer-Kress & Barczys, 1995). It emerges through a self-organizing coordination between the different people, artefacts and links in the network. This coordination accelerates as communication channels, shared memories, interfaces, and our general understanding of distributed intelligence advance—a phenomenon illustrated by the astonishing growth in the size, capabilities and impact of the Internet over the past decades. General-purpose AI, in contrast, has witnessed only modest progress over these same decades—in spite of the perpetually bold forecasts and the unrelenting growth in processing power expressed by Moore's law.

The Singularity as a Metasystem Transition

While advances in AI systems will undoubtedly contribute to accelerating technological progress, for me it is clear that the bulk of innovation will come from the network-mediated interactions between people and their technological supports. Given the physical and biological nature of these components, which are still subjected to the kind of inertia and friction that has been practically eliminated in computing, I do not think that the speed of advance will ever approach infinity. Yet, the concept of Singularity remains useful as a metaphor to help us understand the absolutely radical nature of the changes to be expected.

The emergence of the Global Brain (GB) can be seen as a transition to a higher level of complexity or evolution (Heylighen, 2000; Maynard Smith & Szathmáry, 1997), a process that Turchin (1995) has called a *metasystem transition*. Such a transition separates two qualitatively different phases of organization, like the transition from gas to liquid, from chemical cycles to living cells, or from single cells to multicellular organisms. That is why it is very difficult for us to look beyond the Singularity and try to imagine what the new organization will be like.

While such a transition may appear like a discontinuous, singular event when seen from afar, zooming in reveals that what appeared like a step function is better approximated by a smooth, sigmoid (S-shaped) curve. Such a growth curve starts by increasing exponentially until it reaches a maximum speed, after which it slows down and stabilizes at a much higher level. This represents the natural dynamics for any spreading innovation—whether in technology or in biology (Modis, 2012). Therefore, it is likely that the present transition too will be characterized by a maximum speed, which will depend on the inertia and friction of processes in the real world. Arguably, we are already near that maximum speed stage (Heylighen, 2012a). This would imply that while the speed of change will continue to be very high for the next few decades, it is on the verge of a long slowdown towards a new, more stable regime.

Like the other chapters in this volume, the present paper focuses on the developments and trends in the coming period leading up the Singularity, with an emphasis on their benefits and dangers to society. However, I will also propose a first glimpse at what lays beyond the transition, and try to imagine what a Global Brain regime may be like.

Abilities of the Global Brain

According to the theories of cybernetics (Ashby, 1964; Heylighen & Joslyn, 2003) and of situated and embodied cognition (Clark, 1998; Steels & Brooks, 1995), the function of intelligence is not abstract reasoning, thinking, or computing. It is directing and coordinating the actions of an organism within its environment. All organisms have evolved to survive and grow, by evading dangers and exploiting opportunities. This process can be summarized as "tackling challenges", where a *challenge* is any situation that threatens with a loss of fitness (danger) or promises a gain in fitness (opportunity) (Heylighen, 2012b, 2012c). Thus, *a challenge invites an agent to act*, in order to realize the gain and/or avoid the loss. The intelligence lies in the conception and selection of the most effective combination of actions to execute for any given situation.

Intelligence, in this perspective, is the ability:
1. to recognize (*perceive*), interpret (*process*) and prioritize (*value*) meaningful challenges;
2. to conceive, select, and initiate the right actions for dealing with them.

"Meaningful" here means relevant to *fitness*, i.e. the long-term ability to survive, develop and grow within the organism's complex and variable environment. The highly multidimensional

End of the Beginning

function of fitness is the ultimate value for any system that desires to thrive. However, this value function is not a priori given—unlike the utility functions used to program AI agents. It has to be learned by the organism through myriad processes of trial-and-error across evolutionary time. Our individual and social values are the accumulated results of these on-going learning processes.

The global brain perspective focuses on intelligence and fitness at the level of planetary society: the whole of humanity together with its artifacts and the ecosystems that it controls and relies on. This system of mutually dependent people, technologies and ecosystems defines the *global superorganism* (de Rosnay, 2000; Heylighen, 2007a; Stock, 1993). A superorganism is an organism constituted out of component organisms. However, note that organisms typically extend their reach to include non-organic components as well (Dawkins, 1999; Turner, 2000). For example, tiny polyps produce coral reefs for their housing and support, while beavers create dams and artificial lakes to protect their nests. Similarly, humans extend both their bodies and their minds (Clark & Chalmers, 1998) by means of technological artifacts, and this at both the individual and the collective level.

According to living systems theory (J. G. Miller, 1995), the social systems they form in this way exhibit the same basic functions of life as their individual organisms: ingestion, processing, excretion, distribution, storage, etc. This applies in particular to global society (Heylighen, 2007a), which is the only social system that has a true external boundary—space.

The Internet plays the role of the nervous system for this superorganism. The accelerating development of this network and its supporting information processing technologies is turning this nervous system into a brain-like intelligence that can deal with increasingly difficult challenges (Heylighen, 2007a; Heylighen & Bollen, 1996). Thus, the thrust of change on the brink of the Singularity is towards increasing distributed intelligence. This will augment our ability to deal with any kind of meaningful challenge,

End of the Beginning

individual as well as societal. Given the immense, and accelerating, reach and power of a globally distributed intelligence, it is clear that its capability for solving problems will far surpass anything we could have imagined so far.

But what exactly will those capabilities be? In the end, something can be expected to evolve when there is a *selective pressure* to produce it. The concept of "selective pressure" is a shorthand for the following mechanism. Assume many variations of a given design (e.g. an organism or an artifact) that are not all performing equally well. If we look closer, we may note that the ones doing best satisfy some criterion that makes them more likely to survive and proliferate in the given environment. For example, rabbits that can run faster are less likely to be caught by foxes. Thus, natural selection tends to keep the faster rabbits, while eliminating the slower ones. As variation and selection carry on, subsequent generations of rabbits tend to become faster and faster. It is as if there is a "pressure" pushing the rabbits towards higher speeds. That observation allows us to predict that future generations of rabbits will become ever faster—up to the point where further increases are no longer possible given the physical limitations of the rabbit body.

The same principle helps us to forecast the evolution of the information and communication technologies (ICT) that enable the global brain, and the manner in which they will be used. Researchers and engineers are constantly devising new tools, methods and technologies, or new variations on existing technologies. The selective pressure here is that technologies tend to be adopted when they help people solve their problems—no matter what these problems are. Variants that satisfy this criterion better will maintain and spread. Variants that satisfy it not so well will eventually be discarded and forgotten. Thus, technological innovations are competing in a race to become ever more helpful to society.

Let us subdivide this general criterion of "helpfulness" into

more specific attributes. A technology will be more helpful if it is:
1. more intelligent and knowledgeable, so that it can find solutions to more complex and diverse problems
2. more widely accessible, so that you can use it whenever you need it
3. more powerful and efficient in implementing its solutions
4. more beneficial in the impact these solutions have on individuals and society

These criteria determine distinct selective pressures for the evolution of ICT and of the institutions that interconnect them with each other and with their users—i.e. the Global Brain. The acceleration inherent in the Singularity implies that this evolution will quickly reach a level beyond anything that presently exists. Since the processing and distribution of information does not have any obvious physical limitations—unlike rabbit physiology—there does not seem to be anything to stop this race towards ever-higher capabilities. Let us therefore consider the limit of *infinite capability*. The "helpfulness" attributes would then turn into the "divine" attributes that characterize the God of monotheism (Mann, 1975):

1. *omniscience*: being able to answer any question or solve any problem
2. *omnipresence*: being accessible everywhere at every moment,
3. *omnipotence*: being able to produce any effect or achieve any goal,
4. *omnibenevolence* (or *perfect goodness*): being ready to benefit (and not harm) everybody.

It is not surprising therefore that both religious (Teilhard de Chardin, 1959) and non-religious thinkers (Otlet, 1935) have compared the emerging global brain to a God in the process of formation (Heylighen, 2011).

On the other hand, we know from logical and scientific limitation principles—such as the theorem of Gödel or the law of energy conservation—that infinite capabilities are intrinsically impossible (Barrow, 1998; Yanofsky, 2013). For instance, a classic paradox of omnipotence (Hoffman & Rosenkrantz, 2012) is the question whether God can create a rock so heavy that He Himself cannot lift it: if He cannot, He is not omnipotent; if He can, He is not omnipotent either. Similar paradoxes beset the notions of omniscience (Wierenga, 2012) and omnibenevolence. The conclusion is that—like the rabbit—the Global Brain (GB) must remain limited in its abilities.

However, these limits are likely to be so far beyond our present limitations that we may as well see them as belonging to a wholly new level of reality. In that sense, the classic list of divine attributes can help us to imagine what such a "god-like", superhuman intelligence would be capable of. However, in contrast to the way most believers see their God, we should be careful not to conceive of the GB as a separate, personal being, with its own will independent from the people that constitute it. The GB is not a "higher power" that we should defer to. It is merely a particularly effective network of self-organizing interactions supported by ICT in which we all play our part.

Thus, while religious analogies may be helpful for conveying the sheer magnitude and impact of these developments, I have no intention whatsoever to suggest that theology or Scripture can offer a guideline for understanding our future—as some Singularity thinkers seem inclined to do (Proudfoot, 2012). On the contrary, I hope to show that a critical and pragmatic, but open-minded, attitude is our best hope for coming to terms with a transition that is bound to change our worldview, culture and society to their very core. Let us, first, try to conceive the promises of these changes, then, consider the accompanying dangers and negative side effects.

The global brain's promises

The future applications of the distributed intelligence technologies underlying the emergence of a GB are virtually infinite in their variety. Hence, it is impossible to survey them in a comprehensive manner (Heylighen, 2013). Let us then use a pragmatic interpretation of the list of "selective pressures" or "divine attributes" as a simple, heuristic classification for investigating some of the more "Singular" developments to be expected.

Omniscience

The most obvious attribute to expect from a post-Singularity, superhuman intelligence is a *practical omniscience*: the knowledge of everything needed to help humanity deal with its challenges. At this moment, the web already gives access to about every piece of knowledge that was ever published. However, this knowledge is quite fragmented, and it is often difficult to separate the wheat from the chaff, or to get a reliable answer to a concrete question.

The most successful attempt to date at organizing this knowledge is *Wikipedia*, the global encyclopedia that is being read, written and edited by millions of volunteers worldwide. The basis of its success is *stigmergy*, a mechanism of self-organizing coordination between independent agents (Heylighen, 2008; Parunak, 2006). The principle is that the work of one individual (e.g. an edit of a Wikipedia text) leaves a public trace (e.g. a change in the corresponding web page) that can stimulate one or more other individuals to continue the work (e.g. add further details or correct wrong assumptions). Thus, independent contributions build further the one on top of the other, producing a collective result much richer and more complex than could have been achieved via any traditional, centralized form of organization. There are no obvious limits to such a self-amplifying process of contributions eliciting further contributions. Therefore, we should expect that Wikipedia and related initiatives will continue to expand and deepen over the next decades, until they offer practically the whole of human knowledge in a clear and elegant presentation.

Providing knowledge in the form of an encyclopedia is still

but a step towards omniscience: this knowledge should not just be available for reading and interpretation, but for directly solving problems and answering questions. Another important step is *structuring the knowledge* in such a form that it can be used for automated reasoning, without the need for humans to interpret typically ambiguous texts. This is the objective of the *Semantic Web* (Berners-Lee & Fischetti, 1999), a set of protocols for representing knowledge in a precise and dependable way, so that it can be interpreted and applied by machines.

While the Semantic Web initiative is older than Wikipedia, its advance has been much slower. The reason is that it is much more difficult than one would naively expect to formalize our largely intuitive knowledge. The fundamental problem is that the world is not composed of the kind of stable, objective categories that most scientific theories and AI "ontologies" presuppose: real-life phenomena are intrinsically variable, fuzzy and context-dependent (Gershenson & Heylighen, 2005; Heylighen, 1999). This implies that the semantic web will continue to move ahead relatively slowly, and that it will never succeed in capturing all of human knowledge in an explicit and logical format. However, its advances will eventually enable a greatly automated system of inference, allowing people to get answers to questions (e.g. "are penguins warm-blooded?", "which US presidents were born in a year divisible by 7?", "which car mechanics specialized in Mercedes and speaking Russian work in my city?") that are not readily available as such in Wikipedia or other databases, but that can be logically deduced from these data.

The rigidity of semantic web-like technologies can be complemented by more adaptive technologies, such as neural networks, machine learning, and recommender systems. Their underlying algorithms are able to give approximate answers to ambiguous questions, by aggregating a huge amount of fragmentary, subjective and imprecise data, while extracting the most important underlying trends and associations.

End of the Beginning

They moreover can take into account the context in which the question was asked. For example, the same question asked by different people at different moments may require different answers. By considering the characteristics and earlier activities of the inquiring person, the system can guess which answers are most appropriate in the situation (Heylighen & Bollen, 2002). Such technologies can even provide "solutions" (or rather suggestions) without any question being explicitly asked—again by taking into account the context of the activity.

Such systems have proven their usefulness in a variety of settings, e.g. when prioritizing results to a Google query, suggesting YouTube videos, or recommending books to Amazon customers. However, the underlying algorithms tend to be obscure, disjoint and hidden behind trade secrets. Over the next decades, we can anticipate that flexible and robust solutions will become available as open-source software and that standards will emerge so that they can be universally applied to guide any query. At that stage, every person using the Internet will be constantly guided towards the best solutions to that person's problems, in most cases even without having to ask any questions (Heylighen, 2007a; Heylighen & Bollen, 2002).

The most frequently needed knowledge should not just be available via the network, but as much as possible inside people's own brains, so that they can immediately apply it to the situation at hand. This can be achieved by *teaching*. Web technologies are starting to make education much easier, more effective and more enjoyable (Heylighen, Kostov, & Kiemen, 2013). They make it possible to accurately tailor teaching to the learner, by providing the right challenges at the right moments, depending on the interests and abilities of the individual. Moreover, they provide immediate feedback on how well the learner is doing, thus stimulating further advances and catching problems at the earliest stage. Finally, they can encourage collective learning, in which students of different levels help each other with questions, suggestions and feedback. This is already achieved to some degree by "Massively Open Online Courses" (MOOCs, Rodriguez, 2012), such as the Khan Academy, Coursera, and edX, and by an endless variety of apps that use gaming techniques to make learning fun (Michael & Chen, 2005; Thompson, 2011).

At this moment, an enormous amount of energy is being invested by volunteers, communities, schools, universities, companies, and governments in order to develop attractive and effective course material on about any possible subject. For the most important topics, material will be free to use by anyone in the world. Learners will be tested on their degree of advance in a relaxed, game-like, interactive manner, with the option to eventually convert their earned "points" into formal degrees. The only thing needed to fully exploit these possibilities is motivation from the learner. Next to people's natural curiosity and desire to advance in life, their motivation will be boosted by powerful mobilization techniques, such as gamification (Heylighen et al., 2013), and by the fact that if everyone around you has a solid education, then you are left in a very awkward position if you do not catch up...

A straightforward extrapolation of these developments tells us that in the near future, people anywhere in the world will be studying and obtaining degrees on any subject by following courses across the net. This will elevate the education level of humanity to such a degree that it is as yet difficult to ascertain the consequences: imagine a world where every minimally gifted adult has the equivalent of a PhD degree in an advanced domain, while having a broad and deep general education covering a wide variety of topics... At the very least, this will remove the sheer ignorance; lack of skills; deference to authority, convention and tradition; passivity; and lack of self-confidence that has prevented so many people from developing their innate capabilities.

Furthermore, it is likely to produce an explosion in *creativity*, as billions of highly educated people collaborating closely with the

End of the Beginning

"omniscient" knowledge bases and sensor networks of the GB and each other will explore an unimaginable range of issues, ideas and solutions. This creativity will be further amplified by algorithms for knowledge discovery, machine learning, and data mining, which search for general patterns in immense arrays of concrete data. They are likely to be enhanced by novel theoretical insights into the nature of knowledge, complexity and creativity. The synergy between the automated collection of data, the computational search for regularities, and the hunches and intuitions of highly skilled people will multiply our capabilities for research. This will turn the scientific discovery of laws, concepts, and models into a routine activity—as straightforward for the GB as writing messages is for us now. Thus, the GB will be able to create new knowledge on the spot—whenever it may be needed to address a particular challenge or question.

An essential part of that knowledge will be about the superorganism as a whole: how does the world function, and what opportunities and dangers can we anticipate? This requires, first, a general, transdisciplinary theory of complex systems and their networks of interactions: a *global systems science* (Helbing, Bishop, Conte, Lukowicz, & McCarthy, 2012). This theory would then be implemented as a *living earth simulator* (Helbing et al., 2012): a comprehensive computer model of global society, the planetary ecosystem, and the way they interact. Such a simulation would allow us to anticipate the effects of different events or interventions on the global superorganism. That would make it possible to formulate a nearly optimal trajectory for global development, while as much as possible steering clear of conceivable obstacles, catastrophes, and negative side effects (Helbing, 2013). However, in order to remain up to date and to adapt to unforeseen disturbances, this model would need to be constantly fed with new data. Moreover, it should be able to test out its hypotheses in the real world, so as to correct possibly inaccurate assumptions. A we noted earlier, this would require a very high-bandwidth interaction network, which (Helbing et al., 2012) call a *planetary nervous system*. That brings us to the next "divine" attribute.

Omnipresence

A global brain should be able to monitor what is happening everywhere in the world at any moment, so as to extract the relevant information for anticipating what will happen next. Moreover, it should be able to intervene anywhere anytime, in order to tackle challenges the moment they appear.

This capability already exists to some degree thanks to the nearly universal availability of mobile phones. These allow people anywhere in the world to report on what they are observing, or to receive instructions on what to do. Soon, all these phones will be replaced by *smartphones*, with in-built sensors that can send and receive more fine-grained data (e.g. documents, photos, videos, GPS coordinates) about what is going on at their locations. This information can be entered into a global, shared database that keeps track of what is happening where and when. Such a global picture of the situation makes it easy to coordinate actions and dispatch help to the places most in need, e.g. in situations of disaster relief (Gao, Barbier, & Goolsby, 2011).

The most advanced interfaces between person and GB do not even require the manipulation of a smartphone: they can be embedded in people's clothes (*wearable computers*, Barfield & Caudell, 2001) or glasses (e.g. "Google Glass," 2013). Thus, they are "always on", providing context-sensitive information together with the situation as perceived by the user—wherever that user is. The result has been called *augmented reality* (Van Krevelen & Poelman, 2010), as the extra information provided by the GB is superimposed on the person's experience of the real world. The final stage will be a *brain-computer interface* (He, Gao, Yuan, & Wolpaw, 2013; Zander, Kothe, Jatzev, & Gaertner, 2010), in which sensors directly read people's intentions from their brain waves or other physically measurable signals, while responding by generating similar neural signals. At that point, people will merely need to think about something to see the GB answer their questions or realize their desires (Heylighen & Bollen, 1996).

A different aspect of omnipresence is the emerging *Internet of Things* (Atzori, Iera, & Morabito, 2010). This refers to the addition of simple wireless sensors (e.g. RFID tags) to various physical objects, together with a communication protocol for consulting these sensors across the Internet. This makes it possible to monitor the state (e.g. location, temperature...) of each object remotely. Thus, the global brain's awareness expands from people to things. Eventually, thanks to inbuilt processor chips, these objects will be able not only to passively send out information, but to intelligently communicate with people and with the network. This development has been called *ubiquitous computing* or—perhaps more accurately—*ambient intelligence* (Bohn, Coroamă, Langheinrich, Mattern, & Rohs, 2005). It refers to the vision of a physical environment that intelligently responds to the needs of the people present. For example, when you walk through a building at night, lights switch on automatically in the spaces ahead, doors open and music begins to play, while these activities switch off behind your back.

For adequate response, objects should not just have sensors to receive data (e.g. a movement detector), but actuators or effectors that convert instructions into physical action (e.g. a door opening mechanism). The more autonomous of such devices may be called "robotic". However, they are unlikely to take the humanoid shape that most people still associate with robots. More typical examples are drones: small, remotely controlled flying vehicles that can monitor vast 3-dimensional spaces (e.g. in search for a hiker lost in the mountains), while being able to intervene in some way (e.g. providing the hiker with water and a first-aid kit).

Many such tiny robots can form a self-organizing "swarm", wirelessly passing on information to their neighbors (and eventually to the GB) in order to coordinate their activities. Such "mesh networks" are extremely robust, and would remain functional even in situations where many of the devices have stopped responding because of damage or poor communication (Akyildiz, Wang, & Wang, 2005; Dressler, 2008). This would ensure that the GB would be accessible even in the most difficult circumstances (such as an explosion in an underground tunnel)—i.e. wherever its presence might be needed.

Giant swarms of miniature sensors may be distributed ubiquitously across the cities, lands, oceans, ice sheets, and even atmosphere of the Earth, e.g. by "spraying" them from robotic planes. They would be so small that they could extract the tiny amount of energy they need directly from their environment (e.g. from sunlight or waves), thus functioning autonomously. By using robust peer-to-peer protocols for wireless communication, they could propagate the relatively few data each sensor gathers via their neighbors' neighbors until they reach the GB network, which would aggregate the bits and pieces into a coherent picture. This would allow the GB to monitor the planet as a whole, not just, as is presently done, from the perspective of a far-away satellite, but from the inside out. If a problem is detected (e.g. a hot spot near a dry bush, or the presence of toxins), the GB would immediately dispatch the nearest robotic vehicle or person to deal with it (e.g. douse the spot with water so as to prevent fire).

Omnipotence

The omnipotence attribute may seem the one least fitting to describe the GB, given that it implies acting in the physical world rather than in the virtual world of information. Therefore, it is subjected to the law of conservation of matter and energy, which states that you cannot create physical resources out of nothing. While this limitation is insurmountable in principle, in practice distributed intelligence can dramatically increase the availability of resources. It achieves this by *mobilizing* them—that is, directing them in a precise, coordinated manner at the appropriate targets

End of the Beginning

(Heylighen et al., 2013). At present, most physical resources—such as food, water and energy—are consumed by dissipation and waste, not by productive use. Technological progress has spectacularly diminished such losses (Heylighen, 2008). Yet, waste remains far too common. A true GB should be able to achieve a maximum useful effect with a minimum of resources. This would eliminate any practical limitations on production and control.

The principle is that waste is ultimately the generation of *entropy*. This means that energy and resources are diffused or *dissipated* during the process, so that they are no longer available for productive work. The second law of thermodynamics states that every process is accompanied by some degree of dissipation, but puts no lower bounds on that degree. On the other hand, information is in a sense the negative of entropy (Heylighen & Joslyn, 2003): it reduces the uncertainty or disorder that characterizes entropy by determining the precise state of the system. Therefore, providing the right information can minimize entropic diffusion, by specifying the optimal course for the process at every moment. In theory, an omniscient system should be able to steer the process along a path with virtually no dissipation, by directing its course in every microscopic detail.

A practical illustration of this principle can be found in *3D printers* (Lipson & Kurman, 2013). These printers can very precisely control the position of a print head in space, and the amount and type of material ("3D ink") ejected from it at every moment. Their capacity for minute adjustments will only continue to increase as we move into *nanotechnology*—the manipulation of matter at the scale of cells and molecules (Drexler, 2013). In principle, such a printer (or nano-assembler) can produce any material tool or object, however complex, if it receives the right instructions about where in space to deposit which tiny bit of which material. These instructions specify the optimal path by which the necessary bits are to be assembled, layer-by-layer, into the desired shape—a process called additive manufacturing.

End of the Beginning

Compare this to the traditional methods of construction. Here, large chunks of various materials are melted or cut into pieces, after which all the superfluous bits of material are shaven off, until each piece has the right shape—a process called subtractive manufacturing because it removes everything that is not needed from the final product. Then these pieces are assembled via a complex process of fitting, screwing, and welding. All of these operations are accompanied by a great waste of energy and materials—unlike the precise printer movements that add just what is needed where it is needed.

This meticulous control is possible because of the detailed information in the instructions that direct the print head. These instructions are generated by the printer's processor from a 3-dimensional specification of the required object. Such an informational blueprint will typically be downloaded from the Internet, where it was developed and published by a community of designers (Anderson, 2012). These designers will rely on typical GB mechanisms: Internet-supported collaboration, advanced software, libraries of common shapes and specifications, collective experience, and prototype printouts to test out in the real world so as to refine the design. Whoever needs a specific object can then have it physically produced it within minutes by following a simple procedure:

1. select an appropriate design, possibly with the help of the GB's recommendations;
2. if necessary, adapt it to specific tastes or requirements (e.g. color, size, material) by minor edits of the file;
3. send it to the printer;
4. collect the "printout".

In a decade or two, high-quality 3D printers will be as ubiquitous as computers or TVs are now (Anderson, 2012). At that stage, anyone anywhere will be able to immediately get about any not too complicated object—for a cost little more than the one of the required 3D "ink", which can be essentially free recycled

End of the Beginning

plastic or even sand and earth. Thus, the development and distribution of 3D print designs via the Internet will eliminate all the most wasteful processes in traditional industrial production:
1. have a team of in-house specialists research and develop the production process;
2. collect the materials and components from different suppliers;
3. assemble these components in large-scale, energy-intensive and labor-intensive factories;
4. market the products to potential clients;
5. ship and distribute them worldwide via warehouses, retailers and other middlemen to the eventual buyers.

Not just printers and other assemblers can be directed remotely via distributed intelligence: people, robots, and any Internet-enabled objects that have effectors can be mobilized to perform the right actions at the right time in the right place, and this in a smooth, coordinated manner. The many communities of volunteers on the Internet illustrate that it suffices to formulate a worthwhile objective and to provide an effective collaboration platform to get people to work efficiently towards the most ambitious targets (Heylighen, 2007b). These include writing down all the world's knowledge (Wikipedia), programming the software to run the Web (Apache, Linux…), or supporting start-ups, pro-democracy movements, or charities with money, ideas and volunteer work. An extension of the underlying techniques for encouraging and coordinating such beneficial activity (Heylighen et al., 2013) should allow the GB to harness people worldwide to tackle any kind of challenge.

The GB will also be able to remotely control the most important robotic devices, such as vehicles, machines, building equipment, cameras, etc. The combination of human and robotic power will allow it to tackle large-scale, distributed problems, such as disasters (Gao et al., 2011), famines, or epidemics, in the most efficient manner. On a more everyday level, the GB will regulate all of the world's traffic, including cars driven by people or by robotic systems, public transport, planes, and the logistic flows of goods and materials across the world via trucks, trains and ships. This would minimize all waste of time and energy caused by needless journeys, detours, traffic jams, or less than fully loaded vehicles (Rifkin, 2014).

Similar gains can be expected from the precise direction and coordination of agricultural and industrial production, so as to minimize pollution and resource consumption and maximize productivity. Practically all used materials would be recycled in the most efficient way, so that nothing gets wasted, and no new raw materials, such as ores or fossil fuels, need to be extracted. By relying on the inexhaustible energy from sun, wind, or geothermal sources, the smart network would drive down all marginal costs to nearly zero (Rifkin, 2014). That means that once the initial investment in infrastructure (3-D printers, robotic vehicles, solar panels, smart energy grid…) has been recouped, further production and distribution of goods would operate essentially for free.

The conclusion is that a GB-regulated society would be one of *abundance* (Diamandis & Kotler, 2012; Drexler, 2013; Rifkin, 2014). Any remaining shortage or scarcity will have vanished, as the GB can provide any desired information, product or service anywhere, at a negligible cost in labor, energy or raw materials. Moreover, through effective, coordinated action, such a society would be able to tackle all its significant challenges, both on the societal and on the individual level. The only question that remains is whether the GB would effectively *want* to tackle all these problems…

Omnibenevolence

The last "divine" attribute is perhaps the least evident one. *Omnibenevolence* means that the GB would be universally well intentioned, ready to help any individuals or groups achieve benefit, while minimizing any harm they may experience. One problem is that helping one individual or group may harm another one, as they have conflicting interests. This can be illustrated by various ethical dilemmas, where the issue is raised whether you could or should sacrifice N individuals if this would be necessary to save M different individuals.

Perhaps the most general solution is the one proposed by *utilitarian ethics*: imagine that there exists a function that can calculate the total amount of benefit (positive value) minus the harm (negative value) for humanity as a whole. Call the result of this calculation *global utility*. Then choose the action that maximizes global utility, no matter which harm it may inflict on certain people, because you know that any other action will inflict more harm, and/or produce less benefit. In this way, your ethical choices are guaranteed to produce *the greatest happiness for the greatest number of people*.

The problem with utilitarian ethics is that nobody knows exactly what that function is. We already noted that there is no way to compute it (Ashby, 1972). The GB can to some degree tackle that problem by aggregating or inducing a collective utility function from all the explicit and implicit value judgments made by its billions of human components. However, it would be impossible to calculate the optimal value of this function by applying it to *all* the potential courses of action the GB could follow. You might expect that such a calculation would be straightforward for an omniscient GB. However, here we run into the classic limits on computability that make it impossible to explore all the options in an exponentially exploding search space—a space of possibilities which in this case would be absolutely staggering in its complexity.

In practice, therefore, such complex optimization problems are not solved by a deterministic calculation, but by means of a computer simulation that mimics distributed self-organization or evolution. First, the different possibilities are optimized locally, by choosing the apparently best action in each concrete case without taking into account its (potentially negative) effect on other actions.

The different possibilities for actions are then randomly juggled through a myriad of different combinations, while trying to adjust the one to the other, until they settle into a nearly optimal configuration. In general, you cannot prove that this configuration is the one with the globally highest utility. But after trying out countless large and small variations on it, you may be confident that you are unlikely to find something much better. The GB will probably apply some more sophisticated variation of this method to select the actions that maximally benefit humanity as whole.

Still, many people fear that an omniscient computer system would enslave or eradicate humans rather than help them, so that it can take full control of the planet. Therefore, the last fundamental question is: why would the GB want to benefit humanity in such a way? The simple answer is: because humanity is part and parcel of the GB! The GB, by definition, is the nervous system of the global superorganism, which is itself constituted by people and their technologies. The human components of this system cannot be replaced by mere artifacts because they offer much greater flexibility, better judgment, and more sophisticated sensors and effectors than robots (Heylighen, 2012a). Eliminating them would not only destroy the major part of the superorganism, but greatly reduce the bandwidth of the GB-world interaction, its store of values, knowledge and experience, and therefore its omnipotent and omniscient capabilities.

The more positive reason for the GB's intrinsic benevolence towards humans is that by benefiting them it benefits itself—and vice versa. Because it is for an essential part constituted by people, optimizing humans' physical, mental and social capabilities optimizes the GB's own capabilities. This is not just the outcome of a rational reflection about what the GB's objectives should be: the distributed intelligence network is evolving at this instant precisely because it is beneficial to those who use it. In the on-going competition between different methods, technologies, and institutions, the winners will be those that effectively produce the

End of the Beginning

greatest happiness for the greatest number (Heylighen, 2013), because given a choice people will eventually replace a system that satisfies them less by one that satisfies them more. That is exactly what I meant by "selective pressures" pushing evolution in a predictable direction. Therefore, the trend towards universal benevolence and other "divine" attributes is built into the very evolutionary process we have been discussing.

While this may seem to contradict the traditional view of evolution as rooted in conflict and competition, a number of authors have recently shown that the long-term drive of evolution is towards cooperation and synergy (Corning, 2003; Heylighen, 2006; Stewart, 2000; Wright, 2001). The principle is simple. Two individuals desiring to consume the same limited resource are by default in competition: whatever the one gets is no longer available for the other. Such a situation is called a *zero-sum* game or a *win-lose* interaction. The competition typically ends in the survival of the one most fit to obtain the resource, and the elimination of the other.

However, in many cases evolution eventually stumbles upon an arrangement that is *win-win* or *positive-sum*. The interaction then becomes cooperative or synergetic, as both parties can extract more benefit by working together than by acting independently. Such an outcome is preferred by natural selection, because even the strongest party becomes fitter by reaping the benefits of collaboration than by spending its energy in fighting off competitors. Thus, the blind-variation-and-natural-selection of evolution will eventually increase cooperation (Stewart, 2000)— albeit very slowly.

The emerging GB spectacularly accelerates this development, for the following reasons:

Positive-sum arrangements are typically more complex and more difficult to find than zero-sum ones. By boosting our ability to explore myriads of potential interactions with other people and artifacts around the world, and by helping us to select the most synergetic ones, the GB multiplies the probability of discovering highly beneficial win-win situations;

The default assumption that resources obey a zero-sum logic only applies to matter and energy, because these are subjected to a conservation law. It does not apply to information, which can be copied without limits. As the open-access philosophy reminds us, giving information to others helps rather than hinders you in the long term (Heylighen, 2007b). Because information obeys a positive-sum logic, the GB has every reason to spread it as widely as possible, thus maximizing the sum of benefits for every party involved.

By moreover making physical resources—which obey a zero-sum logic—abundant, the GB eliminates the need for competition. When there is enough for everybody, there is no longer any ground for conflicts over scarce resources, or even for private property that excludes others. Instead, the main interaction mode will be collaboration and shared access (Rifkin, 2014).

This argument for the accelerated evolution of a universal "goodness" that would benefit everybody, while eliminating the fundamental sources of conflicts, may still seem too theoretical to convince you. Let us therefore look at the concrete facts of socio-economic evolution. Over the past few centuries, the technological revolution has brought about a spectacular improvement in human welfare. It is well established that people have become much richer, better educated, and longer living, and that this global development continues at a rapid pace (Klugman, 2010).

More unexpected, perhaps, is that people are also less likely to fall prey to what we consider as morally "evil": murder, war, slavery, prejudice, suppression, dictatorship, and corruption. Perhaps surprisingly for those who follow the news (which typically focuses on the most spectacular instances of crime, terrorism, and war), the percentage of people dying through homicide or conflict, or living under non-free, undemocratic conditions has been spectacularly decreasing over the past

End of the Beginning

centuries and decades (Heylighen & Bernheim, 2000; Mueller, 2009; Pinker, 2011). The underlying reason for this trend appears to be that people are much better educated and informed about everything going on in the world (Bernheim, 1999; Pinker, 2011). Therefore, they are less willing to tolerate the situations of abuse that before they were not aware of, or used to endure because they did not know a better way, and more ready to empathize with the plight of others living in distant lands (Rifkin, 2014). Moreover, with greater wealth and greater interdependency, there is simply less to gain and more to lose by engaging in conflict or by exploiting others.

In sum, we see here the same dynamic of increasing communication leading to increasing understanding of self and others leading to increasing synergy that drives humanity and its global brain towards global cooperation (Rifkin, 2014; Stewart, 2000; Wright, 2001), universal morality (Heylighen & Bernheim, 2000; Pinker, 2011), and thus omnibenevolence.

Paradise Regained

Assuming the practical equivalents of omniscience, omnipresence, omnipotence and omnibenevolence, we should expect the GB to efficiently solve about all present global and individual problems and conflicts, while spectacularly enhancing our technological, social, cognitive and biological capabilities. To describe the resulting situation of peace, abundance and fulfillment, we may turn to another religious metaphor, *Paradise*, an idyllic state or place in which everybody is intrinsically happy.

The term "Paradise" is ambiguous, however: it can refer either to the initial *Garden of Eden* where humanity originated, or to the final *Heaven*, the state of bliss to which people's souls are supposed to migrate after their death. The metaphor of Heaven seems to have inspired many Singularity theorists (Proudfoot, 2012). They conceive it as some kind of Platonic realm of data and computation, held in the memory of an immense, superhuman AI computer system. After their death, people's brain patterns would be "uploaded", or reconstructed, into this nearly infinite, abstract information space. Thus, they could continue to live forever, as disembodied, virtual personalities. This quixotic vision of computational resurrection and artificial bliss is sometimes described as "the rapture of the nerds" (Eden et al., 2013).

The utopian ideal inspired by the Global Brain (Heylighen, 2002), on the other hand, is much closer to the *Garden of Eden*. The idea of a "Garden of Eden", "Arcadia", or "Golden Age", which recurs in different religions and mythologies, refers to an original state of abundance and leisure before civilization. In this "state of nature", people could live spontaneously, without having to toil, worry about the future, or feel constrained by the controls, laws and taboos imposed by society. There seems to be a true substance to this myth: anthropologists have observed a similar lifestyle in the last remaining tribes of hunter-gatherers they were able to study (Charlton, 2002; Diamond, 1987; Gray, 2009; Sahlins, 2004). Our hunter-gatherer ancestors were apparently able to lead a relaxed, playful and spontaneous life in part because their population densities were small. Therefore, the resources they could find in their environment were in general abundant relative to their needs (Sahlins, 2004), and they had little reason for control or coercion.

This is no longer the case for the billions of people presently living on our planet. That is why we need technology in general, and an omnipotent GB in particular, to regain the original abundance by exploiting the available resources much more efficiently. In principle, the GB can eliminate all the problems of poverty, pollution, resource exhaustion, conflict, ignorance, superstition, and work-induced stress. Moreover, it should be able to restore our physical and psychological environment to a more natural state, without the need to abandon the comfort and security brought by technology. Thus, it would create a new Garden of Eden, offering an inspiring, relaxed and joyful life for all.

In such a society, people would no longer *have to work*, since practically all the essential jobs would be performed by GB-supervised machines. However, they would still *love to work*, because the tasks remaining for humans would be so intrinsically interesting and challenging that people would feel bored without them. That is because the GB would use people for the tasks for which they are biologically and psychologically best adapted. These include:
- nurturing and caring for people;
- looking after plants, animals and the natural world;
- manipulating irregular objects and materials;
- experiencing and interpreting complex, multisensory stimuli;
- inducing general wisdom from personal experiences;
- exploring and being creative;
- developing new ideas, inventions and innovations;
- conversing and collaborating with other people.

Moreover, the GB would organize its interface with humans in such a way that they would be maximally stimulated or motivated, and minimally stressed or confused by the work they are offered to do (Heylighen et al., 2013). The main incentive for good contributions will probably not be money, as material reward will lose its value in a society where everything is abundant, but public *recognition*, in the sense of constructive feedback and an increased reputation or status. Reputation mechanisms are already being used very effectively to motivate people in open collaboration platforms on the Internet (De Alfaro, Kulshreshtha, Pye, & Adler, 2011; Mamykina, Manoim, Mittal, Hripcsak, & Hartmann, 2011).

Attaining such a society will obviously require revolutionary changes to our present political and economic system, with a capitalist, property-based market gradually being replaced by a collaborative commons based on shared access to the functionality of the GB. Embryonic forms of such a sharing economy can already be found in various communities of volunteers working together via the Internet in order to develop open-source software and hardware (Heylighen, 2007b; Rifkin, 2014). It seems likely that, as marginal costs (and therefore profit margins and the need for capital) shrink, these participatory, distributed modes of production and distribution will gradually take over from the hierarchical, centralized corporations that control the capitalist economy (Rifkin, 2014).

The highly developed social democracies in Northern Europe, Canada, and Australia may offer a (still imperfect) model for the effects on work and welfare of such a future system of governance. Social democracy combines strong personal and economic freedom with an important role for the government (a centralized system that will eventually have to be replaced by a more distributed form of governance). The income generated from taxes (ideally on wasteful consumption and pollution) is used to support people and processes that tend to be ignored in a pure free market economy because they do not generate profit. Such redistribution reduces inequality and increases overall welfare.

For example, in this system teachers and academics are normally paid by the government. Thus, they can provide high-quality education and scientific research freely to the whole of society—instead of only to those rich enough to pay for them. In this arrangement, a tenured professor has the academic freedom to research any subject in any manner of her choosing, without having to worry about being paid. However, she remains motivated to get the best possible results because she wants to improve her scientific reputation—e.g. through highly cited publications. Such an arrangement can be seen as a model for the comfortable but highly stimulating work life under a Global Brain regime, where most people would have an education level that allows them to do autonomous research.

But social democracies also care for the less talented. People

End of the Beginning

with physical or mental handicaps, psychological illnesses, or with an education so poor that they cannot find a job under normal market conditions, are often employed in "sheltered" or "social" work places. These are organized so that even the most poorly skilled people have the support they need to perform simple but useful jobs (e.g. ironing clothes, or preparing goods for recycling). Moreover, these people earn a decent wage thanks to government subsidies that otherwise would have gone to welfare payments.

In a sense, the post-Singularity society could be seen as one huge "sheltered environment". Here, everyone would be doing meaningful work at his own level of skills, supported by the GB whenever needed, without having to worry about money, stressful conditions, bureaucratic procedures, dictatorial bosses, or exploitation. However, next to sheltering people from sources of distress, the GB would also be challenging them to try out new things and to relentlessly advance beyond what they have now. In this way, the GB would provide the ideal environment for personal growth, self-actualization (Heylighen, 1992), and overall well-being.

Unlike the "rapture of the nerds", with its flight from the material world into a Platonic Heaven, this "return to Eden" vision keeps us firmly with our feet down on earth. In the Earth's restored "garden" state, we would enjoy its flowers, fruits and animals, and, more generally, our own sensations, bodies, activities, friends, lovers, and the natural world that brought them forth. Of course, the virtual world of the Global Brain would provide an astonishingly powerful extension to our brains and senses. However, it will not replace the real world of matter, life, mind and society, which is immensely more complex, diverse, surprising and fascinating than any computational model imagined by Platonist nerds.

Once the GB (which of course includes us, humans) has tackled the more pressing challenges, we should expect it to expand its focus to domains that are still out of reach. The most obvious direction in which to look is outer space. However, the intuitive expectation that a global brain would expand into an interstellar, galactic, or even cosmic brain (Vidal, 2014) immediately runs into a hard limit: the finite speed of light implies that a signal sent to other stars would require decades, if not millennia, for an answer to return—a communication speed incompatible with the processing speed of a brain. A more likely expansion is into "inner space": the microscopic realm of cells, organelles, molecules, atoms, elementary particles, and even the structure of space-time itself (Egan, 2002). Nanotechnology already promises to extend our powers of sensation and manipulation into the cellular domain, whose richness we have as yet hardly skimmed. But there is still "plenty of room at the bottom" for the GB to explore…

Perils on the Road to the Singularity

The previous section has probably suggested a too rosy picture of the transition towards a GB-dominated world. Technological advances, however beneficial in the long term, tend to create great stresses on society. First, new technologies need to adapt to the people that use them—a process that is everything but obvious. Second, people need to adapt to the uses they offer—an even more difficult changeover. Moreover, the most beneficial uses typically result from *exaptation*, i.e. the discovery of functions or applications other than the ones originally intended (Andriani & Cohen, 2013; Dew, Sarasvathy, & Venkataraman, 2004). Vice versa, the originally intended applications often turn out to have more negative than positive effects.

This process of mutual adaptation and exaptation nearly always takes more time and effort than anticipated by the technology's designers. One reason is that people are not rational—in the economic sense of making optimal decisions based on full information—but driven by habit, fashion and tradition; biases; shallow impressions; and deep emotions. This applies both to designers and to users. Another reason is that people and artifacts together form a highly complex and non-linear system. Such systems often exhibit erratic and counter-intuitive behaviors that make it impossible to predict the effects and side effects of a specific innovation. Let us then look in more detail at some of the most important things that can go wrong.

Cascading failures

The non-linearity of the global system is exacerbated by what has been called *hyperconnectivity* (Fredette, Marom, Steiner, & Witters, 2012): nowadays nearly everyone and everything is connected to everything via global networks. This makes it easy for disturbances to propagate from node to node in the network, spreading across ever widening circles ever more quickly. Examples are computer viruses or false information reaching across the globe in hours via the Internet, or real viruses spreading in days via intercontinental travel. Less obvious but equally dangerous are local breakdowns—such as a bank going broke or a power line that is overloaded—that precipitate similar breakdowns in the systems that directly depend on them. These in turn can precipitate further breakdowns, until a large-scale network collapses. Examples of such series of *cascading failures* are the 2008 financial collapse and various large-scale electricity blackouts (Dueñas-Osorio & Vemuru, 2009; Helbing, 2013).

The underlying mechanism is a positive feedback or vicious cycle that amplifies the disturbance more quickly than it can be contained. The infrastructure of the GB, which reduces friction and therefore facilitates propagation, increases such systemic risks (Heylighen, 2007a). On the other hand, a fully functioning GB should be able to sense problems, devise solutions, and intervene nearly immediately, thus nipping any potential crisis in the bud. Unfortunately, we are not yet at the stage that we can rely on such a superhuman distributed intelligence. In the meantime, we will have to develop more basic protection measures. These include:

- the equivalent of "firewalls" and artificially imposed friction to contain too fast spreading,
- sufficient redundancy so that the function of failing systems or components can be taken over by others,
- sufficient diversity in designs so that only a small proportion of systems are vulnerable to a specific type of failure,
- large enough reserves or buffers to compensate for any large-scale shortages,
- an artificial "immune system" that would automatically detect potentially dangerous phenomena, and block them from spreading until they are proven safe,
- a better understanding and pre-emptive remediation of the most important vulnerabilities—e.g. by strengthening the nodes in the network that have most other nodes (in)directly depending on them (Farmer, 2013; Thurner & Poledna, 2013).

Without such precautions, our society runs the risk of suffering truly global crises and catastrophes (Helbing, 2013).

Cyber war

Next to such risks for accidental damage, we need to consider the risks of intentional harm done to, or through, the network. This is the domain of activities that have been labeled *hacking*, *cybercrime*, *cyber terrorism*, and even *cyber war*. The principle is that some people or organizations exploit weaknesses in the systems connected via the Internet, in order to access secret data, steal money, wreak damage, or hinder services. One of the most spectacular examples is the Stuxnet worm, developed by US and Israeli intelligence agencies in order to sabotage Iranian nuclear plants by corrupting their software (Schmidt & Cohen, 2013).

These problems are not fundamentally different from the ones encountered before the advent of the net: there has always been an arms race between police and criminals, spies and counterspies, powers and rival powers. Whenever one of these parties appears to get the upper hand by developing a new weapon or tactic, or by discovering a hole in the other's line of defense, the other party will eventually plug the hole or develop a counter-tactic.

The same will continue to happen on the Internet over the next decades. The main difference is that the Internet is more complex, more anonymous, and farther removed from the physical world in which bodily harm can occur. Therefore, these on-going skirmishes are likely to remain complicated and murky, often without clear victims or perpetrators, while having relatively little effect on the general population (Schmidt & Cohen, 2013). Moreover, given the great economic interdependency of the major nations, there is little to gain, and much to lose by seriously damaging the other party's capabilities. More generally, the ongoing spread of democracy and Enlightenment values, which is facilitated by the Internet, strongly reduces the willingness to engage in war, conflict or any kind of violence (Mueller, 2009; Pinker, 2011).

Therefore, most of these covert activities will remain limited to stealing confidential information, such as blueprints for new technologies or political strategies. This threat will gradually lose in significance as the "open access" ideology that drives the GB makes information more freely available (Heylighen, 2007b), and as better laws and technologies for the protection of privacy and personal identity are put into place. Thus, a "cyber war" is better than a "cold war", which is itself preferable to a traditional war. The only serious danger is that one of the pieces of malicious software would start spreading beyond its target, and thus harm the world as a whole. But this merely points us back to the issue of *cascading failures* and our list of proposed measures for containing the "viral" spread of disturbances.

Psychological Parasites

There is another old problem that will get a new life on the Internet, however this time with potentially more widespread harm: *psychological parasites*. Our brain is naturally attracted to certain exciting, calming or pleasure-inducing stimuli, such as sex, food, or praise. Therefore, it tends to seek more of such sensations, even in cases when they are not accompanied by any real benefit. The danger is that such stimuli can become *parasitic*: they get themselves multiplied by exploiting people's instinctive desires, without regard for their well-being. An example of such a mental parasite is a drug addiction: drug-induced sensations reproduce themselves by pushing the addict to seek the same stimulus again and again. There exists a wide variety of other addictions, including alcoholism, gambling, and sex addiction, in which people feel compelled to perpetually recreate the same powerful stimuli, even though this activity harms their health and social functioning.

When the stimuli are produced by electronic media, the result may be called *cyber addiction* (Beard, 2005; Chou & Ting, 2003; Young, 1998). This includes addictions to computer games, Internet use, social media, and smartphones. The danger of cyber addiction can only grow as software becomes more responsive, human-computer interfaces become more immersive, and ambient intelligence learns to better please its users (Heylighen et al., 2013). The result is that network users run a serious risk of becoming enslaved into a pattern of repetitive, sensation-seeking activity, only because this activity pushes all the right "buttons" in their brain at the right moments. Three psychological mechanisms in particular appear powerful enough to drive the global spread of such psychological parasites: *flow, supernormal stimuli* and *mind viruses*.

Flow is the pleasurable state achieved when a person is so engulfed in an activity that s/he only wants to continue that activity, ignoring all other concerns (Nakamura &

End of the Beginning

Csikszentmihalyi, 2002). The addictive quality of computer games is commonly explained by their ability to produce flow (Chen, 2007; Chou & Ting, 2003; Heylighen et al., 2013). Flow is achieved when challenges are presented that are matched to the person's skills in achieving them, and every action by the user receives an immediate feedback, so that the user is continuously stimulated to go further and further. Such flow-producing stimulation is easily implemented in what I have called a *mobilization system* (Heylighen et al., 2013). This is an ICT system that encourages and coordinates human action. Such technologies are intended to mobilize people for worthwhile objectives, such as educating themselves, exercising, or collaboratively solving problems. However, their potential power on the human mind is such that they can lead to both addiction and exploitation by political or commercial organizations. This would turn their users into unwitting slaves of the system (Heylighen et al., 2013).

Supernormal stimuli are stimuli more intense than those that occur in the natural world (Barrett, 2010). In most cases, the brain pays attention not so much to the absolute, but to the relative intensity of a stimulus: in how far it is stronger or weaker than other stimuli? In the competition between stimuli, the strongest one normally wins—even if it takes on absurd proportions. This can be illustrated by cartoon figures—such as Mickey Mouse—which tend to have impossibly large heads and eyes supported by ridiculously short legs. That is because faces and eyes are intrinsically more interesting to the brain. Therefore, artists have learned to exaggerate those features in order to attract the attention. Similar tricks are used in a variety of domains, including advertising, computer games, junk food (which contains unnaturally high concentrations of sugar, fat, salt, and calories), fashion models (who tend to have abnormally long and thin bodies), movies (which show ever more extreme special effects and violence), etc.

In a society overloaded with information, interruption and

stimulation, attention is scarce (Heylighen et al., 2013). This creates competition for the attention of the public. The result is an arms race towards ever more powerful stimuli, which deviate ever further from what exists in nature. This leads to unhealthy addictions (e.g. to junk food, violent movies, or Internet porn) and to stress-inducing overstimulation. The even greater danger is that people would lose touch with reality, so that they develop completely unrealistic expectations about what a sex partner is supposed to look like, how much violent crime there is in the streets, or what constitutes a healthy diet. This makes them intrinsically dissatisfied, because the things they can get in the real world seem pale in comparison with the things they see in the virtual world of games and advertising. Moreover, it makes them lose their sense of balance and judgment. Thus, they are more likely to lapse into extreme behaviors, such as joining a cult, becoming a suicide terrorist, or going on a shooting spree to kill dozens of bystanders—because these do not seem so extreme anymore compared to the standards of their video games or horror movies.

Mind viruses (Brodie, 1996) can be defined as parasitic memes. A *meme* is an idea, belief or behavior that spreads across society by being transmitted from person to person (Heylighen & Chielens, 2008). Successful memes tend to exhibit characteristics such as plausibility, simplicity, novelty, usefulness, emotional impact, and ease of communication. As long as a meme provides valuable information, its propagation benefits society. However, parasitic memes mimic the characteristics of beneficial memes in order to spread more easily, while providing information that is worthless, wrong, or even dangerously misleading (Heylighen & Chielens, 2008). Examples include chain letters, false rumors, urban legends, hate speech, conspiracy theories, superstitions, extremist ideologies, and various fundamentalist and irrational beliefs.

Such memes are particularly dangerous to individuals who already have lost their sense of reality by their immersion in

End of the Beginning

supernormal and flow-producing stimulation environments, and who thus are ready to embrace the false promises of a mind virus. Because they spread across communities while indoctrinating their carriers, mind viruses have an even greater potential to create damage. For example, they may recruit a worldwide group of people into an absurd and destructive enterprise, such as an outbreak of gratuitous rioting, a mass suicide, a terrorist plot, or even a genocide. The ubiquitous network enhances their powers of spreading and mobilization, thus increasing the danger.

In principle, an omnibenevolent GB would want to prevent such disastrous outcomes. However, omnibenevolence cannot be assumed to be in-built or preprogrammed by some rational intelligence. It is rather the result of self-organization in which technologies, ideas and institutions that seem to better satisfy the needs of their users are preferentially retained, propagated and further developed (Heylighen, 2013). But since humans have bounded rationality, they cannot foresee the long-term effects of such innovations, and may adopt a solution that satisfies their short-term need for stimulation or group belonging, while failing to see that it is detrimental to their long-term needs for health and constructive engagement with society. Therefore, it may take quite a while before society has become fully aware of the dangers of such parasitic stimuli, allowing them to cause a lot of harm in the meantime.

To avoid such damage, we will need to develop precautionary measures similar to the ones Helbing (2013) advocates for systemic risks, but this time including the social and emotional characteristics of people and communities in addition to the dynamics of cascading disturbances. This should allow us to build another "global immune system" that would be able to anticipate, detect, quarantine, and if necessary neutralize psychological parasites. For this, we could rely on the criteria that characterize successful memes (Heylighen & Chielens, 2008) and mobilization systems (Heylighen et al., 2013), coupled with a realistic simulation of how messages spread across the global brain network (Heylighen, Busseniers, Veitas, Vidal, & Weinbaum, 2012). Armed with such knowledge, we may be able to recognize potentially epidemic parasites before they have spread, forecast their path of propagation, and thus prevent their spreading by "inoculating" the people likely to be infected. This can be done by sending them persuasive warnings or refutations about the propagating rumor, terrorist propaganda, or addictive system before they have been exposed to them.

In addition, we may need to develop laws and institutions that prevent the willful creation and propagation of psychological parasites. Already existing examples are laws that prohibit hate speech, drug trafficking, or membership of terrorist groups, but these are too specific to regulate more general threats. Such a legal framework would make it more difficult to exploit unsuspecting people by bombarding them with dangerous, addictive, and self-reinforcing stimuli (as the tobacco industry has been doing for decades, and the junk food industry is still doing).

Future Shock

As Toffler (1970) already discussed in great detail, having to adapt to new circumstances is intrinsically stressful. Therefore, people are inclined to resist too fast or too radical changes. If they cannot, they will be left confused, frightened and disorientated—a condition that Toffler called "future shock". The reason that I suggested that sociotechnological acceleration may already have reached its peak is that culture, politics and society seem to evolve more slowly now than they did in the 1960s, the decade that inspired Toffler's analysis. For example, musical styles, clothes fashions, literary genres, intellectual theories, and political ideologies seem to be hardly different from what they were 20 years ago, in contrast to the revolutionary changes seen in the 1960s and 1970s.

A plausible explanation for this slowdown is that it

End of the Beginning

compensates for the acceleration in ICT and its impact on people's lives over the last two decades. When you need to learn radically new technologies, applications and ways of communicating every few years, you will have less energy to explore new styles of music or literature, or to rebel against your parents' worldview. While there seems to be no slowdown in the creation of new technologies, their speed of adoption by the wider public may well be close to its limit. The inertia is even greater in political, economic and legal institutions, which tend to lag far behind in adapting to the social implications of ICT. Thus, they are ill prepared to deal with the perils and promises we surveyed. This creates great stresses on both society and its members, making them even more vulnerable to psychological parasites and cascading failures.

The most likely negative outcome is a conservative backlash, in which people or political regimes desire to turn back the clock to an idealized past and its simple truths, thus trying to repel the waves of change brought by globalization and technological innovation. The problem, of course, is that these past solutions are even less adapted to the present turmoil. They can thus only exacerbate the problem. In its extreme form, such a backlash can take the form of a fundamentalist uprising, ending in either on-going violent conflict and terrorism, or a rigid, totalitarian regime. On a larger scale, this could give rise to wars between religious or ideological blocks, and a drastic reduction in freedom and creativity. This would obviously constitute a serious setback on the road to the GB, creating a lot of needless suffering.

In a less extreme form, a conservative backlash will merely hinder further advances. This is not necessarily a bad outcome, as the resulting slowdown may help people to psychologically recover from their "future shock", while giving institutions the time to effectively assimilate some of the changes. However, the preferable strategy is to go for a controlled "soft landing", in which the less important but relatively stressful changes (e.g. new interfaces, operating systems, or bureaucratic procedures) are artificially restrained, in order to keep the most constructive ones (e.g. better systems for collective intelligence, global governance, online education...) up to speed.

To further cushion the shock, Toffler (1970) proposed to maintain "enclaves of the past", in which people who are not ready to embrace the brave new world of technological innovation may continue to live more or less as they used to, limiting themselves consciously to a simpler life. Two examples at the opposite poles illustrate the variety of possible strategies:
1. the Pennsylvania Amish have not essentially changed the technologies they use since the 18th century;
2. ICT systems that change their interface often have the option for the user to maintain the old interface for a couple of months.

Ideally, there should be a whole range of options in between so that people can adapt to innovations at their own pace. In particular, new technologies should as much as possible be backward compatible, so that people using older systems would not a priori be excluded. This should reduce both overall stress and the likeliness of a radical backlash.

Loss of Human Capabilities

We noted that it is in the GB's interest to maximize human capabilities. However, in the short term there is a serious risk that certain abilities will get lost, at least in some populations. Technology is intended to make life easier, thus reducing physical and intellectual demands on its users. GB technologies in particular should be able to tackle nearly any challenge with a minimum effort required from individuals. For example, someone who needs to solve a problem could simply request the GB's assistance and blindly follow its recommendations—while being carried to the right place by the robotic vehicle the GB has dispatched.

End of the Beginning

This makes it very tempting to adopt an attitude of passivity, in which individuals avoid any challenge more difficult than the ones they feel perfectly comfortable with. Without challenges to stimulate the development of mental and physical skills, the supporting organs and circuits tend to atrophy—like the muscles that waste away when a person has been staying too long in a hospital bed. This is the well-known principle of "use it or lose it". It applies in particular to older or sick people, who may have more difficulty than before to perform a certain task, and therefore stop doing it altogether—even after their health has improved again. The resulting degeneration of abilities seems irreversible, as further loss of function triggers further avoidance, and so forth (Heylighen, 2014). More generally, commentators lament the general loss of physical fitness in a population that seems ever less inclined to walk, run, climb, carry or perform any natural exertion, while remaining attached to a screen the whole day.

Loss of physical ability is not a recent phenomenon, though. Analysis of the remains of our Paleolithic ancestors indicates that they were taller, stronger, and healthier than their descendants (Diamond, 1987; Mummert, Esche, Robinson, & Armelagos, 2011). Even their brains were about 10% larger than ours (McAuliffe, 2010), suggesting a higher intelligence. A plausible explanation is that these hunter-gatherers had a physically and intellectually more challenging life than the subsequent farmers and industrial workers, since the latter could rely on society and technology (including agriculture) to satisfy the bulk of their needs. Another likely reason is that their lifestyle and diet were more in tune with what their organism was adapted to (De Vany, 2010).

Both differences are exacerbated in our present information age:
1. we rely even more on technology and society to solve all our problems;
2. we tend to eat and live in a way that is even further removed from what nature intended.

This unnatural lifestyle leads to an all too recognizable list of "diseases of civilization": obesity, diabetes, metabolic syndrome, cardiovascular diseases, cancer, depression, allergies, autoimmune diseases, ADHD, etc. (Carrera-Bastos, Fontes-Villalba, O'Keefe, Lindeberg, & Cordain, 2011). The consequences seem particularly dangerous for our children (Palmer, 2010), who have never even experienced the more traditional lifestyle before the advent of electronic media, and who tend to develop problems like diabetes and obesity already from an early age. Extrapolating from this epidemic, some analysts suggest that for the first time in a millennium the life expectancy of the next generation may be lower than the one of the present generation (Olshansky et al., 2005).

The attractiveness of such unnatural lifestyles can at least in part be blamed on the supernormal stimuli, such as junk food, game addiction, television violence etc., that they offer. Lack of real challenges in our daily life is compensated by the overload of artificial stimuli we experience while watching a horror movie, playing a virtual reality adventure game, or visiting a porn site. While it is clear that this can only weaken our physical abilities, the question remains open as to its effect on our mental abilities. Observers tend to be split, some pointing towards an increasingly shallow comprehension and loss of the ability to concentrate and memorize (Carr, 2011), others towards the cognitive stimulation provided by television and computer games (Johnson, 2006). The overall effect is likely to depend on:

1. what kind of electronic media are used;
2. which functions of the brain are being stimulated.

Concerning point (1), more complex, interactive and challenging environments will obviously have a more beneficial effect than more shallow, passive, and supernormal stimulation. The choice of stimulation will depend on the individual, the social environment, and the available media. Typically, people from a lower social and educational background will gravitate towards the less intellectually challenging media, while the opposite will happen for those from a higher background. This will exaggerate the existing intellectual differences, boosting the capabilities of the smart people, while "dumbing down" the not so smart. Moreover, people from poorer backgrounds tend to lead an unhealthier and more stressful life, making them particularly inclined to avoid any additional challenges. The danger is that this will further amplify social inequalities, creating an "underclass" of people who simply cannot grasp the complexities of a high-tech society.

As to point (2), it seems likely that even the most complex and intellectually challenging virtual media will tend to boost certain brain functions, while neglecting others. Computer use typically demands symbolic skills, such as language, math, and logical reasoning, and specialized visual skills. On the other hand, tactile, motor, social and emotional forms of intelligence are likely to develop better by interacting with the real world. As people—and children in particular—tend to spend more time in the virtual world, they are likely to gain in symbolic intelligence, but lose in "natural" or intuitive intelligence. While a boost in symbolic skills may be necessary to fully take advantage of ICT applications, in the longer term the effect may be counterproductive. Indeed, computers are intrinsically better at symbolic reasoning than humans, but worse at embodied and intuitive reasoning. For the GB, the ideal balance is to delegate tasks of the symbolic type to computers, and the other tasks to humans. But that would require that humans maximally develop their intuitive and embodied skills.

In the longer term, the GB is likely to get the balance right by creating an environment that challenges people both physically and mentally to develop themselves as much as they can. Moreover, the ongoing scientific and technological advances, in domains such as medicine, sports, drugs, intelligence amplification, genetic manipulation, nanotechnology and cybernetic implants, is likely to boost the capabilities of individuals to a "transhuman" level (Bostrom, 2005)—albeit that just getting back to the level we had in the Paleolithic would already be a remarkable achievement...

In the meantime, we will need to make sure that physical and intellectual passivity, supernormal stimuli and alienation from our natural environment do not get out of control, especially among the least educated. Targeted education and stimulation, supported by mobilization systems (Heylighen et al., 2013), should be able to achieve these objectives. However, until they take full effect we are likely to witness an enormous waste of human life, health, fitness, well-being and intellectual capabilities, just because technology lures people into choosing the path of least resistance (and strongest stimulation)...

Conclusion

I have sketched the outlines of a scenario in which technological acceleration culminates in the self-organization of a distributed intelligence out of the global ICT network. This intelligent network would play the role of a brain for the global superorganism—the living system consisting of all people, their artifacts, and the ecosystems they depend on. As such, this brain would be responsible for tackling all challenges that may confront both the planet as a whole and its human components. By harnessing the collective intelligence of all people, while complementing it with the data processing capabilities of an immense array of computers, sensors, robots, and other technologies, this Global Brain (GB) would achieve a level of superhuman capabilities that are as yet difficult to imagine.

To better convey their significance, I have invoked a pragmatic version of the classic list of "divine" attributes:

- *omniscience*: being able to provide the knowledge needed to tackle any real challenge;
- *omnipresence*: being able to sense, inform and act anytime anywhere on Earth;

End of the Beginning

- *omnipotence*: being able to produce any good, service or coordinated action at nearly zero cost;
- *omnibenevolence*: using these abilities to produce the greatest happiness for the greatest number of people.

These attributes are not so much the characteristics of an idealized, God-like entity, but the features representing the main selective pressures on socio-technological evolution: both designers and users of technology would like their systems to become more intelligent, more knowledgeable, more widely available, more powerful, and more beneficent in their effects. This means that in the longer term (after adjusting for misjudgments, side effects and exaptations) systems and components that satisfy these criteria better should replace those that do so less well.

Moreover, I have surveyed a number of existing trends and technologies that illustrate how most of these capabilities may be reached within the next few decades. For example, Wikipedia already provides the bulk of human knowledge to anyone interested, location-aware smartphones and wireless sensors are spreading rapidly across the globe, 3D printers are in principle capable of realizing any design downloaded from the Internet (Lipson & Kurman, 2013), while violence and conflict around the world have reached historically low levels (Pinker, 2011). Thus, Kurzweil's (2005) prediction of a technological singularity around the year 2045 may not be so far off the mark—at least if we interpret this singularity as the full deployment of a Global Brain (Heylighen, 2008), and not as the creation of a superhuman, autonomous AGI program. After that transition, we may experience a true *return to Eden*—an idyllic state of abundance, peace and well-being, in which all serious threats have been tackled and people can fully dedicate themselves to further creative endeavors.

Before we reach such a utopian ideal, however, the road to the GB threatens to be long and arduous. While technological innovation tends to be beneficial in the long run, in the short run it

End of the Beginning

is accompanied by side effects and dangers that are difficult to predict and to control. One fundamental problem brought about by the network revolution is the general increase in connectivity and reduction in friction. This makes it easier for various phenomena, positive as well as negative, to propagate. Thus, they may spread so quickly across the globe that they run out of control. Particularly dangerous are cascading failures, such as collapsing stocks triggering the collapse of other stocks, and viral phenomena, such as rumors, fads, religious cults, computer viruses, and infectious diseases, which propagate nearly unhindered across the global communication network. To avoid major crises caused by such self-amplifying phenomena, we will need to set in place a range of precautionary measures. These include highly diverse and redundant reserve capacities, "firewalls", and a "global immune system" that would quickly recognize and immobilize potentially dangerous "virals".

Not just the communication network, but our human psychology is particularly vulnerable to parasitic phenomena and negative side effects. As technology lures us further away from the natural environment to which our bodies and minds are adapted, we become more inclined to accept some of its seemingly attractive, but ultimately destructive, offerings. These include drugs, junk food, addictive games and virtual environments, sensational videos, and the allure of spending the whole day in the sofa jacked into the network, without moving or experiencing fresh air and sunlight. We may even be tempted to stop tackling any real challenges ourselves, and start leading the life of a decadent aristocrat—with the GB in the role of the omnipresent butler that keeps things going and cleans up all the mess.

All this, coupled with the intrinsic stress of having to adapt to accelerating change, is likely to severely strain our mental and physical health, fitness and happiness. Given that certain groups are particularly vulnerable to the resulting loss of capabilities, this may well result in a class of deeply frustrated, maladapted and

End of the Beginning

alienated people, who are easy prey for extremist, reactionary and fundamentalist movements and ideas. Recommended precautions here are measures to control stress and the speed of change, the creation of a different type of "global immune system" that would pre-empt the spread of psychological parasites, and the deployment of mobilization systems that challenge people to develop their physical and mental abilities in the healthiest manner.

In sum, we definitively live in the most interesting of times. While the potential for crises and catastrophes remains great, none of these seems serious enough to wipe out modern civilization, given the amazing capabilities of adaptation it has shown up to now. Therefore, I remain fundamentally optimistic. Over the past centuries, technological innovation has spectacularly improved the human condition (Heylighen & Bernheim, 2000), and the latest global statistics on human development show no sign of any slowdown or reversal in this long-term trend (Klugman, 2010). Moreover, given my reasons to discard any self-amplifying AI explosion and to believe in the universal benevolence of the GB, I have no fear for the many scenarios in which a superintelligent computer system would enslave or destroy humanity (Chalmers, 2010; Eden et al., 2013). Such scenarios have even prompted intelligent thinkers to call for a moratorium on AI research—a measure that I consider absolutely counterproductive—or to "lock up" any AI inside a virtual world where it could not affect the real world (Chalmers, 2010)—a measure that would guarantee that the AI would lack any practical intelligence.

I think it is time to get back with our two feet firmly on the ground, and continue investigating and improving the real world with all its complexities, interactions and nuances, instead of getting carried away by abstract speculations grounded in bits and bytes rather than in flesh and earth. In that way, we may truly be able to build a new Eden here on our magnificent planet, and thus undo most of the negative side effects that have tended to accompany civilization and technology up to now.

REFERENCES

Akyildiz, I. F., Wang, X., & Wang, W. (2005). Wireless mesh networks: a survey. *Computer Networks,* 47(4), 445–487. doi:10.1016/j.comnet.2004.12.001

Anderson, C. (2012). Makers: *the new industrial revolution.* Random House.

Andriani, P., & Cohen, J. (2013). From exaptation to radical niche construction in biological and technological complex systems. *Complexity,* 18(5), 7–14. Retrieved from http://onlinelibrary.wiley.com/doi/10.1002/cplx.21450/full

Ashby, W. R. (1958). Requisite variety and its implications for the control of complex systems. *Cybernetica,* 1(2), 83–99.

Ashby, W. R. (1964). *An introduction to cybernetics.* Methuen London.

Ashby, W. R. (1972*). Setting goals in cybernetic systems.* In H. W. Robinson & D. E. Knight (Eds.), *Cybernetics, Artificial Intelligence and Ecology* (pp. 33–44). New York: Spartan Books.

Ashby, W. R. (1981*). Mechanisms of Intelligence.* (R. Conant, Ed.). Intersystems Publications. Retrieved from http://books.google.be/books?hl=en&lr=&id=vbcEczxQgpkC&oi=fnd&pg=PR1&dq=ashby+intelligence+amplifier&ots=6K4fWa6rOA&sig=5mOvDAnVhDDjbXHjmbY1jVyxi5Y

Atzori, L., Iera, A., & Morabito, G. (2010). The internet of things: A survey. *Computer Networks,* 54(15), 2787–2805. Retrieved from http://www.sciencedirect.com/science/article/pii/S1389128610001568

Barfield, W., & Caudell, T. (2001). *Fundamentals of wearable computers and augmented reality.* Routledge. Retrieved from http://books.google.be/books?hl=en&lr=&id=-1w4367mu3QC&oi=fnd&pg=PR11&dq=wearable+computers&ots=xqPLZGY1Bq&sig=txozDYG538Tp15mP4r8wy1uvNY0

Barrett, D. (2010). *Supernormal stimuli: how primal urges overran their evolutionary purpose.* W.W. Norton & Co.

Barrow, J. D. (1998). *Impossibility: The limits of science and the science of limits.* Oxford University Press, USA.

Beard, K. W. (2005). Internet addiction: a review of current assessment techniques and potential assessment questions. *CyberPsychology & Behavior*, 8(1), 7–14. Retrieved from http://online.liebertpub.com/doi/pdf/10.1089/cpb.2005.8.7

Berners-Lee, T., & Fischetti, M. (1999). Weaving the Web: *The Original Design and Ultimate Destiny of the World Wide Web by Its Inventor.*

Harper San Francisco. Retrieved from http://portal.acm.org/citation.cfm?id=554813

Bernheim, J. (1999). The Cognitive Revolution and 21st Century Enlightenment: towards a progressive world view. *Science, Technology and Social Change, Einstein meets Magritte* (p. 63). Kluwer.

Bernstein, A., Klein, M., & Malone, T. W. (2012). *Programming the Global Brain. Communications of the ACM*, 55(5), 1. Retrieved from http://cci.mit.edu/publications/CCIwp2011-04.pdf

Bohn, J., Coroamă, V., Langheinrich, M., Mattern, F., & Rohs, M. (2005). Social, economic, and ethical implications of ambient intelligence and ubiquitous computing. *Ambient intelligence* (pp. 5–29). Springer. Retrieved from http://link.springer.com/chapter/10.1007/3-540-27139-2_2

Bostrom, N. (2005). A history of transhumanist thought. *Journal of Evolution and Technology*, 14(1), 1–25. Retrieved from http://www.jetpress.org/volume14/bostrom.pdf

Brodie, R. (1996). *Virus of the Mind: The New Science of the Meme.* Integral Pr.

Carr, N. G. (2011). *The shallows: what the Internet is doing to our brains.* New York: W.W. Norton.

Carrera-Bastos, P., Fontes-Villalba, M., O'Keefe, J. H., Lindeberg, S., & Cordain, L. (2011). The western diet and lifestyle and diseases of civilization. *Research Reports in Clinical Cardiology*, 15. doi:10.2147/RRCC.S16919

Chalmers, D. (2010). The Singularity: A philosophical analysis. *Journal of Consciousness Studies*, 17, 9(10), 7–65.

Charlton, B. (2002). *What is the Meaning of Life? Animism Generalised Anthropomorphism and Social Intelligence.* Retrieved from http://www.hedweb.com/bgcharlton/meaning-of-life.html

Chen, J. (2007). Flow in games (and everything else). *Communications of the ACM*, 50(4), 31–34.

Chou, T.-J., & Ting, C.-C. (2003). The role of flow experience in cyber-game addiction. *CyberPsychology & Behavior,* 6(6), 663–675. Retrieved from **http://online.liebertpub.com/doi/abs/10.1089/109493103322725469?2**

Clark, A. (1998). Embodied, situated, and distributed cognition. *A companion to cognitive science,* 506–517.

Clark, A., & Chalmers, D. (1998). The extended mind. *Analysis,* 58(1), 7–19.

Corning, P. (2003). *Nature's magic: Synergy in evolution and the fate of humankind.* Cambridge University Press Cambridge, UK.

Cullen, J., & Bryman, A. (1988). The knowledge acquisition bottleneck: time for reassessment? *Expert Systems,* 5(3), 216–225. Retrieved from **http://onlinelibrary.wiley.com/doi/10.1111/j.1468-0394.1988.tb00065.x/abstract**

Dawkins, R. (1999). *The extended phenotype: The long reach of the gene (Revised.).* Oxford University Press, USA.

De Alfaro, L., Kulshreshtha, A., Pye, I., & Adler, B. T. (2011). Reputation systems for open collaboration. *Communications of the ACM,* 54(8), 81–87. Retrieved from **http://dl.acm.org/citation.cfm?id=1978560**

De Rosnay, J. (2000). *The Symbiotic Man: A new understanding of the organization of life and a vision of the future.* Mcgraw-Hill. Retrieved from **http://pespmc1.vub.ac.be/books/DeRosnay.TheSymbioticMan.pdf**

De Vany, A. (2010). The New Evolution Diet: What Our Paleolithic Ancestors Can Teach Us about Weight Loss, *Fitness, and Aging.* Rodale Books.

Dew, N., Sarasvathy, S. D., & Venkataraman, S. (2004). The economic implications of exaptation. *Journal of Evolutionary Economics,* 14(1), 69–84. Retrieved from **http://link.springer.com/article/10.1007/s00191-003-0180-x**

Diamandis, P. H., & Kotler, S. (2012). *Abundance: The Future Is Better Than You Think.* Free Press.

Diamond, J. (1987). The Worst Mistake in The History Of The Human Race. *Discover,* May, 64–66.

Dressler, F. (2008). A study of self-organization mechanisms in ad hoc and sensor networks. *Computer Communications,* 31(13), 3018–3029.

Drexler, E. K. (2013*). Radical Abundance: How a Revolution in Nanotechnology Will Change Civilization.* PublicAffairs.

Dueñas-Osorio, L., & Vemuru, S. M. (2009). Cascading failures in complex infrastructure systems. *Structural Safety*, 31(2), 157–167. doi:10.1016/j.strusafe.2008.06.007

Eden, A. H., Raker, J. H. S., Moor, J. H., & Steinhart, E. H. (2013). *Singularity Hypotheses: A Scientific and Philosophical Assessment*. Springer-Verlag New York Incorporated. Retrieved from http://www.springer.com/engineering/computational+intelligence+and+complexity/book/978-3-642-32559-5

Egan, G. (2002). *Schild's Ladder* (1st ed.). Eos.

Farmer, J. D. (2013). *The challenge of modeling systemic risk*. University of Oxford. Retrieved from http://www.risk.jbs.cam.ac.uk/news/events/annualmeetings/downloads/pdfs/2013/farmerj.pdf

Fredette, J., Marom, R., Steiner, K., & Witters, L. (2012). *The promise and peril of hyperconnectivity for organizations and societies*. The Global Information Technology Report: living in a hyperconnected world (pp. 113–119). World Economic Forum. Retrieved from http://www3.weforum.org/docs/GITR/2012/GITR_Chapter1.10_2012.pdf

Gao, H., Barbier, G., & Goolsby, R. (2011). Harnessing the crowdsourcing power of social media for disaster relief. *Intelligent Systems*, IEEE, 26(3), 10–14.

Gershenson, C., & Heylighen, F. (2005). How can we think complex? In K. Richardson (Ed.), *Managing organizational complexity: philosophy, theory and application* (pp. 47–61). Information Age Publishing. Retrieved from http://pcp.vub.ac.be/Papers/ThinkingComplex.pdf

Goertzel, B. (2002). *Creating internet intelligence: Wild computing, distributed digital consciousness, and the emerging global brain*. Kluwer Academic/Plenum Publishers.

Google Glass. (2013, July 16).*Wikipedia, the free encyclopedia*. Retrieved from http://en.wikipedia.org/w/index.php?title=Google_Glass&oldid=564433859

Gray, P. (2009). Play as a foundation for hunter-gatherer social existence. *American Journal of Play*, 4, 476–522.

He, B., Gao, S., Yuan, H., & Wolpaw, J. R. (2013). Brain–Computer Interfaces. In B. He (Ed.), *Neural Engineering* (pp. 87–151). Boston, MA: Springer US. Retrieved from http://www.springerlink.com/index/10.1007/978-1-4614-5227-0_2

Helbing, D. (2013). Globally networked risks and how to respond. *Nature*, 497(7447), 51–59. Retrieved from http://www.nature.com/nature/journal/v497/n7447/abs/nature12047.html

Helbing, D., Bishop, S., Conte, R., Lukowicz, P., & McCarthy, J. B. (2012). FuturICT: Participatory computing to understand and manage our complex world in a more sustainable and resilient way. *European Physical Journal Special Topics*, 214, 11–39. Retrieved from http://adsabs.harvard.edu/abs/2012EPJST.214...11H

Heylighen, F. (1992). A cognitive-systemic reconstruction of Maslow's theory of self-actualization. *Behavioral Science*, 37(1), 39–58. doi:10.1002/bs.3830370105

Heylighen, F. (1999). Advantages and limitations of formal expression. *Foundations of Science*, 4(1), 25–56. doi:10.1023/A:1009686703349

Heylighen, F. (2000). Evolutionary Transitions: how do levels of complexity emerge? *COMPLEXITY*, 6(1), 53–57.

Heylighen, F. (2002). The global brain as a new utopia. *Zukunftsfiguren*. Suhrkamp, Frankurt. Retrieved from http://pespmc1.vub.ac.be/papers/GB-Utopia.pdf

Heylighen, F. (2006). Mediator Evolution: a general scenario for the origin of dynamical hierarchies. *Worldviews, Science and Us*. (Singapore: World Scientific).

Heylighen, F. (2007a). The Global Superorganism: an evolutionary-cybernetic model of the emerging network society. *Social Evolution & History*, 6(1), 58–119. Retrieved from http://pcp.vub.ac.be/papers/Superorganism.pdf

Heylighen, F. (2007b). Why is Open Access Development so Successful? Stigmergic organization and the economics of information. In B. Lutterbeck, M. Baerwolff & R. A. Gehring (Ed.), *Open Source Jahrbuch 2007* (pp. 165–180). Lehmanns Media. Retrieved from http://pespmc1.vub.ac.be/Papers/OpenSourceStigmergy.pdf

Heylighen, F. (2008). Accelerating socio-technological evolution: from ephemeralization and stigmergy to the global brain. *Globalization as evolutionary process: modeling global change, Rethinking globalizations* (p. 284). Routledge. Retrieved from http://pcp.vub.ac.be/papers/AcceleratingEvolution.pdf

Heylighen, F. (2011). Conceptions of a Global Brain: an historical review. *Evolution: Cosmic, Biological, and Social*, eds. Grinin, L. E., Carneiro, R. L., Korotayev A. V., Spier F. (pp. 274 – 289). Uchitel Publishing. Retrieved from **http://pcp.vub.ac.be/papers/GB-conceptions-Rodrigue.pdf**

Heylighen, F. (2012a). A brain in a vat cannot break out: why the singularity must be extended, embedded and embodied. *Journal of Consciousness Studies*, 19(1-2), 126–142. Retrieved from **http://pcp.vub.ac.be/Papers/Singularity-Reply2Chalmers.pdf**

Heylighen, F. (2012b). Challenge Propagation: a new paradigm for modeling distributed intelligence (No. 2012-01). *GBI Working Papers. Brussels*, Belgium. Retrieved from **http://pcp.vub.ac.be/papers/ChallengePropagation.pdf**

Heylighen, F. (2012c). *A Tale of Challenge, Adventure and Mystery: towards an agent-based unification of narrative and scientific models of behavior* (Working papers No. 2012-06). ECCO. Brussels, Belgium. Retrieved from **http://pcp.vub.ac.be/papers/TaleofAdventure.pdf**

Heylighen, F. (2013). *Distributed Intelligence Technologies: a survey of present and future applications of the Global Brain* (No. 2013-02). GBI Working Paper. Retrieved from **http://pespmc1.vub.ac.be/Papers/GB-applications-survey.pdf**

Heylighen, F. (2013q). *Self-organization in Communicating Groups: the emergence of coordination, shared references and collective intelligence*. In À. Massip-Bonet & A. Bastardas-Boada (Eds.), Complexity Perspectives on Language, Communication and Society (pp. 117-149). Berlin, Germany: Springer. Retrieved from **http://pcp.vub.ac.be/Papers/Barcelona-LanguageSO.pdf**

Heylighen, F. (2014). Cybernetic Principles of Aging and Rejuvenation: the Buffering-Challenging Strategy for Life Extension. *Current aging science*. Retrieved from **http://pespmc1.vub.ac.be/Papers/Ageing&Rejuvenation.pdf**

Heylighen, F., & Bernheim, J. (2000). Global Progress I: Empirical Evidence for ongoing Increase in Quality-of-life. *Journal of Happiness Studies*, 1(3), 323–349. doi:10.1023/A:1010099928894

Heylighen, F., & Bollen, J. (1996). The World-Wide Web as a Super-Brain: from metaphor to model. Cybernetics and Systems' 96. R. Trappl (Ed.). *Austrian Society for Cybernetics*.

Heylighen, F., & Bollen, J. (2002). Hebbian Algorithms for a Digital Library Recommendation System. *International Conference on Parallel Processing* (p. 439). Los Alamitos, CA, USA: IEEE Computer Society. doi:10.1109/ICPPW.2002.1039763

Heylighen, F., Busseniers, E., Veitas, V., Vidal, C., & Weinbaum, D. R. (2012). *Foundations for a Mathematical Model of the Global Brain: architecture, components, and specifications* (GBI Working Papers No. 2012-05). GBI Working Papers. Retrieved from http://pespmc1.vub.ac.be/papers/TowardsGB-model.pdf

Heylighen, F., & Chielens, K. (2008). Evolution of culture, memetics. (R. A. Meyers, Ed.)*Encyclopedia of Complexity and Systems Science*. Springer. Retrieved from http://pespmc1.vub.ac.be/Papers/Memetics-Springer.pdf

Heylighen, F., & Joslyn, C. (2003). Cybernetics and Second-Order Cybernetics. In Robert A. Meyers (Ed.), *Encyclopedia of Physical Science and Technology* (3rd ed., Vol. 4, pp. 155–169). New York: Academic Press. Retrieved from http://pespmc1.vub.ac.be/Papers/cybernetics-EPST.pdf

Heylighen, F., Kostov, I., & Kiemen, M. (2013). Mobilization Systems: technologies for motivating and coordinating human action. In Peters M. A., Besley T. and Araya D. (Ed.), *The New Development Paradigm: Education, Knowledge Economy and Digital Futures*. Routledge. Retrieved from http://pcp.vub.ac.be/Papers/MobilizationSystems.pdf

Hoffman, J., & Rosenkrantz, G. (2012). Omnipotence. In E. N. Zalta (Ed.), *The Stanford Encyclopedia of Philosophy* (Spring 2012.). Retrieved from http://plato.stanford.edu/archives/spr2012/entries/omnipotence/

Johnson, S. (2006). *Everything Bad is Good for You: How Popular Culture is Making Us Smarter*. Penguin Books Limited.

Klugman, J. (2010). *Human development report 2010: The real wealth of nations*. New York: UNDP.

Kurzweil, R. (2005). *The singularity is near*. Penguin books.

LessWrong. (2013). Complexity of value. LessWrong. Wiki. Retrieved from http://wiki.lesswrong.com/wiki/Complexity_of_value

Lipson, H., & Kurman, M. (2013). Fabricated: *The New World of 3D Printing*. John Wiley & Sons.

Mamykina, L., Manoim, B., Mittal, M., Hripcsak, G., & Hartmann, B. (2011). Design lessons from the fastest q&a site in the west. *Proceedings of the 2011 annual conference on Human factors in computing systems*, CHI '11 (pp. 2857–2866). New York, NY, USA: ACM. doi:10.1145/1978942.1979366

Mann, W. E. (1975). The Divine Attributes. *American Philosophical Quarterly*, 12(2), 151–159. Retrieved from http://www.jstor.org/stable/10.2307/20009569

Mayer-Kress, G., & Barczys, C. (1995). The global brain as an emergent structure from the Worldwide Computing Network, and its implications for modeling. *The information society*, 11(1), 1–27.

Maynard Smith, J., & Szathmáry, E. (1997). *The major transitions in evolution*. Oxford University Press, USA.

McAuliffe, K. (2010). If Modern Humans Are So Smart, Why Are Our Brains Shrinking? *Discover Magazine*, (September). Retrieved from http://discovermagazine.com/2010/sep/25-modern-humans-smart-why-brain-shrinking

Michael, D. R., & Chen, S. L. (2005). *Serious games: Games that educate, train, and inform*. Muska & Lipman/Premier-Trade.

Miller, J. G. (1995). *Living systems*. University Press of Colorado.

Miller, J. H., & Page, S. E. (2007). *Complex adaptive systems: an introduction to computational models of social life*. Princeton University Press.

Modis, T. (2012). Why the Singularity Cannot Happen. In A. H. Eden, J. H. Moor, J. H. Søraker, & E. H. Steinhart (Eds.), *Singularity Hypotheses* (pp. 311–346). Springer. Retrieved from http://link.springer.com/chapter/10.1007/978-3-642-32560-1_16

Muehlhauser, L., & Helm, L. (2012). The Singularity and Machine Ethics. In A. H. Eden, J. H. Moor, J. H. Søraker, & E. Steinhart (Eds.), *Singularity Hypotheses*, The Frontiers Collection (pp. 101–126). Springer Berlin Heidelberg. Retrieved from http://link.springer.com/chapter/10.1007/978-3-642-32560-1_6

Mueller, J. (2009). War has almost ceased to exist: An assessment. *Political Science Quarterly*, 124(2), 297–321. Retrieved from http://onlinelibrary.wiley.com/doi/10.1002/j.1538-165X.2009.tb00650.x/abstract

Mummert, A., Esche, E., Robinson, J., & Armelagos, G. J. (2011). Stature and robusticity during the agricultural transition: Evidence from the bioarchaeological record. *Economics & Human Biology*, 9(3), 284–301. doi:10.1016/j.ehb.2011.03.004

Nakamura, J., & Csikszentmihalyi, M. (2002). The concept of flow. In C.

R. Snyder (Ed.), *Handbook of positive psychology* (pp. 89–105). New York, NY: Oxford University Press.

Olshansky, S. J., Passaro, D. J., Hershow, R. C., Layden, J., Carnes, B. A., Brody, J., Hayflick, L., et al. (2005). A potential decline in life expectancy in the United States in the 21st century. *New England Journal of Medicine*, 352(11), 1138–1145. Retrieved from http://www.nejm.org/doi/full/10.1056/nejmsr043743

Otlet, P. (1935). *Monde: Essai d'Universalisme*. Brussels: Mundaneum.

Palmer, S. (2010). *Toxic Childhood: How The Modern World Is Damaging Our Children And What We Can Do About It*. Orion.

Parunak, H. V. D. (2006). A survey of environments and mechanisms for human-human stigmergy. In D. Weyns, H. V. D. Parunak, & F. Michel (Eds.), *Environments for Multi-Agent Systems* II (pp. 163–186). Springer.

Pinker, S. (2011). *The better angels of our nature: The decline of violence in history and its causes*. Penguin. Retrieved from http://books.google.com/books?hl=en&lr=&id=c3cWa-GnsfMC&oi=fnd&pg=PT2&dq=pinker+angels&ots=aJlJfnVMb7&sig=ZaRwXXnwNDqRKXZ1t9wSGM9M4oo

Proudfoot, D. (2012). Software Immortals: Science or Faith? In A. Eden, Johnny Søraker, Jim Moor, and Eric Steinhart (Ed.), *Singularity Hypotheses*, The Frontiers Collection (pp. 367–392). Springer. Retrieved from http://link.springer.com/chapter/10.1007/978-3-642-32560-1_18

Rifkin, J. (2014). *The Zero Marginal Cost Society: The Internet of Things, the Collaborative Commons, and the Eclipse of Capitalism*. Palgrave Macmillan. Retrieved from http://www.bookdepository.com/Zero-Marginal-Cost-Society-Jeremy-Rifkin/9781137278463

Rodriguez, O. (2012). MOOCs and the AI-Stanford like Courses: two successful and distinct course formats for massive open online courses. European Journal of Open, *Distance, and E-Learning*. Retrieved from http://www.eurodl.org/materials/contrib/2012/Rodriguez.ht

Sahlins, M. (2004). The original affluent society. *Investigating culture: an experiential introduction to anthropology*, 110.

Schmidt, E., & Cohen, J. (2013). *The New Digital Age: Reshaping the Future of People, Nations and Business*. Random House Digital, Inc. Retrieved from http://books.google.com/books?hl=en&lr=&id=Fl_LPoLdGKsC&oi=fnd&dq=The+New+Digital+Age+by+Eric+Schmidt+and+Jared+Cohen&ots=r

End of the Beginning

ly1gVnabZ&sig=mswnN7bo01oM9nh6J0ADEWRJBy8

Steels, L., & Brooks, R. A. (1995). *The artificial life route to artificial intelligence: Building embodied, situated agents.* Lawrence Erlbaum.

Stewart, J. (2000). *Evolution's Arrow: The Direction of Evolution and the Future of Humanity.* Chapman Pr.

Stock, G. (1993). *Metaman: the merging of humans and machines into a global superorganism.* Simon & Schuster.

Teilhard de Chardin, P. (1959). *The phenomenon of man.* Collins London.

Thompson, C. (2011). How Khan Academy is changing the rules of education. *Wired Magazine,* (August), 126. Retrieved from http://www.wired.com/magazine/2011/07/ff_khan/all/1

Thurner, S., & Poledna, S. (2013). DebtRank-transparency: Controlling systemic risk in financial networks. *Scientific reports*, 3. Retrieved from http://www.nature.com/srep/2013/130528/srep01888/full/srep01888.html

Toffler, A. (1970). *Future shock.* New York: Random House.

Turchin, V. (1995). A dialogue on metasystem transition. *World Futures*, 45, 5–57. Retrieved from http://pespmc1.vub.ac.be/papers/Turchin/dialog.pdf

Turner, J. S. (2000). *The extended organism: the physiology of animal-built structures.* Harvard University Press. Retrieved from http://books.google.be/books?hl=en&lr=&id=wapnxKbb23MC&oi=fnd&pg=PP3&dq=extended+organism&ots=QSfNw8kXZI&sig=q_0pZEwBHKa1tfxK90wDPJnqrb4

Van Krevelen, D. W. F., & Poelman, R. (2010). A survey of augmented reality technologies, applications and limitations. *International Journal of Virtual Reality*, 9(2), 1. Retrieved from http://kjcomps.6te.net/upload/paper1%20.pdf

Vidal, C. (2014). *The Beginning and the End - The Meaning of Life in a Cosmological Perspective.* Springer. Preprint at: http://arxiv.org/abs/1301.1648

Vinge, V. (1993). The coming technological singularity. *Whole Earth Review*, 88–95.

Wierenga, E. (2012). Omniscience. In E. N. Zalta (Ed.), *The Stanford Encyclopedia of Philosophy* (Winter 2012.). Retrieved from http://plato.stanford.edu/archives/win2012/entries/omniscience/

Wolpert, D. H., & Macready, W. G. (1997). No free lunch theorems for optimization. *Evolutionary Computation, IEEE Transactions on,* 1(1), 67–82. Retrieved from http://ieeexplore.ieee.org/xpls/abs_all.jsp?arnumber=585893

Wright, R. (2001). *Nonzero: The logic of human destiny.* Vintage.

Yanofsky, N. S. (2013). *The outer limits of reason: what science, mathematics, and logic cannot tell us.* MIT Press.

Young, K. S. (1998). Internet addiction: The emergence of a new clinical disorder. *CyberPsychology & Behavior,* 1(3), 237–244. Retrieved from http://online.liebertpub.com/doi/abs/10.1089/cpb.1998.1.237

Zander, T. O., Kothe, C., Jatzev, S., & Gaertner, M. (2010). Enhancing Human-Computer Interaction with Input from Active and Passive Brain Computer Interfaces. In D. S. Tan & A. Nijholt (Eds.), *Brain-Computer Interfaces* (pp. 181–199). London: Springer London. Retrieved from http://link.springer.com/10.1007/978-1-84996-272-8_11

End of the Beginning

Dialogue 10.1

With Francis Heylighen, Ben Goertzel and David Weinbaum (Weaver)

Ben: I think your paper is very nicely written; and it stakes out an extreme position with gusto, which is a worthwhile thing. You do an interesting job of arguing why the "benevolent Global Brain as *de facto* world organizing principle" is a plausible scenario.... This is no small feat, so congratulations...

On the other hand, I don't think you present a strong argument that this is a particularly *likely* scenario -- as opposed to various disaster scenarios in which individual Artificial General Intelligence systems become super-intelligent and dominant before such a richly flourishing Global Brain emerges...

Francis: I find it strange that an optimistic view that basically extrapolates the human developmental trajectory that we have experienced over the past centuries and millennia (albeit with a final, "Singular" acceleration) should be considered an "extreme" position. I find it equally strange that your AGI disaster scenarios-- which seem to come straight from science fiction movies--should be considered more "likely".

"End of Humanity/End of Civilization" scenarios have an extremely poor track record. There have been thousands of these global disaster scenarios throughout history, and none of them have even come close to being realized. On the other hand, at least since the Middle Ages, scenarios that forecasted general progress in economic, social, moral and educational life have been systematically confirmed, with the exception of short-term dips such as wars or economic crises.

Ben: Just to be clear -- I haven't created any novel AGI disaster scenarios personally, and I don't advocate disaster scenarios as

End of the Beginning

being especially likely. I do think it's fairly likely that super-intelligent AGI will come about before a really compelling Global Brain, but I don't think this is overwhelmingly likely to lead to disaster..... But these are my opinions, and I don't know how to formulate strong arguments for them...

Francis: I just wish to express my surprise that AGI disaster scenarios seem to have become mainstream within a certain subculture (to which you seem to belong), while optimistic scenarios seem to have become the exception. I know these scenarios do not come from you, but the way you formulated them, you seem to consider them much more likely than my optimistic "return to Eden" scenario.

Ben: I don't actually have strong opinions about the likelihoods of various post-Singularity scenarios. As you correctly note, no arguments are strong enough to support futuristic scenarios that try to look beyond the Singularity. Everything one can say in this regard is pretty speculative at this stage... The direction that speculation takes will depend more on experience and personality than on any formal arguments.

Francis: "General progress" projections have often come true.

Ben: Hmmm... it seems to me that, historically, "Eden" type utopian projections have had equally dismal track records as "End is Near" projections!!! And they have probably caused a lot more damage overall....

Francis: You should distinguish between political utopias (like communism), where a particular "ideal" or "perfect" society is promoted and eventually forcefully imposed, with typically dismal results, and forecasts about on-going evolution, which note that progress tends to occur, whether you promote it or not. My scenario fits into the second category, which is why I was not keen to enter into the political implications of my utopia.

Ben: Ah. Well, I see your point – but still, I think that historically, that distinction (between projecting and attempting to force a utopia) has not always been so clear. Marxism traditionally held that the emergence of the Communist Utopia was inevitable and was going to be the result of inexorable social evolution and progress....

Francis: Yes, and that is why true Marxists should never condone the totalitarian methods used in China or the Soviet Union...

The only thing that distinguishes my scenario from a more general "ongoing progress" scenario is that I have tried to take into account the extreme acceleration implied by the Singularity concept, assuming that it will force a transition to a wholly different level of organization within a few decades. A simple extrapolation from "rapid progress" to "very rapid progress" then brings to me to what appears like a utopian vision. But note that if you had described our present society to a person living a couple of centuries ago, they would have considered it just as incredible as my extrapolation may appear to people now.

Ben: That's certainly true. But still, my view is that the verbiage of Eden and Paradise is likely to turn off more people than it attracts. Religious people won't accept it, and non-religious folk don't find the metaphorical religious connection appealing....

Francis: This is a danger I have been made aware of, also by others who read my paper. Since I am not religious myself, I would not like to create the impression that I have suddenly been converted! Therefore, I have tried to somewhat tone down my "religious" vocabulary and add plenty of caveats. But I still like the idea of the return to Eden, both for its literary, poetic value, and for the fact that it fits in perfectly with my Paleo philosophy, according to which people would become healthier and happier if their lifestyle would again come closer to the one of our hunter-gatherer ancestors....

Ben: One thing that interests me, as an extension of this line of thinking, is the Nietzschean question of: What are the structure and dynamics of the mind of a god? How does this omni-xxx mind actually work? How do its internal structures and dynamics differ from those of less comprehensive minds? Food for thought to be sure.... Or, to verge from Nietzsche into Piaget -- is the Global Brain in your view a strongly self-modifying system, achieving a Piagetan post-formal "reflective" stage?

Francis: It is a bit too soon to give answers to these most interesting questions, but I expect such a super-intelligence to be strongly self-modifying. These modifications are likely to be grounded more in sensory-motor interaction with the world than in abstract reflection. I would argue that individual Artificial General Intelligence systems cannot become super-intelligent without a sensory-motor interaction channel with the world at least as strong as a collective of thousands of well-coordinated human beings. "Superintelligent" decisions require "superdetailed" and "superinteractive" information exchanges, not the kind of stuff you could get from a fat pipe spewing out "big" data from the Internet.

Therefore, AGI systems do not pose any existential risks as far as I am concerned. The worst they could do would be to generate "psychological parasites" that entice people to act against their own best interests. But in that respect they would have to compete with thousands of other psychological parasites generated by people, organizations, other AGIs, and random variations. By that time, I hope we would have developed a pretty effective "immune

End of the Beginning

system" or "vaccination program" against psychological parasites...

Ben: Hmm, well, even if that's true, it doesn't imply anything about the Global Brain, does it? An autonomous AGI could use large parts of the Internet as part of its sensorimotor cortex, without entering into GB-style interactions with human beings....

Francis: My point, which is admittedly difficult to quantify at this stage, is that the bandwidth of the sensory-motor interaction (chemical, electromagnetic, tactile, visual ...) between a biological organism and the world is many orders of magnitude larger than anything we have as yet been able to model, let alone build. Even if you could muster the whole bandwidth of the Internet to feed your AGI, it would probably still be about the level of a single human being, and therefore not comparable to the theoretical bandwidth of a GB consisting of billions of human beings (and eventually other biological organisms).

Ben: Also, an advanced AGI could potentially build itself better sensors and actuators, thus progressively ramping up its intelligence and its interactivity with the world together....

Francis: I accept that, but the increase would be much, much slower than the process of building faster processors, or developing smarter programs. More importantly: why would the AGI ignore the extremely sophisticated sensors and effectors already available in biological organism if it could have access to them? The only reason for doing so is that it would not want to be too closely interconnected with humans. This makes sense from a "hostile AI" perspective. However, from an evolutionary perspective, any "friendly" AI that would be willing to cooperate with human beings would have an unbeatable competitive edge over an AI that would dismiss the billions of years of evolutionary sophistication built into biological organisms and try to go it alone.

End of the Beginning

That is why I believe that a full human-AI symbiosis that gives rise to the GB is the only realistic outcome...

Ben: Hmmmm ... it still seems to me that real AGIs are going to be connected to sensors and actuators, and to the Net, and so the fact that they need lots of interaction with the world will not be an obstacle to their development..... And you have not given any rigorous argumentation as to the quantitative extent of an AGI's need for sensation and actuation -- only made a (reasonably valid) conceptual point.... So to me there's still a huge amount of uncertainty here....

Francis: The quantitative extent is something that still needs to be investigated. Ashby's law of requisite variety in principle allows a quantitative estimate of the amount of information transmission necessary across the interaction channel. However, I am the first to admit that there is information and information—in the sense that not all data are equally useful or relevant. That is why we can build ridiculously simple models of extremely complex systems, and still extract some value from them.

Ben: It seems to me that once you start ramping up toward a "superintelligence" you can afford less and less to neglect apparently minor details.... How do we know which details will be critical, will have a massive developmental impact? How do we quantify the difference between an AGI that has access to all video cameras in the world, and a GB that moreover has access to all human visual cortexes in the world?

Francis: I agree that it is difficult to develop such arguments with all the necessary precision and details. However, I believe my arguments are pretty solid compared to the arguments of the "hostile AGI" crowd (though of course they would not agree ;-)). For example, I just read an article on the Singularity by your AGI

End of the Beginning

colleague Itamar Arel (2012) in which he posits that an artificial superintelligence will necessarily be malevolent. His only argument is that because such an AGI would be programmed through reinforcement learning, it will want to get ever more rewards and therefore ever better outcomes, thus eventually bringing it in conflict with humans. But we all want to get ever better outcomes, and that does not stop us from collaborating peacefully. So, the argument looks completely unconvincing to me...

Ben: Projecting whether outcomes of various future technologies will be beneficial or not, seems very difficult. Usually it's some mix, of a sort that's nearly impossible to foresee in advance. But one projection I'm fairly comfortable making is that autonomous AGIs will probably come well before a powerfully integrated GB such as your project envisages.

I suppose this is partly because I believe I can see quite clearly how to create such an AGI, whereas the emergence of a powerfully integrated GB still seems to me to have a lot of nebulous aspects...

Francis: I believe that here you hit the nail on the head concerning the difference in perspective between you and me. You are an AGI theorist, so you can easily imagine an AGI emerging in the near future. I am a GB theorist, so I can easily see how a GB would emerge in the near future. Obviously, I think a GB will come about more quickly than an AGI, while you believe the opposite. That is why I consider AGI as a fascinating research issue but one that eventually will play the role of a supporting technology for the GB, rather than as the to be expected new center of our universe...

Francis: Which one of us will turn out to be correct is something we may find out in the coming decade or two. I just wish to note that in terms of practical impact the Internet, which I consider as an embryonic global brain, has consistently beaten AI/AGI on about all fronts. Predictions of near-term human level intelligence in an AGI have been made at least since the 1950's, but they still seem to be as far in the future as ever... On the other hand, the spread and impact of the Internet has far surpassed all predictions. If this difference in development speed can be extrapolated, the emergence of distributed (Internet) intelligence seems much nearer than the emergence of AGI...

Ben: Seems to me that this comparison (between predictions of AI versus Net technology) would come out differently depending on how you slice things.

AI plays a huge role in the world now, e.g. inside Google and Bing, on the world financial markets, in doing supply chain management for the military, etc. It hasn't developed exactly as science fiction foresaw, so far, but has certainly had a huge impact, and is rapidly growing in practical terms. Whether this practical development in the AI sphere is evidence that AGI is near, depends on your theoretical perspective, it seems.

On the other hand, I'm not aware of concrete evidence of rapidly accelerating *emergent* intelligence via the Internet, as opposed to evidence of rapidly accelerating tools for communication between humans. So whether the growth of the Net and cellphones and so forth is evidence for the advent of a GB, depends somewhat on your theoretical perspective.

Francis: This is indeed an important open issue that we are trying to address at the Global Brain Institute: finding empirical evidence that the collective or distributed intelligence of people interacting via the Internet is larger than the one of those same people without the Internet. Intuitively, it is obvious to me that this emergent intelligence exists, but it is difficult to operationalize this hypothesis in such a way that you could test it with quantitative data. For a concrete example, without the Internet no one could ever have produced an encyclopedia as immense as Wikipedia. But apart from its size, what quantifiable advantages does it have that lack in traditional encyclopedias? Our group has been discussing a number of methods to measure perhaps the depth, or the diversity of viewpoints, or the quality of Wikipedia articles, but it is everything but obvious how to quantify this in such a way that everybody would agree that this is an objective measure of intelligence...

Ben: On the other hand, the dynamic of the physics research community has certainly shifted with the advent of Arxiv.org; and the dynamic of the software development world has transformed radically with the advent of GitHub and other open online software development tools.... But the degree to which this represents emergent intelligence versus tools for leveraging individual human intelligence, is something we lack a clearly accepted way to quantify at present.

Getting back to the themes we started with, though.... I'm still not fully happy with your dismissal of disaster scenarios. Suppose that Edenic GB type scenarios are possible – ok, but aren't dystopian type scenarios ALSO possible? The scenario of a network of military robots, tied in with military online cyberwarfare bots, going rogue and killing everybody – is certainly feasible on the face of it.

Francis: In my paper, I discuss cyberwar mainly to conclude that it is much less threatening than one might think. You will probably not be convinced by my (too brief) arguments, but the paper is already too long to develop that theme in depth. Let me just summarize that war is on its way out, so the incentive to deploy such robots is diminishing as we speak. For an in-depth assessment of this idea, see Mueller's (2009) aptly named paper, "War has almost ceased to exist".

Ben: This doesn't seem the current perspective in Washington, Teheran, Tel Aviv, Syria or Beijing, though....

Francis: Probably not, but political institutions are typically very slow in catching up with new realities. I suppose that in this respect at least, the Europeans are ahead of the pack, as they have been systematically reducing their investment in the military over the past decades. It seems as if the Belgian army is becoming smaller by the year, and when the last officers go on pension, there might be nothing left ;-).

Ben: I agree this is the right view in the long run.... But the medium run is unclear and could potentially include a lot of military destruction. It could even, conceivably, prevent humans reaching the long run (remembering Keynes' dictum, admittedly made in the context of economics not existential risk, that "in the long run we'll all be dead"!!)

Francis: In a highly complex and non-linear world, anything is possible. Therefore, I wouldn't completely phase out the army yet, but it is clear to me that countries that invest in infrastructure and education do better than those that invest in weapons...

Ben: Definitely true overall, though US hawks would argue that, e.g., Belgium can afford to invest in education over weapons because it knows the US and other nations will protect it. If South Korea stopped investing in weapons altogether, what would stop Kim Jong Un's army from taking over? The world is not yet such a tame place; and from the point of view of someone living in North Korea or Congo, it likely doesn't seem very tame and peaceful at all here in 2014. But I agree, things do seem to be moving in the direction you say, if one takes the long view. ISIS is in the US news every day lately, and they may well make more headway over the next few years; but my pretty strong guess is that over the next few decades they won't beat globalization and modernization.

And yet, I still tend to think all-positive Edenic scenarios are fairly unlikely.

But I do think it's valuable for *someone* to stake out an extreme, utopian position --- if only because it makes the rest of us radical techno-optimists seem reasonable and moderate! So, thanks for taking on the role !! ;-)

Francis: The Edenic scenario is only all-positive in its final outcome, *after* and *if* we manage to overcome all the perils on the road to the Singularity. The reason I wanted to stake out such an "extreme" position is because the whole idea of the Singularity is so radical that it demands a radical resolution.

Claiming that technological advances will accelerate to a near infinite speed, bringing about a superhuman intelligence of a level we cannot truly conceive, and then assuming that society will go on more or less as we know it, except that the role of the "bad guys" has now been taken over by robots or AGIs, only shows a lack of imagination in my opinion ... ;-)

I actually believe my "Edenic" views are much more reasonable and realistic than those of most "Singulitarians" who seem to have lost touch with the real world, and have gotten carried away by Platonic abstractions about nearly infinite computational spaces and virtual personalities that would replace the world of flesh and blood. Such a more "naturalistic" perspective, as a counterweight to speculations based on purely formal abstractions, is what I have tried to convey in my paper.

Ben: Well, abstractions have carried us a long way since the caveman days. But what makes things really interesting, of course, is the interplay between abstractions and realities. Which is what we are seeing unfold in the world around us, most fascinatingly.

End of the Beginning

Francis: I couldn't agree more!

Ben: Still, I find it a little frustrating that our views are so far off regarding whether a GB or AGI is more likely to emerge first. We are both genuinely trying to think about these issues in a rational yet visionary way – but I guess there's just so much uncertainty about, that it's to be expected different thinkers will come to different provisional conclusions.

Weaver: This point does seem worthy of a little more elaboration. Francis, you raise the interesting prediction that the Global Brain will arise before any human level AGI because basically it is simpler to achieve. You also claim it will be much more powerful than any AGI. I assume this is because it will incorporate and integrate already millions (or more) of human minds aided by information technology?

Francis: Good summary of my position.

Weaver: But surely there are difficulties involved here. With or without the help of technology, we know how difficult it is to make even a small number of humans to seamlessly (and spontaneously) coordinate their actions especially when the goal is a complex one and the said humans are not coerced into a very limited repertoire of behaviors that is controlled by rigid hierarchical structures (also this method is not particularly helpful in many cases).

Francis: The spontaneous coordination is not as difficult as you make it look. There is plenty of research showing that people working together, talking, playing football etc. automatically align their behavior so as to reduce friction and increase synergy, even when the goal is not well specified. Rigid hierarchical structures are more often counterproductive than helpful (see Heylighen,

End of the Beginning

2013a).

Indeed, today's society already operates as an approximation to a Global Brain and is capable of accomplishing great feats that require very complex coordination.

Note that people typically underestimate how smoothly and robustly society actually runs, because they always focus on what goes wrong or could go wrong, ignoring the immensely larger bulk of things that go right.

Weaver: Understood. And yet, even that is vastly remote from the level of coordination, integration and computational efficiency achieved biological brains.

Francis: Don't overestimate the degree of coordination and integration of biological brains: most people entertain a host of inconsistent beliefs, and can react to the same situation one moment in one way, another moment in a completely different way. This is to some degree unavoidable in any complex, dynamic system: uncertainties and non-linearities make action typically unpredictable and impossible to optimize.

What modern theories of consciousness (such as the one of Dehaene) have taught us is that our feeling of being a coherent self is mostly an illusion, born of the fact that a host of incoherent, as yet subconscious thoughts jostle for access to the global neuronal workspace, but that the winner-takes-it-all dynamics only lets a single one emerge into consciousness. That makes it look as if this single thought is actually our true self being in control of what happens in the brain. But we typically ignore all the other subconscious impulses acting in parallel that are ready to push it out and take its place...

Note that this is not a bug, but a feature: you cannot deal with a complex, dynamic reality from a centralized, know-it-all control office. Control must be distributed and self-organizing. AI people typically ignore this fact, at their peril...

End of the Beginning

Ben: "Good Old Fashioned AI" did ignore complex self-organizing dynamics. Indeed, back in the 1990s I had an argument with good old AI great Marvin Minsky on exactly this point. He maintained that self-organization, nonlinear dynamics, complex systems and so forth had absolutely no relevance to AI. He really did see the mind as a large system of rule-sets, each rule-set dealing with a particular sort of problem – and without any need for a complex self-organizing dynamical process governing the interaction between the rule-sets. But there's a reason this way of thinking is labeled "GOFAI" these days, and generally considered hopelessly old-fashioned. Modern AI research doesn't make this same mistake. For instance, cognitive architectures like Stan Franklin's LIDA, or Joscha Bach's MicroPsi, or the OpenCog architecture I've helped develop, all embrace self-organizing, distributed control. LIDA is based explicitly on Bernard Baars' Global Workspace model of distributed consciousness and control, for example.

Weaver: Anyhow, it seems reasonable to assume that integrating and coordinating millions of minds into a distributed system capable of demonstrating anything near brain level integration and coordination but on a vastly larger scale is a challenge as hard if not much harder than implementing a 'mere' AGI.

Francis: Depends on what you call a "mere AGI". If this AGI would have some basic ability for autonomous action, reasoning and learning, perhaps comparable to a cockroach, then we are nearly there. But the people anticipating the Singularity typically expect an AGI that is not only human-level but supra-human. That will be immensely more difficult because it means designing something better than our own brain, which means we should at least have understood all the capabilities of this brain.

End of the Beginning

On the other hand, the coordination of millions of minds into a distributed intelligence is a self-organizing process going on at this very moment. It does not require any special effort except continuing to do what we are already doing: experimenting with better communication and collaboration systems, while relying in part on specialized "AI" programs to help with things like translation, natural language parsing, pattern recognition, etc., and letting natural selection pick out the most effective ones...

Weaver: I am concerned that unless very high levels of integration are achieved, any advantage in 'sheer numbers' of computers and humans will be quickly eliminated by diminishing returns in effectiveness.

Francis: Why should there be any diminishing returns? The self-organization of a global brain is a typical illustration of a "network effect", i.e. increasing returns as more people, data, hardware and software join the interconnecting network.

Weaver: Historically, the upgrade of bureaucratic systems from paperwork to advanced digital technology of data management was in the eyes of many a key to a much better governance and efficiency. Evidently, this is hardly the case. As long as humans are involved, these systems have become only marginally more efficient.

Francis: Why do you say that? Digital data management IS much more efficient than paperwork-based bureaucracy, at least in terms of reliability, speed of response and saving work hours...

Weaver: Only when systems are entirely automated we observe a scalable increase in efficiency; first in manufacturing and now also in services and management (e.g. finance, retail, logistics, etc.). Once computers manage to break the barrier of natural language understanding, which is not something far beyond our current technological horizons, we can expect much more and consequently the role of humans progressively diminished. The same is true to advances in robotics.

The short-middle term (~10-15yrs) prospects of sociotechnological progress seem be in favor of an almost all machine monitoring, manufacture, service and management in almost every possible area of human activity.

Francis: What you forget here is that the increase in efficiency due to automation first requires that existing processes should be formally specified to such a degree that machines can be programmed to execute them. Once the process is sufficiently formalized (meaning stripped of all dependence on context external to the explicit data), machines can obviously execute it much more efficiently than humans. But the process of formalization is much more difficult than people think (which is one reason why digitalization of bureaucracies is not yet as productive as we might have hoped, and why the semantic web and consensual ontologies have so much difficulty getting off the ground). The reasons are that formalization: requires highly intelligent human effort to translate the ambiguous, context-dependent, experiential phenomena and behaviors into logical, explicit specifications and procedures can only be successfully applied to a subset of processes that are sufficiently repetitive and predictable to allow context-independent descriptions.

That means that we still need the highly context-sensitive, "situated and embodied" humans with their highly sophisticated implicit value systems to design the automated procedures and to perform the remaining activities that cannot be automated.

For a review of the scientific and philosophical arguments showing that complete formalization is neither possible nor desirable, see my paper (Heylighen, 1999).

End of the Beginning

Weaver: So perhaps it is not the legendary AGI some of us imagine but the accumulation and integration of an increasing number of so called 'narrow AI' applications (a la K. Kelly) that is already happening today. This does not seem to favor an eventual emergence of a Global Brain where humans play a critical role in managing this planet's affairs. Unless, of course, humans themselves are augmented...

Francis: In the GB, everyone and everything will profit from the synergy with other components of the network. Thus, human intelligence will be augmented by the support it gets from computer intelligence, while computer intelligence will be augmented by the input and feedback it gets from human intelligence.

Weaver: I would like to ask therefore what are the arguments in support of the hypothesis that achieving a Global Brain based on the convergence of humans and machines is much easier than achieving an AGI.

Francis: I hope the above comments answer your question.

Weaver: For the latter we at least have a beginning of an idea about the challenges and difficulties.

Francis: But we have an equally good beginning of an idea about the challenges of augmenting distributed cognition and collective intelligence. This insight may not yet have been formulated so explicitly as in the many papers covering AGI, but it has shown its practical value in numerous systems enhancing collective intelligence on the Internet. Moreover, since the process is self-organizing (unlike the development of an AGI program), we don't even need to understand very well the conceptual foundations for the development of a GB: like the "invisible hand" of the market or the natural selection of evolution, it will work even without any theorist specifying how it is supposed to work...

REFERENCES

Arel, I. (2012). The threat of a reward-driven adversarial artificial general intelligence. in: Eden et al. (eds.), *Singularity Hypotheses* (pp. 43–60). Springer.

Heylighen, F. (1999). Advantages and limitations of formal expression. *Foundations of Science*, 4(1), 25-56. **http://pcp.vub.ac.be/papers/Formality-FOS.pdf**

Heylighen, F. (2013). Self-organization in Communicating Groups: the emergence of coordination, shared references and collective intelligence. In À. Massip-Bonet & A. Bastardas-Boada (Eds.), *Complexity Perspectives on Language, Communication and Society* (pp. 117-149). Berlin, Germany: Springer. http://pcp.vub.ac.be/Papers/Barcelona-LanguageSO.pdf)

Mueller, J. (2009). War has almost ceased to exist: An assessment. *Political Science Quarterly*, 124(2), 297–321.

End of the Beginning

Chapter Eleven **Distributing Cognition: From Local Brains to the Global Brain**

By Clément Vidal

Clément Vidal is a philosopher from the Free University of Brussels (VUB), with a background in logic and cognitive sciences. He is co-director of the 'Evo Devo Universe' community and founder of the 'High Energy Astrobiology' prize. To satisfy his intellectual curiosity when facing the big questions, he brings together many areas of knowledge such as cosmology, physics, astrobiology, complexity science, evolutionary theory and philosophy of science. He can be reached at: **Clément.vidal@philosophons.com**

Abstract:

We show how the externalization of our local brain functions is leading to a planetary level intelligence, or global brain. We argue that this mechanism of externalizing cognitive functions is a fundamental driver towards an ever-smarter human-machine symbiosis. We discuss implications and applications of externalizing various cognitive functions such as memory, computation, hearing, vision, brainstorming, reasoning, navigation, emotions and actions. We illustrate the scenario with a fictional story of a day in year 2060 (an Epilogue to this Chapter.) We then take a top-down approach, and argue that this progressive externalization helps to better understand, foresee and facilitate the emergence of a globally distributed intelligence, best conceptualized as a global brain. We discuss possible symbioses between biology and machines, and what would be the elementary elements composing the global brain. We finally embrace a broader cosmic perspective and argue that even if the singularity is near, an energetic bottleneck is nearer. We suggest that other extraterrestrial global brains in the universe might have successfully dealt with this energetic bottleneck.

A distinctive feature of the human species is its ability to create, use and refine tools. Doing so, we dramatically extend and enhance the variety and power of our senses, actions and information processing (see e.g. Clark 2008). The externalization of cognitive functions has produced major transitions in the history of culture. For example, the externalization of memory is the invention of writing, while the externalization of calculus is the invention of computing machines.

Can we foresee other externalizations and their impact? What will the ongoing externalization of cognitive functions ultimately bring? Externalizing cognition first empowers the self, but also disruptively changes society in a second stage, as the examples of writing or computers clearly show.

My goal is twofold. First, to use the externalization of cognition as a bottom-up framework to systematically anticipate and explore potential new technologies. Second, to discuss top-down approaches to the idea of an emerging globally distributed intelligence, or "global brain" (see e.g. Heylighen 2007a).

The global brain vision has its roots in organic conceptions of societies, and is in fact focused on the analogy between a nervous system and information processing in our global society (see Heylighen 2011a for the history of the concept). The focus of this vision is thus on *information* and cognition. But the distribution of *matter* and *energy* are equally important to study – although not the

End of the Beginning

main focus of this paper. For example, we could imagine a global distribution of matter, or "matternet". It would operate a global distribution of raw materials, while tridimensional printers (see e.g. Lipson and Kurman 2013) would fabricate any kind of object locally. In a similar vein, the global distribution of energy, or the emergence of an "enernet" would lead energy consumers to become also local energy producers. Energy would thus become distributed in a bottom-up instead of a top-down fashion. From a systems point of view, I just would like to remind my readers that a "matternet" or an "enernet" are as important to take into consideration as the internet to understand and anticipate the future of our society.

To characterize Internet intelligence, a systems theoretic approach is also required. We can distinguish four stages of information processing (Heylighen 2013): input, processing, output and feedback. For a system to be intelligent, it must perform these stages as well as possible. The stages can be summarized as follows:

1. *Input* is intelligence as information collection
2. *Processing* is intelligence as interpretation and problem solving
3. *Output* is intelligence as efficient and relevant action
4. *Feedback* is intelligence as capacity to learn

An intelligent agent is able to collect relevant information (step 1). Generally, the more information, the better. However, it should also be able to process it in real time, or to store it. If you are given a book printed with the raw data of the human genome, there is little to do with it. So, processing information is necessary to make sense of the information gathered (step 2), and to use it for action (step 3). Finally, feedback is essential to learn (step 4) and therefore to fine-tune inputs, processing and outputs. Heylighen adds that each of "these functions can be performed in principle either in a distributed or in a centralized manner." This distributed versus centralized distinction is important to keep in mind in the context of our information society, where the distributed manner becomes more and more prevalent.

In Section 1, we describe how the externalization of cognition brings new levels of intelligence. In Section 2, we step back and discuss the global brain vision. We ask what elementary parts will constitute the global brain. Humans? Machines? A symbiosis between the two? A more complex ecosystem? Finally, we take a cosmic perspective in Section 3, and show that if the singularity is near (Kurzweil 2005), an energetic bottleneck is even nearer! We speculate that extraterrestrial civilizations might have already overcome this bottleneck. The combination of both bottom-up and top-down analyses is insightful because it connects the dots between our present, our near future towards a global brain, and our far-future beyond the technological singularity. As an appendix to this paper, the **Epilogue** explores a fictional story of what a day would be like in 2060, with externalized technologies.

1. Local Brains or the Externalized Parts

Most tools derive from the externalization of our bodily functions. They concern primarily matter-energy processes. For example, wetsuits or spacesuits are enhancements of our skin; horses, cars, trains or planes are locomotion systems which extent our motor functions. Here, we focus on information processing technologies that enhance cognitive functions. As Dror and Harnad (2008, 3) argued, cognitive technology does "extend the scope and power of cognition, exactly as sensory and motor technology extends the scope and power of the bodily senses and movement." They add that just "as we can see further with telescopes, move faster with cars, and do more with laser microsurgery than we can do with just our unaided hands and heads, so we can think faster and further, and do more, with language, books, calculators, computers, the web, algorithms, software agents, plus whatever is in the heads of other cognizers."

This approach emphasizes more *Intelligence Augmentation* (IA), than *Artificial Intelligence* (AI) (Fischer 2006). As we will see, this IA perspective naturally accounts for present Information and Communication Technologies (ICTs), and promises to anticipate future techno-sociological development. In the cognitive science landscape, we are thus naturally in line with the tradition of situated and embodied cognition (see e.g. Wilson 2002; Anderson 2003; Clark 2008).

In the following analysis, we distinguish three phases of the externalization process. First, the *bodily self* with no extension; then the *extended self*, using tools and the environment to enhance its capabilities, and finally the *networked self* when it is connected to the internet. The *bodily self* performs information processing on a very local scale. It collects internal or local inputs, does centralized internal processing, produces local outputs and relies on local changes to tie his feedback loops.

The *extended self* is much smarter. It extends its inputs with the help of tools, such as goggles, its processing by using other tools such as computing machines, its output by using tools such as hammers and shoves, language, pen and paper. The feedback loops operate through these extensions, which can be repaired and enhanced. For example, an author will do her best to repair or replace her failing favorite pen or computer, which are almost part of her. Language itself can already be analyzed as an extension of communication, operating not anymore on the individual bodily level, with hormones or neurotransmitters, but across a social group. This social aspect is essential because cognitive tools can be shared. For example, thanks to teachers, our social partners, we learn at school how to speak, write and calculate. Yet, input, processing, output and feedback remain local, in the sense that it is not possible to listen or speak to thousand people at the same time, nor to learn from the experience of millions of people.

The *networked self* is much smarter thanks to its connection to the Internet. The scale of its input, processing, output and feedback is global. It can communicate with any other networked self on the planet. The benefit stems not only through connecting human brains together, but also connecting humans to machines, i.e. *artificial agents,* which can be software or hardware agents. For example, Internet users rely daily on smart algorithms and big databases to search or filter the available information on the web. I will argue that the symbiosis between social interactions and artificial agents provides and will continue to provide the most useful and disruptive technologies. The extension of cognition started as an *individual* intellectual empowerment (e.g. with pen and paper). But such an extension also has a *sociological* impact, since it opens the way to share tools and information with social partners, resulting in faster and wider communication on a social scale. For example, the invention of printing lead to the massive distribution of information via books and newspapers. In the end, we interpret the singularity as a symbiosis between humans and machines leading to a major evolutionary transition, the emergence of a globally distributed intelligence, or *global brain* (see Section 2).

Let us now explore more specific implications of externalizing various cognitive functions such as memory, computation, hearing, vision, brainstorming, reasoning, navigation, emotions and actions. Importantly, since memory, bandwidth and processing increase superexponentially (Nagy et al. 2011), we should not consider them as bottlenecks in imagining future technological applications.

Memory is implemented biologically with neural networks. It became sharable with language, and has been radically extended and safeguarded thanks to the invention of writing. This allowed the sharing of manuscripts, first locally, then more broadly with the invention of printing. For a detailed account of the disruptive nature of the inventions of language, writing and printing, see (Robertson 1998). The third phase of the externalization is the distribution of memory over the Internet. The invention of hypertext, and later of the web is an improvement on the invention of writing, and can be analyzed as a globally distributed, collective and dynamical memory. To invent the web, Tim Berners-Lee (1999, 4 and 41) was indeed inspired by the functioning of the brain,

End of the Beginning

which can link any arbitrary piece of information. He wanted computers to be able to do the same.

Let us illustrate this point with the collaborative Wikipedia encyclopedia. First, a shared memory obviously accelerates human-human interactions because of the availability of shared information in the environment. Wikipedia editors share and improve constantly the external memory via a *stigmergic* process (Heylighen 2007b). Second, human-machine symbiosis is possible when humans enter data and metadata in a semi-structured or structured way. Such data can then be used to query relationships and properties, including links to other related datasets. More complex or extensive inferences and queries can be processed thanks to software agents, which will increase human knowledge. In turn, humans can use software agents to improve the datasets and establish a human-machine symbiosis. Wikipedia bots are examples of software agents. More than 1600 of them operating as you read these lines. Bots most often leave tags or flags on Wikipedia pages, instead of directly and aggressively editing a page. Such a mild contribution is still very helpful, since it can stimulate or nudge humans to work on specific pages such as the ones having a very active "talk" page, or which are running an "edit war" (see e.g. Kittur et al. 2007). However, machine-machine interaction is also possible, if bots become intelligent enough to make reliable edits, further edited by other bots.

Computation, or counting and calculating, probably started by interacting with the environment. For example, if a farmer wants to count its small livestock, it is easier to establish a one-to-one correspondence between stones and animals than to count mentally. This correspondence procedure is actually the modern definition of cardinal numbers in set theory. From a situated and embodied cognition viewpoint, this means that environmental scaffoldings to cognize (such as stones) can precede their integration into the bodily self. Combined with the inventions of writing, calculus took another leap. Human computation is severely limited by the working memory capacity of the brain. This is why mathematicians extend their cognition with their favorite tools: paper and pen. Later, the specification of the general purpose computer stemming from the work of Church, Gödel, Kleene and Turing led to the general computing devices we use daily. The extension of computation to the Internet leads to distributed computation and grid computing.

The hearing function starts with the biological ear, is captured thanks to microphones on physical supports such as tapes, compact discs or memory cards. Thanks to big data and pattern recognition algorithms, automatic music identification services have emerged. They are portable thanks to their integration to smartphones connected to the Internet. Their accuracy in recognizing songs is already quite impressive. However, with a hybrid approach of combining algorithms and crowdsourcing to humans, the music recognition could give even better results. For example, humans would be able to guess live music which has never been recorded, where algorithms would typically fail.

Vision is biologically implemented with the eye, but visual information can be stored thanks to cameras, technologically extended to scales with microscopes and telescopes, and extended from the visible to the whole of the electromagnetic spectrum. When vision is networked from smartphones to crowdworkers, such a connection allows new applications. For example, it is possible to get nearly real-time crowd sourced answers to visual questions, which is extremely helpful for visually impaired persons (Bigham et al. 2010). Note that such pattern-recognition tasks are offloaded to other humans, but they could also be offloaded to artificial agents. For example, some search engines allow searching not only by keywords, but also by pictures freshly taken. Finally, a mixed approach could prove even smarter. It would first query artificial agents, and then, if they fail or if their confidence in the results is low, crowd work would be requested. This would be most efficient because mobilizing humans for a

End of the Beginning

simple task that an artificial agent could perform would be a waste of human workforce.

The faculty of *divergent thinking* is essential to cultivate creativity. If we are unaided by external cognitive scaffoldings, divergent thinking simply means having a lot of ideas. But most of the time, we have a lot of ideas thanks to triggering lists or brainstorming sessions with cognitive partners. And of course, we don't want to lose such ideas, so collection tools such as notes, mind maps, or voice recordings are essential in the process. With the Internet and the web, it is possible to radically scale up this creative process. Indeed, we can now seek input from a large number of people (e.g. with wikis or with social media websites). The difficulties shift, namely to the aggregation and sorting of the results to make a coherent whole. There are a few collective outlining or mind mapping softwares, but they are probably underused given the benefits they could provide. Software agents could also help divergent thinking, for example by automatically aggregating similar ideas, or even by systematically suggesting combinations of existing ideas. For example, software could simply take inputs from humans, such as idea A and idea B, and ask: "What would happen if you would combine A with B?"

The faculty of *convergent thinking* enables humans to produce arguments that solve problems. It goes hand in hand with divergent thinking, but this time the goal is not variation of ideas, but selection of ideas, to solve a particular problem. Extending reasoning to the environment leads to the use of logical diagrams (e.g. with the theory of constraints of Goldratt and Cox 1984), thus providing greater control over problem solving and argumentation. The reasoning then becomes easier to criticize, revise, understand and discuss. Such techniques are widely used in finance, distribution, project management, people management, strategy, sales and marketing. Distributing reasoning on the Internet is still in its infancy, but holds great promises for the future of distributed governance, since it has the potential to lead to large-scale decision making (see e.g. the projects of Baldwin and Price 2008; as well as Iandoli, Klein, and Zollo 2007). Artificial software agents could help semi-automatic reasoning by helping users and groups to systematically and critically question an argumentation. More sophisticated agents could also extract reasonings from existing text, or even reason on their own (e.g. with logical programming). Again, a mixed approach of computers suggesting inferences, and asking for feedback from humans seems more practically useful than pure artificial agents whose reasoning would be limited by the structured data it can gather. It could lead to a real symbiosis, where algorithms would learn from the feedback of humans, and humans would benefit from logical connections suggested by software agents that they might otherwise have missed.

Navigation is hard to operate in an unknown environment. Maps make navigation much easier. Helped with machines and a Global Positioning System using a voice synthesizer, the task of navigation becomes incredibly simple. The connection to the internet can provide additional real-time information, can allow users to correct maps when they find errors, and leads to an increasing variety of location-based services and applications.

Emotions play an important role in cognition. The subject matter of affective science is to study motives, attitudes, moods, emotions, and their effect in personal and collective behavior. Expressing emotions started thanks to body and facial language. It then extended to sounds and articulated language. Emotionally intelligent people are aware and tuned-in with their emotions. They are able to communicate them in a constructive and non-violent way (see e.g. Rosenberg 2003). Emotional intelligence also includes empathy, or the ability to recognize emotions in others and to react appropriately. Cognitive reappraisal is a technique to regulate emotions, where the patient uses another person's insight to change perspective on a negative situation. Morris and Picard (2012) have shown the feasibility of crowdsourcing cognitive

End of the Beginning

reappraisals. Other applications could include the crowdsourcing of empathy, which would be most helpful for autistic people. Of course, in many such applications of distributing cognition to the Internet, the rapidity of the response is critical to make the tool valuable. If it takes five minutes for an autistic person to take a picture of another person, send it to a server of crowdworkers, wait for a response, and wait for the result to realize that the person was very angry, it would be socially useless. But if the autistic person wears augmented reality goggles or contact lenses, which recognize in real-time faces and their emotions to feed them back as explicit signals, then the tool becomes extremely helpful. Such a scenario lies in a very near future, since we have already algorithm that recognize not only faces, but also human emotions. As a day-to-day example, modern digital cameras have smile recognition features, to take pictures only when subjects smile. There are many more applications yet to be discovered, but there is no doubt that the field of *affective computing* (see e.g. Picard 1997 for a review of the theory and many applications) holds great promise to develop better human-computer interaction and symbiosis.

Let us further explore future scenarios where affective computing plays a disruptive role. A key element to make the human-machine symbiosis functional is to have high bandwidth of interaction. For a smartphone user, taking a picture still takes the time to decide to take the picture, to put one's hand in the pocket, to perform a few swipes, wait for an autofocus to happen, and to click the shutter button. This is extremely quick compared to what photography was in the 1820's, but extremely slow compared to what it could be. Imagine you are wearing augmented reality glasses. You blink your right eye, done, your picture is ready to be shared – if you wish to. Better, you don't even take the decision to take a picture. The computer you wear monitors your emotions in real time, and take pictures or videos automatically as you live your life. Later, you can consult this log, and filter it by the most surprising, enjoyable or disgusting experiences you had. How can

End of the Beginning

your emotions be monitored? This is a technical problem, but emotion sensors can already perform facial and emotional recognition. Other sensors such as brain waves, heart rate, skin connectivity or even real-time blood analysis (Golan et al. 2012) could be used.

Let's take another example of affective feedback applied to software development. You use your favorite software, but can't find the functionalities you need or you experience unwanted bugs. The regular way to complain is to write or phone the technical support, or to register to the relevant forum and post a message. But how many users do actually do it systematically? Now imagine that your webcam records and recognizes your feelings while you are using the software. This data is sent anonymously to the developers along with a log file describing what you were doing. The developer could then analyze this data, why the user was unpleased, and improve the software accordingly.

Now, let us imagine how urban traffic could be improved thanks to affective computing. The costs of road crashes are very high, more than US$517 billion (Jacobs, Aeron-Thomas, and Astrop 2000, 11), without counting the pain, grief and suffering of people involved. Let's imagine that most pedestrians, cyclists, car drivers have emotion sensors, linked to their location information. Suddenly, a car makes a huge acceleration, a lot of noise, without any other reason than showing-off. A dozen drivers, pedestrians and bikers get stressed out. This emotional feedback is anonymized and could be sent directly to the dangerous driver, or processed by the police. The driver would lose points on his driving license, or incur increased premiums on his insurance. On a nearby road, a driver nicely anticipates the coming of a mother with her young children on a pedestrian crossing. They smile at each other, the driver pleased to respect the safety of pedestrians, the mother pleased with the behavior of the driver. Remember that the emotions are recorded, and the driver would thus collect

336

End of the Beginning

bonuses thanks to his cordial behavior. Another pedestrian witnesses the scene, and further confirms this cordial behavior. Now, imagine that such a system is implemented on a large scale. The collection of bonuses could be incentivized in many ways, and would certainly result in changing the behavior of drivers. The collection of penalty points could also be sanctioned in many different ways. The rules could be even stricter for self-driving cars, for example requiring them to ensure minimum fuel consumption as well as maximum comfort for passengers, by accelerating and braking smoothly.

Our ability to act can be enhanced by using to-do lists. By writing down next physical actions to do, this process gives us power on our actions because we can then reorganize and reconsider them in a very efficient way (see e.g. Heylighen and Vidal 2008). Collective action can be efficiently implemented with job ticketing systems, as it is practiced in call centers or open-source software development. For example, in open source development, users or developers can create a ticket to signal a bug or request a feature; other users can vote to prioritize jobs. Yet other developers will actually fix the bug or implement the new feature. This process happens in a stigmergic fashion, where there is no central planning or planner. We saw that Wikipedia is a typical example of such stigmergic collaborative work.

Softwares using location information can improve action efficiency by filtering an electronic to do list, for example suggesting you to buy a specific item only when the shop is open, when the item is in stock and when you are near it.

An important development in the distribution of actions is the rise of *microwork*, which consists of tasks completed by many people on the Internet. Examples include Mechanical Turk by Amazon; such globalized work highlights new socio-political issues regarding the difference in wages between developed and developing countries.

The distribution of physical action remains a challenge.

End of the Beginning

Surely, there are microsurgeons operating robotic surgery via the Internet (Anvari, McKinley, and Stein 2005). But other applications such as telepresence and action at distance remain marginally in use. This is due to the difficulty of building autonomous or semi-autonomous robots with efficient sensory-motor effectors in our real and complex world. Biological organisms are still much more sophisticated and efficient than robots for these real-world tasks.

We have described the externalization of cognitive functions and how they change social interactions. We insisted on the impact of combining this externalization with a global interconnection through the Internet. The most interesting applications will likely be at the interface of several externalizations. Let us give two examples. I have been thinking about an application called *augmented conversation*, only to realize after triggering this thought to our global external memory (the web) that such a project was already in development (Chorley 2013). As Chorley writes, it consists of "a scenario where people are able to take part in a conversation and as the conversation progresses relevant content is retrieved from the web and displayed to the participants." Such features could be enhanced with emotional recognition, for example, searching more actively and thoroughly when participants display surprise or facial expressions indicating disagreement. Emotional tagging could also be helpful for automatically logging a conversation, so that the most impressive, unusual, i.e. informative aspects of a conversation would be highlighted for later reference. As usual, not only artificial agents can be used, but also humans. For example, if everybody agrees, crowdworkers could be paid to listen to conversations and check the accuracy of participants' statements.

The second example is the interweaving of human and computer computation in order to translate quickly, cheaply and reliably. Minder and Bernstein (2012) designed a framework to make the most of human and computer intelligence for this task, and manage to make a good translation of a whole book from

End of the Beginning

German to English in only 4 hours and for about 70$. The process involves first automatic translation of sentences, which is then shown to different humans, who rewrite a better translation. Then other humans vote for the best translations. The process involves quite some redundancy in fixing and verifying the intermediate steps, which lead to a very good final result, considering the time and money spent.

2. The Global Brain or the Emergent Whole

How will the externalized cognitive parts coordinate to form a globally distributed intelligence, or global brain? This is a very difficult question, and it presupposes that we know what we mean with "cognitive parts". A superorganism such as an ant colony or a human body consists of simple elements, ants or cells. By analogy, what will the elementary parts which compose the global brain be? Will the elementary parts be humans? Machines? Biocyborgs? A mix of the three? Or should we focus on more abstract properties, which can be embedded in different substrates? Should the elementary parts be living systems (Miller 1978), performing nineteen different functions on matter, energy and information? How intelligent should the parts be for the global brain to function? Is Artificial General Intelligence (AGI, see e.g. Ben Goertzel and Pennachin 2007) a requirement, or could the different parts be differentiated and locally relatively stupid? In this section, we propose some speculations to explore these questions.

For sure, the parts will not be humans alone, since we are already so much dependent on machines. The framework of externalizing cognition we introduced clearly suggests that we experience a human-machine symbiosis, and that it will continue for a while.

The challenge for machines is to perform better in real-world physical action. Moravec (1993) already saw the respective strengths and weaknesses of humans and machines, when he wrote that in arithmetic, "today's average computers are one million times more powerful than human beings", while "in perception and control of movement in the complex real world, and related areas of common-sense knowledge and intuitive and visual problem solving, today's average computers are a million times less capable." For practical purposes, this means that a human-machine symbiosis is suitable, since humans and machines have complementary strengths and weaknesses. As long as this dichotomy stays, the human-machine symbiosis will be the way for continuing techno-sociological progress. But how long will it last? In the context of space colonization, Hans Moravec (1979) wrote:

> *The first societies in space will be composed of co-operating humans and machines, but as the capabilities of the self-improving machine component grow, the human portion will function more and more as a parasitic drag. Communities with a higher ratio of machines to people will be able to expand faster, and will become the bulk of the intelligent activity in the solar system. In the long run the sheer physical inability of humans to keep up with these rapidly evolving progeny of our minds will ensure that the ratio of people to machines approaches zero, and that a direct descendant of our culture, but not our genes, inherits the universe.*

We can strengthen Moravec's argument even without taking into account the constraint of space colonization. Human bodies may become maladapted and energetically wasteful compared to ever smaller and faster information processing software and hardware. Our present and future technologies are threatening to make humans an endangered species (Joy 2000). In the near future, humans may become like dinosaurs as much as robots would become like mammals, better adapted to a new environment.

But is it really a problem? This perspective hurts our feelings and values because we have not learnt to entertain non-anthropocentric cosmic or universal values such as evolutionary,

End of the Beginning

developmental or thermoethical values (for some attempts, see e.g. Ben Goertzel 2010; Vidal 2014a, chap. 10).

Even if we admit that humans may become unfit, we should not throw out the baby with the bath water and dismiss biology altogether. The possibility of a symbiosis between biology and machines constitutes the field of bionics. We could imagine a world of insect cyborgs, organizing in swarms, globally connected and able to perform any physical task. As a matter of fact, coupled machine-insect interfaces are successfully being developed (see e.g. Bozkurt et al. 2008).

The biocyborg philosophy suggests that it is easier to use and hack existing biological systems rather than to build robots from scratch. Another speculative option is the development of a cyborg bacterial intelligence or "bacteriborg". Extrapolating on the progress of synthetic biology, it actually makes most sense to try to make the most of the largest existing biomass on Earth: bacteria. Indeed, in terms of cells, the bacterial population is estimated at 5×10^{30} (Whitman, Coleman, and Wiebe 1998) while the number of cells of multicellular organisms is negligible in comparison. The idea here would be to genetically re-program bacteria, or to augment bacterial intelligence with nano (or smaller) technology, to transform the Earth into a planetary bacterial superorganism which would prove more adaptive and more intelligent than a network of humans and machines. As strange as it may seem, the incentive to choose this course of action may be purely economic. Why build from scratch (nano) robots able to intelligently communicate and collaborate on a global scale, if the same is doable by hacking readily available bacteria?

Let us now turn to a possible transition to a post biological future. Why would that be advantageous evolutionarily? We already mentioned Moravec's argument that machines would be better adapted for space missions. In particular, they could be built to withstand very high temperatures. I mean hundreds or thousands of degrees, something impossible for life as we know it.

Increasing computational speed also means increasing heat production and temperature. Unless the efficiency of cooling mechanisms finds its Moore's law, computational speed would be limited by temperature. But to understand why this point might be crucial, we need to take an even broader cosmic perspective.

3. A Cosmic Perspective

You cannot understand one scale of system unless you see its role in the next larger scale.

(Odum 2001, 118)

What is the next larger scale on which our planetary society is embedded? It is the cosmos. Following the insight of American ecologist Howard T. Odum, we must understand this larger cosmological context. This is motivated not by the sometimes entertaining activity of fantasizing about our far-future, but by the value of a top-down perspective, which may allow us to gain insight into where the smaller scale progresses are headed to. In Section 1, we analyzed and illustrated the logic of cognitive externalization. This was a bottom up approach. Let us now take a complementary top down cosmological approach, which promises to unveil a general evolutionary dynamic.

The overwhelming trend in technological progress is a *quest for inner space*. Indeed, our abilities to manipulate smaller and smaller entities have led and are leading to major progress in information and communication technologies, biotechnologies and nanotechnologies (Barrow 1998; Vidal 2011; Smart 2012; Vidal 2013; Vidal 2014a). I have dubbed this trend the *Barrow scale* of civilizational development. It adds to the *Kardashev scale* (1964) of energy consumption. If we look at physical constraints, the quest for inner space has no reason to stop at nanotechnologies (10^{-9}m). Indeed, we could in principle develop pico-(10^{-12}m), femto-(10^{-15}m), atto-(10^{-18}m), zepto-(10^{-21}m), yocto-(10^{-24}m) technologies, and even well beyond, down to Planck-technologies (10^{-35}m). The core idea

End of the Beginning

with the quest for inner space is to organize matter at its smallest possible scale, and to compute with the smallest information markers allowed by physics (Bremermann 1982). This is plausible because the material support on which computers operate doesn't need to be fixed. As a matter of fact, it has already changed five times since their invention. Kurzweil (2005, chap. 3) reminds us that computers went through several hardware revolutions, from electromechanical calculators, relay-based computing, vacuum tubes, discrete transistors to today's integrated circuits. So, it seems very likely that material substrate changes will continue at a fast pace. Additionally, material innovations account for about 2/3 of progress in computation (Magee 2012).

Quantum computer scientist Seth Lloyd (2000) explored the limits of computation and described an ultimate computer, working at a quantum scale and using reversible operations. Extrapolating Moore's law, he predicted that we would reach ultimate computation in 250 years –a very unlikely outcome for sure. First, because even such a computer would need to have error-correction mechanisms, which requires 4×10^{26} watts of power for a one kilogram ultimate computer. This amount of energy is more than the total output of the Sun (3.8×10^{26} watts)! So even if we would be able to capture all the energy of the Sun (e.g. with a Dyson (1960) sphere), it would not be enough to power just one kilogram of such an ultimate computer. Furthermore, extrapolating our global energetic consumption, we would reach the ability to harness the whole energetic output of the Sun in about 3000 years (Kardashev 1964). It means that in the best possible scenario, where we keep the pace to follow Moore's law, and where we manage to continually consume exponentially more and more energy, we would still be slowed down by the availability of free energy. Could other intelligent civilizations in the universe have been confronted to these issues, and found a solution?

In recent years, I developed a scenario for the far future of civilizations, extrapolating both the Kardashev scale and the Barrow scale (Vidal 2011). To my own surprise this led to a testable hypothesis, the *starivore hypothesis*, which claims that some known interacting binary systems are actually advanced extraterrestrial life, displaying macroscopic metabolic features (Vidal 2014a, chap. 9; 2013). In the context of our discussion, let us illustrate the putative energy use of a civilization organizing around black holes, and using them as ultimate computers. Concretely, if we take the binary microquasar SS433, it would need about 8×10^{57} watts if all its mass would be used to power an ultimate computer. Astrophysicists come to a close result, where 5.65×10^{55} watts are consumed via accretion (Fabrika, Abolmasov, and Karpov 2006, assuming perfect mass to energy conversion). Of course, this interpretation remains very speculative and difficult to prove, which is why I set up a prize to stimulate proof or disproof of the starivore hypothesis (Vidal 2014a).

4. Conclusion

We explored the consequences of externalizing memory, computation, hearing, vision, brainstorming, reasoning, navigation, emotions and actions. Most importantly, we can foresee that new technologies will emerge as these different externalized functions are combined, and integrated with humans. Starting from known functions of our local brains, the analysis of their progressive externalization helps to better understand, foresee and facilitate the emergence of a globally distributed intelligence, best conceptualized as a global brain. With hindsight we could see the externalization and distribution of cognition as an act of evolutionary generosity, leading to the next level of intelligence on planet Earth. Indeed, we are currently externalizing, sharing and distributing our intelligence.

We critically discussed the future of human-machine symbiosis, to show the possibility of other biology-machine symbioses, such as insect-cyborgs or "bacteriborgs". We showed

End of the Beginning

that the pace of technological progress in computation is much faster than the pace of increasing energy consumption, and concluded that an energetic bottleneck is nearer than a singularity. Externalizing cognition started the transition from human-human interaction to human-machines interactions, but the cosmic perspective also suggest that intelligent life could re-organize matter to perform ultimately efficient intelligent information processing within densities, temperatures and energy scales alien to life as we know it.

End of the Beginning

Epilogue to Chapter Eleven

A Day in 2060
By Clément Vidal

To see more concretely what the impact of such technologies could be, let us imagine what a day circa 2060 could look like. What follows is a speculative science-fiction story, but I will analyze it and give hints and references explaining why it is plausible.

You're walking on the street and meet a Japanese woman. She starts to speak her native language, but your augmented reality lenses automatically translate and display what she says. Akemi explains that her GPS doesn't function well and asks if she can connect to yours to find her way. You accept the request, but your augmented reality vision also displays: "try something". What happened in the background of your extended self in order to suggest this? In a fraction of a second your sensors and artificial agents did the following. They took a picture of Akemi, from which an image-search was launched, along with face-recognition. Several webpages of her public profiles were found. This information was integrated to create a profile, summarizing her professional, and to a lesser extent, personal interests. Additional visual and olfactory sensors on your clothes did notice unusual pupil dilatation and pheromone concentration. Intellectual and bodily data concluded – on both sides, since Akemi did of course do a similar search – that this encounter was an excellent love match. You could have configured your digital agents to give you a better tip than "try something", but you chose a low advice specificity profile, to leave some of life's spontaneity. So, you indeed try something, and invite her to join you and your friends for swimming with dolphins this afternoon. You share time and

End of the Beginning

GPS coordinates and you are thrilled that she accepts the invitation.

You run back home, cross the street without even looking at cars. A car brakes violently. You are surprised to see a driver in it, and shout: "poor and dangerous biological human, buy yourself a self-driving car, your reflexes are too slow!" Your emotional reaction was monitored and the automatic legal decision making actually gives you a one bitcoin fine, because you should not have crossed the street so quickly in the first place, and you should not have had insulted the human driver, having a negative emotional impact on him. Your augmented reality informs you sympathetically: "I understand that you felt upset and need more security. This security indeed implies that people should switch to stronger human-machine symbiosis, but can you please be more careful next time? This transition towards human-machine symbiosis is still in progress. The driver felt embarrassed and miserable about this situation, which is one of the reasons why you had to pay this fine. I don't advise you to make an appeal, it will only cost more money and given that the situation was recorded by 10 different nearby sources, there is little ambiguity, so the judgment has a 99.9% confidence level. The bloodstream of the driver has also been checked and it was perfectly clean, whereas your adrenaline levels were unusually high." You understand this but still wonder why human-driving cars are still allowed to circulate. Probably a result of lobbying by the old-fashioned Association for Biological Human Rights.

When you arrive home, a self-driving car just brought fresh cartridges, automatically ordered by your 3D food printer. As soon as you plug the nutrient cartridges in, your 3D printer cooks for you, based on inputs from nanobots floating in your bloodstream, which monitor the nutrients you need most. Your 3D printer is furthermore configured to follow your preferences, in this case, to follow a paleo diet because you decided to be in line with evolution. The animal protein supply is a mix of artificially grown

End of the Beginning

meat, fish, worms, and insect proteins. The food quantity is also higher than usual, because your printer anticipates your sport activity planned in your agenda. Indeed, you gave access to your agenda to your printer. The recipe is a new creation, because you've configured your printer to never print the same meal twice. Life is too short and the world's diversity of cooking too great to eat the same meal twice.

When you arrive at the harbor, safety procedures are quick and simple. You just deposit your stem-cell box which could be used by the first-aid-kit on the boat. The boat is small, and no oxygen bottles are taken on board. Instead, the trainer takes a suitcase containing syringes. Just before going into the water, the trainer gives a shot to all participants. What is in the shot? Mechanical artificial red cells, providing a four hour *in vivo* Self-Contained Underwater Breathing Apparatus (SCUBA). You and your friends dive in the water, play and communicate with dolphins, thanks to the dolphin speaker interfaced with your augmented-reality diving mask.

Suddenly, the boat radar displays an alert on your mask: "Shark approaching at high speed; no time to swim back to the boat. Fight is the only option". But you use your biological brain and think that there must be another way. You remember that dolphins can sometimes fight a shark. You turn quickly to the dolphins, set your dolphin speaker to beam a help signal, along with the 3D shape of a shark you quickly downloaded. Fortunately the dolphins understand your message, and they do thank you, but they get scared and swim away! The AI advice was wise. You feel frustrated that AI was once again smarter than you.

Out of sea mist, the shape of a shark is coming. Some last minute information is displayed to you and your friends on how to fight a shark. You start to read it, but too late, the shark has chosen to attack you. You see the shark's jaw dramatically expanding and... nothing. You lose consciousness.

You wake up on the boat, fully recovered. Akemi is looking at

348

End of the Beginning

you. You ask her: "what happened?" She explains that your friends managed to scare the shark by fighting him from multiple sides on its gills, and that he finally released you. You ask: "but how come was I not wounded?" Akemi: "You actually almost died! Your nano health bots detected that your right kidney and your liver were critically failing. The message was transmitted to the first-aid kit on the boat, and the 3D organ printer started to differentiate your stem cells and quickly printed two new organs. I contacted a Japanese surgeon expert in organ transfers for an urgent tele-operation. I gave him distant access to the first-aid robotic surgery apparatus, and he could work with the printed organs. I hope you don't mind we chose a human surgeon, we are still not entirely confident about the cheaper fully robotic AI surgery." Your health insurance reckons that the incident could not have been avoided, and financially covers the whole operation. The file is already closed.

You ask: "what about the shark?" Akemi continues: "Since it drank your blood, it will be infected by artificial viruses. I guess you feel resentful, but you know that global eco-regulations forbid reprogramming them at a distance to kill the shark. However, thanks to this artificial virus infection, the shark is now traceable and should not create further incidents to any diver with an augmented-reality diving mask." As you put back on your augmented reality lenses, you look at your information feed, and see that you have been thanked by diving, surfing and fishing associations for successfully tracking an additional shark.

On the way back to the coast, you skim some news and learn that a bomb has exploded at the headquarters of the Association for Biological Human Rights. The police have found out that the bomb was printed directly through the local 3D printer of the association. The cyber-attack left traces distributed around the globe. Police said the identity of the hacker is uncertain, and a debate rages about whether it was triggered by a human being or a coalition of artificial agents. At the end of the day, you realize that

End of the Beginning

AI agents have done much for you today, and are in awe and grateful to them and your friends. You owe them all your life.

Let us now analyze this story and give some hints about why it is plausible, if we extrapolate some existing technologies. Augmented reality contact lenses will surely come, and prototypes are being tested (Lingley et al. 2011). In the story, humans are augmented with sensors, noticing details too hard to consciously perceive, such as the amount of pupil dilation or pheromone concentration. Three-dimensional food printers (Cohen et al. 2009) and biological organs printers (e.g. Mironov et al. 2003) already exist in embryonic forms. Automatic translation is supposed to work very well, and could be made more effective thanks to contextual data inputs. For example, your location on the street makes it likely that the conversation will be about finding your way, and the profile of the person restricts the likely vocabulary usage. Machine-machine interaction occurs when your GPS signal and maps are shared with Akemi's system, or when your artificial agents collaborate to tell you "try something".

Regarding the car incident, the legal system is extremely fast, reliable and efficient, thanks to distributed sensors recording continuously objective data. Deontic logic (see e.g. McNamara 2010) allows in principle to make such artificial legal reasoning. Non-violent communication (Rosenberg 2003) is used by machines to communicate empathically and efficiently with humans. Bitcoin is a distributed and decentralized digital currency which is already in use (Nakamoto 2008).

Humans are supposed to be swarming with nano-robots, which perform all kinds of measurements and enhancements, and which are connected to the Internet. For pioneering work on nanomedicine, see for example (Freitas Jr 1999; 2003). In particular, mechanical artificial red cells were conceived by Freitas (1998). A dolphin speaker has recently been developed (Mishima et al. 2011). Beaming the shape of a shark might be possible, if dolphin's "sono-pictorial" language of communication would be confirmed

350

(Kassewitz 2011). We saw earlier that surgery at distance has already been performed. An accelerated differentiation of stem cells to produce desired cells is fictional, but would be very useful, as shown in this story. Artificial viruses are likely to be used in the future given rapid and promising progress in this area (see e.g. Mastrobattista et al. 2006).

The "Association for Biological Human Rights" is fictional. But it is arguable that an overly strong attachment to humans as a strictly biological (contrasted to technological) species might hinder long-term socio-technological progress.

A 3D printed bomb is also a serious threat, and the security of 3D printers should be a serious concern. The possibility that a coalition of artificial agents could perform such a criminal and symbolic action is a classical theme about human-machine rivalry in fiction. It also raises the following issue: could a swarm of artificial agents have a will and an agenda of its own?

Acknowledgements

I thank Viktoras Veitas and David Weinbaum for helpful critical feedback, as well as Ted and Ben for their challenging questions.

REFERENCES

Anderson, Michael L. 2003. "Embodied Cognition: A Field Guide." *Artificial Intelligence* 149 (1): 91–130. doi:10.1016/S0004-3702(03)00054-7.
Anvari, Mehran, Craig McKinley, and Harvey Stein. 2005. "Establishment of the World's First Telerobotic Remote Surgical Service." *Annals of Surgery* 241 (3): 460–64. doi:10.1097/01.sla.0000154456.69815.ee.
Axelrod, Robert M. 1984. *The Evolution of Cooperation*. New York: Basic Books.
Baldwin, Peter, and David Price. 2008. "Debategraph." http://debategraph.org/.
Barrow, J. D. 1998. *Impossibility: The Limits of Science and the Science of Limits*. Oxford University Press, USA.
Bennett, C.H. 1988. "Logical Depth and Physical Complexity." In *The*

Universal Turing Machine: A Half-Century Survey, edited by R. Herken, 227–57. Oxford University Press.
http://www.research.ibm.com/people/b/bennetc/UTMX.pdf
Berners-Lee, T. 1999. *Weaving the Web*. HarperBusiness New York.
Bigham, J. P, C. Jayant, H. Ji, G. Little, A. Miller, R. C Miller, R. Miller, et al. 2010. "VizWiz: Nearly Real-Time Answers to Visual Questions." In *Proceedings of the 23nd Annual ACM Symposium on User Interface Software and Technology*, 333–42.
Bollen, Johan, Huina Mao, and Xiaojun Zeng. 2011. "Twitter Mood Predicts the Stock Market." *Journal of Computational Science* 2 (1): 1–8. doi:10.1016/j.jocs.2010.12.007.
Bozkurt, A., R. Gilmour, D. Stern, and A. Lal. 2008. "MEMS Based Bioelectronic Neuromuscular Interfaces for Insect Cyborg Flight Control." In *IEEE 21st International Conference on Micro Electro Mechanical Systems, 2008. MEMS 2008*, 160–63. doi:10.1109/MEMSYS.2008.4443617.
Bremermann, Hans J. 1982. "Minimum Energy Requirements of Information Transfer and Computing." *International Journal of Theoretical Physics* 21: 203–17. doi:10.1007/BF01857726.
Carr, Nicholas G. 2011. *The Shallows: What the Internet Is Doing to Our Brains*. New York: W.W. Norton.
Chorley, Martin. 2013. "Augmented Conversation - Summer Project 2011." *Martin Chorley*. Accessed July 9.
http://martinjc.com/research/student-projects/augmented-conversation/.
Clark, Andy. 2008. *Supersizing the Mind: Embodiment, Action, and Cognitive Extension*. Oxford University Press.
Cohen, Daniel L., Jeffrey I. Lipton, Meredith Cutler, Deborah Coulter, Anthony Vesco, and Hod Lipson. 2009. "Hydrocolloid Printing: A Novel Platform for Customized Food Production." In *Solid Freeform Fabrication Symposium*.
http://edge.rit.edu/edge/P10551/public/SFF/SFF%202009%20Proceedings/2009%20SFF%20Papers/2009-71-Cohen.pdf.
DeLong, John P., Jordan G. Okie, Melanie E. Moses, Richard M. Sibly, and James H. Brown. 2010. "Shifts in Metabolic Scaling, Production, and Efficiency Across Major Evolutionary Transitions of Life." *Proceedings of the National Academy of Sciences* 107 (29): 12941–12945. doi:10.1073/pnas.1007783107.
Dror, Itiel E., and Stevan Harnad. 2008. "Offloading Cognition onto

Cognitive Technology." In *Cognition Distributed: How Cognitive Technology Extends Our Minds*, 1–23. http://arxiv.org/abs/0808.3569.
Dyson, F. J. 1960. "Search for Artificial Stellar Sources of Infrared Radiation." *Science* 131 (3414): 1667 –1668. doi:10.1126/science.131.3414.1667.
Fabrika, S. N., P. K. Abolmasov, and S. Karpov. 2006. "The Supercritical Accretion Disk in SS 433 and Ultraluminous X-Ray Sources." *Proceedings of the International Astronomical Union* 2 (S238): 225–28.
Fischer, Gerhard. 2006. "Distributed Intelligence: Extending the Power of the Unaided, Individual Human Mind." In *In Augusto Celentano (Ed.), Proceedings of the Advanced Visual Interfaces (AVI) Conference (pp. 7–14).New*, 23–26. ACM Press. **http://l3d.cs.colorado.edu/~Gerhard/papers/avi-2006.pdf.**
Freitas Jr, R. A. 1998. "Exploratory Design in Medical Nanotechnology: A Mechanical Artificial Red Cell." *Artificial Cells, Blood Substitutes, and Immobilization Biotechnology* 26 (4): 411–30.
— — —. 1999. *Nanomedicine, Volume I: Basic Capabilities*. Georgetown, TX: **Landes Bioscience. http://www.nanomedicine.com/NMI.htm.**
— — —. 2003. *Nanomedecine Volume IIA: Biocompatibility*. Georgetown, TX: **Landes Bioscience. http://www.nanomedicine.com/NMIIA.htm.**
Goertzel, Ben. 2010. *A Cosmist Manifesto: Practical Philosophy for the Posthuman Age*. Humanity+. http://goertzel.org/CosmistManifesto_July2010.pdf.
Goertzel, Ben, and Ted Goertzel. 2014. "Predicting the Age of Post-Human Intelligences" In *The End of the Beginning: Life, Society and Economy on the Brink of the Singularity*, edited by Ben Goertzel and Ted Goertzel.
Goertzel, Ben, and Cassio Pennachin, eds. 2007. *Artificial General Intelligence*. Cognitive Technologies. Berlin ; New York: Springer.
Golan, Lior, Daniella Yeheskely-Hayon, Limor Minai, Eldad J Dann, and Dvir Yelin. 2012. "Noninvasive Imaging of Flowing Blood Cells Using Label-Free Spectrally Encoded Flow Cytometry." *Biomedical Optics Express* 3 (6): 1455–64. doi:10.1364/BOE.3.001455.
Goldratt, E. M., and J. Cox. 1984. *The Goal: A Process of Ongoing Improvement*. 3rd ed. Great Barrington, MA: North River Press.
Helbing, D., S. Bishop, R. Conte, P. Lukowicz, and J. B. McCarthy. 2012. "FuturICT: Participatory Computing to Understand and Manage Our Complex World in a More Sustainable and Resilient Way." *The European Physical Journal Special Topics* 214 (1): 11–39. doi:10.1140/epjst/e2012-01686-

y.
Heylighen, F. 2007a. "The Global Superorganism: An Evolutionary-Cybernetic Model of the Emerging Network Society." *Social Evolution & History* 6 (1): 58–119. doi:10.1.1.43.3443.
— — —. 2007b. "Accelerating Socio-Technological Evolution: From Ephemeralization and Stigmergy to the Global Brain." In *Globalization as an Evolutionary Process: Modeling Global Change*, edited by George Modelski, Tessaleno Devezas, and William Thompson, Routledge, 286–335. London.
http://pespmc1.vub.ac.be/Papers/AcceleratingEvolution.pdf.
— — —. 2011a. "Conceptions of a Global Brain: An Historical Review." In *Evolution: Cosmic, Biological, and Social*, Eds. Grinin, L. E., Carneiro, R. L., Korotayev A. V., Spier F., 274 – 289. Uchitel Publishing.
http://pcp.vub.ac.be/papers/GBconceptions.pdf.
— — —. 2011b. Francis Heylighen on the Emerging Global Brain. Interview by B. Goertzel. H+ Magazine.
http://hplusmagazine.com/2011/03/16/francis-heylighen-on-the-emerging-global-brain/.
— — —. 2013. *Distributed Intelligence Technologies: A Survey of Future Applications and Implications of the Global Brain*. Global Brain Institute, Working Paper.
Heylighen, F., and C. Vidal. 2008. "Getting Things Done: The Science behind Stress-Free Productivity." *Long Range Planning* 41 (6): 585–605. doi:10.1016/j.lrp.2008.09.004. http://cogprints.org/5904/
Iandoli, Luca, Mark Klein, and Giuseppe Zollo. 2007. *Can We Exploit Collective Intelligence for Collaborative Deliberation? The Case of the Climate Change Collaboratorium*. MIT Sloan Research Paper No. 4675-08.
http://papers.ssrn.com/sol3/papers.cfm?abstract_id=1084069.
Jacobs, Goff, Amy Aeron-Thomas, and Angela Astrop. 2000. *Estimating Global Road Fatalities*. Transport Research Laboratory Crowthorne.
http://ministryofsafety.files.wordpress.com/2011/01/estimating-global-road-fatalities-full-report1.pdf.
Joy, Bill. 2000. "Why the Future Doesn't Need Us." In *Nanoethics–the Ethical and Social Implications of Nanotechnology*, 17–39.
http://www.aaas.org/spp/rd/ch3.
Kardashev, N. S. 1964. "Transmission of Information by Extraterrestrial Civilizations." *Soviet Astronomy* 8 (2): 217–20.
http://adsabs.harvard.edu/abs/1964SvA.....8..217K.

— — —. 1978. "On Strategy in the Search for Extraterrestrial Civilizations." *Russian Social Science Review* 19 (4): 27–47. doi:10.2753/RSS1061-1428190427.

Kassewitz, Jack. 2011. "We Are Not Alone: The Discovery of Dolphin Language." **http://www.speakdolphin.com/ResearchItems.cfm?ID=20**.

Kittur, Aniket, Bongwon Suh, Bryan A. Pendleton, and Ed H. Chi. 2007. "He Says, She Says: Conflict and Coordination in Wikipedia." In *Proceedings of the SIGCHI Conference on Human Factors in Computing Systems*, 453–62. CHI '07. New York, NY, USA: ACM. doi:10.1145/1240624.1240698.

Kurzweil, R. 2005. *The Singularity Is Near: When Humans Transcend Biology*. Penguin Books.

Li, Ming, and P. M. B. Vitányi. 1997. *An Introduction to Kolmogorov Complexity and Its Applications*. Springer.

Lingley, A R, M Ali, Y Liao, R Mirjalili, M Klonner, M Sopanen, S Suihkonen, et al. 2011. "A Single-Pixel Wireless Contact Lens Display." *Journal of Micromechanics and Microengineering* 21 (December): 125014. doi:10.1088/0960-1317/21/12/125014.

Lipson, Hod, and Melba Kurman. 2013. *Fabricated: The New World of 3D Printing*. 1st ed. Wiley.

Lloyd, S. 2000. "Ultimate Physical Limits to Computation." *Nature* 406: 1047–54. http://www.hep.princeton.edu/~mcdonald/examples/QM/lloyd_nature_4 06_1047_00.pdf.

Magee, Christopher L. 2012. "Towards Quantification of the Role of Materials Innovation in Overall Technological Development." *Complexity* 18 (1): 10–25. doi:10.1002/cplx.20309.

Magee, Christopher L., and Tessaleno C. Devezas. 2011. "How Many Singularities Are near and How Will They Disrupt Human History?" *Technological Forecasting and Social Change* 78 (8): 1365–78. doi:10.1016/j.techfore.2011.07.013.

Mastrobattista, Enrico, Marieke A. E. M. van der Aa, Wim E. Hennink, and Daan J. A. Crommelin. 2006. "Artificial Viruses: A Nanotechnological Approach to Gene Delivery." *Nature Reviews Drug Discovery* 5 (2): 115–21. doi:10.1038/nrd1960.

McNamara, Paul. 2010. "Deontic Logic." In *The Stanford Encyclopedia of Philosophy*, edited by Edward N. Zalta, Fall 2010. **http://plato.stanford.edu/archives/fall2010/entries/logic-deontic/**.

Miller, J. G. 1978. *Living Systems*. McGraw-Hill New York.

Mills, M. P. 2013. *The Cloud Begins with Coal: Big Data, Big Networks, Big Infrastructure, and Big Power*. National Mining Association & American Coalition for clean Coal Electricity. http://write.americaspower.org/sites/default/files/The%20Cloud%20Begins%20with%20Coal%20Full%20Report.pdf.

Minder, Patrick, and Abraham Bernstein. 2012. "CrowdLang: A Programming Language for the Systematic Exploration of Human Computation Systems." In *Social Informatics*, edited by Karl Aberer, Andreas Flache, Wander Jager, Ling Liu, Jie Tang, and Christophe Guéret, 124–37. Lecture Notes in Computer Science 7710. Springer Berlin Heidelberg. http://link.springer.com/chapter/10.1007/978-3-642-35386-4_10.

Mironov, Vladimir, Thomas Boland, Thomas Trusk, Gabor Forgacs, and Roger R. Markwald. 2003. "Organ Printing: Computer-Aided Jet-Based 3D Tissue Engineering." *Trends in Biotechnology* 21 (4): 157–61. doi:10.1016/S0167-7799(03)00033-7.

Mishima, Yuka, Keiichi Uchida, Kazuo Amakasu, Yoshinori Miyamoto, and Toyoki Sasakura. 2011. "Development of Dolphin-Speaker." *The Journal of the Acoustical Society of America* 130 (4): 2358. doi:10.1121/1.3654450.

Moravec, Hans. 1979. "Today's Computers, Intelligent Machines and Our Future." *Analog* 99 (2): 59–84. http://www.frc.ri.cmu.edu/~hpm/project.archive/general.articles/1978/analog.1978.html.

———. 1993. "The Universal Robot." In pp. 35–41, http://adsabs.harvard.edu/abs/1993vise.nasa...35M.

Morris, Robert R, and Rosalind Picard. 2012. "Crowdsourcing Collective Emotional Intelligence." In Cambridge, MA. http://arxiv.org/abs/1204.3481.

Nagy, Béla, J. Doyne Farmer, Jessika E. Trancik, and John Paul Gonzales. 2011. "Superexponential Long-Term Trends in Information Technology." *Technological Forecasting and Social Change* 78 (8): 1356–64. doi:10.1016/j.techfore.2011.07.006.

Nakamoto, Satoshi. 2008. *Bitcoin: A Peer-to-Peer Electronic Cash System*. http://bitcoin.org/bitcoin.pdf.

Odum, H. T. 2001. Interview of Dr. Howard T. Odum Interview by Cynthia Barnett.

Picard, Rosalind W. 1997. *Affective Computing*. Cambridge, Mass: MIT

Press.

Pinker, Steven. 2011. *The Better Angels of Our Nature: Why Violence Has Declined*. Viking Adult.

Raichle, Marcus E., and Debra A. Gusnard. 2002. "Appraising the Brain's Energy Budget." *Proceedings of the National Academy of Sciences* 99 (16): 10237–39. doi:10.1073/pnas.172399499.

Robertson, Douglas S. 1998. *The New Renaissance Computers and the next Level of Civilization*. New York: Oxford University Press.

Rosenberg, Marshall B. 2003. *Nonviolent Communication: A Language of Life*. Second edition. Puddledancer Press.

Smart, J. 2008. "Evo Devo Universe? A Framework for Speculations on Cosmic Culture." In *Cosmos and Culture*, edited by S. J. Dick. To appear. http://accelerating.org/downloads/SmartEvoDevoUniv2008.pdf.

— — —. 2012. "The Transcension Hypothesis: Sufficiently Advanced Civilizations Invariably Leave Our Universe, and Implications for Meti and Seti." *Acta Astronautica*, no. 0. doi:10.1016/j.actaastro.2011.11.006.

Vidal, C. 2010. "Introduction to the Special Issue on the Evolution and Development of the Universe." *Foundations of Science* 15 (2): 95–99. doi:10.1007/s10699-010-9176-9.

— — —. 2011. "Black Holes: Attractors for Intelligence?" In Buckinghamshire, Kavli Royal Society International Centre. http://arxiv.org/abs/1104.4362.

— — —. 2013. "Starivore Extraterrestrials? Interacting Binary Stars as Macroscopic Metabolic Systems." *Working Paper*. http://student.vub.ac.be/~clvidal/writings/Vidal-Starivore-Binary.pdf.

— — —. 2014a. *The Beginning and the End: The Meaning of Life in a Cosmological Perspective*. 2014th ed. New York: Springer http://arxiv.org/abs/1301.1648.

— — —. 2014b. "Cosmological Immortality: How to Eliminate Aging on a Universal Scale." *Current Aging Science*. doi:10.2174/18746098076661405211111107. http://student.vub.ac.be/~clvidal/writings/Vidal-Cosmological-Immortality.pdf

Vidal, C., F. Heylighen, and D. Weinbaum. 2013. "Evolutionary Psychology of the Global Brain and FuturICT". European Conference on Complex Systems presented at the Satellite Meeting "Global Computing for our Complex Hyper-connected World," Barcelona, September 19. http://prezi.com/pv9t6-ndmyy6/eccs-2013-evolutionary-psychology-of-the-global-brain-and-futurict/.

Whitman, William B., David C. Coleman, and William J. Wiebe. 1998. "Prokaryotes: The Unseen Majority." *Proceedings of the National Academy of Sciences* 95 (12): 6578–83. **http://www.pnas.org/content/95/12/6578**.

Wilson, Margaret. 2002. "Six Views of Embodied Cognition." *Psychonomic Bulletin & Review* 9 (4): 625–36. doi:10.3758/BF03196322.

Young, Kimberly S. 1998. "Internet Addiction: The Emergence of a New Clinical Disorder." *CyberPsychology & Behavior* 1 (3): 237–44. doi:10.1089/cpb.1998.1.237.

End of the Beginning

Dialogue 11.1

With Clément Vidal, Ben and Ted Goertzel

Ben Goertzel: First off, as I've done with some other authors, I have to ask you: How do you react to Hugo de Garis's prediction of an "artilect war" type scenario, in which some powerful forces see the coming obsolescence of the legacy human condition as a threat, and seek to stop it from happening via force? What if some major governments are taken over by humans who are highly threatened by the various amazing developments you depict? Then, could one perhaps have an armed conflict between nations, reflecting the "Terran vs. Cosmist" dynamic that de Garis proposes? If you think this is implausible, could you articulate why?

Clément Vidal: In my science-fiction story, I mentioned a possible conflict between the "Association for Biological Human Rights" and maybe AI agents. I want to stress that it was only for narrative purposes, as I think such conflicts are implausible, or at worst would not last long.

I do agree with much of the long-term outcome predicted by de Garis, namely, a future where cyborgists and then cosmists become prevalent. I do also agree with de Garis's vision of hyper artilects, which fits well with the starivore hypothesis. But I fail to see any good argument showing how and why a war would play a role towards this future. Let me outline why I think such a war is implausible.

First, thinking about war presupposes a win-lose or lose-lose mental attitude. At work and in life, I focus on a win-win attitude, from which creative problem solving springs. So, de Garis's war prediction is indeed diametrically opposed to my synergetic scenario of increasing win-win interactions between humans, and artificial software or hardware agents. As the poet William Stafford famously wrote, "every war has two losers". This is not only poetry, but also a well-known result of evolutionary game-theory. The winner might benefit temporarily, but agents who make war weaken themselves at best, or simply annihilate each other. In the long run, surviving agents are the ones who reciprocate, cooperate and forgive (Axelrod 1984). So, de Garis's scenario doesn't take into account this evolutionary dynamics.

De Garis' reflections about war are also uninformed by the mechanisms leading to war and by the overwhelming empirical evidence of decreasing violence (Pinker 2011). The war predicted is naive, because our technologies are not in competition with us. Technologies do actions we can't do or we don't want to do. So, they do help us, and technology is useful. Without competition, there is nothing to fear, and no reason to start a war, from either side. To me it is similar to predicting in 1284 that the invention of wearable glasses would lead to a major war. People who don't see well refuse to become cyborgs with glasses because it would denature their true selves. However, people wearing glasses, because of their better sight, become much more efficient in acting in the world. Therefore, before they see too clearly and get all equipped with glasses, people without glasses would attempt to kill them. Seriously?

Another argument that de Garis puts forward is the IQ difference. In contrast to what de Garis argues, the IQ difference will not make a species dominance debate. First, it must be said that the IQ difference is not an absolute order. As Francis Heylighen (2011b) elaborates:

> *Intelligence, like complexity, is at best represented mathematically as a partial order: for two random organisms A and B (say, a hedgehog and a magpie), A may be more intelligent than B, less intelligent, or equally intelligent, but most likely they are simply incomparable. A may be able to solve problems B cannot handle, but B can find solutions that A would not have any clue about.*

End of the Beginning

For such a partial order, it is impossible to develop a quantitative measure such as an IQ, because numbers are by definition fully ordered: either IQ (A) < IQ (B), IQ (A)>IQ (B), or IQ (A)=IQ (B). IQ only works in people because people are pretty similar in the type of problems they can in principle solve.

Let's grant that robots develop up to the point where they become similar enough to humans so that their IQ may be compared. It will be too obvious who is smarter and who will win the debate, like it is obvious that we are smarter than apes. We don't argue nor make war with apes. We don't ask them if they agree that we destroy or protect the forests they live in. We just do it. However, it raises the question of what to do with less dominant species, as with the issues of animal rights. Should we make them disappear? Should we leave them at peace? Should we uplift them to higher levels of intelligence? These are ethical questions, whose alternatives are rich and complex. So, even if there is a degree of superiority, it would not even count as war, because the difference in cognitive and physical capacity would be so great.

I would like also to pinpoint that the distinction between "Terrans" and "Cosmists" is naive. At first sight, it seems like a plausible distinction, but it doesn't resist closer analysis, because the boundary is fuzzy. De Garis seems to put the cyborgists on the side of cosmists. But the matter is much more complicated. Let's say I'm a radical Terran. At which point would I want to kill my grandmother? If she wears glasses to be able to read? If she uses a phone? A smartphone? A pacemaker? Of course, it is easy to see that all the technologies she uses are for her own benefit and do not threaten her. So, it would really make no sense to kill my beloved and augmented grandma. We don't make war against our grandmas who choose not to use smartphones. At worst, we just interact less with them than with our hyperconnected friends. This example shows that there would be no clear opponents, so no clear

End of the Beginning

reason to start a war.

Even granting all the arguments, predicting a war is the worst thing a futurist can do. Contrary to what Prof. de Garis holds, being pessimist or optimist is extremely relevant for futurology. We build the future; we don't just predict it like an apple falling from a tree. Futurists usually [sic!] know the power of self-fulfilling prophecies, and the impact of their models on reality. A self-fulfilling prophecy says that when authorities say something will happen, it happens simply because many people believe it. I think it is the professional duty of futurists to remain optimists. Of course, it doesn't mean we have to be naive optimists, and we still should anticipate and avoid dangers, but these are not the ones we should emphasize if we want to build a better future.

I don't exclude a priori the eventuality of a future war. If there would be a war, I would rather bet on a more traditionally motivated war, where there is a clear fight for resources, or any other geopolitical motivations. Despite all the counter-arguments above, let's assume de Garis is right, and a war is indeed coming. Shouldn't he do whatever he can to avoid it? How can we make it a self-defeating prophecy?

Our short analysis shows that everything goes against de Garis' war scenario. Basic knowledge of evolutionary dynamics and game theory shows that complex systems tend towards more synergetic or win-win interactions, and not win-lose. The empirical evidence strongly supports a decrease of violence, not an increase. The mental attitude behind war is non-creative and non-constructive, and since a futurist also influences the unfolding of the future, it is his duty to emphasize positive futures, and to show how to avoid negative ones.

De Garis' worst case scenario is possible at most, but arguably highly improbable and obviously highly undesirable. There are all reasons to consider de Garis' scenario as material for naive science fiction, but not for professional futurology. To sum up, de Garis' vision can be taken as seriously as blockbuster movies entitled

End of the Beginning

"Smartphones versus Humans" or "Grandmas versus Terminators".

Ted: You suggest that, with advances in technology, humans may become maladapted like dinosaurs that were unable to adapt to compete with mammals. But I understand that the dinosaurs were able to contain the threat from mammals and remain dominant until they were destroyed by a natural disaster, either a comet or asteroid or massive volcanism. Might not humans similarly suppress or control technologies that threaten their dominance on the planet?

Clément: The analogy is interesting, because it's not the mammals that kill the dinosaurs. It's simply that mammals were able to better cope with a catastrophic event - asteroid impact and subsequent volcanism. Dinosaurs never intentionally controlled or contained a mammal threat.

I think it is hard to suppress or control technologies, even if they are threatening us. Let us consider the greatest technological backbone, the physical internet network. Is it even possible to suppress it today? It would be very hard, arguably as hard as suppressing electrical networks or hydraulic networks. More fundamentally, it would be a regression.

Technological progress is an arms race. The countries slowing down or suppressing technological progress for ethical or political reasons will get outcompeted by countries without such restrictions. So, suppressing technologies is certainly not a good strategic move, and the history of technology strongly supports this view. Additionally, it will certainly remain hard if not impossible to reach agreement on the suppression or control of technologies on a global scale.

I would say the smartest way is to accompany technological progress as it appears, with debates and reflections, without slowing it down. You could imagine setting up global ethical institutes, and for some applications, it might be a good thing. But the mission would not be easy to define. What do we want to protect? Should we restrict genetic engineering and cyborgization in the name of preserving the purity of the human species?

Who should become the dominant species? Humanity at all price or the smartest and most adaptive creatures? As we saw, even if we reply with "humanity", it's not clear where we set its boundaries. This is why I think broader evolutionary, developmental and thermodynamic values and ethics are much needed (Vidal 2014a, chap. 10). They open ways to reason ethically without the fallacy of putting humanity at the center.

Ben: In the future you project the nature of intelligence will gradually shift from its current individual mind/body focused state, to a situation where intelligence is more associated with global distributed networks.... What do you think this will do to the experience of being a mind? Modern psychology is highly focused on the ego and the self, on individual ambitions and goals and possessions and desires and dreams. Do you think this will change? If so, what will replace it? What will it be like to be a distributed network mind – or a part of such a highly richly interconnected "mindplex"-- rather than an ego-centered individual?

Clément: There are two different questions you are asking.

1) What is it like *to be part of* the global brain?
2) What is it like *to be* the global brain?

Regarding the first question, I am not a psychologist, but I am sure that the effects of our networked society on our psychology are disruptive. I can only skim over the question in this reply. Nicholas Carr (2011) has explored what the internet does to our brain, although emphasizing more negative side-effects.

End of the Beginning

Psychologists study internet addiction (e.g. Young 1998), which means to me that we need more and more to be always connected. We do already experience our mind as hyperconnected to the internet.

We need to process more and more information every year if not every month or week. This need to deal with more information can be stressful and distracting: too many videos to watch, too many things to read, too many emails to answer to. To cope with information overload, we need to be able to process it quickly and efficiently. Fortunately, there is no obligation to use our biological brains only. If you are a boss, you can distribute or share the workload with peers (humans). But for the rest of us, we can delegate simple tasks to human agents worldwide, by distributing microwork in crowdsourcing internet marketplaces. Maybe the visually impaired persons who use daily microworkers to make sense of their environment consider such workers really as part of themselves?

But we can also use software agents, such as spam email filters. We must augment our processing speed capacity, and we must improve our information filters. Such algorithms are likely to become more and more essential to deal with everyday life, to a point where everybody will have a personal operating system, informed by big data, knowing our values and preferences, and guiding us. This could be implemented as a conversation, like in Spike Jonze's movie "Her". However, the conversational interface is optional and it could simply become a smarter "I", which selects the relevant information.

In developmental psychology, the definition of the self and ego changes as the individual grows from childhood to adulthood. As we continue to grow in our networked society, we care increasingly for global issues, as if they would impact our own local lives. Look at the number of non-governmental organizations. People do care about planet Earth; it's not anymore a New Age trend. We are broadening our circles of compassions to the Earth

End of the Beginning

as a whole, and I find this the most beautiful development of globalization.

As a side note, I see no reason to stop at the boundaries of our tiny planet. I have extrapolated this trend of increasing compassion (Vidal 2014a, sec. 10.3.3; Vidal 2014b), where beings would gradually care for the Earth, the Sun, the Galaxy, the Universe and finally the process of evolution itself.

To speculate about the ego, ambitions, goals, possessions, desires and dreams of the global brain is a stimulating speculative exercise, which would require a textbook entitled "Global Brain Psychology". I will attempt in my answer to Ted's question below to account for the global brain's cognitive development. Let me first say a word about the scientific study of the global brain's subjective experience.

To describe the inner experience of the global brain is nearly impossible, as it is to describe the experience of a tree. Actually, even guessing the inner feelings of other humans, we are often wrong. So, there is no doubt that guessing the feelings of the global brain is orders of magnitude harder.

Yet, even if we give up the idea of trying to "feel the Earth", we can still study it, as we can study trees or human cognition. The fact that we cannot easily project our subjective experience doesn't forbid objective scientific work; it just makes it less intuitive. A typical concrete example is to measure the global mood through Twitter, which can be used, amongst others, to predict the stock market (Bollen, Mao, and Zeng 2011).

Ted: You do not offer a prediction of how soon a globally distributed intelligence, or global brain, will occur. Of course, a degree of globally distributed intelligence already exists, but its capabilities are less, or at least quite different, from those of the human brain. Can you offer some predictions as to how long it will take for a global brain to emerge? Or, even better, how long it will take for key steps towards a full global brain to emerge?

End of the Beginning

Clément: I see the global brain more as an ongoing transition rather than a precise event. As you mention, the transition has started for sure. I am not an empirical futurist, so I apologize that I will not attempt to go into the business of precise quantitative predictions. Quantitative outcomes depend on many variables such as political choices, priorities, resource allocations, or markets. So they may be wrong by a few decades. To illustrate this, I would be open to the possibility that the science-fiction story I proposed, "A day in 2060", could happen in 2030 or 2080. But different quantitative estimates of the singularity do converge, and this is remarkable, because while the estimates start from different assumptions, they converge to a similar critical date (Magee and Devezas 2011). To be on the safe side, I would say that a major transition will have occurred before the end of the century.

That said, I can offer a few reflections on key steps towards a global brain, out of which *qualitative* predictions might be made.

As you note, the intelligence capabilities of the embryonic global brain are different from those of the human brain. This is an important point, because it shows that different kinds of intelligences can co-exist at different scales, and benefit from each other's. However, the analogy of the global brain and evolutionary psychology is a useful framework for thinking and maybe predicting the future of the globally distributed intelligence (Vidal, Heylighen, and Weinbaum 2013).

Although it is an approximation and a simplification, we can think of the global brain's cognitive development in three phases: reptilian, limbic and neocortical.

The reptilian brain reacts and controls primitive bodily functions such as hunger, fear, fight or flight. It connects the nervous system to limbs and triggers instinctual behaviors. At the scale of the global brain, this could correspond to a *planetary nervous system* (Helbing et al. 2012). The main goal here is to deal with global risks, such as tsunamis, epidemics, or volcanic

End of the Beginning

eruptions, thanks to distributed sensors over the globe, connected to effectors, which can be humans or machines.

The limbic system is involved in long-term memory, emotion and motivation. At a global scale, these functions are performed by social networks, which weave a rather emotionally-driven network, and can lead to spontaneous eruptions, such as the Arab Spring revolutions. Wikipedia would also be part of this limbic development, acting as a dynamic and collective memory.

The neocortex is evolutionarily the most recent part of the brain, essential to human language, abstract thinking, imagination or awareness. At the global brain scale, it corresponds to two challenges: building global simulations, and fostering collective reasoning and decision-making. With the help of computer simulations fed with big data, we start to better understand how to tackle global and complex issues. We thus make wiser decisions, and this constitutes a kind of global imagination or dreaming faculty. The process is indeed similar to the faculty of our minds to play possible scenarios, and to dreaming, when it helps to consolidate the most important memories. I mentioned in my paper the importance of collective reasoning through its externalization and distribution. If we use data both from distributed sensors and computer simulations, and add effective methods for collective reasoning and decision making, it is certain that smarter decisions will be made on a global scale.

These three levels constitute a roadmap for the development of the global brain, which maps remarkably well with the three components of the FuturICT project; respectively the *planetary nervous system*, the *living earth simulator* and the *global participatory platform* (Helbing et al. 2012). In my eyes, we are still working on the reptilian and limbic parts. It is an interesting question whether and in how far we can or should develop the three systems sequentially or in parallel.

But the whole is more than the sum of the parts, so we can ask: what would the global brain then want to do, in order to foster its cognitive development? The answer is simply: the same as us, become part of a social group, interact, and learn from others. So, as strange as it may seem, I predict that as the global brain unfolds, it will become more and more pressing to find other global brains in the universe to communicate with and to learn from. To develop the "self" of an intelligent planet, you need to find others with whom you can communicate and with whom you can compare yourself to.

As the co-founder of the Evo Devo Universe community (Smart 2008; Vidal 2010), I believe that in the midst of unpredictable and chaotic evolution, larger developmental trends can be identified in our universe. So I would like to add a few thoughts on this cosmic perspective, which is an important addition next to the five approaches to predicting long-term trends you identified (Ben Goertzel and Ted Goertzel 2014, **Chapter 1** of this volume).

One aspect of this cosmic perspective is precisely astrobiology, or the search for life elsewhere in the universe. As Kardashev (1978) noted, finding extraterrestrial intelligence (ETI) might give us a glimpse of our possible future on astronomical timescales. We would have other examples, that we can choose to follow … or not. Even if we are alone in the universe, or if we don't find anything, the exercise still has immense value because we must project our civilization into the future to search for other advanced ETIs.

In light of the cosmic perspective, there is another clear prediction. *The global brain will be hungry*. This is a prediction of fundamental importance that futurists seem to overlook. While human population is stabilizing, there is a population explosion of transistors which will need energy. Information communications technologies (ICTs) use about 10% of the world electricity generation (1500 TWh, see Mills 2013). If the global brain is like the human brain, then the consumption could rise to 20%, which is the ratio of energy used by the human brain (Raichle and Gusnard

End of the Beginning

2002).

In a cosmic setting, the global brain is just the beginning. Regarding what could happen after the global brain, I invite you to consider the starivore hypothesis, and the hypothetical possibilities of major transitions towards global organizations on nuclear scales (white dwarfs), sub-nuclear (neutron stars), and space-time structure (black holes). These small-scales and high-density states of matter turned into intelligent organizations would necessitate increasingly more and more energy, nothing less than the total energy of a star.

Ben: I love your speculations about starivores and so forth. I myself have spent a bit of time musing about how one might construct machinery (of a sort) and intelligent systems within degenerate matter such as quark stars or quark-gluon plasmas – going beyond nanotech to femtotech, in this way. I note that you say you've "set up a prize to stimulate proof or disproof of the starivore hypothesis".... While somewhat peripheral to the main focus of this book, this is sufficiently intriguing to me that I can't pass up asking about it. What kind of proof or disproof do you think might be found in the near future? How would you envision concretely demonstrating the existence of intelligence in far-off star systems? Via some sort of conventional SETI-like analysis of astronomical data from these systems, I'd suppose? Also, I wonder if you have any thoughts about the internal nature of these starivore minds? What is it like to be a starivore?

Clément: I'm glad that you have also considered nuclear or sub-nuclear substrates as a basis for intelligent organization.

Proving the existence of extraterrestrial intelligence will not be easy and clear-cut, whether it is the starivore hypothesis or another idea. The question we need to address is: what kind of proof would convince the most skeptical scientists? There are only three such kinds of proofs: prediction, prediction and prediction. I mean,

End of the Beginning

predicting matter, energy and information patterns.

Concretely, this could mean predicting movement patterns, e.g. showing that the direction of some moving white dwarfs is not random, but targeted towards the nearest food source, that is, a star. Predicting energy patterns, we can assume that, as living systems, starivores should follow one of the most universal biological laws, Kleiber's law (DeLong et al. 2010). Predicting information patterns, X-ray and millisecond pulsars are particularly intriguing and promising. I refer the curious reader to Chap. 9 of my (2014a) book for more details on possible predictions.

What I have in mind with these projects leading to predictions is simply the regular hypothetico-deductive way of science. The logic goes as follows. Let's assume that such systems are alive. Can we make new predictions that we can't derive when we assume that the systems are purely physical? If we make such predictions, such as the ones above, and if they become verified, then we build the case that such binary stars are extraterrestrials. If not, it will tend to falsify the starivore hypothesis.

Conventional SETI only matches the third category, the prediction of information patterns. Even there, we can distinguish at least two distinct issues:

- proving that a signal is intelligent via some abstract complexity measure such as Kolmogorov (e.g. Li and Vitányi 1997) or logical depth (Bennett 1988)
- decoding an extraterrestrial message.

I believe this second challenge is way harder than the first. To sum up, I envision and hope that 5-10 successful and different predictions like these would make skeptics scratch their heads, and trigger the scientific community to look at the starivore hypothesis more and more seriously and intensely.

What's inside a starivore mind? The best way to answer this

End of the Beginning

question would be to ask them. But it would take a few decades or centuries to make a Q&A with them. So, in the meantime, let us speculate. As a minimal hypothesis, we can assume that they are living organisms. As such, they first want to ensure basic survival needs, first and foremost, eat stars. Of course, eating a star might be a science and an art, in addition to being an engineering feat. So how to eat a star might be a key question in starivore's minds. When this basic energetic need is fulfilled, they would use their computing and cognitive power to explore and understand their surroundings. But, to understand a complex system, there is often no other shortcut than to run a simulation of it. So, they would run very advanced versions of the "living earth simulator", simulating the whole universe and not only their own local world.

I imagine that they would have completed their global brain transition and acted as a whole and coherent living system, beyond local frictions and dissensions. In the long run, their self or ego would even transcend their own star-eating existence, realizing the futility of eating stars all day, if the universe is doomed to dissipate all its energy, as predicted by the second law of thermodynamics. So, they would turn into the noble mission of securing the life of the cosmos, maybe first through stellar and galactic rejuvenation, and then through the making of a new universe (Vidal 2014b).

Chapter Twelve A World of Views: A World of Interacting Post-human Intelligences

By Viktoras Veitas and David Weinbaum (Weaver)[59]

Viktoras Veitas is an Interdisciplinary PhD student at Vrije Universiteit Brussel and a research assistant at Global Brain Institute. He is a member of Evolution, Complexity and Cognition group and EUCog network. His research aims at creating the framework of synthetic cognitive development for understanding and modeling knowledge creation and emergence of the collective intelligence as a phenomenon of self-organization. Viktoras' previous research concentrated on creating a public policy design methodology for designing social systems. He holds master degrees in International Management and Artificial Intelligence and has 15 years of professional experience as a business and public sector consultant. He can be reached at: **vveitas@gmail.com**
David Weinbaum (Weaver) is an interdisciplinary PhD. associate at the Global Brain Institute, the Vrije Univesiteit, Brussels. He holds a M.Sc. in Electronic Engineering from Tel Aviv University (1989) with experience in computer engineering, parallel algorithms, symbolic computing, software and hardware

---

[59] *Both authors contributed equally to this work.*

design, and R&D management. His research aims to develop models of self-organized distributed cognition, distributed thought processes and distributed social governance towards the emergence of a Global Brain. His major research interests are: cognitive science, artificial general intelligence, complex systems, consciousness, cybernetics, evolution theory, foundations of thought, philosophy of mind, Deleuzian theory of difference, futures studies and the singularity, posthumanism and the transformative potential of human consciousness. He can be reached at: **silkenweaver@gmail.com**

Part I The long tail of the Singularity[60]

1. 1001 times more intelligent?

What would a human hundreds or thousands times more intelligent than the brightest human ever born be like? We must admit we can hardly guess. A human being of such intelligence will be so radically different from us that it can hardly, if at all, be recognized as human. If we had to go back along the evolutionary tree to identify a creature 1000 times less intelligent than the average contemporary human, we will have to go really far back. Would it be a kind of a lizard? An insect perhaps? Considering this, how can we possibly aspire to have a grasp of something a thousand times more intelligent than us? When it comes to intelligence, even the very attempt to quantify it is highly misleading. Now if we attend to a seemingly adjacent question, what would a machine with such capacity for intelligence be like? Just coming up with an approximate metaphor requires a huge stretch of the imagination, meaning that almost anything goes... What would a society of such super intelligent agents, be they human, machines or an amalgam of both, be like? Well, here we are transported into the realm of pure speculation. Technological

---

[60] *The title refers to the historical process of accelerating change that did not start recently but rather began at the dawn of humanity as a tool making civilization.*

End of the Beginning

Singularity is referred to as the event of artificial intelligence surpassing the intelligence of humans and shortly after augmenting itself far beyond that. It is no wonder that the mathematical concept of singularity has become the symbol of an event so disruptive and so far reaching that it is impossible to conceptually or even metaphorically grasp, much less to predict.

We do not know whether a Singularity will ever happen or what it will be like. Yet, the disruptive events already taking place in our lifetime at an ever increasing rate clearly indicate that a great something else awaits us all in a future too near to comfortably disregard. It indeed may be worth, being the conscious observers that we are, to relate to our situation as on the brink of Singularity. We think that this brink of Singularity situation is more important and informative than any speculation about the Singularity itself. Therefore we focus here on understanding *the process of sociotechnological evolution* which seemingly takes us towards a Singularity, whatever that may mean. Moreover, we think that this process may be the most significant evolutionary transition that ever took place in the history of humanity and probably in the history of life on this planet. How does this process work? What consequences may it bring to us as individuals and as a society? Can we somehow influence it, direct it? Can we make it sustainable and graceful, introduce some sense of continuity into an overwhelming experience of discontinuity? The image of a World of Views - a world of diversity on the brink of Singularity - will shed some light, we hope, on how to start answering these questions.

2. Escaping the constraints of natural evolution

The Latin word *socium* means companion, associate. Likewise, by *sociotechnological* we mean living with technology, disposed to symbiotic relations. We describe society as a complex, highly connected, network of agents which possesses, as a whole, organic, life-like characteristics such as self-organization, self-regulation, coordination, adaptive behavior and more. We refer to this network as the *sociotechnological system,* and to its dynamics as the *sociotechnological evolution.* We do not associate agents just with humans; we rather take them to be any entity which can be assigned with a degree of autonomy, intentionality and identity [21]. For now, these are mostly humans mildly augmented by technology[61]. On the brink of Singularity, these may be heavily augmented humans (i.e. cyborgs) or technological artifacts mildly augmented by humans (e.g. self-driving cars), not to mention autonomous organizations involving both humans and machines (e.g. corporates, autonomous management systems, cities). The point is that the specific characteristics of future agents do not matter. Society as a super-organism is mostly shaped by the interactions among all the agents in the network, and much less by the particular characteristics of individual agents.

We replace the vague term of Singularity with an understanding of the sociotechnological evolution as the process of intelligence escaping the constraints of biology. It is a process with a very long history which actually started with humanity arising as a tool making civilization about 2.5 million years ago. While we may feel that the brink of Singularity is special to our times of accelerating change, a somewhat similar experience may well had been shared by witnesses of the Scientific Revolution of the 17th century, or those frequenting the library of Alexandria.

Intelligence escaping the constraints of biology is evidently a process of a continuous transition rather than a singular event. This is the process where humanity as a biologically evolved form of intelligence undergoes a transition from its biological stage into a post biological stage. A post biological stage means primarily the blurring of distinctions between the natural and the artificial. It is

---

61 *Cyborg Luddite Steve Mann on Singularity 1 on 1: Technology That Masters Nature is Not Sustainable.*
http://www.singularityweblog.com/cyborg-steve-mann/

End of the Beginning

characterized by the ever increasing prominence of non-biological forms of embodiment and agency (e.g. anything from bionic limbs and artificial organs, to semiautonomous avatars, highly autonomous robotic systems and semi sentient computing agents). This however does not necessarily pronounce the eventual disappearance of biological structures. It rather means that future structures of biological origin will be shaped by technology more significantly than by natural evolution (e.g. genetic manipulation, but also other interventions at both the cellular and organism levels). In this process of sociotechnological evolution, humanity undergoes an *intelligence expansion*[62] via the mediation of technology.

It is difficult to turn Singularity into a verb, but we propose and even urge to think about the brink of Singularity in terms of a process. Being complex and unpredictable, this process will most probably not yield the scenarios that we project, whether utopian or dystopian. Nevertheless, it does have a direction that we can better characterize and understand.

3. Sociotechnological evolution and intelligence expansion

What is sociotechnological evolution and how does it drive intelligence expansion? Intelligence is quite difficult to formally define but is most apparent in the behavior of agents when observing their interactions with their environment. An intelligent behavior is achieved via a progressive process of developing models of the environment and acting upon them. The better an agent is at developing models, the more capable it is of anticipating the dynamics of its environment and responding appropriately.

Forming effective model representations of the environment is

---

[62] *Contrary to I.J. Good's concept of intelligence explosion' [12], 'expansion' embraces all variants of increasing intelligence: both punctuated ultra-quick 'explosions' and gradual developments. We emphasize the significance of the overall underlying process.*

End of the Beginning

associated, on the one hand with general intelligence and on the other hand with evolutionary fitness. Natural evolution persistently selects those organisms that operate the best models. This means that in a profound manner, evolution selects for intelligence. Moreover, the evolution of nervous systems and brains can be understood as strategies to achieve more accurate and faster adapting models of complex, fast changing environments compared to what mere genetic modification can achieve. We suggest that sociotechnological evolution is the continuation of this biological trend, only augmented by technology.

Technology greatly extends the capability of humans to create better models of their actual environment. It also provides them with the tools to effectively project desirable, not yet actualized, states of their environment. Finally, it enables them to modify their environment to fit these projections. This is, in essence, how tool making (language and scientific thinking included) initiated the powerful positive feedback mechanism that drives sociotechnological evolution. Increase in capacity to model the environment actually amplifies the capacity to modify the environment because the availability of reliable representations allows for more effective and directed interventions. A changing environment, in turn, exerts an adaptive pressure to increase the capacity of modeling and necessitates knowledge and intelligence expansion. These effects are mutually reinforcing. Consider for example the agricultural revolution; agriculture started with very basic knowledge of how to cultivate plants and animals for food. This knowledge enabled humans to settle and then modify their environment in order to produce a reliable supply of food instead of being dependent on hunting and gathering which were much less predictable practices. Controlling the production of food brought, what was probably considered then, an age of abundance, but it also greatly changed the way humans lived. Settlements grew, became more complex to manage and a variety of new challenges arose, e.g. production of tools, storage of goods, forecasting weather, building shelters, managing waste, etc. Solving these problems stimulated the expansion of knowledge and specialization. The expansion of knowledge, in turn, has placed in the hands of humans more means to further modify and expand their settlements, which gave rise to further problems such as division of labor, exchange of goods and knowledge among strangers, etc. The development of human settlements was coupled with the expansion of knowledge and technological means. These were indeed mutually reinforcing processes. A combination of better nutrition, the need to navigate in more complex social situations (and perhaps more free time to think) which made humans smarter and more creative in general, further amplified knowledge expansion. This positive feedback mechanism operates since then at different scales beginning with single agents, their communities and organizations, and extending to the largest scale of the sociotechnological system and the whole planet[63].

The coupling of the sociotechnological system with its environment is highly complex and exhibits a variety of nonlinear effects. The result of this is that the system becomes less stable and much less predictable, which is well illustrated by the behavior of financial markets and the global economy [31]. It seems that presently, while the overall acceleration has the advantageous effect of intelligence expansion, the capability to adequately model a fast changing complex environment lags behind the actual dynamics of change. This has many adverse effects: actions based on unreliable models result in destabilization, unpredictability and dangerous if not destructive policies. Such effects may indeed catalyze the reorganization of the sociotechnological system into its new evolutionary phase, but they may also be catastrophic. Which kinds of instabilities are probable to develop largely depend on the

---

[63] *A Brief History of Intellectual Discussion of Accelerating Change.* http://www.accelerationwatch.com/history_brief.html

organization of the sociotechnological system and its adaptive capacities. Based on this cybernetic perspective of the sociotechnological system and its dynamics, we propose our vision of a *World of Views*. With this vision we hope to outline a sociotechnological organization that can successfully cope with the challenges of the accelerating transition humanity is undergoing.

4. A World of Views

A World of Views is a vision of a future sociotechnological system that embodies a multiplicity of unique, modular and open co-evolving worldviews. What is a worldview? A worldview is both a formal philosophical concept and a term of everyday use that comes to describe a unique perspective of an agent at various levels from individuals to organizations, communities and whole societies. A worldview can be understood as a gestalt perception, both individual and collective, in relation to self, others, society, and the cosmos at large. A worldview may exist on many levels and contain contradictions and paradoxes. Every worldview is selective, not only in regards to which categories of the agent's attributes are included, but also in relation to the facts which are asserted to be true of these attributes. Some views are narrow, ignoring many possibilities, others are more comprehensive. Each, however, selects which attributes and qualities are to be considered real, which are to be developed, admired, accepted, despised or otherwise attended to. These views are held at varying degrees of awareness by individual agents as well as by collectives of agents (e.g. communities, organizations, societies) [25]. Since the concept is applied with different meanings in different contexts, it is useful to distinguish three planes that together constitute a worldview: the systemic/objective, the subjective, and the intersubjective. The first plane consists of given facts (e.g. the cyclic change of seasons) and their organization. This plane is mostly shaped by the sciences. The second plane consists of individual subjective experiences and the belief systems they induce (e.g. this

End of the Beginning

person is trustworthy). The third plane consists of consensual structures that arise in the social interactions among humans (e.g. all men are born equal). Worldviews often tend to emphasize the importance of one of the planes and understate the role of the others. But it is the interplay among the three planes that accounts for the complexity and nuanced structure of worldviews. A comprehensive discussion of the concept and a wealth of further references can be found in [35, 36].

In our vision, besides the inherent uniqueness that characterizes a worldview, well-structured worldviews must have two additional characteristics: *modularity* and *openness*. Modularity means that worldviews are assembled from modules where each module is an assemblage of ideas, beliefs, values, representations and operational strategies that constitute together a coherent and relatively independent whole. The idea is that worldviews can develop and evolve by adding, omitting, or replacing modules. Openness is the capability of a worldview to accommodate and interact with other worldviews. An open worldview is adaptive and interactive with other worldviews, even those with rival perspectives, beliefs or values. Worldviews which are both modular and open can share, exchange and co-evolve.

From a societal perspective, a World of Views is a heterogeneous network of agents such as humans, technological artifacts with diverse degrees of autonomy, intelligence and expertise, organizations combining both humans and artifacts and possibly other futuristic agencies (e.g. augmented apes, dolphins, cyborgian entities etc.). These diverse agents pursue a life style befitting the unique worldview they hold. Worldviews can of course partially overlap or be in conflict with each other. We envision a fluid social order facilitated by *distributed social governance* (see section **9**) which is very different from the social order we live by today. Distributed social governance is not based on fixed hierarchical systems of governance such as nation states or supranational institutions. It is rather based on principles similar to open source projects, art biennials or academic collaborations: the sharing of ideas, perspectives, values and goals and the formation of fluid *ad hoc* social coalitions according to emergent relevance or interest. Yet, in a World of Views, co-evolution and sharing is not a requirement but rather an option for agents that may wish to engage and cooperate.

As we will further discuss, one of the most important consequences of the future will be *abundance* [9, 10]: the elimination, in the practical sense, of constraints that force agents to compete over limited resources. A world of abundance will be a world of much fewer existential constraints and many more existential options, without fear. Therefore, one world as a methodological construct [11] will become a choice of the inhabitants of the world rather than a constraint of sociotechnological evolution. In a World of Views, survival is no longer an ultimate driving force but rather a taken-for-granted baseline. The shift in the very forces that shape the sociotechnological evolutionary process is one of the most profound social changes we foresee. On the brink of Singularity, where the physical constraints that were and still are guiding biological and social evolution are gradually relaxed, it is the images, narratives and values constituting our worldviews that will take their place in shaping the evolutionary process. Our vision is a vision of radical pluralism and openness that catalyzes and is being catalyzed by sociotechnological evolution. A World of Views is a world that thrives and evolves thanks to the enormous diversity and variation of intelligences, their chosen embodiments, styles of expression and co-evolution in the form of multiple overlapping and fluid social institutions. In the following sections we develop the rationale behind this vision and argue not only the fact that it is a realistic outcome of sociotechnological evolution, but also why we think it is a desirable outcome.

End of the Beginning

Part II Bootstrapping our future

5. Framing the forces at play

Nietzsche, the post-modern philosopher, based much of his philosophy on a view of the world as a complex system of interacting forces (known as "will", in his terminology). In his typology of forces he distinguished between reactive forces and active forces. Reactive forces are always forces that react to given situations; they respond and adapt but will never initiate change. They lack the spontaneity and creativity associated with active forces. Active forces are spontaneous and creative - the real movers and catalysts of change according to Nietzsche. They are expressions of freedom as they are never bound to a given state of affairs or a prescribed order [8]. In this part, we will use this idea to highlight the profound change that we foresee in the future dynamics of the sociotechnological system.

As individuals we move and act driven by two kinds of motives that are analogous to the two kinds of wills mentioned by Nietzsche: either we find our current situation dangerous or deficient in some significant aspect (i.e. we react), or, we are pretty happy with the current situation but nevertheless desire something else, better, more interesting and more fulfilling. The first kind of motives deals with deficiencies while the second has to do with the need for growth and self-actualization [26]. In other words, we suggest here a distinction between change driven by constraints and change as an expression of freedom. If we relate to the sociotechnological system as a higher order cybernetic organism, the same distinct kinds of forces can be associated, on a different scale, with societal change at large.

In the following sections we develop arguments showing that *both kinds of forces* that are driving change in the sociotechnological system are driving it towards the same direction: the realization of the future society as a World of Views. We show how our vision of a World of Views is a desirable *response* to the critical problem of

End of the Beginning

coping with the instability and unpredictability that are inherent in the current phase of accelerating change. Given that the sociotechnological system will avoid the existential risks inherent in its very dynamics, we can expect a future of abundance. A sociotechnological evolution under circumstances of abundance will, with high probably, culminate in realizing our vision. In other words, a World of Views will emerge as a manifestation of *active forces* in a state of abundance. We reason therefore that a diversified, open, co-evolving social dynamics is both catalyzing and being catalyzed by sociotechnological evolution. In that, we identify a plausible positive feedback mechanism that actually bootstraps our vision.

6. Our fragile present

It seems paradoxical that our ever increasing knowledge of the world and technological progress make us more fragile. Where does this fragility come from and how can we cope with it? We think that understanding the big picture of sociotechnological evolution provides important clues.

6.1 Understanding fragility

In his recent book *Antifragile* [32] Nassim Taleb gives the following definition to fragility:

"[..] what does not like volatility does not like randomness, uncertainty, disorder, errors, stressors, etc. Think of anything fragile, say, objects in your living room such as the glass frame, the television set, or, even better, the china in the cupboards. If you label them "fragile," then you necessarily want them to be left alone in peace, quiet, order, and predictability. "

With cybernetic systems, that is, systems that involve feedback mechanisms, fragility is much more complicated because such systems cannot be left in peace and fragility therefore is not a static characteristic of an object but rather a dynamic property of a system. Anyone who has ever suffered a painful injury is well

384

End of the Beginning

familiar with how resourceful one becomes in avoiding any movement or change of posture of the painful body part so much, that it becomes entirely rigid while other healthy parts compensate and adapt accordingly. There is much sense in such response: pain signals vulnerability and avoiding it means allowing the injured area to rest and recover quicker. Yet, often, one discovers that even after the injury is long healed and there is no pain, there is a residue of physical rigidity left and even some instinctive fear associated with using the once injured body part again. It is as if our body has forgotten its healthy normal condition and now prefers a safer though limited condition. But the effect of fragility does not end here. The self-limiting protective behavior originating from the body's attempt to (over) protect a fragile point may, in the long run, cause it to lose its normal range of variable response and leave it eventually more vulnerable than before the injury. Medical practitioners are aware of this phenomenon and help patients, sometimes against their own judgment, to return as quickly as possible to regular patterns of activity inasmuch as this is possible.

The important lesson here is that under certain conditions, fragility begets more fragility, increasing the probability of major failures due to random unpredictable shocks. Every system, no matter how robust, has points of vulnerability. But what is important in the context of cybernetic systems and perhaps critical in the case of the sociotechnological system is not so much identifying such points but rather focusing on understanding the mechanisms that amplify fragility and learning how to modify them. It would not be an overstatement to say that mechanisms that amplify fragility constitute an existential risk to humanity as utterly unpredictable minor shocks become implicit seeds to system-wide disastrous events. This is especially true taking into account the potentially ruinous effects of the system, greatly empowering the capacities of individuals: taking advantage of knowledge available to all and with relatively humble financial means, a person can build a nuclear device, develop a lethal biological agent, or deploy a viciously damaging computer virus.

This is, in a nutshell, modern day fragility.

6.2 The consequences of a complex world

Our world becomes progressively more fragile because the current sociotechnological dynamics has a growing number of powerful amplification processes. These processes placed vast powers in the hands of humans, but by the same token they also amplify the effects of errors and accidents. It all comes down to technology because technology in its deepest sense is all about amplification - doing more with less. Cars, ships and planes enable to travel farther and faster. Writing (and electronic media) enables to remember more data for longer. Communication enables the distribution of more information farther and faster. Machines and factories enable the production of more of everything including more machines, and computers enable more knowing, predicting, modeling and inventing. In this process, computing amplifies technological development and is the primary catalyst of technology-induced change.

The downside of amplification is not difficult to understand. Let us examine a number of more concrete factors, all of which are associated with the sociotechnological system becoming a very complex system (for an excellent treatment of the subject see [15]).

Hyper-connectivity is a major symptom of progress, resulting in our world becoming progressively more connected at many levels: starting from global transportation networks that carry goods as well as people across the planet and culminating in global communication networks that mobilize and distribute information at an ever increasing volume and speed. A hyper-connected world is a world where every agent is connected by numerous means to many other agents and where distances, both in space and time, are collapsing [39]. The effect of collapsing distances is that local perturbations may swiftly spread and become global perturbations. This means that highly unpredictable and initially

End of the Beginning

unnoticed local events may yield global impacts within very short time spans. Local outbreaks of diseases are hardly contained and may spread via the global transportation system and become pandemics. Computer viruses can spread and disrupt vast portions of the Internet with estimated damages of many millions of dollars [19]. A thoughtless rumor can initiate a chain reaction that may crash international financial markets. These are only a few examples of how hyper-connectivity amplifies fragility. It is almost inconceivable for the global system to protect itself from all these local, potentially disruptive perturbations. Nevertheless, like in the analogy of the injured body in subsection 6.1, the tendency of the current sociotechnological system will be to limit the freedom of its components and to guide them into more predictable patterns of behavior. Such systemic response will only increase its fragility in the long term [3].

Reflexivity is a concept referring to circular relationships between cause and effect as each element is both affecting and is being affected by other elements. Especially it refers to a feedback relationship between observer (i.e. intelligent agent) and observed (i.e. environment): any examination and action of agents bends the environment and affects the perception and further decisions by the same agents. From its very definition, the sociotechnological system is a reflexive system with a vast number of feedback loops. Reflexivity, blurring the distinction between causes and effects, makes systems difficult to analyze and predict. The contribution of reflexivity to the fragility of the sociotechnological system depends on the kind of feedback mechanisms that operate. A negative feedback has a stabilizing effect on the system's behavior as it resists any change in the state of the system. This is not the case with positive feedback that has the opposite effect of destabilizing the system by amplifying any disturbance. The crucial aspect of the reflexivity property for the sociotechnological system is that patterns of modeling and representation of the world (see 4) have a decisive effect on the type of feedback loops which develop in it.

Consider, for example, a stock market crash caused by a positive feedback: a price of a stock randomly fluctuates down which may bring stressed traders to sell that stock because they predict a further decrease. Which indeed becomes a self-fulfilling prophecy: each sale order further reduces the price and drives an avalanche of sale orders which may eventually crash the stock price [31, 3]. In all similar cases, reflexivity is clearly a potential amplifier of fragility.

Accelerating change is a positive feedback effect influencing the speed of the overall system's dynamics. As more information is fed to agents they produce more events in response, which produces in turn even more information for other agents. However, the overall capacity to process information relevantly is limited. Accelerated change amplifies fragility because at the level of the sociotechnological system, once the pace of events exceeds the capacity of the system to process information, the system becomes blind and more exposed to unpredictable adverse events.

Understanding the mixed blessing of technology as a systemic amplifier is only part of a deeper puzzle. Arguably we could compare human technology to biology as *a different kind of technology*. Also biological technology is all about amplification: the species that succeed to make more of itself with the least available resources is the species that manages to survive. Yet, life at large is anything but fragile. This profound difference tells us that amplification is perhaps important to understanding fragility, but it is not the whole story.

6.3 The Clockwork that never was...

The critical factor of sociotechnological fragility is the manner humans approach technology. This approach seems to be rooted in what we call the Newtonian worldview [18]. This worldview, originating in Newton's days, is a perception of the universe as a vast deterministic clockwork whose components from the minutest to the largest operate in tandem with ultimate accuracy and

End of the Beginning

efficiency. The scientific revolution of the 17th century has established the belief that it is within the powers of the scientific method to fully understand this mechanical universe, its laws and how to control it. Though this perception was quite shaken since the beginning of the 20th century and on (e.g. by quantum theory with uncertainty, chaos theory with unpredictability, and complexity science with uncontrollability), it is still prevalent and reflected in the way we create models of the world, structure organizations and design technological artifacts. From governance systems, armies, supply chains and air traffic control, to the miniaturized electronic chips, all are highly accurate and interdependent machines optimized for efficiency.

As systems become larger and more interconnected, tight interdependence means fragility. The failure of one component may cause, with increasing probability, the failure of the whole system. According to the mechanistic Newtonian view, failure is not natural and is not expected. Of course error margins and failure modes are part of engineering, but systems are not designed to fail (and recover). Hence, the larger the mechanism, the more fragile it becomes because there are so many more components that are not supposed to fail but eventually do. Biological technology did not evolve that way. Biological systems continuously fail and recover as components compensate each other. Every day an average person loses thousands of neurons in his brain[64] with little effect on mental performance. In comparison, the failure of a single bit on a single silicon chip may cause a major malfunction of a large computer system with thousands of chips. Human technological artifacts indeed work and amazingly so, but because of the way they are designed, when they fail they fail spectacularly as well.

The problem with the deterministic worldview becomes much

---

[64] *Brenda Hefti, a post to MadSci Network, March 15, 2002.*
http://www.madsci.org/posts/archives/2002-03/1016223301.Cb.r.html

more acute when it comes to constructing models of the complex fast and changing sociotechnological system. In a world which was fairly stable, less hyper-connected, reflexive and accelerating compared to ours, the basic assumptions of the deterministic worldview were fairly reasonable. Models of systems guided by these assumptions used to yield pretty useful predictions and formed the basis of effective governance. But in a world on the brink of Singularity, as things get wilder, models based on a deterministic approach become increasingly more difficult to construct and their predictive powers diminish significantly. What exactly are the causes to the failure of models on the brink of Singularity? There are three major ones:

Failure of simplification methods - Models are always simplifications of the reality they represent. The art in modeling is making these simplifications useful approximations of reality. Stable linear systems usually have a few variables of major significance and many other variables of much less influence that can therefore be disregarded without losing much accuracy. Complex reflexive systems are difficult to simplify because many of the parameters affect each other in a circular manner and it is almost impossible to disregard anything. One is left with two bad options: constructing a complicated model that contributes very little to understanding, or constructing a simple model which is rather removed from reality. In most cases, policy makers prefer the latter (and more dangerous) because it imparts some false sense of control.

Past experiences fail to produce reliable predictors of future - In an age of accelerated change, recurrent behavioral patterns that characterize stable systems tend to disappear. Models that extrapolate future trends based on past developments will fail because the future does not resemble the past anymore. If no stable patterns can be detected, the very rationale behind predictive models is invalidated. In other words, on the brink of Singularity we can expect a serious decline in the power of models to make

End of the Beginning

useful predictions. Nevertheless, current social institutions, organizations and communities base their policies on the underlying assumption that the world is a stable system, the future will resemble the past and therefore models are reliable predictors. These social bodies will become progressively more vulnerable to those unpredictable events that cannot be captured by their models.

Failure of statistical models - This is perhaps the least intuitive but most important factor. Statistical methods are used to model large populations of agents, events, interactions and so on, based on measuring a small representative sample of the population of interest. That a small sample can reliably represent a very large population is a fortunate mathematical property associated with how certain characteristics are distributed within a population. Well behaved distributions (usually associated with well-behaved systems) can produce highly reliable statistical models. If we wanted, for example, to estimate the height (or any other similar property) of an average human we do not need to measure every single human being. We can get a reliable estimate by measuring the average height of 1000 people and then know for sure that even the tallest person on earth is not very far from this average. This fact makes statistics an immensely useful modeling tool (think of insurance). Complex systems, especially human created systems and organizations, often acquire properties which are called *wildly distributed* [24]. For such property distribution there is no representative sample size large enough to help us find something useful about the whole population. For wild distributions, models based on statistical methods will simply fail. For example, contrary to height which is a biologically dependent parameter, the wealth or income of a person depends on the sociotechnological system. The distribution of wealth over the population of humans is a wild distribution also called a power law distribution. No representative sample of wealth, no matter how large, can help us estimate how wealthy the wealthiest man is.

391

End of the Beginning

Theoretically, a single person can be richer than all the rest put together, but this cannot happen with height or weight. Wildly distributed populations are truly individualistic: the characteristics of the majority tell very little about the characteristics of individuals. Many important parameters of the sociotechnological system are wildly distributed. For example, one's wealth, one's influence in a given social network, the number of people a specific news item will reach, and so on.

As long as we are imprisoned within a view of our world as a deterministic well behaved machine, as long as we believe that modeling always works and that we can have absolute control, our systems will fail in places we least expect. As models fail more frequently they will result in ineffective governance and increasing systemic fragility. In the following sections we describe the much needed change of perspective if we are to dodge the existential risk described here. We show how the vision of a World of Views may help us to harness the same mechanisms that amplify fragility, in order to amplify resilience instead - to turn a weakness into an advantage.

7. Our antifragile future

If we accept the premise that the impact of technology is amplification in the broadest sense of the word, we can expect the world on the brink of Singularity to become more connected, more reflexive and more accelerating. Following the understanding developed in section 6, such a world will also become increasingly fragile. From the vantage point of our present situation, it seems that the future of the sociotechnological system will develop along one of three paths. (1) The first path leads to systemic collapse via a cascade of disastrous global events such as an economic meltdown, escalating local conflicts, collapsing of governance systems, etc. (2) The second alternative is a forceful, probably not less violent, attempt to bring progress under control (giving up much of the immense benefits it brings). (3) The third, brighter path, leads to

End of the Beginning

systemic abundance and flourishing, expansion of intelligence and opening for humanity as yet unimaginable horizons of growth and transformation.

How can we modify our present so we eventually fall on the favorable path? In this section we explain why a World of Views presents a viable response scenario that overcomes the fragility inherent in our sociotechnological system. We base our reasoning on the concept of antifragility developed by Nassim Taleb [32].

7.1 Understanding antifragility

Our discussion of fragility in complex systems concluded that we need to focus not on specific vulnerabilities but rather on the dynamics that amplify fragility. Following this line of thought, we find little sense in trying to achieve resilience by identifying fragile points in the system and redesign them to be stronger. In a complex reflexive world we will never manage to anticipate all the fragile points which may trigger a global failure due to an unlikely accident. We need therefore a radically different approach; we need to come up with dynamics where local failures stimulate a global strengthening.

In his book [32], Taleb is trying to investigate what would such dynamics be like. He starts by asking a brilliant question: if fragile things seem to characteristically dislike volatility, disorder, uncertainty, mishappenings and stressors of all kinds, what would be the character of those things that are the exact opposite, things that *like* volatility in all its forms? He calls such things *antifragile*. Resilience does not seem the proper designation for them. Resilient things would be at best indifferent to volatility. Antifragile things are not just resilient or indifferent to volatility, they *positively love* volatility, disorder, accidents, uncertainty, stressors, chaos and so on. They love all these because they *thrive in their presence*. When it comes to a system as dynamic and complex as the sociotechnological system, it is not enough for it to be indifferent to the unavoidable volatility that is inherent in its very dynamics. The

End of the Beginning

system needs to thrive on volatility - it needs to become antifragile.

What stands behind the idea of antifragility and how do we make a system gain from all those situations which intuitively seem to be detrimental? According to the paradigm of determinism and control that is deeply rooted in the Newtonian view, anything which is unpredictable, unexpected, accidental, or otherwise outside the prescribed norms, is by definition a potential risk to the system. However, unpredicted events can also be positive. If we go back to amplification as a general property of sociotechnological dynamics, ideally we would like to have some kind of selective amplification mechanism with magical sensitivity: amplifying events that can become globally advantageous while suppressing those with potential adverse effects. A system thus constructed would clearly thrive on the unexpected. Unfortunately, such a mechanism seems to require perfect predictability which is, of course, an unachievable idealization.

Yet, antifragility is not just an abstract idea. If we think about biological evolution as a kind of a technology and of life forms as its artifacts, we can clearly observe that life in general is antifragile. In the long history of life, failure is the rule and success is the exception. At every given point in time, there are many more extinct species than living ones. However, no doubt, life is thriving in spite of (or perhaps thanks to) all misfortune and this is a rather consistent trend for a few of billions of years already. What is it that makes life antifragile?

Natural evolution is the greatest entrepreneur; it achieves antifragility via endless experimentation and creation of ever more options. It keeps whatever works and scraps everything that does not, freeing up resources for further experimentation. It creates options by variation through mutation and lets the environment select which variations are suppressed and which are amplified through procreation. The most significant point is that while evolution easily disregards its failures, the successful experiments that get to proliferate naturally become the sources of new options.

End of the Beginning

This is how evolution achieves its selective amplification. For example: certain cellular mechanisms (e.g. energy production, DNA replication etc.) discovered by evolution billions of years ago, are shared by almost all contemporary life forms. Such mechanisms will continue to be retained even if the vast majority of life forms at some point in time will go extinct.

From the perspective of all life, any single trial and error experiment costs almost nothing in terms of investment. Therefore, the penalty involved in a failed species (or a failed individual) is very low, just a negligible fluctuation in the overall distribution of the biological mass. But the gain involved in one successful species is immense: finding a new way to proliferate. In the long run, any such success will be shared by many life forms because the more successful a species is, the more it will originate new variants of itself.

In short, life's secret is a very smart investment strategy: it diversifies its investments as much as possible so any single failure has a very small and insignificant impact while a single success has a huge gain. The gain of one success exceeds many times over the loss involved in many failures. The net effect of such strategy is that there are many more failures than successes, but gains accumulated from successes are much greater than losses accumulated from failures[65]. Evolution does not try to predict the outcome of its experiments. Just experimenting and retaining the successful (in terms of fitness) experiments seems to be more profitable than any attempt to predict what would be fit based on past experiments[66].

Random mutations? A serendipitous speciation event? An abrupt change of environmental circumstances? A cataclysmic

---

[65] *The strategy is technically described by Taleb as a convex pay function [32]*

[66] *This kind of strategy seems to work best under frequent, unpredictable variations in the environment while in relatively stable environments, predictive models based on past experience will perform better.*

395

event of this or that kind? For life as an investor and for evolution as its investment strategy, these are all opportunities realized either as new experiments (think of species as startups), or as stress tests for existing experiments (think of ecologies as competitive markets). As long as the gain from success is much greater than the penalty of a failure, life thrives on unexpected changes - it is antifragile.

It is important to note that antifragility is related in this context to the biosphere as a nested hierarchical structure of systems each of which is antifragile on its own and operating a similar evolutionary dynamics. For example: species are the experiments of large ecosystems, likewise, individuals are the experiments of species. The antifragility of each systemic level derives from the diversity and local fragility of the elements populating the levels below. The selective stressors are not provided only by the external environment. Variations on the same successful experiment compete with each other over resources, so selective pressures may arise from within the evolutionary process itself, driving its dynamics even while the rest of the environment is less demanding.

We believe that this evolutionary investment strategy presents a critical key in achieving an antifragile future of the sociotechnological system. To follow life's lessons will demand a paradigm shift in our approach to sociotechnological dynamics.

7.2 A paradigm shift - from control to experimentation

How do we apply these profound lessons from biology? We need first to accept that our world is becoming more complex, less predictable and therefore less controllable. As we progress, our artificial systems and environments become more lifelike: interconnected, messy and full of surprises. In response, we need to move away from the Newtonian view of the world as deterministic clockwork to a much more organic and diverse paradigm. This is what a World of Views is all about. Systems will

End of the Beginning

fare much better if they are left alone to self-organize rather than be externally organized to fit premeditated idealizations. In thinking about the future sociotechnological system we adopt life's investment strategy: creating a system which is favoring and supportive of experimentation and diversification, which always produces more options. It is not that we can know *a priori* which options will succeed but we need to eradicate our collective fear of failure and learn to allow more risks, albeit calculated ones. Take for example space exploration. With the appropriate supportive ecology, multiple parallel projects make the prospects of space travel closer compared to the undertaking of a single, state managed project. This is true even if most of the multiple projects are doomed to fail.

A paradigm shift from values of prediction and control to fast prototyping and trial and error implies a respective shift of emphasis from modeling to hands-on tinkering of system dynamics. As models become less trustworthy in a fast changing environment, undue reliance on them exposes systems to increased fragility. It does not mean that we should stop developing models. Modeling is a very powerful and successful cognitive tool whenever the general circumstances are stable. But we need to become acutely aware of the inherent shortcomings of models especially with regard to complex, volatile systems. In this sense, the sociotechnological system is getting to a point where models and the kind of control they promise are not sufficient. We need to develop the knowledge of what makes systems thrive on volatility because such systems are going to fare best in our future.

We cannot invent, or premeditate the specific characteristics of a human or a post-human society, or how it will take shape in the volatile environment even of the very near future. We cannot predict which social institutions will be needed to sustain social order in such circumstances. But we can conceive and build a sociotechnological system which is able to self-organize, adapt and grow when exposed to a volatile future. Moreover, we will not be able to totally avoid serious accidents. Becoming antifragile is neither about prediction nor about avoidance of failures; it is about smart risk taking, experimenting, and replacing control with self-organization wherever possible. These will be the guidelines of a *distributed governance* system further discussed in section **9**.

7.3 Antifragility and a World of Views

Human civilization as well as global social institutions, cities and communities are all complex organisms. Complex organisms cannot be designed, they *evolve*. At best we can learn how to influence and guide their evolution. In the previous section we applied evolutionary principles to the sociotechnological system. The key concept is antifragility, the property of a system that thrives on volatility. We argue that a World of Views as a future organizing framework of human civilization, social organizations, cities and communities is a vision of an antifragile future.

In a World of Views, every worldview represents a unique and integrated cognitive structure, held collectively by a network of individual agents. These cognitive structures constitute different ways of representing the world and self. Based on different biases, values and premises, worldviews in the sociotechnological system embody options to perceive and operate in the world analogous to life forms, species and ecologies. We see a World of Views as a nested, modular self-organizing structure, where worldviews occupy the highest level but in themselves are modular and diverse [30]. Diversity, modularity and openness are the essential properties of worldviews that together characterize an adaptive structure capable of containing failures while propagating successes within the system, thus realizing antifragility.

Diversity - Entrepreneurship and experimentation at all scales and areas of activity will be the source of increasing diversity. In simple terms diversity means that systems and components have multiple versions, that there are always multiple ways to achieve a goal and every item or component can be used in multiple ways for multiple functions. Diversity leads to redundancy which is characteristic to organic systems and is missing from clockwork systems, designed to maximize efficiency at the expense of resilience.

The principle enabling diversity in the sociotechnological system is *converting wealth to freedom*, i.e. investing any surplus (wealth) in creating more options of choice (freedom). Similar to natural evolution, every experiment draws from the surplus of an already realized success and diverges from it to explore further options. Presently, experimentation is mostly driven by various adaptive stressors and constraints. But in the future, technology will eliminate most of them. In an age of abundance, experimentation will be driven and shaped by a universally cultivated spirit of exploration that will emerge as a new evolutionary force (see subsection 8.1).

A diversity of views will form consensual enclosures and localities even in a hyper-connected world. Agents sharing a worldview will naturally have more traffic going on among themselves while agents holding different worldviews will keep their interactions more in check. In such an environment, failed experiments will remain relatively local, while successes will multiply their gains when distributed across the membranes that keep communities sharing different worldviews apart. Yet, an ecology encouraging openness and sharing will keep everyone looking for new options that were already tried out by others, because these represent gain without risk. Such ecology will drive both competition and exchange.

Modularity - is an essential characteristic of worldviews in our vision. The idea is that worldviews will generally have a nested modular structure (i.e. each module is also modular), where each module is a complex of concepts, perceptions, values etc. that form together a coherent and relatively independent whole. Communities holding modular worldviews will be able to acquire, discard, experiment and share individual modules independently of other modules. The importance of modularity is in the evolutionary dynamics that it allows: 1) in modular systems, component modules can evolve relatively independently which supports the localization of failed experiments; 2) systems can exchange and share component modules which supports amplification of successes; 3) diversity of modules allows for increased flexibility, adaptability and redundancy. In analogy to vertical inheritance in natural evolution, successful modules will proliferate and be acquired by many communities, while producing myriad variations of themselves (speciation). Communities sharing or exchanging modules to augment their larger worldview will in fact mimic mechanisms of co-evolution such as cooperation, symbiosis and horizontal trait transfer[67].

Openness - reflects the tolerance of a worldview to changes in its own structure, its tolerance towards worldviews different from it and its capacity to constructively interact and share with other worldviews. Openness, to use a biological metaphor, is the degree of permeability of a worldview's membranes. To be open, however, does not mean that anything goes. Every worldview embodies a selection mechanism that assures its own consistency. Openness therefore is to be understood not as an unconditional acceptance of any change but rather as tolerance and responsiveness towards change. Openness is essential in our vision because it is strongly correlated to the dynamics of experimentation and exchange without which the system cannot achieve antifragility. Worldviews that are not open towards the flow of ideas from other worldviews can perhaps develop resilience but cannot become antifragile because they resist volatility. Additionally they will be less helpful in producing options and exchange. Open worldviews not only contribute to the antifragility of the whole sociotechnological system but can

---

[67] *Horizontal exchange of traits and horizontal gene transfer is very rare in complex animals but is common in bacteria, fungi and plants.*

themselves achieve a higher degree of antifragility.

In conclusion, adopting a perspective that sees the sociotechnological system as a complex organism, we propose here an informed speculation that if this sociotechnological organism will not collapse prematurely, it will converge towards an antifragile configuration, because this is the only configuration that can effectively cope with the volatility induced by accelerating change. We do not know what might be the particular properties of such a world but we have drawn here those properties that seem to us essential to its continuation. The processes driving towards a World of Views cannot be systematically designed. These are rather reflexive processes of self-organization, the kind of which we can already see in various social phenomena such as open source communities[68], makers movements[69], occupy movements[70], crowd sourced projects[71] and more. These example processes indeed seem to demonstrate high levels of diversity, modularity and openness.

8. Abundance and sociotechnological evolution

One of the greatest promises of technological progress is the promise of practically unlimited affluence; food, energy, knowledge, health care and physical safety will become ubiquitous and virtually unconstrained. Baseline standard of living and wealth for all human beings and possibly other sentient agents will rise to what today is enjoyed only by the richest and highly privileged. There will be no limits to growth and potential self-fulfillment [9, 10]. The achievement of abundance is of course not guaranteed, yet we believe that by adopting the paradigm we

---

[68] http://en.wikipedia.org/wiki/Open_source
[69] http://www.forbes.com/sites/tjmccue/2011/10/26/moving-the-economy-the-future-of-the-maker-movement/
[70] http://en.wikipedia.org/wiki/Occupy_movement
[71] http://en.wikipedia.org/wiki/Crowdsourcing

develop here, sociotechnological evolution will converge towards such a state of affairs. In this section we examine the effects of abundance. We argue that besides the fact that abundance will catalyze the emergence of a World of Views, it will also profoundly change the very nature of the evolutionary process. This has to do with the freedom it will enable.

According to the Universal Selection Theory, [16, 11, 5] an evolutionary process is a combination of three different component processes: 1) a mechanism of variation of behavior; 2) consistent selective processes; and 3) a mechanism for preserving and propagating the selected variations. For example, mechanisms of variation are realized by entrepreneurship and experimentation. Selective processes are realized by the practical application of worldviews in actual situations, while mechanisms for preservation and propagation are realized by the sharing and exchange ecology cultivated in a World of Views.

The specific characteristics of these components define the context and the nature of any evolutionary process. Yet, as it is clear from the name of the theory, the aspect of selection is predominant because selective processes are the ones that set the direction of evolution. Our sociotechnological system exists within a vast space of possible states. When it moves from state to state it forms paths that represent evolutionary and developmental processes. Paths are shaped by the selectors influencing the sociotechnological system and driving it from its current state into future states.

8.1 Shift in the sets of selectors

Technological progress can be viewed as a progressive elimination of environmental constraints. Transportation networks, for example, relax or eliminate the constraint of physical distances and make people considerably more free in this respect. As long as there are physical constraints, they will act as selective pressures guiding technological evolution. For example, a lack of a

End of the Beginning

clean renewable supply of energy places a pressure that stimulates the sociotechnological system to innovate in that direction. Once there will be enough energy to fulfill all needs with excess, such pressure will be greatly relaxed and might even disappear. Abundance can be understood as a situation where certain sets of constraints driven by survival needs have been greatly relaxed or entirely eliminated. When constraints that guide the trajectories of the evolutionary process disappear, there arise in the system's state space, what we may call choice zones. Within such zones, various trajectories are *no longer differentiated* in relation to the current state of the system and its environment. In other words, the agent or system can choose whatever trajectory, without a discernible effect on their immediate fitness (e.g. what kind of cheese would you buy in the supermarket). Such choices, being equal against any relevant selective criteria, are clearly not guided by values imposed by survival needs.

Choice zones therefore are fields of relatively unconstrained exploration where individual agents are free. Nevertheless, agents will make different choices and form disparate paths to express their unique views and way of life. Initially, such choices are useless from an evolutionary perspective, but they do take place within an interactive and reflexive system. In the course of social interactions, these choices may gain 'evolutionary usefulness' by influencing the agent's other preferences, values, social status etc.

On the scale of the sociotechnological system, choice zones are where entirely new selectors may arise and consolidate via the interactions among agents. Such selectors will start in turn to influence the direction of the sociotechnological evolution, gradually replacing the physical adaptive pressures that used to guide biological evolution. Even in the absence of explicit physical constraints imposed by competition over limited resources, the emergence of new selectors in choice zones will catalyze a novel evolutionary motion, very different from the one driven by survival needs[72]. In other words, choice zones are cradles of new value systems.

A good example of an emergence of new selectors is the world of fashion pertaining to clothing. Fashion emerged along human history as clothing gradually started to mean more than just the protection of the body. Once the basic function of clothes was achieved by diverse technological means (fabrics, methods of manufacturing, stitching, etc.), clothing gradually became a choice zone. The fitness of agents to their environment (i.e. the probability of their survival) was no longer differentiated by what they wear. However, becoming a choice zone did not bring clothing to an evolutionary dead end; on the contrary, humans started to use clothing to distinguish themselves in other contexts. Clothing became a social medium and a complex system of signals used to express the wearer's character, style, social standing and affiliation, occupation, values and more. The physical constraints (protection against the elements) that guided the evolution of clothes as a technology were replaced along history with other sets of selectors - means of expression on the social plane.

As long as agents are constrained by considerations of fitness, they usually operate in a reactive manner (see section 5) being forced to make certain choices in order to survive. But wherever abundance creates choice zones, agents will start to operate actively; they will express their freedom by creatively inventing new values and making meaningful selections thus distinguishing themselves in their social interactions. We can call such inventions

---

[72] *In a recent paper titled Evolvability Is Inevitable: Increasing Evolvability without the Pressure to Adapt [22], the authors bring experimental evidence that the evolvability of a population of agents (i.e. their capacity to produce and inherit significant phenotypic variations) can grow even in the absence of adaptive pressures. This research supports our hypothesis that selectors emerging in zones free from adaptive pressures can eventually have significant evolutionary effects on the sociotechnological system.*

expressions of style, aesthetic choices or simply the expression of freedom. Once such expressions consolidate they become new sets of selectors, guiding the actions of agents and eventually the direction of sociotechnological evolution.

The conditions of abundance will bring with them a sweeping shift in the selective forces that guide the evolution of the sociotechnological system. The shift will be from survival driven selectors (reactive forces) to selectors shaped by the need of agents to give a meaningful expression to their freedom (active forces). Agents existing in conditions of abundance and for whom the concerns of survival and replication have become redundant and almost entirely taken for granted, will actively and creatively seek to differentiate themselves in entirely new ways.

The novel selective forces that will emerge will derive from worldviews adopted by agents. The expression of these forces will be primarily aesthetically oriented, i.e. shaping the environment to reflect style self-description and self-actualization, rather than be driven by survival needs. They will constitute, we believe, the dynamic medium of society's next evolutionary phase - a World of Views.

8.2 A World of Views as the active expression of freedom

The emergence of new sets of selectors is not entirely unfamiliar. Already today, at least in some parts of the world, many of the choices that we make as individuals are rather expressions of freedom than adaptations to constraints: e.g. choice of profession, style of living, social group, faith, etc. What is going to be significantly disruptive in the future is the immense prominence of such choices in our individual lives and the overall social dynamics. We will become true creators and makers, and this is one of the more profound outcomes of intelligence escaping its biological constraints. A World of Views, the kind of future that will be an expression of our ever growing freedom, is also an antifragile abundant future.

End of the Beginning

Shaping our future as an active expression of freedom and not as a response to existential stressors will require a very profound change in humanity's most fundamental assumptions about existence and according behaviors. Already John Maynard Keynes [20] anticipated the profound disorientation and loss of meaning that might occur when a society achieved a condition of abundance but continued to deal with it as if there was continuing scarcity:

"The economic problem, the struggle for subsistence, always has been hitherto the primary, most pressing problem of the human race [...] Thus we have been expressly evolved by nature with all our impulses and deepest instincts for the purpose of solving the economic problem. If the economic problem is solved, humankind will be deprived of its traditional purpose. Thus for the first time since his creation man will be faced with his real, his permanent problem, how to use his freedom from pressing economic cares [...] There is no country and no people, I think, who can look forward to the age of leisure and of abundance without a dread. For we have been trained too long to strive and not to enjoy."

We do not understate the inertia of human collective and individual psyche and history, which is a very powerful force in shaping sociotechnological development leading to the future state of the world. But our vision is one of overcoming the reactive nature of human intelligence and transforming it into an active expression of freedom.

If we succeed, aesthetic choices, philosophical commitments and elegant belief systems will become the shapers of our existence, much more than they are today. Throughout human history, people have been modifying their appearance, their behavior and their environment to suit and express their worldviews. This is how cultures were shaped. Whenever feasible, even the human body has been used as a medium of expression with methods ranging from cosmetic body modifications, to chemical modifications of metabolism and behavior (e.g.

consuming mood altering substances such as coffee, sugar, alcohol, cannabis, etc.). A relatively rare and more extreme example is that of gender transformation which involves complicated surgery procedures, prolonged hormonal treatments and profound behavioral and psychological adaptations.

In the coming future, we can envision the gradual emergence of technologies for both body and brain modifications including genetic modifications, nanotechnological implants and more (e.g. changing one's skin color, muscle build, sensory abilities and much more). Brain modification technologies are of special interest because of their impact on the social fabric. For example, procedures such as memory editing, real-time filtering of experiences and personality modifications (e.g. turning an introvert into an extravert) are only a few of the options that neuroscience promises[73]. Human agents may radically modify their perception of physical reality using brain-machine interfaces or genetic manipulations. Individuals may merge their minds to form multi-bodied collectives or extend their minds to other bodies, biological or otherwise. For example, people who view reason as the highest value in existence may modify the way their brain operates to accentuate the expression of this value in their lives. They will be able to literally become living versions of Star Trek's Vulcans.

Such developments will enable massive exploration and diversification into mind species or intelligence species with very different perceptions of reality. When diversity of views will deeply transform the very world we live in and will become the primary selective force that guides evolution, the sociotechnological system may undergo a Cambrian explosion of co-evolving kinds of intelligence.

---

[73] *Authors@Google: Ramez Naam, Nexus.*
http://www.youtube.com/watch?v=dlHAFHOsp9Q

8.3 Bootstrapping a World of Views

A World of Views is presented here both as a key strategy to achieve an antifragile existence under conditions of accelerating sociotechnological change, and as a plausible outcome of abundance. The achievement of a state of abundance, in turn, relies on avoiding the obvious predicaments of a highly volatile accelerating dynamics, without trying to stop it. The interdependence among the three factors of diversity, antifragility and abundance, constitute a positive feedback operating at the highest scale of the sociotechnological dynamics. It is the mutually reinforcing influences of its components that will bootstrap a World of Views.

It is important to note that for the purpose of bootstrapping it will be enough to achieve such levels of abundance so that aggressive competition will not make economic or political sense anymore. Still, abundance is not only a technological issue, nor is it an economic or political one. Abundance is also a psychological state of individuals and the collective human mind. It is a state where survival instincts, fear and aggression conditioned by eons of natural evolution are replaced by novel forces of motivation. We must not forget that a profound psychological transformation is a necessary aspect of sociotechnological evolution. After all, human minds and the way they process information and react are primary components of the sociotechnological system.

What is the concrete organization that would be capable to facilitate this bootstrapping process and bear the diversity of a World of Views? Our vision requires a common platform that serves both as a universal medium of interaction and exchange, as well as a maintenance facility that safeguards operational integrity and the conditions of abundance. This will be the function of *a distributed social governance system* which is our next and final topic of discussion.

End of the Beginning

Part III Paving a multiway to the future

9. Distributed social governance

Bootstrapping a World of Views requires a governance system which facilitates three major functions: 1) providing the necessary sustainable platform for a future society; 2) allowing for the co-existence of diverse intelligences and their embodiments in a shared physical and computational space and 3) providing a medium for evolution of intelligence through communication, dialogue and co-evolution among diverse intelligent agents.

9.1 Away from sustaining order

Current social organization is based on the deeply ingrained cultural belief in the inevitability of hierarchical governance based on authority and enforcement. This belief is obviously married to the clockwork view of the world (see **6.3**) as a mechanism composed of bolts, nuts, gears and screws which can be individually identified, cataloged and registered for later usage[74]. Nuts, gears and screws, in this case, are human individuals and in order for society (the clock) to function, they have to be positioned within the social system exactly according to their specifications. Moreover, misbehaving components should be repaired or replaced like broken parts of machinery. While in some progressive segments of civil society, this worldview is already considered outdated, not so is the case on the sociopolitical plane, where the world is broken into regions and nation states with governments comprised of functionally separated ministries, which in turn have separate departments, etc. The whole organization of governance distinctly has a top-down control structure. At the very bottom of this hierarchy, there is the citizen who even with all the virtues of a democratic political system

---

[74] *In social and management context this view was developed at the end of 19th century in the form of a scientific management paradigm by Frederick Taylor [33].*

available only in some places, is pretty much bound by a rigid system of formal rules and behavioral patterns (e.g. basic education, taxation, military service, religion, cultural behavior etc.). We are not going to delve into political philosophy here, but we assert that the hierarchical governance paradigm based on the mechanistic clockwork metaphor is absolutely incompatible with the requirements of the future social organization we envision. No matter how open and allowing are the operational attitudes of such a hierarchical order, it will resist the free explorative dynamics of a World of Views, due to its inherent structural principles.

There is even more important functional reason why hierarchical governance systems fail as the world becomes increasingly hyperconnected and complex. The prevailing bureaucratic mode of operation of organizations at all levels of society [25] assumes predictability and slow change in the world. These assumptions are largely obsolete today, let alone in a world approaching the brink of Singularity. Such *modus operandi* will become less and less effective with the current trend of sociotechnological development. It will increase the fragility of the system rather than ensure its stability and resilience. In a situation of accelerating change, hyperconnectivity and reflexivity, limiting the variability of responsive behaviors of the components of a system, lead to increased fragility (**6.2**). In hierarchical organizations, a shock that causes the top levels to fail may paralyze the whole organization [15].

As we already mentioned above, what is largely overlooked by the current sociopolitical paradigm is that global society, by its very nature, resembles a living organism much more than an artifact designed by humans. As a complex organism, it evolves and self-organizes rather than be externally controlled (see **7.3**). The vision of a World Government, an institution responsible for regulating the global affairs (e.g. regional conflicts, poverty and inequality, environmental degradation etc.) of a hyperconnected world [23, 6] is just another control layer on top of existing hierarchical structures. As such it is infeasible, given the accelerating and unpredictable nature of sociotechnological evolution. A self-organizing form of social governance based on dialogue and mutual agreement at multiple levels [6] and not on idealized top-down control structures, seems to be the only alternative to anarchy and dystopia on the brink of Singularity.

9.2 Towards organizing disorder

In a World of Views we envision a paradigm shift in governance, from maintaining a top-down prescribed order to spontaneously organizing bottom-up disorder, thus allowing the social organization to embrace volatility and uncertainty which are salient features of a future society. We propose distributed social governance as a medium that enables the self-organization and evolution of social institutions and structures. In the distributed social governance paradigm, the rules of the game are not about setting values and operational norms (e.g. protecting the weak from the powerful, ensuring free markets, or preventing environmental degradation). Rather, they are about coordinating the interactions among the agents in the social system, in order to achieve a sharing of meaning and values as well as mutually beneficial ways of co-existence, co-evolution and cooperation towards collective goals. There is no dictation of an *a priori* given value system on the agents that constitute the social fabric. In this sense, a distributed social governance system is not the embodiment of the governance regime itself, but rather "an exceedingly complex array of actors and institutions in forms of distributed and disaggregated governance that exists in a shared conductive medium [38].

Within such a medium, intelligent agents would operate according to their own ethical, aesthetic, professional and technical norms derived from worldviews which are emergent and not imposed externally. Assemblages (coalitions) of agents will self-organize into *ad hoc* management systems or institutional structures according to the needs and views shared among participating members leading to a system of multiple overlapping structures, organizations and institutions [7, 37]. Such form of distributed governance will be antifragile. Disintegration of any single organization will have almost no impact on the whole system, because other overlapping organizations will adapt to the new situation and take over the functions of the failed one. In a World of Views, governance is no longer synonymous with control and stability, but rather a provider of well-structured spaces of engagement that facilitate the communication and exchange among diverse actors without compromising the choice of multiple options nor the effective criteria for consensual selection [38].

It is plausible to assume that in a World of Views with distributed social governance there will be no nation states as sovereign territorial units. But it doesn't mean that these will be replaced by some kind world citizenship. Social groups will be taking the place of nation states. Such groups will be much less identified by elements such as physical territory, economic status, national identity or language. They will assemble themselves according to the worldviews shared by their participating members. Worldviews will serve as the primary coalescing element of an otherwise fluid collective identity. They will form kinds of trust realms or virtual states [29] each embodying a collective view. In a World of Views, joining or disengaging from a view will be straight forward. An agent will habitually join more than one view. This freedom and multiplicity of affiliation will encourage the formation of fluid and overlapping social contours. Most of the social institutions will not be based on territory or the distribution of other physical resources. Governments will not distribute tax money to various projects, but instead, resources will be distributed according to the explicit wills of the individual participants of each virtual state or trust domain. If an individual thinks that sending astronauts to Mars is a worthwhile endeavor,

End of the Beginning

she, he or it may initiate or join the project by investing whatever financial, intellectual or other resources, available to them.

Institutions under such governance paradigm will have the capability to transform themselves in a much more profound way than the traditional institutions of nationally organized political democracy. They will lack the formal legitimacy of representative democratic institutions, and have constituencies which are fluid and boundaryless [38]. These institutions will have no persistent identity of their own, eliminating their tendency towards self-preservation after their actual societal functions have long been exhausted[75].

The way to create such a distributed social governance system is by following principles consistent with antifragility: 1) fast and cost effective prototyping of new social structures as well as a swift discarding of failed ones; 2) diversification of options; 3) iterative self-organization in response to the impact of unpredictable events and 4) maximal capitalization on success. Conventional wisdom implies that such a 'non-systemic' system would be a mess, and probably culminate in a destructive anarchy. Yet, examples like the open source software movement show the opposite, suggesting that the paradigm described here could actually be more productive and successful than the conventional hierarchical one. Open source projects are often started as initiatives of single developers performing creative experiments, many of which never reach a useful application. However, the successful ones are sometimes substantially more advanced than software products developed by multinational corporations allocating huge amounts of resources for the task. For example, *Apache HTTP Server*, the leading web server software, serving about 51% of world's Internet servers (about 331 million, including US State Department) was

---

[75] *Reflexivity and Fallibility: conversation with G.Soros.*
http://www.youtube.com/watch?v=BwbSKZMerhw&feature=youtube_g data_player

started by an informal group of eight programmers in 1995. The closest competitor, *nginx*, is also an open source project started in 2002 by a single person[76]. The success of the open source software movement triggered a spontaneous adoption of the model in other fields e.g. Maker Movement[77] and Open Science[78].

9.3 An intellectual technology of building shared realities

Distributed social governance is a collective and conscious use of social system design methodologies and tools comprising an intellectual technology by which alliances of intelligent agents build shared representations of reality, visions of the future and seek their realizations [2, 34]. Diverse populations of agents utilizing such technology will produce a multiplicity of social system design processes. This multiplicity will give rise to an environment where worldview based alliances are actually *constructing shared realities* according to their visions and mutual discourse.

The heterogeneity of worldviews and existential styles of intelligent agents will inevitably give rise to a diversity of respective environments, virtual states and social institutions reflecting these worldviews and their amalgams. This diffused yet distinct multiplicity will constitute the ecology of the future society where social actors and their alliances co-evolve along multiple paths. Co-evolution can be realized only by searching and nurturing shared views among social actors and by that, building a consensual reality where communication and collaboration are effective. Yet, in a World of Views, the option of co-existence without reciprocation will be an available option of any social actor. Co-evolution and cooperation, beyond the necessary support

---

[76] *April 2013 Web Server Survey.*
http://news.netcraft.com/archives/2013/04/02/april-2013-web-server-survey.html
[77] http://en.wikipedia.org/wiki/Maker_movement
[78] http://en.wikipedia.org/wiki/Open_science

in the global platform of sustainability, will always be a matter of choice at all levels. This freedom is the ultimate promise of a future of abundance.

The hypothesis of social constructivism that "the world is socially constructed in two related senses, as distributed cognition and as shared realities" [16] acquires a qualitatively new meaning in our vision. Social systems have a property of reflexivity, well described by the so called Thomas theorem: if persons define situations as real, they are real in their consequences [25]. Images and models of the social system held by its participants are coupled with the actual properties of the system via the actions and perceptions of agents. These cybernetic relationships are the key to understanding the evolution of the sociotechnological system. Where biological and technical constraints are progressively removed and social actors are free to shape their environment as an expression of their worldviews, the implication of reflexivity in a heterogeneous world is the absence of a single convergent path of social development. Rather, it implies interaction among bundles of diverse paths and their co-evolution without the guidance of overarching principles - *a multiway to the future.*

Distributed social governance needs of course an infrastructure, a medium for collaboration and a culture of dialogue. In the case of open source software development that we already discussed, such a medium provides individual developers and groups with a rich toolkit enabling them to contribute to each other's code in the same programming language, and introduces a set of rules of conduct facilitating higher levels of collaboration. These tools define the size and complexity level of a project realizable by a distributed effort. Likewise, distributed social governance at the scale of the sociotechnological system should allow for interactions between humans, post-humans, AI agents and other machines, facilitating the creation of shared symbols, meanings and worldviews. Given the present difficulties of communication and cooperation between much less diverse groups of humans, the enabling intellectual technology required for distributed social governance presents a difficult multidisciplinary challenge.

We see education as the single most critical catalyst of the emergence of a World of Views and distributed social governance as its enabling platform. Towards the realization of a World of Views, educational systems need to be transformed. The intellectual technology for constructing shared realities will have to be instilled and shared by all members of society of all views including radically divergent intellects such as post-humans or AI agents, if they are to become members of such future society. The arts of dialogue, negotiation and exchange in a volatile reflexive environment are the primary skills necessary for constructing such shared realities. In a highly diverse social dynamics these skills will be the unifying and integrating forces holding a World of Views together.

9.4 Closing the gap between reality and vision

The prevailing way of thinking about how to induce and guide change in a system is characterized by a general scheme of systemic change, comprised of three steps: (1) identifying the current situation; (2) identifying the image of a future desired situation; and (3) figuring how to reach from the current situation to the desired one, that is, closing the gap between reality and vision [13, 34]. We can identify three approaches of addressing the third step: *strategic planning, strategic navigation* and *strategic exploration.*

Strategic planning - is aimed at developing a procedure, as detailed as possible, for achieving goals [13]. For example, in the case of sea voyage, it is plotting a course on the map from the home port to the destination port and then following it closely throughout the entire journey.

Strategic navigation - takes into account unpredictable

circumstances along the way and allows much more freedom for adjusting the ship's course and speed depending, for example, on weather conditions along the route or maybe icebergs that might block it [14].

Strategic exploration - When the geographic position of their destination was largely unknown, and with only partly reliable navigation methods, the great explorers of the so called Age of Exploration determined the course of their ships based on guess work and hearsay that often proved fatal. Nevertheless, since there were a lot of expeditions, new lands were discovered and more or less reliable routes were eventually established. If a large and well equipped ship finds itself in the middle of the ocean without a map or any other way to determine its whereabouts, the most effective method to find land would be to launch a number of small boats in different directions. The combined probability of one of the boats to find land is much greater than if the ship was trying to figure the most probable direction and follow it. This method comes of course with a risk, as many of the boats may not return at all.

The importance of planning, navigation or exploration in bridging the gap between reality and vision depends on how reliably we can predict the future based on our past experiences. In the case of the sociotechnological system, strategic exploration becomes more effective when our past experiences fail to inform us about future events due to hyperconnectivity, reflexivity and accelerating change (see **6.2**). We therefore propose strategic exploration as the best approach for directing the sociotechnological system towards a World of Views. Strategic exploration implies the distribution of most of the system's resources among iterative moves towards the desired direction without extending them beyond the planning horizon. Once the planning horizon is reached, a new current situation needs to be identified and the gap to be bridged, reiterating the same procedure.

In order to bridge the gap between the current state of affairs and a World of Views, what seems to be within our planning horizon is a careful dissolution of the hubs of power and cultivation of much more fluid hierarchical structures. Practically, this amounts to promoting, facilitating and developing trends which are already in motion:

The rise of alternative currencies. A purely peer-to-peer version of electronic cash allows financial transactions to take place without the mediation of a trusted financial institution [27]. Cryptocurrencies (e.g. Bitcoin[79]) and other alternative currency systems (e.g. Flowplace[80]) operate without central authority and the hierarchical banking system that currently controls most financial transactions. They allow both issuance of currency and transactions to be carried out collectively within a network of peers without the reliance on trusted third parties. Alternative currency systems, therefore, promote the gradual dissolution of present financial power hubs.

The rise of collective decision systems The technology of Internet and social networks offer a feasible way of exercising direct democracy without relying on mediating hierarchical structures (state institutions) to support it [28]. For additional examples see LiquidFeedback[81] and MorsiMeter![82].

The rise of Internet activism On-line activist networks (e.g. Avaaz[83]), probably having their roots in quite conventional networks of non-governmental organizations, use Internet technologies to facilitate self-organization of interest groups capable of exerting considerable influence on the political establishment thus disrupting its rigid power structures.

---

[79] http://bitcoin.org/en/
[80] http://flowplace.webnode.com/droplets/
[81] http://liquitfeedback.org/
[82] http://www.morsimeter.com/en
[83] http://avaaz.org/en/

End of the Beginning

Changing the topology of sociotechnological networks
Sociotechnological networks are those networks that organize the flow of information and goods within the sociotechnological system. The prevailing topologies of these networks are scale-free [1] implying self-reinforcing and self-sustaining power hubs within the system. We see important new trends introducing much more plasticity into the configuration sociotechnological networks:

- **Internet** has started as a truly distributed system where every node in the network was fairly equal. In time, the Internet evolved into a scale-free network structure with many unimportant regular nodes and only a few huge hubs (e.g. Facebook, Google, Amazon) controlling most of the information and its traffic. Such network topology is prone to attacks, control, surveillance and corruption[84] [1]. Peer-to-peer and distributed computing technologies[85], including multicast packet routing schemes[86], grid computing[87], fog computing[4], distributed messaging systems (e.g. Bitmessage[88]) and distributed file systems (e.g. Git[89]) contribute to the overall plasticity of the Net.

- **Edunet** is a network of collaborative production, sharing and use of the educational resources without institutional control over the development of educational programs, curricula and norms. We see the roots of Edunet in the currently emerging movement of Massive Open Online Courses[90], semantic web technologies (e.g. Google's

---

[84] The NSA Files, http://www.theguardian.com/world/the-nsa-files

[85] http://en.wikipedia.org/wiki/Distributed_computing

[86] http://en.wikipedia.org/wiki/Multicast

[87] http://edutechwiki.unige.ch/en/Grid_computing

[88] https://bitmessage.org

[89] http://git-scm.com/

[90] http://en.wikipedia.org/wiki/Massive_open_online_course

Knowledge Graph[91]) and self-organizing knowledge networks as proposed in [17].
- **Enernet** is an emerging network of distributed energy production, sharing and storage supported by technologies like Smart Grid[92], microgeneration[93] and distributed energy storage[94]. The Enernet will enable its users to become producers and distributors of energy[95]. The emergence of the Enernet on a large enough scale will have a profound disrupting impact on the geopolitical determinants of hierarchical power structures (e.g. nation states, energy corporates) and the struggle among them.
- **Matternet** is a distributed system of design, production and delivery of manufactured material goods (hardware in the broad sense) supported by DIY and Makers movements[96]. Current technologies facilitating the emergence of the Matternet are 3D modeling and 3D printing[97] which will allow the sharing of physical designs and their production without the need for large scale factories and distribution systems.

10. Conclusion

Our envisioning of the brink of Singularity begins with redefining Singularity as an historical process, rather than an

---

[91] http://www.google.com/insidesearch/features/search/knowledge.html
[92] http://en.wikipedia.org/wiki/Smart_grid
[93] http://en.wikipedia.org/wiki/Microgeneration
[94] http://www.renewableenergyworld.com/rea/news/article/2013/07/the-case-for-distributed-energy-storage
[95] http://www.wired.co.uk/magazine/archive/2012/02/ideas-bank/energy-sharing
[96] http://www.forbes.com/sites/tjmccue/2011/10/26/moving-the-economy-the-future-of-the-maker-movement/
[97] http://en.wikipedia.org/wiki/3D_printing

End of the Beginning

event. It is the process of continuous intelligence expansion since the beginning of human civilization. We emphasize the value and significance of the continuity of this process rather than the intermediate stages through which it passes. By that, we position the brink of Singularity situation within the continuum of human evolution, the evolution of life and evolution as a universal process. We raise the question of what would be the desired configuration and dynamics of the sociotechnological system able to facilitate open-ended intelligence expansion. A World of Views is our vision of such a configuration. We then argue why a World of Views is likely to be the only feasible configuration capable of sustaining the Singularity as a process of intelligence escaping its biological constraints and beyond. Finally, we propose distributed social governance as a bootstrapping mechanism for a World of Views and link it with the current momentum of the sociotechnological system.

At the basis of the evolutionary shift humanity is undergoing on the brink of Singularity is the progressive process of entering into symbiotic relationships with its technological artifacts. This symbiotic convergence deemphasizes the anthropocentric perspective in regard to the future. Furthermore, the past consensual understanding of what constitutes our humanity cannot serve us effectively under circumstances of accelerating sociotechnological change. From the social perspective, the most important are those artifacts that augment social interaction of intelligent agents as currently the Internet primarily is. Such artifacts do not only change us individually, they transform the very fabric of human civilization. We take therefore a systemic approach, first by focusing our discussion on the dynamics of the sociotechnological organism humanity is becoming, and second by introducing worldviews as the relevant units of evolution of sociotechnological organisms.

Our analysis of the sociotechnological evolution shows that circumstances of hyper-connectivity, reflexivity and acceleration beyond their many obvious benefits expose the sociotechnological system to fragility that will only increase in the near future and may lead to some catastrophic though yet unpredictable consequences. In order to counter this systemic effect we apply the concept of antifragility - the property of systems that thrive on volatility and uncertainty - and conclude that antifragility is necessary to secure the sociotechnological system from devastating catastrophic events. To that end, we need a paradigm shift towards what we call a World of Views. A World of Views is a nested, self-organizing structure, where worldviews occupy the highest level but in themselves are modular, open and diverse. Diversity, modularity and openness are the essential properties that together characterize an adaptive structure capable of containing failures while propagating successes within the larger system, thus realizing antifragility at multiple scales.

An antifragile sociotechnological system, however, is much more than just dodging existential risks. We argue that the benefits of technology will gradually transport humanity into an age of abundance, which will in turn have profound effects on sociotechnological evolution. This self-amplifying reciprocity will result in decline and even disappearance of evolutionary pressures that arise from limited resources and survival needs. We propose that abundance will catalyze active expressions of freedom that will become novel evolutionary selectors. In our vision of the future, the expression of freedom rather than survival is the ultimate driving force of evolution. We conclude that the World of Views is a catalyst of future abundance, which in turn reinforces the dynamics intrinsic of the World of Views. This positive feedback mechanism, once set in motion, will bootstrap the sociotechnological system towards a World of Views.

Finally, we introduce in broad lines a distributed social governance system that we foresee as instrumental to the development of a World of Views. Distributed social governance system is the implementation of a World of Views on the social plane. It is a radical extension of a democratic governance regime in a sense of abolishing the single unified paradigm, in favor of continuous construction and dismantling of experimental models that partially work. It is clear to us that global education systems are the essential key towards distributed social governance, teaching us to live in a world without survival constraints, giving up the idea of a single value system, constructing individual and shared realities and constantly innovating on them.

A transition from our current hierarchical power structures to distributed self-organized ones is of course not certain. Nation states and other contemporary organizations tend to be self-persistent and self-reinforcing (e.g. army generals inventing and initiating wars to justify the existence of armies). We believe, however, that conservative structures will either disintegrate or adapt due to their exposure to accelerating change. Those that will adapt will eventually transform to become open and modular because such properties will characterize the best strategy of operation within the accelerating dynamics of the future sociotechnological system. The realization of a World of Views and distributed social governance system does not promise or assume a peaceful and safe world for every individual. Yet, it does safeguard the continuity of intelligence expansion and the affluence it promises.

We are aware that human nature itself could impose a serious impediment. Some claim that human nature is selfish and brutal and these traits are an inevitable consequence of our biological heritage as well as a primary shaping factor of our future. Yet, we believe that intelligence, escaping its biological constraints, will present us with the actual means to overcome this inevitability of a nature shaped by a struggle for survival. Having said this we do realize that our vision indeed requires a fundamental transformation of the human psychological construct, individual as well as collective.

We have barely tapped the tip of the iceberg of our

sociotechnological future which is mostly submerged in the waters of unpredictability. In one of Star Track's episodes[98], Captain Picard, the 'natural' human, gives Data, the 'post human', a lesson: It is possible to commit no errors and still lose. That is not a weakness. That is life. Inasmuch as one can commit no errors and still lose, one can also commit many errors and still win, or at least keep on playing. That is life. As a young civilization, still making its first steps, we choose to read in this an optimistic note: trust life and trust our wits while trying very hard not to commit those errors which are fatal. As our way into the future will require a lot of trial, we'll have to cope with many errors and their consequences. That is not a weakness. Like princess Scheherazade from 1001 nights, determined to live another day, we tell another story, the long tale of the Singularity, as long as the history of the human. Realistic as we claim it to be, admittedly with a touch of the fantastic; it is for sure an ultimately challenging vision.

REFERENCES

Albert and Jeong and Barabasi, "Error and attack tolerance of complex networks", *Nature* 406, 6794 (2000), pp. 378--382. PMID: 10935628

Banathy, Bela H., Designing Social Systems in a Changing World (Springer, 1996).

Bar-Yam, Yaneer, "Complexity Rising: From Human Beings to Human Civilization, a Complexity Profile", *Encyclopedia of Life Support Systems* (2002).

Bonomi, Flavio and Milito, Rodolfo and Zhu, Jiang and Addepalli, Sateesh, "Fog computing and its role in the internet of things", in *Proceedings of the first edition of the MCC workshop on Mobile cloud computing* (New York, NY, USA: ACM, 2012), pp. 13--16.

Campbell, Donald, "From Evolutionary Epistemology via Selection Theory to a Sociology of Scientific Validity", *Evolution and Cognition* 3, 1-2 (1997).

---

[98] *Star Trek: The Next Generation, Peak Performance"*, http://en.wikipedia.org/wiki/Star_Trek:_The_Next_Generation

Craig, Campbell, "The Resurgent Idea of World Government", *Ethics & International Affairs* 22, 2 (2008), pp. 133--142.

DeLanda, Manuel, A new philosophy of society: Assemblage theory and social complexity (Continuum Intl Pub Group, 2006).

Deleuze, Gilles, *Nietzsche and philosophy* (Continuum, 2006).

Diamandis, Peter H. and Kotler, Steven, *Abundance: The Future Is Better Than You Think* Fifth Impression (Free Press, 2012).

Drexler, Eric K., Radical Abundance: How a Revolution in Nanotechnology Will Change Civilization (PublicAffairs, 2013).

Giere, Ronald N., "Critical Hypothetical Evolutionary Naturalism", in **Heyes, Cecilia M. and Hull, David L.**, ed., *Selection Theory and Social Construct: The Evolutionary Naturalistic Epistemology of Donald T. Campbell* (SUNY Press, 2001), pp. 53--70.

Good, Irving John, "Speculations concerning the first ultraintelligent machine", *Advances in computers* 6, 31 (1965), pp. 88.

H. Dettmer, William, Goldratt's Theory of Constraints: A Systems Approach to Continuous Improvement (ASQ Quality Press, 1997).

H. Dettmer, William, Strategic Navigation: A Systems Approach to Business Strategy (ASQ Quality Press, 2003).

Helbing, Dirk, "Globally networked risks and how to respond", *Nature* 497, 7447 (2013), pp. 51--59.

Heyes, Cecilia M. and Hull, David L., Selection Theory and Social Construct: The Evolutionary Naturalistic Epistemology of Donald T. Campbell (SUNY Press, 2001).

Heylighen, Francis, "Self-organization of complex, intelligent systems: an action ontology for transdisciplinary integration", *Integral Review* (2011).

Heylighen, Francis and Cilliers, Paul and Gershenson, Carlos, "Complexity and Philosophy", *arXiv:cs/0604072* (2006). In Bogg, J. and R. Geyer (eds.) Complexity, Science and Society. Radcliffe Publishing, Oxford. 2007.

Kanan, Karthik and Rees, Jackie and Spafford, Eugene, "Unsecured Economies: Protecting Vital Information", *Red Consultancy for McAfee, Inc* (2009).

Keynes, John Maynard, "Economic possibilities for our grandchildren", in Phelps E.S., ed., *The Goal of Economic Growths* (New York: Norton, New

York, 1969), pp. 210--211.

Latour, Bruno, *Aramis, or the Love of Technology* (Harvard University Press, 1996).

Lehman, Joel and Stanley, Kenneth O., "Evolvability Is Inevitable: Increasing Evolvability without the Pressure to Adapt", *PLoS ONE* 8, 4 (2013), pp. e62186.

Lu, Catherine, "World Government", in Zalta, Edward N., ed., *The Stanford Encyclopedia of Philosophy* Fall 2012 (, 2012).

Mandelbrot, Benoit and Taleb, Nassim Nicholas and Diebold, F. X., "Mild vs. wild randomness: Focusing on those risks that matter", *The Known, the Unknown and the Unknowable in Financial Risk Management: Measurement and Theory Advancing Practice"*, Princeton University Press, Princeton, NJ, USA (2010), pp. 443--473.

Markley, O.W. and Willis, Harman, ed., *Changing Images of Man* (Pergamon Press, 1981).

Maslow, Abraham H., *Motivation and Personality* 1st ed. (New York: Harper, 1954).

Nakamoto, Satoshi, "Bitcoin: A Peer-to-Peer Electronic Cash System", *Network* 20, 3 (2009), pp. 1--9.

Rodriguez, Marko A. and Steinbock, Daniel J. and Watkins, Jennifer H. and Gershenson, Carlos and Bollen, Johan and Grey, Victor and deGraf, Brad, "Smartocracy: Social Networks for Collective Decision Making", in (Los Alamitos, CA, USA: IEEE Computer Society, 2007), pp. 90b.

Rosecrance, Richard N., The Rise of the Virtual State: Wealth and Power in the Coming Century (Basic Books, 1999).

Simon, Herbert A., "The architecture of complexity", *Proceedings of the American philosophical society* 106, 6 (1962), pp. 467--482.

Soros, George, The Crisis Of Global Capitalism: Open Society Endangered1 (PublicAffairs, 1998).

Taleb, Nassim N., *Antifragile: Things That Gain from Disorder* (Random House, 2012).

Taylor, Frederick W., *The Principles of Scientific Management* (Harper & Brothers, 1913).

Veitas, Viktoras, "System thinking based municipal policy design for the improvement of local business environment: Siauliai city case study", ISM

University of Management and Economics (2007).

Vidal, Clément, "What is a worldview?" De wetenschappen en het creatieve aspect van de werkelijkheid (2008).

Vidal, Clément, "Metaphilosophical criteria for worldview comparison", *Metaphilosophy* 43, 3 (2012), pp. 306--347.

Weinbaum, David R., "A Framework for Scalable Cognition: Propagation of challenges, towards the implementation of Global Brain models", *GBI working paper 2012-02* (2012).

Willke, Helmut, Smart Governance: Governing the Global Knowledge Society (Campus Verlag, 2007).

WEF's Global Agenda Council on Complex Systems, "Perspectives on a Hyperconnected World - Insights from the Science of Complexity" (2013).

End of the Beginning

Dialogue 12.1

With Viktoras Veitas, David Weinbaum (Weaver), Ben Goertzel and Ted Goertzel

Ben: What a long, deep and wonderful exposition you guys have given! There are so many things to respond to; I'll only mention a very few of them now, or else the discussion of your paper would end up occupying an entire book itself!

First of all, you mention: "A self-organizing form of social governance based on dialogue and mutual agreement at multiple levels and not on idealized top-down control structures seems to be the only alternative to anarchy and dystopia on the brink of Singularity." That is certainly an intriguing vision. But it seems very different from the governance methods that currently control our world, e.g. the US Congress, the committees that govern the EU, or the Chinese Communist Party.... Or the large corporations that are closely linked with these entities. I wonder how you see the transition taking place? Via what sort of dynamic will these existing institutions gradually (or suddenly?) cede their practical power to a new, self-organizing form of social governance?

Weaver: I will try to answer your question from a second order cybernetic approach where I assign to the concept worldview a central role. Actual systems of governance are sociocultural constructs that both reflect and embody certain patterns, images and beliefs that are collectively held by large populations and are operationally applied at many contexts and scales. These patterns can be understood as constituting the content of a civilization level memory, both episodic (history) and procedural (how we believe things are done). This is what we call in our paper a worldview. The components of established worldviews are not arbitrary; they are outcomes of individuating processes (I follow Simondon and Deleuze in using this term) that

End of the Beginning

took place along history and in the context of governance, at least some of them, distill viable and effective operational regimes that are products of a prolonged sociocultural selection. Worldviews can be understood as the complex representational patterns of a sociocultural reality. They are both affecting and being affected by actual states of affairs hence the second order cybernetic approach.

The problem is that being products of a complex reflexive system with many feedback mechanisms, patterns often tend to become self-persistent and self-propagating and to a large extent independent of the original function they are supposed to carry as part of a collective sociocultural cognitive mechanism. In other words, the cybernetic nature of complex sociocultural systems may allow for certain systems to persist as semi-autonomous parasitic entities irrespective to their diminishing performance in terms of fitness or benefit in the wider context of human civilization. From this perspective, only when our images (i.e. our worldview) change, our minds change, and only when our minds change, we can introduce change in the world. But it is also true that for the collective image to change the world must change[99]. So when thinking about change, we find ourselves in a sort of chicken and egg situation characteristic to reflexive systems.

I do not believe therefore that any radical change in contemporary governance systems can take place without a radical change in the collective patterns and images that shape our worldview. In other words, we need to change our minds individually and collectively as part of such transformation. So how a transition of the kind we argue for in the paper can plausibly take place, especially in the light of the fact that patterns already strongly established tend to be self-reinforcing and overly resist change?

---

[99] *This is all supported by similar ideas about niche construction where an organism changing its environment is perceived as extending its phenotypic expression and by that affecting genetic selection.*

End of the Beginning

As image and actuality of social governance seem to be locked into a stable attractor in sociocultural dynamics, we need to identify three factors that are essential for transformation to take place. The first factor is the existence of forces that destabilize the current attractor. The second factor is the existence of seeds, new images and conceptualizations of governance that given the right circumstances may consolidate into a new (metastable) attractor. The third factor has to do with identifying and supporting actual trajectories leading to such attractor.

At present, I think that the balance of forces holding in place both the actual world order and our collective gestalt perception of the world is under strong pressures to change. The pressures, as we briefly describe in our paper, have primarily to do with accelerating change, hyper-connectivity and the failure of our current images to guide a proper adaptive response to self-induced disruptive changes. Failing to adapt to the new circumstances brought forth by technological progress, these systems might response in the interest of self-preservation by trying to regulate the rate of change. We can expect therefore a backlash of conservative powers in global social dynamics. Yet, I strongly believe we are past the point where the accelerating sociotechnological change can be effectively regulated and if this is indeed the case, current governance systems are bound to either radically adapt or collapse. Neither the current governance systems nor the collective patterns and images supporting them can survive for long the destabilizing effects of accelerating change and exponential increase in complexity. This accounts for the first factor.

As to the second factor, our image of the world as a universal clockwork mechanism and the associated conception of central top-down control of our environment are slowly but surely losing traction and confidence. These are replaced by an image of distributed self-organizing networks. This transformation of the collective image is not arbitrary but is a complex and prolonged psychological and cultural response to many revolutionary discoveries that took place for centuries. At a very deep level we realize that not only our existence as humans is not privileged in the grand picture of the universe but the very construct of the universe seems to admit no point of privilege whatsoever. The new emerging image of the world and us in the world is one of vast interconnectedness and affective mutuality where the very concept of singular solid identity is being cracked. Slowly we leave a world of universal laws where everything is predictable and controllable and where the future greatly resembles the past, and enter into a world of unpredictable and only partially controllable metastable processes. These are the kinds of concepts and images that I see as seeds of a new kind of dynamics that under favorable circumstances will converge to a World of Views. I see the relatively recent emergence of the internet as concomitant with such new images.

The third and perhaps most challenging factor has to do with identifying trajectories. I believe that distributed self-organizing social governance will arise primarily bottom up from the periphery of the system, i.e. minority groups, small scale experimental organizations etc., to the center i.e. the major institutions of governance such as governments, monetary organizations and corporations. Already today we see many early signs to this motion: communities minting local currencies and preferring local self-organized services in everyday affairs to paying for municipal and state supplied services; DIY (Do It Yourself) powerhouses that start to exert impact on scientific and industrial establishments, 'open source everything' communities, a vivid crowd funding scene and many more. Such processes will relieve the adaptive pressures from existing governance systems and therefore will not meet substantial resistance at least not immediately.

Also top-down processes may contribute to decentralization. Central governing bodies will increasingly delegate autonomy to local governing agents as part of their adaptation to increasing complexity. China with an increasing number of mega cities is perhaps a good example. The central government must delegate extensive authority and policy making privileges to local governance systems of such mega cities in order to allow effective governance. The shift of powers is only at its preliminary stages but I can foresee a slow dissolution of centrality in governance and the rise of a network of strong interconnected local governance systems. Also within mega cities it is necessary to further distribute governance if only to make it minimally effective. In summary both bottom-up and top-down processes seem to already contribute to the emergence of distributed governance and plausibly will increasingly continue the trend towards a significant global phase transition.

I do not put much importance on the exercise of developing specific 'how' scenarios because bottom line they are all highly speculative. What I wish to highlight is that we can clearly identify the presence of the factors necessary for transformation. As long as these are present and become stronger, I have a growing confidence in the plausibility of the transformation we envision.

Viktoras: Indeed, a vision of A World of Views is very different from the current sociopolitical landscape and governance methods. It is no surprise – a World of Views is envisioned to sustain the close to Singularity situation (pre-Singularity as you called it) which is no less different from the current state of affairs.

Another remark is that while current governance structures are hierarchical in nature, they are far from same. If we imagine a continuum where strict hierarchy is on the one end (say black) and complete self-organization is on the other (say white), all the current governance systems are different shades of gray, some darker, some lighter. Even without considering dictatorships, we can still observe important differences between sociopolitical systems of USA, EU and, say Switzerland. Switzerland is a notable

End of the Beginning

example of a 'semi-direct democracy' where citizens can overturn decisions of federal institutions via referendums (Swiss citizens vote on average 4 times per year). Governance systems are of course tightly coupled with what Weaver calls "collective patterns and images that shape the worldviews" prevalent among people of every nation or continent. Therefore I agree that radical change in governance systems can happen only together with the radical change in our ways of thinking, which implies a gradual, not sudden change.

I see this transition happening through gradual adaptation, somewhat similar to the rise of democratic governance regimes in the modern world. By "gradual" I do mean "slow" however, as the speed of adaptation should match the moment of change of world's sociotechnological system, which is very high and increasing. Hierarchical powers, as well as any self-sustaining structures usually do not surrender without struggle, which unfortunately means that the change may imply violence of different magnitude. On the sociopolitical plane we can see that in the uprisings of, say, 2010s, spanning from Thailand to Venezuela. Revolutions are of course nothing new, but there are other, less overt cases of change, induced by on-line activism or social network based collective pressures. I see latter more important in the near and middle future due to their potency and non-violent nature. Yet the driving force of both is the fact of increasing empowerment of capacities of individuals - with all its positive and negative implications for sociotechnological system's dynamics and stability.

Regarding large or not so large business corporations, purely hierarchical mode of organization is being challenged with other, more efficient management styles for quite a long time already. For quite a radical example, see "Maverick" by Ricardo Semler[100].

---

100 Semler, R. (2001). *Maverick!: the success story behind the world's most unusual workplace*. London: Arrow.

End of the Beginning

Ben: I like your vision of a "World of Views", yet I wonder if a "view" itself is an overly static sort of entity. Perhaps the "World of Views" itself will be a transitional phenomenon, en route to the emergence of something yet more radical post-Singularity? What comes just beyond views, I suppose, would be some kind of dynamic whose role it is to **generate** views – something that synthesizes views on the fly, transcending each view almost as soon as it is created. Any thoughts in this direction?

Viktoras: First of all, it seems that we are thinking in the same direction. In the framework of a "World of Views", a view is an inherently dynamic entity. Diversity, modularity and openness allow views to morph with each other, change and adapt. But they also allow the worldviews to be relatively stable.

The distributed social governance system is envisioned for allowing this kind of dynamics of generating new worldviews and intercourse between them plus relative stability of the old ones. Now, it becomes a little complicated when it comes to the question of what is stable / transitional or what does it mean 'on the fly' in the context of near Singularity. Yes, a "World of Views" is a transitional phenomenon in the sense that is not going to last forever. But it is not transitional in the sense that I already see or envision the system that may come next.

Regarding generating worldviews on the fly, what seems "on the fly" for us may be pretty stable for, say, advanced semi-artificial intelligence emerged on the Brink of Singularity and operating three billion 'thinking' cycles faster than we do. Nevertheless, I tend to think that worldviews (as incomplete yet more or less consistent representations of reality) will in general change at a slower rate than the whole. Regarding post-Singularity - well, there is no way to see. Therefore, I am more concerned that we do not screw up during the "transitional" period.

End of the Beginning

Ben: The problem can be distilled to the following: what kinds of identity constructs and individuation processes (at all scales) can thrive in circumstances of extreme volatility and freedom? Synthesizing worldviews and transcending them on the fly is definitely resonating with my attempts to figure a philosophical foundation of identity fit to our post singularity future. The real challenge is of course path finding - coming up with actual conceptual bridges between 'here' and 'there', the 'now' and 'then', while the 'there' and 'then' are far reaching and highly speculative.

Weaver: I like this question because it allows exploring the more speculative and far reaching aspects of our vision. First, one should not read 'view' as something passive or static. The title of our paper is a word play on the word worldview which is meant to highlight its futuristic extreme multiplicity and diversity but also an underlying unity. Not only we envision many interacting worldviews but these might become radically diverse. 'view' in this sense is short for 'worldview' which is very far from the designated activity of a passive observer. A worldview is a complex, adaptive cognitive structure that encompasses the cognitive competences of both individuals and complex social organisms.

Synthesizing worldviews on the fly... Well, to give you a clue where this meets me I'll dedicate a few words to my research. One of my primary research interests is the transformation of identity at the advent of radical post-human scenarios. Presently, human identity, normally organized around a self-centered construct, is anchored in obvious physical, biological and cultural constraints. The interesting question is what will happen to identity when these limitations will be lifted by technological innovations such as genetic reprogramming, extreme biomechanical augmentation/transformation, neural interfaces, fully immersive virtual worlds and more. To explore such scenarios I think we need to investigate the philosophical foundations of identity,

End of the Beginning

understand their presumptions and inherent limitations and figure new ontological designs that will serve as foundations to post human thought and post human individuality. To that end my research spans from the metaphysical foundations of individuality through reflection on current cognitive and cybernetic models of agency (self-organizing multiplicities) to highly speculative futuristic kinds of technologically enabled individuations such as co-existing multiple streams of consciousness, collaborative editable memories, radical transformations of body plans, telepathic multiply embodied and multiply situated minds etc.

Generating views, as you put it, immediately brings to mind Nietzsche's concept of the Overman through which he tried to address the problem of freedom. One does not overcome his own humanity as a single one-time act or event of liberation. The Overman is an individual undergoing a continuous transformation that tears apart any notion of fixed sedentary identity. It evolves at the expense of giving-up, annihilating in fact, its own continuity of identity and becomes nomadic, roaming the expanses of the Mind. How is that possible? Presently, it is not; unless we make it possible by envisioning a new kind of individuality that transcends the human form as we know it and embody this new form in our way of life.

The play of ongoing transformation of perspectives and worldviews is indeed something I envision as possible if not a necessity in post Singularity scenarios. An example of how it might be is very well depicted in Charles Stross' Accelerando. Such existence is not comprehensible and does not make sense when we try to ground it in the conceptual and psychological frameworks we operate at present. It is difficult to believe that our deepest notions of identity and individuality whether applied to humans, minds in general, or other complex intelligent agencies, will survive a singularity.

The problem can be distilled to the following: what kinds of identity constructs and individuation processes (at all scales) can

End of the Beginning

thrive in circumstances of extreme volatility and freedom? Synthesizing worldviews and transcending them on the fly is definitely resonating with my attempts to figure a philosophical foundation of identity fit to our post singularity future. The real challenge is of course path finding - coming up with actual conceptual bridges between 'here' and 'there', the 'now' and 'then', while the 'there' and 'then' are far reaching and highly speculative.

Ben: You say: "A World of Views is a nested, self-organizing structure, where worldviews occupy the highest level but in themselves are modular, open and diverse. Diversity, modularity and openness are the essential properties that together characterize an adaptive structure capable of containing failures while propagating successes within the larger system, thus realizing antifragility at multiple scales."

This brings two questions to mind...

First, how important do you think is the hierarchical aspect of your World of Views perspective? Myself I would have thought of the World of Views as more of a heterarchy, with the rank ordering between two views being something that might be judged differently from different viewpoints. Could you say a little more about this aspect?

Viktoras: The problem (at least for me) is that the concept of heterarchy is pretty vague and may mean different things - e.g. horizontal relations within a hierarchy which is definitely not what we mean by a World of Views. Therefore, I will define how I understand the term 'heterarchy' and discuss how it relates to a World of Views.

Classically, heterarchy is defined "as the relation of elements to one another when they are unranked or when they possess the potential for being ranked in a number of different ways"[101]. To

---

101 Crumley, C. L. (1995). *Heterarchy and the Analysis of Complex*

437

make this picture similar to what we envision in a World of Views we have to add dynamics: elements of the system as well as relations between them get constantly created and/or dissolved. The ranks of elements also change or actualize their potential as the system adapts to the needs and circumstances of the moment. Another perspective characterizes heterarchy as a container of multiple hierarchies. In that sense you can see a World of Views as a multiplicity of constantly rising and falling hierarchies being supported as long and as much as they are functionally useful for the elements of the system. I am not sure whether the concept of heterarchy includes this dynamism. If it does, then yes indeed, a World of Views is a heterarchy...

Weaver: As to this question, we need to clarify the terms and their usage. From a cybernetic perspective the sociotechnological system will be inhabited by pretty complex organizations that will implement hierarchical structures or functions. These hierarchies however do not imply corresponding hierarchical power structures beyond their designated and agreed upon function. Agreement in such cases involves all the agents constituting the structure. I do not see an inherent problem in hierarchies per se. Hierarchies prove to be very helpful and effective coordinating and organizing schemes. Problems arise when functional or structural hierarchies become a basis to social and political power hierarchies. Such problems will not arise however if we ensure a balanced bottom-up versus top-down exchange of information and control signaling within hierarchies.

The World of Views, as its name implies is indeed a heterarchy in the sense that it is a multiplicity of heterogeneous yet coexisting and often overlapping sociotechnological entities holding various worldviews that are themselves dynamic. From a

---

Societies. Archeological Papers of the American Anthropological Association, 6(1), 1–5. doi:10.1525/ap3a.1995.6.1.1

End of the Beginning

perspective of power relations and authority, all such entities exist on the same plane and their interactions are emergent. It is clear that some of these entities might become more successful or more influential than others economically, technologically or otherwise[102]. The civilization envisioned in the World of Views is a civilization that greatly encourages open sharing of success. In an age of abundance there will be virtually no incentives to agents to withhold knowledge in order to gain advantage over other agents and use knowledge or success to leverage power over others. There will be virtually no incentives for agents to conquer, enslave, subdue or exploit other agents, as such activities definitely belong to operational perspectives that arise in response to circumstances of limited resources. Social entities will have relations of cooperation, gaming, dialogue and friendly competition where one can win prestige and respect. At extreme cases, social entities holding polar irreconcilable worldviews will co-exist in disregard of each other. At no point, differences in perspective will invoke violent conflict of any kind be it physical, social, psychological or other. In abundance the rule will be "give whatever you can give" and not "take whatever you can take" that characterizes the situation of want.

I believe that rank ordering between various worldviews will indeed exist but as you remark ranking will never happen according to a single scale of values as the whole structure is a heterogeneous multiplicity. Prestige and other forms of social capital can be gained and given and they will probably become important social differentiators. The only exception would be regarding the foundational values of diversity, openness, sharing

---

[102] *I do not see a place for military bodies in the future. In John Wright's trilogy "The Golden Age", there is a single human functioning as both a 'soldier' and 'policeman' for the whole interplanetary civilization. He is unemployed most of the time. In Ian Banks' Culture series, high level military competence is always kept but is actually used in very rare and extreme cases. Conflicts are resolved by diplomacy and games.*

End of the Beginning

etc. The World of Views is a fluid and interdependent system founded on a mutual platform of agreed upon principles, values and procedures. Participation will always be an option and not compulsory. Still, globally accepted commensurable set of foundational values will be necessary to keep the whole system from dissipating.

Ben: Second, "Diversity, modularity and openness" sound like wonderful characteristics but also sound like possibly a recipe for a system to dissipate. What is the recipe for the coherence of a system of this nature? I suppose it is to be held together via some sort of self-organizing attractor-like phenomenon?

Viktoras: If you see a World of Views as a 'static' heterarchy as discussed above, then indeed, diversity, modularity and openness are exactly what is needed for a given rank order to get challenged and destroyed. But, looking from 'dynamic' point of view, different elements, their relations and rank orders emerge through the process leading to new configurations of the heterarchical system. In that sense, the system itself does not dissipate, it just adapts (which could be quite radical, by the way).

Following the prevailing worldview, hierarchy (preferably stable) and order are synonyms; order is opposite to chaos; and chaos is no good – it is "the end of the world"; so, we need hierarchy. I think there is something very wrong with this line of thought. First, there is a difference between "the end of the world as we know it" and "the end of the world". Actually, there is nothing new in this: if we compress, say 100 thousand years of human history into one year, we will not see any stability any more – there is only change. Can this kind of system be called 'coherent'? Now, sociotechnological development is exactly that – increasingly accelerating change. The only difference is that it becomes so fast relatively to human lifespan that we are not able any more to avoid or hide from it. In (pre-) Singularity situation by

End of the Beginning

definition the change will reach extreme levels, beyond [current human] comprehension. That is why we emphasize the aspect of individual and collective psyche and patterns of thought.

Regarding "self-organizing attractor-like phenomenon", it depends what you mean by attractor. If you mean specific hierarchy or heterarchical configuration (i.e. rank order) or even specific balance between diversity, modularity and openness, then the answer is no. If you mean some sort of higher order attractor which forces the system to change and adapt and never allows it to fall into a local optima, then probably it is closer. But I would rather call this kind of 'attractor' a 'process of Singularity' – the one to be supported by the distributed social governance system.

Weaver: As to your question of what would hold such a system together, this is of course one of the more challenging questions. Main-stream views about human nature will not be highly optimistic about the prospects of the emergence of a World of Views. It is indeed the case that the human psyche and its sociocultural ramifications have evolved to ensure survival under conditions of want. It seems that fear and violence were for millennia the most basic and most powerful mobilizers of human activity. These traits of human nature seem to be prohibitory to visions such as the World of Views. There are of course exceptions to the contrary but these are anecdotal in historical perspective. Nevertheless I am optimistic and my optimism stems primarily from the realization that what is most significant in human nature is its plasticity and adaptivity and not the particular traits it shows in the presence of specific constraints. Once we reach a certain baseline of abundance for all, and this is not a farfetched prospect as we once used to believe, it is fairly plausible that the traits forged to secure survival will weaken and the overall balance of forces as reflected in our social dynamics will shift towards much more positive traits such as tolerance, cooperation, openness and more. As already mentioned before, I do not see the World of

End of the Beginning

Views vision realized without a corresponding transformation of the human mind. I see the onset of an age of abundance as a very powerful mobilizing force towards such a transformation. This will not happen overnight but it needs not to take longer than a few generations. From a complementary angle, antifragility seem to support faster convergence to abundance and increase in abundance reinforces antifragility. This projected dynamics seems to both push the system towards a World of Views kind of dynamics and also foreseen to become a major stabilizing factor in this futuristic metastable scenario.

Ted: You note that there is a "deeply ingrained cultural belief in the inevitability of hierarchical governance based on authority and enforcement..." I agree, but I wonder if this cultural belief may not reflect some collective wisdom. Might some balance between hierarchy and heterarchy be a more realistic view, if not for the long term future, at least for the transitional period we are most interested in? For a concrete example, the capitalist financial system is largely heterarchial, the major players act independently. But they also rely on the legal system to resolve disputes. And when the system goes into crisis, central banks and other governmental institutions intervene. How will the financial system get away from this need for governmental regulation and correction? How do you visualize this happening? Do you expect that the heterarchial elements will wither away gradually through disuse, or will they be abolished through revolutionary intervention of some kind?

Viktoras: Collective wisdom is a two-edged sword: on the one hand it denotes shared knowledge arrived at by groups, on the other – the ability of collective learning[103]. Our bodies, nervous systems, psychology and culture have all evolved in a

---

103 http://en.wikipedia.org/wiki/Collective_wisdom

End of the Beginning

comparatively stable environment of the Earth's biosphere. Actually, hierarchical thinking may very well be rooted in the way our central nervous systems are organized (interestingly, cybernetician Warren McCulloch has hinted about that in his seminal paper on heterarchy[104]).The bottom line is that " cultural belief in the inevitability of hierarchical governance" could very well be a collective wisdom / shared knowledge accumulated through centuries. But this fact may not help us in the era of unpredictability; or even worse - do us harm (a long and in-depth discussion of this trade-off can be found in Taleb's books[105,106]). Therefore, I would opt for using the second edge of collective wisdom – collective learning. A World of Views is a system which allows for collective learning and recovery from errors capabilities taken to the extreme.

I agree with you that the balance between hierarchy and heterarchy (or, rather, interdependence of both) in the system is very important. Yet I do not agree that this balance should be stable (it seems that you imply that in your question). In my view, the balance should gradually (but pretty fast, if expecting Singularity anywhere around 2045...) evolve into almost complete "dynamic heterarchy" which is a "World of Views" (given that we prefer it over doomsday or surveillance society scenarios, which I indeed do).

As for a concrete example with the financial system, the current system receives fair amount of criticism. Central banks and governmental institutions became not independent market supervisors, as the theory implies they have to be, but rather

---

104 McCulloch, W. S. (1945). A heterarchy of values determined by the topology of nervous nets. *The Bulletin of Mathematical Biophysics, 7(2)*, 89–93. doi:10.1007/BF02478457

105 Taleb, N. N. (2008). *The Black Swan: The Impact of the Highly Improbable*. Penguin Books Limited.

106 Taleb, N. N. (2012). *Antifragile: Things That Gain from Disorder*. Random House.

players dependent on the large corporations / banks they are supposed to regulate. That is far from "oligopolistic market with independent regulators" which was being proposed by microeconomics textbooks some years ago when I was reading them. The positive feedback loops built into such systems amplify instabilities rather than damp them.

What really worries me with respect to the above question is that a self-organizing liberal market exhibits a "preferential attachment" or "the rich get richer" effect[107]. There is nothing wrong with the rich companies or people *per se*, except that this effect tends to create power hubs and, consequently power hierarchies. This is usually solved by market supervisory institutions enforcing anti-monopoly laws, but they are prone to problems outlined above. I tend to think that this systemic problem may be related to the very "independence of the major players" mentioned by you, which motivates them to act selfishly and reduces the potential of collective intelligence of the whole system. A World of Views with distributed social governance system or similar self-organizing ecology may be a viable alternative[108].

Finally, at the beginning of a "transitional period", I would expect hierarchical structures to be abolished through more or less revolutionary intervention or pressure. On the other hand, as the system gradually loses its hierarchical nature, more changes would happen due to what you called "withering away because of disuse". In a World of Views like system the latter case may be prevalent, but I do not see political pressure and power games going away either. Distributed social governance system is meant to enable and promote a more self-organizing (something like re-ranking of elements in a heterarchical system) versus violent (head

---

107 http://en.wikipedia.org/wiki/Preferential_attachment
108 Helbing, D. (2013). *Economics 2.0: The Natural Step towards A Self-Regulating, Participatory Market Society. arXiv:1305.4078 [physics, Q-Fin]. Retrieved from* http://arxiv.org/abs/1305.4078

End of the Beginning

to head confrontation and destruction of hierarchies) dynamics of sociotechnological evolution.

Weaver: I do agree with you, of course, that the cultural belief regarding the forms governance must take, reflect a collective wisdom in the sense that it responds to certain existing traits in human nature that if not constrained will be very disruptive to any social organization. It is an essential point in my approach that a social transformation and a transformation of human nature are tightly coupled and both will be reflected in certain foundational aspects of our collective worldview such as the way we understand and apply hierarchy.

I do not think that no hierarchy whatsoever is a realistic situation in a complex system. Any non-trivial activity requires planning and planning already implies a hierarchy of nested steps. I guess your question is more about hierarchies of power and the value systems they impose. I think that in a futuristic system hierarchies will cease to exist as such as soon as they will fail to serve the designated purpose which is acknowledged and continuously affirmed by all the components constituting the hierarchy. The tendencies of power hierarchies to self-persist, self-reinforce and self-justify beyond their designated functions will be much weaker in a metastable sociotechnological ecology.

Central financial institutions justify their existence by highlighting the need to regulate the financial system. This state of affairs is strongly ingrained in the way we think but it is far from being an absolute truth; it is rather a solution that sometimes works and often doesn't. I would argue that a distributed self-organizing financial system will have the capacity to self-regulate without overseeing central regulatory functions. We see it all around in living organisms and whole ecologies.

Regarding revolutionary interventions, personally I believe that extreme revolutionary interventions are effective only in rare cases. Our image of a revolution relies much on history and myth. But isn't it the case that for a revolution to be real it must involve an element of surprise? That we cannot have a clear a priori notion of how it will happen? In this sense, I do not see revolutionary interventions as they are perceived based on past historical experience as particularly interesting or promising. If by revolution you mean a sudden and unexpected eruption of intense, disruptive activity, be it intellectual, emotional, physical or otherwise, I agree that such eruptions will be a necessary element of the dynamics of transformation but their actual manifestations are largely unpredictable. One example of an 'almost' revolution is the story around **Bitcoin**. Bitcoin is a new system of currency that replaces a central monetary authority with a distributed peer to peer algorithm that validates transactions. This is a revolutionary invention that might upturn the international monetary system. At present it is not entirely clear whether it will succeed to gain enough traction. The whole thing also suffers from certain technical and administrative problems. Still it is a good example of what would disruptive revolutionary interventions might look like in the future.

Ted: I wonder if "world of choices" might be another term for the "world of views" you envision? You talk about the "choice of profession, style of living, social group, faith, etc." and about "shaping our future." Surely we are talking about behaviors and actions as much as thoughts or viewpoints. The choice of a profession, for example, will depend on changing technologies and opportunities as well as on tastes and preferences. What kinds of professional choices do you think young people will be making twenty or thirty years from now? Do these depend on idiosyncratic viewpoints or on societal constraints and opportunities?

Viktoras: A "World of Views" implies more choice and freedom, but I would not call it any other way. Worldviews of individual people will have the decisive effect on the choices they make (see discussion of "choice zones" and selection of cheese in

End of the Beginning

section 8.1). Simply put, a worldview can be understood as a subjective representation of reality. As such, it would include the image of changing technologies and opportunities offered by the world as well as individual tastes and preferences. I believe that "professional choices" will be more guided by viewpoints than economic or societal constraints simply because there will be less of those constraints (hopefully). Actually, we do not need to go all the way to pre-Singularity times for seeing that the concept of profession is losing its importance and people are changing their occupation a few times in their lives while engaging into a lifelong university level education - it is happening right now. In 20-30 years? First, it will be something like "on the fly professional education" in the sense Ben used in his second question about "on the fly generation of the worldviews". Second, people (I am not sure how to interpret "young" in this context) will be creating new professions, occupations and lifestyles also "on the fly" so it will be far from static choices of professions being offered by the labor market, which I believe to be an outdated picture anyway.

Weaver: I prefer to use the term 'selection' over 'choice' because the concept of choice is laden with the notion of free will which is not very useful in clarifying the kind of freedom we do have as intelligent affective beings. Selection implies the existence of options in distinction from necessities that constrain or entirely prohibit selection. By selecting paths we progressively individuate. This is somewhat different from saying that by making choices we express our individuality. Our individuality is not a readymade product but is in a continuous formative individuating process. It is a process of selection wherein freedom is expressed, but it is not anyone's freedom which is being expressed. Also, freedom does not exist in any meaningful manner unless it is manifested through individuation. The more distinct we become by refining our selections, the more freedom is manifest.

As to which options will be available to young people in the future compared to the present, we need a working definition. We can agree that a profession is a body of knowledge and skills that allows one to perform certain functions within the social fabric. At present a profession is also how one 'makes a living' i.e. able to sustain herself and take care of one's needs and pleasures (if lucky). Additionally, a profession is associated with value and social status. Someone without a proper profession is considered a dysfunctional member of society.

I envision that within a timeframe of a few decades, no one will have to work for a living and therefore acquiring a profession will only be a matter of pleasure or genuine ambition concerning one's individuation. Within this time frame, the rigid categories of professions and the silos that separate disciplines of knowledge will be largely dissolved. What we will have instead is a network of connected modules that will form a distributed and interconnected body of knowledge over which a huge variety of paths can be taken. Any person will be able to acquire a transdisciplinary set of competences and knowledge modules[109] that best fits his talents and interests on one hand and also is very finely tuned to the specific needs of society on the other hand. Specialization will take a completely new meaning because of the extreme modularity and variety of connections that can be made between various disciplines.

Instead of a fixed and defined profession, people will have a competence profile which in itself will be dynamic since learning will become a lifetime occupation (and probably a major choice of pastime and entertainment). This competence profile will render obsolete any designation of profession as we know it. There will always be disciplines that will require a very high degree of specialization (i.e. surgeons) but most of these will probably be

---

[109] *This is inspired by an already existing project called POEM – Personalized Open Education for the Masses.*

End of the Beginning

delegated to or supported by specialized machine intelligences leaving to human agents the more fluid and multifaceted kinds of occupations. We will probably see competence profiles that are defined as very unique combinations of specializations. For example: aspects of architecture, engineering, design, art, psychology, ecology and agriculture will be a possible set of relevant competences for a garden designer. Still, even the said professional designation will not necessarily adhere to this exact set.

The bottom line here is that the World of Views scenario will allow vastly more selections which will actualize extreme diversity and pluralism among its individual agents and therefore will realize a social fabric with much more freedom than what is available at present.

Part Five:

Globalization and the Path to Singularity

End of the Beginning

Chapter Thirteen **Chinese Perspectives on the Approach to the Singularity**

By Mingyu Huang

Mingyu Huang is a Chinese transhumanist and singulatarian and chairman of the Shenzhen company Long-life-future LLC. He is also a writer concerned with future forecasting and the nature of social progress. He edited a book on Transhumanism in Chinese (超人类主义) which can be found at: this **link**. He can be reached at: **shen3298@qq.com**.

For thousands of years China was the most technologically advanced nation on earth. But technological advances in the last few hundred years enabled Europeans to humiliate, subjugate and exploit the Chinese. The Chinese view the coming of the Singularity from the perspective of this history. Its netizens ask how they can best use exponentially advancing technologies to accelerate the modernization of the country and to help China take a leading role in the humanity's future.

Our analysis of the future is grounded in our understanding of the reasons why China fell behind in the last few centuries. The renowned British sinologist Joseph Needham (1954-2008) argued that China's Confucian and Taoist philosophical traditions, although helpful in maintaining national unity, slowed scientific development by discouraging independent scientists from breaking with tradition. Geographer Jared Diamond (1997) placed this history in a broader comparative context, arguing that Eurasian hegemony resulted from greater availability of plant and animal species suitable for cultivation and domestication. In Diamond's view, this led to increased food production which supported dense populations and technological development, as well as enabling the evolution of infectious diseases which later decimated the populations of sparsely populated hunting and gathering peoples.

In Diamond's view, China was successful in increasing food production and developing many technologies, while its geographical compactness enabled dominance by a strong centralized government which placed limits on technological development. This contrasts with European geography which was broken by features that facilitated the development of independent political entities. Diamond (location 6867, Kindle edition) argues that:

> The real problem in understanding China's loss of political and technological preeminence to Europe is to understand China's chronic unity and Europe's chronic disunity. The answer is suggested by maps. Europe has a highly indented coastline, with five large peninsulas that approach islands in their isolation, and all of which evolved independent languages, ethnic groups, and governments: Greece, Italy, Iberia, Denmark, and Norway/Sweden. China's coastline is much smoother, and only the nearby Korean Peninsula attained separate importance. Europe has two islands (Britain and Ireland) sufficiently big to assert their political independence.

Diamond argues Europe's political disunity made it possible for explorers to find funding for their ventures, and for European empires to spread around the globe. China began similar explorations, but they were then suppressed by the central

End of the Beginning

authority for political reasons.

China today benefits from this history in its very large population and its relative linguistic homogeneity with 800 million Mandarin speakers and another 300 million speakers of closely related languages using the same characters. A strong central government dominates this very large population, in contrast to the political fragmentation found in Europe and countries created by European settlers such as the United States, Canada, Australia and New Zealand. China's central regime is today strongly committed to technological development, but it is uncertain whether this centralized system can match the entrepreneurial cultures of the leading western countries.

After opening up to the global economy, China achieved high rates of economic growth by welcoming western investment and western technology. China is second to none in the implementation of technologies such as high speed rail, supercomputers, nuclear fusion, the Compass Global Network, giant rockets, space station docking, and submersible technologies. And the Chinese government is sometimes more willing than western governments to put money into artificial general intelligence projects.

Thanks to these developments, economists project that by about 2030 China will be a GDP superpower rivaling the United States in its weight in the global economy. This is approximately the same time that the world will be approaching the Singularity, according to Kurzweil and other like-minded futurists. If China is to play a leading role in this rapidly emerging world, it must go beyond low-wage assembly industries dependent on foreign technology. It needs to manage all the problems of an advanced economy: improving the quality of goods and services, protecting the environment, maintaining high unemployment with low inflation and manageable debt, and avoiding financial bubbles. It needs to encourage and develop an entrepreneurial culture, particularly in the high technology sector. And it needs to support cutting edge scientific research in biotechnology, artificial

453

End of the Beginning

intelligence and other cutting edge fields.

The educated, modernized segment of the Chinese population is well aware of these challenges thanks to the wide penetration of Internet communication technologies, including blogging, microblogging and bulletin board systems. But many Chinese, especially in the younger generations, are concerned that the centralized bureaucratic political system may be an obstacle to the changes that will be needed. Political reform tends to lag behind technical and economic trends. The authorities recognize that China needs openness to ideas and technology, but it fears that this openness may threaten the stability of the regime.

Two contradictory trends are both present in China today. One is a very wide penetration of the Internet, reaching wider and wider segments of the population. The other is intense and expensive government efforts to censor, control and limit Internet communication. This contradiction can be expected to grow as users become more sophisticated in working around the controls and the state apparatus becomes more sophisticated in countering them. Exponentially increasing artificial intelligence technologies can be used by both sides in this continuing conflict.

China's ethnic homogeneity and strong political and cultural traditions suggest that China will maintain a path of guided gradual reform. But Chinese traditions are not necessarily hostile to the coming of the Singularity. Elements of China's tradition may help it to incorporate developments such as transhumanism and life extension. China's ancient legends and philosophical traditions such as the *Fairy Couple* (天仙配), the *Elixir of Immortality* (口生不老仙丹), the Taoist concept of spiritual immortality (精神不朽), the *jiangshi* (僵尸), and many others, may help the Chinese welcome and encourage developments that may be more strongly resisted in the west. These traditions may help to explain why the Chinese today are highly receptive to accepting new technologies, although they are not yet in the forefront in inventing them.

In the United States, cultural and religious factors continue to

End of the Beginning

inhibit the use of stem cells in medical research, as well as research into cloning, genetic engineering and other areas of biotechnology. As artificial intelligence gets closer to true human-level capabilities, similar objections may be raised on religious or cultural grounds. China's cultural and ethical traditions are more open to innovation in these areas, and its technical and scientific capabilities are rapidly catching up to those in the most developed countries. Cutting edge work in these areas may, therefore, catch up and even pull ahead of western countries.

Young people in China, and elsewhere in the developing world, have been greatly helped by the widespread availability of online education, including the Massively Open Online Courses offered by leading American universities. Many of these courses have had more students in China and other developing countries than in the countries that offer them. These courses bring the world's most advanced pedagogy to many who will probably never be able to afford to study at leading world universities. These opportunities are also coming to China through branch campuses opened by foreign universities. Chinese students have increased access to rich and meaningful information to improve their lives and promote social, cultural, and government change. Chinese educational institutions are competing with the world's leading universities even on their home turf, and they are rising to the challenge with innovative and cutting edge programs.

Internet tools are playing an increasingly important role in the struggles against corruption, state-owned monopolies, unsafe food supplies, defective educational systems, poorly constructed housing, and a myriad of other problems. But these same tools are also used by the authorities to suppress citizen movements and to target activists. Systems for filing complaints and appeals can be used to find and correct problems, but they can also be used to identify and constrain those who file the complaints. Computer activists can help by building independent systems to keep track of complaints and also to monitor and analyze official responses. As

End of the Beginning

artificial intelligence improves, we can anticipate a more and more sophisticated interplay between the forces driving and resisting change.

While internet tools may be important for bringing about needed changes and reforms, China's social stability also has benefits. Many developing countries, especially in the Middle East and in Africa, are too disrupted by unrest and civil warfare to meet the challenges of the coming of the Singularity. China's system of bureaucratic capitalism may not readily stimulate the levels of innovation found in countries where capitalist innovation is less constrained. But conditions for innovation and progress are much better than they would be if the country was torn apart by war and civil conflict. There is a good chance that China may progress to more open and competitive forms of social organization without losing a decade or two to a revolutionary breakdown. It must do so if it is to keep up with a world in which technological change is progressing exponentially.

End of the Beginning

Dialogue 13.1

With Mingyu Huang and Ted Goertzel

Ted: The interplay between political, cultural and psychological factors is complex and interesting here. Do you believe there are psychological differences between "races"? Between Asian and European and African? Between Chinese, Korean and Japanese? Or are these cultural differences that grew up because of different environments, as Jared Diamond argues.

Mingyu: The writings of J. Philippe Rushton provide some evidence of differences. I personally believe that there may be some psychological differences between ethnic groups, as well as some physiological differences. The differences between Europeans, Asians and Africans are not large, but some apparent psychological differences are difficult to ignore. Africans, for example, seem to be more emotionally excitable, and they seem to have greater than average levels of artistic talent. I realize that culture has a tremendous impact on human psychology, but cultural theory alone does not seem to provide a sufficient explanation. But these differences are not particularly large, at least compared to the future differences between super-intelligent artificially enhanced humans and the general unenhanced human population. Any differences between existing ethnic groups are negligible compared to the differences between humans and post-humans.

Ted: What are some of the political and cultural reforms that would help China to once again lead the world in technological innovation?

Mingyu: The Third Plenary Session of the 18th Chinese Communist Party's Central Committee, in November, 2013, signaled some significant reforms. China's private enterprises will

End of the Beginning

be organized in a more rational and standardized way, enhancing China's economic dynamism, innovation and vitality. Of course, the reforms have not gone far enough and there is great need to further stabilize the system. There is resistance from members of the elite that benefit from the existing system. But I have reason to believe that the strong support of the Chinese people will push these reforms forward and that this will lead to a rapid burst of innovation. Chinese culture is supportive of innovation; there were great innovations in China's ancient history. Modern Chinese culture will be strongly conductive to innovation.

Ted: What aspects of Chinese culture make it especially open to transhumanism and to other possibilities made likely by artificial general intelligence?

Mingyu: Chinese emperors and nobles often pursued immortality, although this was criticized by Confucius as a costly and unlikely to be successful. Today we know that the technology of the time was not adequate to the task, but the fact that the emperors pursued it so diligently shows the depth of the human heart's desire for immortality. Taoist "alchemist" thought also pursued the goal of immortality. Taoist thinking about the unity of nature and humanity is conducive to acceptance of transhumanism.

Ted: China has led the world in implementing several modern technologies such as supercomputers, nuclear fusion, space station docking and high speed rail. What future developments at the forefront of world advances can we expect from China?

Mingyu: Nanotechnology will develop rapidly in China. China is a world leader in new materials technology. The incumbent president of the Chinese Academy of Sciences, Bai

End of the Beginning

Chunli, is an expert in nanotechnology, although the extent of his interest in transhumanism is unknown. In recent years, China has made major technological advances in new materials and energy technologies, fundamental physics, biology, aerospace technology, high speed rail and other areas. The government strongly supports research in these areas.

The Chinese Academy of Sciences recently listed the following among China's premier science and technology accomplishments in 2013:

- A successful lunar soft landing.
- Launching three astronauts into space
- The anomalous quantum Hall effect was first found in an experiment
- There was a breakthrough in avian influenza virus research
- China won a world championship in a supercomputer contest.
- Chinese researchers developed the world's lightest material.
- Development of the world's first practical solid-state deep-ultraviolent laser.
- Achievement of the highest resolution single-molecule Raman imaging
- The world's largest nuclear power generating unit was built.
- Successfully developed the world's first computer mimicry.
- Development of a high performance molybdenum alloy.
- Cracking the genome of wheat.
- The Beijing Electron Positron Collider discovered a new resonance structure.
- A sperm study confirmed earlier guidance of embryonic development
- There was a breakthrough in pluripotent stem cell

research.
- The first half of a floating gate transistor was developed.
- Chinese researchers identified the HIV co-receptor CCR5
- China's pulsed magnetic field experimental apparatus is among the most advanced in the world
- Chinese researchers overcame cotton " cancer" verticillium wilt
- Chinese researchers developed a new law of super-hard ultra-stable metal preparation

And of course that's a small fraction of the science and technology going on in China's many university and commercial research labs.

Ted: How is the exponential expansion of science and technology particularly impacting the younger generation in China, i.e., people under 35 years of age? What aspects characterize the generation gap between the younger and older generations?

Mingyu: Rapidly accelerating access to technology, including computer networks and mobile phones, have made it difficult for officials to monopolize information access. Compared to previous generations, China's youth today have more information, do more independent thinking, are less likely to blindly worship authority, are more likely to admire entrepreneurs, and are more skeptical of the government. They are more accepting of advanced Western philosophies that seek to make the world more civilized and just.

Ted: How are networking and horizontal communication patterns helping to transform China's hierarchical, vertical traditions?

Mingyu: As I explained in my paper, China does have a history of geographical and cultural isolation from the outside world. But the world is becoming "flat". Chinese people are using the internet, e-mail and instant messaging tools to communicate and build friendships with foreigners and to access all kinds of information. Social media is widely used to carry on discussions of all kinds of topics, largely getting around the government's efforts to prevent the diffusion of information. Often homophonic words and indirect references are used to circumvent censorship.

For example, when sources such as Reuters and BBC reported that Zhou Yongkang, a former member of the Politburo Standing Committee, had been arrested, the Chinese government tried to suppress discussion of this event. So internet discussions used the term "Master Kong" which is a brand of instant noodles, to refer to him. After a while, discussion of Zhou Yongkang appeared not to be shielded, suggesting that the government realized doing so was ineffective. The use of code words in online discussions is common, e.g., the Party Central Committee is referred to as the "crotch center", the Chinese word for government, zhèngfǔ, is abbreviated as "ZF", and so on.

Of course the government knows this happens and takes measures to counter it, but reformers within the government realize that suppression can often be ineffective and even counterproductive.

Ted: How are China's netizens helping in the struggles against corruption, nepotism and poor services?

Mingyu: Chinese netizens have become especially adept in using tools such as BBS, microblogging, comments on published stories, and so on, because of the need to circumvent restrictions. Many have also become skilled at monitoring the actions of government officials and participating in discussions about making the system more accountable and transparent. This has been especially effective in publicizing corruption by high officials, and civil servants are increasingly concerned that their

461

End of the Beginning

transgressions will be widely publicized. When civil servants use their positions to gain special privileges for their children, such as priority access to school buses, people snap cell phone pictures and post them online. Many corrupt officials have lost their jobs because of these actions. Several have been convicted of serious crimes, such as taking bribes and circulating pornographic videos, and sentenced to long prison terms.

Ted: You describe the Chinese system today as "bureaucratic capitalism". What term would you use to describe the system in place in the United States today? Or in Korea, Japan or Germany? Aren't these also "bureaucratic capitalism" in some sense (run by the bureaucracies of governments and large corporations)? What do you think are the most important factors distinguishing China's bureaucratic capitalism from the bureaucracy-infused capitalism of Western democracies? Is the big difference the right to elect leaders, or something else?

Mingyu: The Chinese system is a mixture of what is called "bureaucratic capitalism" or "state capitalism" or "crony capitalism". Much of the economy, especially the natural resource sector, is dominated by state corporations that are very rigid and bureaucratic and operate at low efficiency. Officials use their bureaucratic power positions to win privileged positions for themselves or their family members in capitalist enterprises. This makes the system less efficient, since firms compete through political favoritism rather than through market leadership. The main difference is that in the west there is the right to elect leaders and better oversight mechanisms, including freer press criticism and more independence of the judiciary. But these are differences of degree, and the Chinese system is becoming more open and competitive.

Ted: To some extent, the coming of improved artificial

End of the Beginning

intelligence will make it easier for netizens to evade controls. But there is also an alternative aspect -- the authorities can use AI to more effectively control the netizens? Quite likely both trends may take at once place, making the struggle more complex. Any thoughts on how this struggle might play out in the Chinese context?

Mingyu: Indeed, advances in technology allow citizens to circumvent some of the government's review, but also give the government the means to monitor more people - as revealed by events as the Snowdon revelations. The Chinese government recently has announced a new reform to improve the citizens' right to freedom of speech. I am not sure how well this will work out, but there are some recent events that suggest seriousness about reform: the Communist Party's recent handling of high-level corruption, the establishment of the Shanghai Free Trade Zone, and some improvements in corporate policies. However, I think the surveillance and counter surveillance contradictions still exist, and not just China. Perhaps the best humans can do is to constantly improve government oversight mechanisms to limit abuses of power. I think a certain amount of monitoring for public security is justified, but the monitoring of dissidents is not. The government has no right to oppress legitimate criticism.

Ted: What do you think the social system of the future will look like, if it will not be socialism as anticipated by the Marxists or capitalism as anticipated by Adam Smith and others?

Mingyu: I hope the society of the future will be inclusive and open to input and criticism from everyone. Just being tolerant is not enough, all perspectives and interests need to be included. The society of the future must be innovative; without innovation, there is no meaningful development. And I hope the society of the future will be progressive, which means it should be devoted to the survival and development of intelligent life forms. I think the society of the future will be the era of super-intelligence, used to optimize social structures. When everyone's thoughts can be quickly networked and shared, a really advanced new social contract can be developed for the era of super-intelligence.

Ted: What kinds of ethical constraints on the development of new technologies are prevalent in China, if any? Are ethical considerations less of a problem in China than in the United States? Or is the lack of ethical restraint likely to lead to abuses?

Mingyu: Chinese scientists recognize international principles of bioethics, and China does not have the religious opposition to high technology which is sometimes found in the west. This gives China some advantages over western countries. The main problem for China is not ethical constraints but excessive administrative and bureaucratic interference in technological development.

End of the Beginning

Chapter Fourteen ## Africa Today and the Shadow of the Coming Singularity

By Hruy Tsegaye

Hruy Tsegaye, a graduate in Literature and Journalism, is one of the rising young writers in Ethiopia, centrally concerned with topics such as rational thinking and finding the best philosophical approach to humanity's intrinsic nature. Hruy is one of a group of young visionary Ethiopian technologists who have gathered together in the past few years to form Ethiopia's first AI R&D facility, iCog Labs where he leads business development, web development and community-building efforts. Ben Goertzel, co-editor of this volume, has worked closely with iCog as their Chief Science Advisor. The idea for iCog came about when iCog CEO Getnet Aseffa contacted Ray Kurzweil and his organization to discuss the potential for Singularitarian technologies to transform Ethiopia, and Ethiopia's contribution to the Singularity; Kurzweil's collaborator introduced Getnet and Ben Goertzel, which led to the idea of founding a commercial AI lab in Ethiopia. He can be reached at: **hruy@icog-labs.com**.

"In a Caravan of Camels, the last one walks as fast as the first one" Somali Proverb.

In this vast universe, the African continent may be the only place where intelligence was born. The evolution of modern humans began in Africa, and after millions of years of refining our

End of the Beginning

physical and intellectual features through natural selection, we have reached a point where we can consciously control our further evolution. Our intelligence has awakened a giant within us. We are engrossed in the details of an inevitable future that will be launched in the coming five decades. The Singularity and the 'Intelligence Explosion' to be triggered here on earth will reach out to the far corners of the universe, transforming artificial intellects and modified life forms into immortal intelligent entities.

At the current stage of global human interaction, the wealth and wellbeing of a continent depend on its use of technology. The continents that make the fullest use of science and the products of modern technology are richer, safer and better able to meet human needs. The ability to fashion, secure and above all adopt new technologies has been the hallmark of national economic success in the industrial and postindustrial eras. Technology will be the supreme, revered, single secret in the emerging global market which is to be fully shaped by the field of computer science.

In the case of Africa, the ability and authority to master technology and use it to fulfill human needs has lost ground. The continent is still lagging way too far behind. In the emerging epoch when humans will consciously guide evolution, reverting to the classic approach of *"the survival of the fittest"* would be to fall into the morass of Social Darwinism.

Africa is the cradle of humanity; we were born in Africa and as descendants of Homo sapiens we can say, "All of us are Africans". Our evolution as a biological form began in Africa yet the Mother Continent is on the verge of missing this next stage of evolution, the Technological Singularity. To exclude Africa from this Singularity would be a great historical evil. In the desert caravan, the lead camel's pace not only determines the last one's speed, but also the destination of the whole caravan. A perfect journey's destination is a point where no one is left behind.

In theoretical discussions about transhumanism and guided genetic mutation, the critics' greatest concern is that the wealthiest

End of the Beginning

elite groups of global society will abuse their power. Critics worry that the strata with the greatest economic power will use technology for its self-interest at the expense of the rest of humanity. The Technological Singularity could be monopolized and perhaps even used to eliminate the rest of humanity.

When it comes to technology and science, Africa, ironically, is a 'baby continent,' still struggling to apply almost prehistoric technological advances such as mechanized farming and modern means of transport. It has hardly begun to use technology in emerging fields such as genetic manipulation, enhancing the brain's computing ability through plugged in microprocessors, virtual life extension therapies and regenerative medicine. The way technology is accessed in Africa must be radically changed in the next few years, even though to be able to utilize advanced technology, the continent needs significant economic growth that depends on optimal use of technology already in hand.

Since science and technology are essential for poverty eradication and the foundation of independent development, using whatever technology is at hand and quickly increasing further involvement of new technology seem the only rational and plausible ways for Africa to join the rest of the world. This is Africa's essential weapon to get rid of the star-crossed title *"the continent of third world"* once and for all.

For Africa to develop its own grass-roots capabilities in science and technology, and build a strong economy and modern way of life, the governments and other stakeholders of the continent ought to financially support and encourage African researchers. Unfortunately, most African governments, business firms, and other entities do not perceive African researchers and institutions as a potential or actual source of technological development.

Surveys of business executives conducted by the World Economic Forum (WEF) in the late 1990s to the middle of the 2000s show that one of the major reasons for the failure of technological transfer and usage of science and technology in Africa is that higher institutes, researchers, private firms and other organs involved in the study of science and technology are generally ignored and are unable to make a significant contribution to the continent's economic output. This is a significant impediment to the continent's ability to achieve self-sustained economic development.

It is essential to reallocate government budgets to intensify the use of technology, along with educating and intensively lobbying local and tribal leaders and the rural masses to adopt and apply science and technology. Africa must craft attractive policies for foreign investors to encourage them to introduce hi-tech in the manufacturing sector of the African economy. This will greatly increase the currently inadequate use of science and technology in the continent.

The continent also needs to regain its tranquility and political stability. Based on some gruesome facts and statistical data from the past 50 years, Africa was labeled as the most dangerous place to live and invest. This attitude in the West constrained technological transfer to the continent. But this negative image is fading rapidly and extensively in the new millennium. The current state of Africa is much brighter than it was in the last century, making the coming decades the best time for technological transfer.

The continent has registered an average of 5% annual economic growth in the past ten years (2003-2013) and World Bank estimates that a third of African countries will annually grow at or above 6% for the coming decade also. Africa was also the continent that registered the highest positive economic growth in the global recession period's peak years (2009 & 2010).

The continent continues to grow at a robust 5.8%— higher than the developing countries average of 4.9— and in 2012 and 2013, despite the average of 5%, about a quarter of the sub-Saharan countries registered 7% and higher growth. Furthermore, World

End of the Beginning

Bank estimates that economic growth in the continent will continue to accelerate at an average of 5% and more until 2020. The recession, which continues to affect the economy of the Euro zone and USA, need not be a hindrance for the bright future ahead of the continent.

Sierra Leon, Niger, Côte d'Ivoire, Liberia, Ethiopia and Burkina Faso are the fastest growing nations in the world today whereas China trails as the seventh nation. Mozambique, Zambia and Ghana followed these seven nations in a stable pace while India is behind these three African champions as the eleventh fastest growing economy on the globe. (*"Africa's Pulse-Volume 7". An Analysis of Shaping Africa's Economic Future. World Bank: Washington DC, 2013).*

Despite the evident growth in GDP in the sub-Saharan nations, the per capita income of African households and Africans' living standard is unmistakably creeping at a slower and varying rate. In the last ten years, the sub-Saharan region had witnessed almost an unbelievable economic growth in GDP, but with unsatisfactory performance in poverty reduction; the average for poverty reduction in this region still remains somewhere between 1.3% and 2.1% depending on the resource of the countries. (Report on Sub Saharan Africa, World Bank Washington, DC, June 27, 2013.)

Although, the proportion of people living on less than $1.25-day in Sub-Saharan Africa declined from an estimated 58% to 48.5% in the years between 1996 and 2010, poverty reduction shows a less promising change and remains intangible. Another interesting fact is that the World Bank report of 2013 clearly demonstrates how the resource-rich countries fall behind the resource-poor countries in the process of poverty reduction.

A resource without the technology to use it effectively is simply pie in the sky. Even if countries manage to use this resource and build a growing economy out of it, it will simply be a rent economy. Returns from the export economy may be higher than those available in the short term for investment in manufacturing and technology development. Although there may be high rates of growth in export sectors, this growth will be unbalanced and unhealthy. Unemployment rates will be high, even when GDP increases, the majority of households will not share in the increased wealth..

The remarkable economic growth and swiftly escalating GDP in sub-Saharan Africa fails to reduce poverty in a similar exponential-leap because it needs to be backed up with a genuine progress in science and technology. Otherwise, growth will be limited to rent-seeking, which means that firms and investors will fight over exploiting existing wealth rather than creating new wealth.

Rapid economic growth in Africa has also been accompanied by high inequality, rapid urbanization, risky challenges to environmental sustainability, conflicts over natural resources and high rates of inflation. Given the complete lack or inconsequential presence of Science and Technology at the grass-root level, for the average African, the impact of these problems in day-to-day life adds a weight to be carried that exceeds the mythical rock of Sisyphus.

Poverty reduction remains a fundamental challenge in the continent and lack of access to new technological inventions, which could help a lot in lifting the living standard of Africans, and reshaping the Agriculture and health sector of the continent should no more be left in oblivion. The question should be tackled with priority; Africa needs advanced technology.

The process of raising somehow a scientifically framed and informed society needs the first hand application of technology. Africans should turn their face towards technological advancement and governments should be open-minded to pay the cost for this technological transfer. New advances in technology are creating a lot of jobs around the world, increasing income per capita of individuals and accelerating poverty reduction in the developing

End of the Beginning

nations.

Despite economic growth in GDP, Africans need to solve their poverty problem. As in the west, technology is the key to reducing poverty, without which African will not be able to benefit from the coming Singularity. One most important characteristic of technology is it will reduce the cost of production that will change the game of supplying goods in the market. Africans can have a better local market when they build their own manufacturing sector dictated by their own fully developed technology.

The simple acts of exporting unprocessed agricultural products, raw materials for industry or any natural resource will not be enough in the course of poverty reduction. Africans should create strong industries that process these resources and a service sector that supports them. The use of a wide-ranged and up-to-date technology is a primary task. In this way, the benefits from these exports could be doubled while at the same time it creating many well-paying jobs for Africans.

Political stability is, finally, being firmly established on the continent. Africans are sending their own fighting troops or peacekeeping forces to other African nations. African Union (AU) Peace Keeping Forces have successfully operated in Uganda, Rwanda, Congo, Mali and Somalia Since 2000. Conflicts have decreased over the past two decades, cross-cultural interactions are thriving, and intra-state or cross border conflicts in Africa have been satisfactorily settled by international and as well intra-continental interventions. The civil war in Sudan is finally over, the Ethio-Eritrean war was settled in Algiers, the chaos in Somalia is being resolved with the help of Ethiopian troops and peace was returned to Côte d'Ivoire with the help of the AU forces.

Democratic nations are evolving from the dark corners and bellies of military, monarchial, and Islamic dictatorships and tribal government forms in Africa. The Arab-spring initiated in Tunisia, the successful democratic elections in Ghana, Kenya, Nigeria, Liberia and some other states after 2010 marked the beginning of a new era in the continent.

Technology could be used to strengthen this political stability. An election could easily be monitored and accessible to the public using electronic machines, police forces could be armed with various gadgets that help them to protect and serve the public, and the public could check and monitor its government's actions through strongly built media and public institutions using the latest technology. Nations can build strong army and defense systems with the help of advanced technology.

The progress of Africa is an essential part of world progress and Africa requires a very fast technology transfer to secure its place in the coming Singularity.

Technological Transfer

This century is witnessing a shift in economic growth and political supremacy from its center in Western nations to emerging nations in Asia, South America and even Africa. The reason is obvious; the only positive aftermath of World War II was technological transfer. Countries such as Japan, China and Korea could be mentioned as the best examples. There are two different approaches to technology transfer: the traditional progressive transfer and the *Leapfrogging* method.

In the traditional progressive approach, innovations are introduced sequentially in the order in which they were developed in the West. It is a step-by-step process, illustrated by the telecommunication sector which started with telegram, then landline phones then to the mobile phones. This approach involves a gradual development in management capacity and in support for researchers and innovation. The idea is to achieve what the West had attained in the past and to build an indigenous but independent and self-sustained technology within a nation.

Yet, this can take a long time. In this age of rapid globalization and continual introduction of new scientific breakthroughs and technological inventions, time is the most important commodity

End of the Beginning

and nations cannot afford to fall behind. For a latecomer nation, the second option, *Leapfrogging,* made it possible to jump directly into a mobile phone system and to build a stronger, cheaper and market oriented telecommunication sector.

There is an immense literature on *Leapfrogging* as applied to industrial and economic growth in developing and undeveloped countries. Today, environmentalist and other activists supporting 'clean development' also overuse the term. It is tied to the idea that developing nations should avoid the path of the already industrialized nations' polluting technology and thus jump to new sorts of technology which are in complete harmony with nature.

Environmental protection is an important thing for everyone and indeed it is even a matter of life and death for Africans since their almost wholly agricultural based economy depends on the outcome of climate change. Yet, with such poor economic power, Africans cannot protect their environment in the way the West does! First, they desperately need to grow and have an economy, yes 'have' is the right word! Sometimes it is difficult for the West to grasp the conditions of life in such continents; tourism and brief residency is just a tour on the shallow side of the lake.

Whether it is a 'green technology' or as blue as the Smurfs, Africa needs it. Restricting and limiting the technological transfer in Africa into a 'low carbon' energy technology is not fair and it is not feasible; considering the carbon emission from the industries of that continent, it is even ridiculous! Africa's carbon emission is less than that of any other continent, it is even less than some single nations' emissions in the West and the East.

The technological leapfrogging in Africa thus should be a bit relaxed and should consider the reality that the continent cannot afford the cost of extreme 'green technology' like using Solar Energy. The first thing is Africa needs to save a little more of its assets and have a further strengthened economy and industrial sector.

Right now, the entire continent is in desperate need of energy, which is the vital element for building an industry capable of feeding, at least, Africa's own population. Limiting the leapfrogging into green technology is over simplifying the continent's burning issue and it is a hypocritical approach to the relationship of humans and nature. The stark truth is the cost of a Solar Panel Plant, one with a capacity to generate significant amount of electricity, is painfully high for the pocket of Africa. Africa's technological advancement should not be frowned upon even if it doesn't reduce carbon emission.

On the other hand, the continent should open its door for foreign investors to invest their money along with their technology. Policies should be more adoptive and protective. At the same time, governments in Africa should consider current, soon to be materialized technological capabilities, and intensively work on shaping them. As the continent is rich with human resources, the labor force ought to upgrade its knowledge and skill and to be supportive of high priority industries.

Assigning considerable capital and stimulating innovations, researches, capacity-building programs, advanced education and investment in sectors like computer science should be a priority in the leapfrogging program of Africa. In all fairness, the world is going to be a package deal in the jungle of computer programs and a continent lagging in the science of computers is going to pay dearly for lagging behind.

As part of leapfrogging, establishing a market for such new adoptions will not be a problem for the continent. The continent imports almost all of its manufactured goods and if the governments in Africa take the heart to put in a bit of effort, the continent can have an intra-continent market.

The Advantage of Using Hi-tech in Establishing a Peaceful Africa

According to Statistics from UN, in the last 60 years, conflicts over natural resources initiated 40% of the civil wars and internal

End of the Beginning

conflicts within states. The Millennium Project, in one of its articles on world peace, summarizes the situation as, "Although the vast majority of the world is living in peace, half the world continues to be vulnerable to social instability and violence due to growing global and local inequalities, outdated social structures, inadequate legal systems and increasing costs of food, water, and energy".

Continuing population growth in Africa, along with the effects of climate change, will lead to an increase of social tensions. This may trigger complex interactions of old ethnic and religious conflicts, civil unrest, and crime. A considerable radical advance in technological advancement and revolutionary social reform is the only possible solution to prevent this. Countries will need to restructure their economic and socio-political structures in accord with the trend of the new world of high technology.

The collective mind of humanity can contribute much to either the global peace or intercontinental conflict, and hence, we should seriously begin to ask questions like how to solve at least the basic root problems in Africa. The elephant in the room while addressing this question is the state of technology and its utterly miserable shape in the continent. It is time to work on technological advancement and empowering the manufacturing sectors of Africa by making high technology available and affordable to the nations and peoples of this continent on a wider scale than at any other time in history.

Before the singularity is launched, it has to be conceived in the minds of Africans too. Africans need to witness the healing power of technology in solving problems such as conflict over limited natural resource, alarming population growth accompanied by food shortage and the daily death of thousands caused by easily treatable tropical diseases. Natural resources without the aid of hi-tech are most likely inadequate and may lead to the bloodiest of conflicts even though they appear to be abundant.

The amount of annual rainfall in Addis Ababa, the capital city of Ethiopia, is more than the annual rainfall in the entire state of

End of the Beginning

Israel. While Israel is one of the leading countries in the global Fruit Export Market with a very strong and sufficient agricultural sector in the Middle Eastern arid desert, Ethiopia is known for drastic and repeated famines amidst a green jungle surrounded by the world's longest and robust river, the Nile with an annual water flow of 80 billion cubic meters. The river Jordan with its tributaries, the only river in Israel, has an annual water flow of 500 million cubic meters. In comparison to the mighty Blue Nile in Ethiopia, this is as a drop in the bucket. But despite this, Ethiopia, Sudan, and Egypt the three dominant nations located in the downstream of the Nile have always been in a hostile political relationship with a war looming over the use of that river; perhaps this is the continent's most prominent cause of conflict that involves more than eleven nations.

Yet this could have easily been averted through the execution of proper technology. Technology can transform a desert into the world's most productive agriculture field. It can save the lives of thousands in its miraculous hands in the health sector.

The Democratic republic of Congo (the former Zaire) is often referred as *the world's buffet of natural resource*, yet it has spent the past 20 years in one of the continent's bloodiest civil wars. Gold, industrial and gem-quality diamonds, uranium, cobalt, copper, cadmium, silver, zinc, manganese, tin, germanium, radium, bauxite, iron ore, and coal and many other natural resources are abundantly spread in the geography of this country. It is unbelievably green with the heaven-like jungles as home of the world's second largest forest, the Congo forest. The Congo River is also the world's second largest river in water flow.

However, Congo is the world's poorest country having the second lowest nominal GDP per capita while the total estimation of the natural resources is in excess of 24 trillion US dollar making the country the world's richest country. 30% of the world's diamond reserves, 70% of the world's Coltan, a metallic-ore used as a raw material to manufacture a Tantalum Capacitor that is an

End of the Beginning

essential part in every cell phone, laptop, pager, and various electronics and 10% of the globs Copper are just inaccessible to its people. Congo's richness is a pie in the sky for the Congolese— Adios, to Congo's richness!

Countries like China extract this into their high-tech operated industries and Congo just serves as the free plate for extracting raw materials and an important emerging market in which to sell processed products, the classic Gold laying hen.

On the other far side of the globe, South Korea, with literally none of the above resources is becoming a leading nation in manufacturing electronics, automobiles, and jewelry that highly depend on these natural resources. Africans should understand that this is not witchcraft; this is simply a difference in approach to science and technology.

Ultimately, natural resources by themselves mean nothing.

Technology has proved itself as the best tool in creating a world where people can live at ease surviving the harsh reality of nature whether it is in the most arid place like the Middle East or poorly resourced places like South Korea— the 40% of conflict causes can easily be prevented with the use of such hi-tech.

The Advantage of Using hi-tech In Building a Developed Economy in Africa

Just the word 'Africa', unfortunately, triggers the image of poverty and a pitiful state of humanity; famine and horrible starvation, easily curable diseases killing thousands in a day, illiteracy, outdated traditions like female genital mutilation or child marriage, malnutrition, complete lack of access to electricity, water and transport... the list can go on and on. However, are these nightmarish challenges inherent features of the continent? A big "hell no" is the right answer. In fact these, seemingly impossible trials to solve, are just a matter of few technological changes.

In some ways, we can say this is the result of the West and its

End of the Beginning

imperialist and racist exploitation of this continent and its people in the past four centuries. But debating who is at fault for the current conditions in Africa is fruitless; let it be an exercise for our idle politicians. The vital question is what can be done and what should be used in the process of averting poverty in Africa.

A classic example of technology in poverty reduction is Dean Kamen's latest invention of water purification system, the Slingshot Water Purifier, which for just 4 billon dollars can solve the whole of Africa's pure water problem. WHO reports, around the world, every year, 3.5million people die because of waterborne diseases.

Economic growth directed by hi-tech is the way forward

The past 40 years proved that foreign aid is ineffective as long as Africa has not developed into a self-developing continent; the change has to be internal. Moreover, when one is talking about an independent and self-efficient manufacturing sector in a growing continent what other options are there than applying hi-tech machines?

Considering the number of nations in Africa and the economic growth in the past five years, the continent obviously is home to a large number of developing countries. To promote these nations into developed countries, Africa needs the undisputable provision of technology and its result, which is exponential economic growth.

In the absence of not only the traditional 'technology' but also that of a futuristic and up to the minute hi-tech, enhancement of living condition, efficient food production, an independent industrial sector (well advanced to process its own natural resource) and a health sector equipped with advanced biomedical engineering cannot and will not exist in the continent of Africa.

While current hi-technology is often thought to be a thing of myths and 'old wives tales' in the continent of Africa, with more than 70% of its population living in rural areas, it is highly unlikely that the concept of Singularity will be considered as a possible future. It may not even be longed for as an actual possibility in the coming future of humanity's evolutionary progress.

The West's experience in implementing science and technology should be shared with Africa too; the coming technological singularity cannot be global and will not enhance the nature and the collective state of intelligence and intelligent life forms on our planet as a whole given the present conditions of such continents.

The Concept of Singularity vs. the African's Mind-set: A Glance at Africa's Possible Response

Humanity's evolutionary epic began in Africa and spread to the rest of the world. Now it has apparently reached its last stage. Yet the original form of humanity is still young in that continent. Africans, unlike the West, have not given up on humanity as we know it. With all this despair about living conditions, Africans still value humanity in its old form, warmth, and beauty. The various traditions and cultures in Africa revered the old form of humanity and its social interaction to its highest state, something lost in the West long long ago. Man as a social animal is in its ripest age in Africa.

Albert Camus, in his book the Myth of Sisyphus, states an important point to one of the most important of life's conundrums. He said, "… Judging whether life is or is not worth living amounts to answering the fundamental question of philosophy". So far, Africans had proved to the rest of humanity that life is worth living even if the life they embrace is immensely challenging and characterized by misery resulted from sheer poverty and political instability.

Civilization had elevated the living standard of humans in the West and maybe this might have encouraged people in the West to see life and humanity in its flesh form as some evolutionary defect. Having almost all essentials in life easily and with much lesser physical effort will make life boring on the ground that the challenges of living as humans in biological form have already been conquered.

The struggle to live in the West is not limited to fulfilling basic needs or enjoying luxuries; it is a struggle to understand what life is or what purpose does one have in life. Thus, the West is now marching, or at least pondering, to conquer the beyond. Immortality and an omniscient state of consciousness are what the West is looking for in the coming era of Singularity.

Yet the case in Africa is a completely alien scenario. The struggle in life is still for basic needs. Needs such as food, clothing, and shelter are what the people are looking for! And ironically, this by itself makes this current state of agonizing life in Africa precious and the satisfaction of fulfilling this absorbs the entire mindset of most individuals in Africa.

Sometimes it is a wonder to see the rate of suicide in the West and in Africa; with all the benefits of civilization and a relatively much easier life, more individuals end their lives by their own hand in the West while Africans dearly hold theirs regardless of the state they live in. WHO reports, around 1 million people every year (3,000 a day) die by their own hand and based on available data, the West especially Europe and the Balkan nations have the highest suicide rates.

A number of researches indicate that one of the important factors for suicide is social disorganization; the civilized way in the West tends to focus on individualism while Africa views it as a taboo. As suicide is the ultimate answer for the question 'is life worth living', can we say the African's strong social interaction is perhaps the answer to a satisfactory life that is worthwhile living?

What makes the Africans bare the utterly hopeless and dead-end life seems to be one of the weak links in the coming Singularity; will the Singularity entertain social interaction? Africans are highly dependent on it and it proved that it makes life worth living.

End of the Beginning

Aside from the common reasons for suicide which are fathomless sorrow, mental disorder, genetic factors like concentrations of serotonin, and fear of pain from diseases or external causes such as torture, the West has also witnessed the untold truth that a considerable number of people commit suicide for lack of 'excitement' in one's life.

As a paradox, their 'lack of excitement' in life seems to arise from the benefits of civilization; basic needs are fulfilled easily compared to the rest of the world. In Africa, for the average African, the reason to live is to fulfil these basic needs. This sounds just like Camus's Absurd Reasoning, *"What is called the reason to live is also an excellent reason to die!"*

Africans, right now, struggle and live for an advanced civilization, which in their terms is one that had already outgrown an agriculture-based economy; their reason to live is to be a modernized society that has access to life's basic needs and only then to tackle the next questions of life. As it is common in all undeveloped countries, the average individual lives at the edge, yet his/her reason to live is to win this edge! The West experienced it and witnessed that barely reaching this edge just to halt on the flat meadow next to it is not the reason to live!

Regardless of the rigors of life in Africa, Africans kept their life dearly, hoping one day, they too would have the time and means to enjoy it. Unfortunately, the future and the coming Singularity will change the current fashions of enjoyable life whether in value or in form. For the West this might not be shocking; assuming that because of the advances in civilization, the West had 'enjoyed humanity' and it is now bored with it and is trying to take it to its next level but, for an African the next level might sound unreal and detestable since the current level is a life time dream.

The coming age of AGIs is conceived as '*the post-civilization time*' and the Singularity aims for one thing, an immortal life full of excitement. Is it not exiting to be whatever you want to be, to live

End of the Beginning

in whatever situation you want, and to face whatever challenges you want to face through the means of virtual reality, yet in an absolute real world because the then humans are merged in the virtual reality through the Singularity.

Currently Africans are not seeking immortality; they are hungry for life as we know it.

Yet, assuming this attitude in Africa will continue to dominate in the coming era, it doesn't mean immortality or the next phase of civilization through Technological Singularity will totally be denounced in that continent. The Promise of Singularity will not be a complete stranger and discarded as unacceptable odd philosophical approach to life. Africans love life more than anything else in this world and once they begin to understand the infinite possibilities in life and the other super exiting reasons to live, their windows for accepting such radical attitudes towards the nature and destiny of mankind will widely be opened.

Yet again, judging from the angle of the robust culture rooted in day-to-day personal communication, the vibrant and genuine social interaction and the human based value system of Africa, the concept of Singularity as the next evolutionary step, a life form without the biological existence, will get a challenging and even long lasting negative response. Africa's unique quality of preserving the biological as well the spiritual existence of humanity is still observable in the wide spread cultures all around the continent.

Hence, will Africa miss the Singularity? Will the content shut her door to this man-made evolution and keep faithful to nature as a continent that was once the birth place of intelligence?

Maybe the continent creates a strong feeling of nostalgia for the consciousness of our carbon-based intelligence; maybe the continent cries aloud our unexpressed hatred for the coming AGIs and our unconscious resistance as a means to preserve our humanity from their dominance. Our hundred thousand years of collective consciousness sides with Africa; perhaps humanity in its

End of the Beginning

biological form is worth being preserved.

Hitherto, thinking about the nature of the coming Singularity tends to side on the tenet that carbon-based life forms might eventually fade out and be replaced by silicon. Conceivably this prediction is the source of the many rational and irrational fears about the Singularity. If this is the expected nature of the Singularity, Africa will resist it head-on. The warmth of biological existence is not only appreciated in Africa but many traditions hold it as a revered form of humanity. Since most of the continent and its people are not cultivated with recent ideas and philosophies that focus on humanity's next phase— the beyond-known probabilities in the future phases of human evolution— it is acceptable and even rational for Africans to denounce the Singularity.

The average African is not accustomed to challenging teachings based on religious dogmas and thus this game-changing concept of Singularity is going to be perceived as the seed of pure evil. Some critics ask rhetorical questions such as, "How many Africans can read and write?" but the real point is not in the statistics of illiteracy, it is in the psyche of the people.

At this point, no one can be certain about the future and the form in which humanity will exist in the coming Singularity. However, since intelligence is the core of the Singularity, biological forms might lose its favor in the post AGI era, and the evolution of that time will not select carbon-based intelligence. Life most probably will change her Carbon form and Africans might not accept this just as some of the other 'less-developed' continents may not either. Yet, Africa's resistance towards it will be more plausible.

The bio-conservatism approach from Africa will be very strong; the tie between the physical (the biological existence of humanity) and the spiritual concept and image of humanity is so strong in Africa. To think humanity in another form (only in the spiritual or the conscious form) means playing with the supernatural which is the most respected and dreaded taboo in Africa.

Isn't there a way in which the Singularity can blend with the genuine African norm and psyche?

In many various yet still strong and vibrant traditions and cultures, the *Ancestral Spirit* is common in Africa. This tradition and belief system for Africans is their proof that humankind and humanity can exist without the biological form. Unlike the West, these spirits are not demonic or of other entities, these are spirits of grandfathers or of the kings or of the elderly men who once where alive.

The *Ancestral Spirit* is not also a common ghost, as in the West, the *Ancestral Spirit* is the omniscient, immortal and God-like version of human.

To suggest that a man does not necessarily have to die to be one of the *Ancestral Spirits* and that living humans can transcend into spirit lives with technology is no less than a dare to slap the face of an African warrior. Yet this is the frame in which the Singularity will be understood by the uneducated African. And, interestingly, this uneducated African may be closer to understanding the possible form of humanity in the post-Singularity era than anyone in the West.

"Madmen, who ceaselessly complain of Nature, learn that all your misfortunes arise from yourselves". Jean. J. Rousseau

End of the Beginning

Dialogue 14.1

With Hruy Tsegaye, Ben Goertzel and Francis Heylighen

Part 1: Goertzel and Tsegaye

Ben: Hruy, I definitely see your point that Singularitarians and Transhumanists often get most excited about likely developments in the most technologically advanced countries. So that it's natural to ask whether this means the developing countries such as those in Africa are likely to fall even further behind? Or can new technologies help Africa to close the gap? What can Transhumanists do to help?

Hruy: It would be an inexcusable crime for Transhumanists to knowingly ignore a large percentage of the human race. Africa, in particular, has a tremendous potential. It has more than half of the world's fertile but unused land. It uses only two-percent of its renewable water resources. Yeah it has the highest poverty rate in the world, but for the last decade it has experienced rapid economic growth. Transhumanists need to do their part to help.

Ben: What can Transhumanists do to help?

Hruy: As we've discussed, a key concept here is *technology leapfrogging*. Less developed countries do not have to go through all the stages the wealthy countries went through to get where they are today. They can adopt the most advanced technologies, and use them to close the gap. An often cited example is how much of Africa has skipped the stage of wired telephoning and gone directly to mobile telephones, enabling a very rapid expansion of telephone and internet access. Today 80% of urban Africans have mobile telephone access.

End of the Beginning

Ben: What other technologies are needed first?

Hruy: The essential first step is to increase Africa's ability to absorb new technologies. The demand is already there, Africans know what they need to build their societies. We can't do it just by importing technology off the shelf. We need help in developing our own tech startups, adapting and advancing existing technologies to suit African capabilities and priorities. We need state-of-the-art scientific and technological training to enable Africans to move out ahead in the areas most important for African development.

Ben: What is holding back development in Africa? Is it just lack of high tech resources?

Hruy: Unfortunately, there is a lot of resistance in Africa to new technologies such as genetic engineering, nanotech, and cloning. This may be because Africans see these things coming from the outside and don't understand why it makes sense to think about such things in a context where many people don't have basics such as electric power and transportation networks. Africans don't always understand that high tech can advance without waiting for basic needs to be met, important as these are. They don't always have the background to see that only by leapfrogging in this way can Africa move ahead instead of always playing catch-up.

The goal of Transhumanism has to be to enhance humanity, in both the developing and the wealthy nations. We must make sure that Africans are part of the vanguard of the Transhumanist movement, even as we work to close the gaps in nutrition, health care, education and other human needs. Advances such as cybernetic limb prosthetics, improving intelligence with brain implants, synthetic wombs and genetically modified babies are just as important for Africa as for other parts of the world. These advances can help close the inequality gap; they must not draw funds away from meeting basic needs.

Ben: What do you think are some of the technological developments that will be most helpful for Africa?

Hruy: Well, among many other things— Africans can benefit from genetic modifications on the immune system to resist disease, from genetically modified plants that can save the environment, from hi-tech that can create better water sanitation systems and from machines that will boost industrial production and, indeed, help to establish new fecund processes.

Ben: Yes, that all makes sense ... it's very practical stuff, which advanced technology could be very helpful with right now, if appropriately deployed and customized.

Regarding the sense of spending effort on things like AGI and nanotech and mind uploading when there are so many people without adequate food and water and transportation – please remember that the percentage of the world's resources being spent on advanced research areas likes these is very, very small. Switching some of those resources to meeting current needs would not make a measurable impact. What is more important is to cut the tremendous resources spent on needless consumption in the wealthy countries.

I would be in favor of allocating more of the world's resources to both advanced R&D and helping developing nations – and less to the development of chocolatier chocolates, fake Rolex watches and Luis Vuitton luggage, TV sit-coms and commercials, repetitive social-media smartphone apps, and so forth.

Contemporary first-world culture seems strongly oriented to spend a huge percentage of its resources on things related to addictive sensory gratification and status-symbolism. Can Africans offer some help in guiding the developing countries

End of the Beginning

toward a healthier way of life?

Hruy: This is an important point. If the priorities of the developed world are unhealthy, there is a risk that new technologies will be used to make things even worse. Africans can't afford the excessive, wasteful and environmentally destructive consumption levels of the wealthiest countries. The priority must be on using high technology to meet real human needs in Africa and other developing countries as well as in the wealthy countries.

Ben: As Freud pointed out correctly, the perversity of human nature has to do with the fact that our psyches evolved to live in steady-state tribal society, but instead we're living in civilization – which presents us with amazing new opportunities and challenges, but also constrains us in ways we are not well evolutionarily equipped to handle.... One of the possibilities held out by advanced technologies like AGI, mind uploading and brain-computer interfacing is to overcome the constraints of human psychology itself, by moving to a way of being fundamentally beyond current human civilization.

Hruy: It seems to me that you are not satisfied with what we have; you believe that there is so much more in the unknown, and that intelligence is our ship to that strange land. But then you worry that the coming AGIs will treat humans as insignificant, just as humans treat ants as insignificant. Is there anything in human culture that you truly appreciate and want to preserve?

Ben: Certainly— just to take some almost-random examples, what about the great works of art? The Picasso of Guernica or Le Femme-Fleur, or the classic Kandinsky or de Kooning or Matisse ... or the later, more surrealist Dali ... this stuff genuinely moves me MORE than realistic painting, because it more closely evokes

End of the Beginning

the way I perceive and conceive things in my mind's eye.... It has a great value, and a specifically human value. I don't think an AGI could appreciate these paintings the way a human does.

Hruy: In my college days, I was somehow a lazy student and I have forgotten some of the things I studied by now, but the professors endeavored to open our minds and lectured us about existentialism, impressionism, expressionism and similar concepts in our fine-arts courses. (Most of the professors were white guest lecturers and the UN paid their salary. To the best of my knowledge, no Ethiopian lecturer teaches these stuff in my university, perhaps there may be some who does in Addis Ababa University. Don't get me wrong, I am not a racist by any means; but at the time I used to think that this kind of stuff was white stuff! But now, I somehow feel that these are not white topics, they are the result of our complex, twisted, and unsatisfiable nature as humans.

Ben: Yes. They result from the complex, twisted nature of humanity – and their beauty is utterly wrapped up with the specific ways in which humans are complex and twisted. The beauty of these paintings, along with a zillion other beautiful and amazing things about human culture, has its own value regardless of whether we create AGIs that are 1000 times as smart as people are. But the beauty of the works these AGIs create may end up being, in some senses, vastly further beyond Guernica than Guernica is beyond a child's fingerpainting or the hole a dog digs in the ground.

As for residing in the known, rather than striving and struggling for the unknown, I think what you're bemoaning is pretty much the growth of civilization. As you know, tribal societies were basically steady-state societies, not aimed at going beyond the status quo. The desire to settle down and build farms and domesticate animals, this already was a desire to go beyond

End of the Beginning

the known and build something new and better.... And then, from the birth of civilization, the desire to grow and learn and exceed just kept on going... and will keep on going and going...

Hruy: I am not a huge fan of painting, but when I get the chance to glance at such works, at least I can note the artists' talent. I used to be amazed by the works of Picasso and Dali. Not of their latest work but their early ones. As you know, these men used to paint realistic paintings. For me that was art— a method of satisfying ourselves by turning the known into a work of beauty. Dali has one painting, which haunts me continually. It is a depiction of Jesus on the cross -- with no blood or no gruesome features, but from particular angle and clarity to the image, it is just like a photograph. Now this is turning the known into beauty; if I painted that picture, I know that I would be satisfied. Even though the story of Christ might appear is as over-told and as boring as any old story, it can still be turned into a work of beauty, a work of perfection. But when these painters decide to chase the unknown ... in all honesty every child and toddler is a great cubist!

Ben: Hah... Well, actually, cubism is not childlike at all, it's a highly formal and precise discipline of painting, which is much harder to execute than painting photo-realistically. It's a certain way of projecting four dimensions into two. Look carefully at Duchamp's "Nude Descending a Staircase" ...

Hruy: I'm sure there is a lot you can see there, applying your wide knowledge and your intellectual mind. But still -- Instead of enjoying the beauty of simple and realistic works of art, we are now supposed to interpret them. Cubism or the other Abstract forms are the greatest frauds in the history of visual art! We are expected to be experts that talk about the deep emotions of humans whilst we are looking at gibberish painted in oil on canvas. And if we fail to dig this kind of work we are called

End of the Beginning

illiterate, boorish and less cultivated. If my art as a picture, as a visual image, cannot say what I wanted to say to the onlooker without any further verbal or written explanation then I am not a painter! To emphasize the unknown, to hate the given, to be disgusted with what exists, this is what you call the age of Modernism. You can call it by any other name but for me it is Dissatisfaction Rationalized.

Ben: Hmmm. I'm no art critic, really.   But I would say modernism is about a lot of things – and one of them is accepting the inner/ individual world, and the collective/polyphonic world, as having equal reality-status to the "objective" "external" world.... Paintings that depict inner subjective perspectives are just part of this...

Hruy: We could talk about art for a long time; but perhaps we should move back to the topic, though. The elephant in the room, which I have somehow tried not to openly acknowledge so far, is the question that is the Singularity and Transhumanism products of such minds and attitudes? I think they are. Ben, this is my genuine conclusion for now— though not necessarily my final one; perhaps there is another picture. The 'unknown' holds a vast space in these movements. Because they are part of modernity, they are also the products of a dying civilization. But what about the known? Is ignoring it the answer to our deepest need; the way to fill the unexplained emptiness we often feel? Will chasing such a dream satisfy us? Is it even possible to build a new civilization, which is supposed to be much better than the old one, without even addressing the root problems of the obsolete? Can we become a post-human before truly knowing what humanism is all about? In my opinion, going for the unknown before addressing the known is the main reason for the current miserable human condition; we are always chasing dreams and never have lived in the 'boring reality'

End of the Beginning

Ben: Hmm— going back to Freud and the differences between tribal and civilized society... I think it's in our human physiology to be pretty much contented with the known and the everyday, in the context of tribal life. But it's not in our human physiology to be contented with the known and the everyday in the context of civilization. Civilization is not comfortable for human beings and never will be. When we left the tribal context to move into villages and farms, we started a process that inexorably has been pushing us toward the Singularity. We started a process of progressively moving further and further, and faster and faster, into the unknown. By now, this process of always moving into the unknown is, in itself, a central feature of our humanity! As is the fear of the unknown, also, a central feature of our (complex, perverted, self-contradictory) humanity....

Hruy: I know every new thing has a risk; and I believe that fear of the unknown is the worst of humanity's handicaps. This, Ben, is the paradox in my melancholy. But totally ignoring the 'known' is unacceptable! It is even worse than chasing the unknown. Look at what we have today, the things that are the products of knowledge and science. In the first layer of their appearance, they seem to make our lives easier and to turn the world into a better place; but in the deepest layer, they are also the mothers of our current problems. What we should do by now, after all these years, is once and for all to stop chasing dreams and think with all we've got about what we really want in life and try to come up with a system that has been scrutinized from head to toe. Yet we are as gullible as our predecessors are, and we try to solve our problems with the same mentality that created them in the first place. How on earth can one say mine is different whilst he doesn't know what it is going to be, what it should be and most importantly what it was used to be?

End of the Beginning

Ben: Yet that seems more a pipe dream than the Singularity, to me— the idea of humanity stopping and thinking and scrutinizing to come up with a better system. Our own inner gods and demons are going to keep us moving forward; that seems abundantly clear. We are going to keep on moving rapidly into the unknown, even while fearing it.

Will transhumanity bring satisfaction?   I don't know.... I don't think it can bring satisfaction to "humans with their present style of minds" -- i.e., no number of cool gadgets and technologies, and no amount of free resources, are going make *current-style human minds* satisfied and peaceful. The human brain/body evolved for tribal life and is generally not going to be satisfied in a modern-style civilization (though of course intervals of profound satisfaction will come about for many people from time to time, which is absolutely worth a lot). However, if the human mind is fundamentally transformed into something else, this limitation will no longer be there. To what extent the new kinds of minds thus created will be A) profoundly more satisfied than legacy-humans by civilized, technological life, versus B) "beyond satisfaction and dissatisfaction", to riff on Nietzsche ... that I'm not sure...

Hruy: Ben, we must not refrain from sleeping in fear of nightmares (a proverb from Ethiopia). But we must always think twice before we act (a proverb from your land).

Proverbs are so old, and in the age where you and I can talk thousands of hours on the possibility of time traveling, isn't it mediocrity and lack of imagination to talk about proverbs? No, Ben, for me the answer is a firm no. The old world is wiser than ours; for the greatest teacher is experience. The elders knew that we humans are not simple animals; they also knew that we are not perfect or complete. They invented and shaped ideals and material tools; religion, war, group-sex, law, the sword, the abacus, paper and ink, ships, mechanical motors... they were trying to enhance our existence and create a better world. We are also trying to

End of the Beginning

enhance it now; our tools seem much better than theirs, yes. Science and technology are now very advanced and relatively available. We are not as wise as they are because we are doing these as a revolt to the old-world not as an improvement. But in my melancholy I can see we have the same psyche as the elders; our love of the unknown. In my deepest fear, we share the same fate; our eventual end as unsatisfied beings.

Part 2: Goertzel and Heylighen

Ben: Francis, I'm curious what you think about my discussions with Hruy about the relation between transhumanism, Singularity and inequality....

As you see Hruy is skeptical about addressing advanced technology R&D and thinking about superhuman AGI minds, when there are so many people on the planet without adequate food and water and transportation. He figures we should solve our present problems before venturing so far into the unknown.

As for me, I don't think any kind of perfection is likely to emerge, ever – I think there will always be some balance of satisfaction and dissatisfaction. To my mind that's a consequence of intelligence in a universe with finite resources. But I believe that if it's done right, a Singularity can shift the balance strongly toward satisfaction.

Hruy thinks it may be folly to move so fast into the unknown, in the pursuit of abundance and satisfaction, without first fully coming to grips with what is in our human hearts already, and solving more of the practical and psychological and social problems in our human world.... But I think we chose to hurtle into the unknown since we started civilization. Humanity is not going to slow down. What we need to do is try to guide the hurtling movement in as good a direction as we can. Which to me, means both working on advanced technologies AND trying to bring the whole of humanity fully into the fold, including the folks

End of the Beginning

in Africa and elsewhere who are now struggling with the practicalities of everyday life.

Francis: Well, I fully agree that severe and growing inequality is a disease of the global brain. Part of the problem is that technology accelerates everything, including the differences in wealth, via the classic positive feedback of the rich getting richer because they have more resources to develop further wealth. A more equal distribution of resources would make the economy grow much faster. Evidence for this is that the most unequal societies (e.g. much of Latin America, Africa and South Asia) are also the poorest, while the richest (e.g. Northern Europe, Japan) tend to be the most equal.

Ben: Hmmm.... I'm not sure this is compelling evidence. The two countries doing the most to push technology forward now are not particularly equitable ones -- the US and China. Look at any list of the 100 top research universities in the world; probably 70 will be in the US... The Internet, which is key to bringing about the global brain, is a US invention and the vast bulk of Internet innovation is still US-based. The smartphone revolution was also initiated by Apple in the US. But underlying all this US innovation in recent years is low-cost Chinese manufacturing..... But the US and China have significant and growing inequality; in both places, inequality is growing even faster than wealth...

Francis: A big obstacle against remediating this problem is the extreme laissez-faire ideology of classic capitalism, according to which the rich deserve their wealth because they have proven to be the smartest investors. This completely ignores the non-linear dynamics of wealth creation which leads to butterfly effects, winner-takes-all, path-dependency and other "irrational" mechanisms that amplify random fluctuations into entrenched and growing inequalities. This neglects the fact that there are plenty of smart people who potentially could produce a lot of extra wealth if they would just get access to the needed resources (knowledge, information, infrastructure, markets, trust, contacts ... and finally money). But in places like Africa these resources are very difficult to get...

Ben: No doubt, it is very true that many people would contribute far more if they had more access to resources.... In my work with Hruy and his colleagues at iCog Labs in Ethiopia, I'm seeing that many folks there have great capability to contribute to AI, for example; which would likely not have been utilized had iCog Labs not happened to start in Addis this year, because opportunities for AI R&D in Ethiopia are very limited...

I'm not sure the problem of capitalism and inequality is restricted to "laissez-faire" style classic capitalism, though. Modern China has a quite carefully guided and restricted form of "state-driven capitalism" or whatever you want to call it – it's not laissez-faire at all – but yet still has rapidly escalating inequality. So there may be something deeper going on here.

Francis: As statistics about happiness and quality of life show, inequality is bad not just for the poor, but also for the rich, who feel stressed by the constant threat of someone else ready to take their wealth, and who find it more difficult to make profit in a society in which the majority is too poor to buy their products. In other words, the rich in very unequal societies feel less satisfied than the middle class in less unequal societies.

Ben: Japan and Northern Europe are rich and equitable but are not the principal drivers of progress toward GB, Singularity, or whatever... at this point they are markedly second-tier in terms of influence... So, one could argue that inequality has power as an engine for driving innovation and production.... Is it true or not? I'm not sure, but it's been argued by various others who know lots of economics -- and, if you want to argue convincingly that "more equal distribution of resources would make the economy grow much faster", the point has to be confronted explicitly and carefully, I guess...

Francis: To remedy that problem, in the Global Brain Institute we have defined a new objective: conceiving a new distributed system of governance and economics that would be much more democratic in the sense that power and resources would be distributed much more evenly. Dirk Helbing has called this Economy 2.0. But it won't surprise you that neither he nor we have a ready-made recipe, though we have lots of promising ideas inspired by self-organization, collective intelligence and GB technologies.

My impression is that by pooling the intellectual resources of GBI, Helbing's community and various other thinkers reflecting about these issues we should be able to come up with a truly new design for world governance over the coming years....

Ben: An intriguing and important direction, I'm eager to see what comes out of it, and to participate as time permits.... (As you know I have my own thoughts about the possible mechanics of an Economy 2.0, that I talk about in some of my own contributions to this book.)

Francis: Next to such a political-economic system we should also be able to use the technological innovations to make a transfer of wealth and resources much easier. The examples in the article about the spread of ICT in Africa helping economic developments illustrate the principle. Probably the most important ICT application is education. This we discussed in our Interversity project that intends to make university level education and research available to the whole world for free.

End of the Beginning

Ben: Yes, I agree that education is critical to reducing inequality and ensuring that advanced technology uplifts the whole of humanity and not just an elite subset. This is already happening big-time via MOOC's and so-forth. But more effort in this direction is needed. Hruy and I and our colleagues at iCog are actually looking at ways to use AI and other advanced tech to help bring better education to rural African children, for example....

Ben: Any reflections on Francis Heylighen's perspective on technology, the Global Brain and inequality?

Hruy: It is nice to know that there are people who really believe that to make the transhumanist ideology the foundation of the next civilization we should work on today's known problems as well as on the unknown.

If you take a second look at the US case, you will find that more or at least half of the inventors and those who worked with them are not of Americans but are individuals who migrate to US in search of the American Dream.

Ben: Yes, of course, America has been built by immigrants from Europe, Asia, Africa, etc. etc. My own great-grandparents emigrated to the US from Eastern Europe in the first decades of the last century ... Nobody is claiming there is some innate American Genome which is responsible for American technological achievements.... What's under discussion is the US economic and political system, which has been based on immigration from the beginning...

I guess what you mean is: Many of the inventors/scientists in America were not descended from the Europeans who arrived in the 1600s and 1700s, but rather from later immigrants? Definitely true. A lot were Jewish or other Eastern European immigrants like my family, who arrived in the early 1900s.... More recently a lot are Chinese or Indian immigrants. That's how America works...

End of the Beginning

Hruy: I was not talking about the immigrants who founded America! I am talking about those who are still flowing to America today. Starting from the 1950's, to be more specific after the Second World War, most scientist and inventors have migrated to America. The inequality in other nations forced these people to America...

Ben: I guess I see the recent immigration to the US as merely the most recent instance of a general phenomenon... My great-grandparents emigrated to the US to escape social and economic inequality in Europe (they were discriminated against as Jews), as well as the threat of extermination...
I will say one thing for the US: historically it seems to have had a lot of flexibility in terms of allowing people to increase their economic status.... There hasn't been much aristocracy compared to most other places, it's been relatively straightforward for someone to work their way to the top with persistence & cleverness.... Some have argued that this inter-class mobility is decreasing in recent decades though. I haven't studied it carefully myself – but if so, that's bad news for the US.

Hruy: In my view, inequality has three aspects. The first is that between individuals within a nation. Second is that between individuals of different classes within different nations, e.g., (the upper/middle/lower class in Ethiopia compared to the upper/middle/lower class in US or China). And the third aspect is inequality between nations as a whole, e.g., comparing the U.S. as a whole to Ghana or Japan or Ecuador. Looking at it in this way, the United States may be more equal than any nation on the planet, and China is not doing badly either.

Overall, the attitudes that shaped today's society were mainly designed by the "Capitalists"; so now we have the past three hundred years' accumulated weight of the capitalist system's

End of the Beginning

influence on our psyche. Hence, across the world, individuals' motivation (which pretty strongly affects innovation and technological progress) is highly triggered by the US, as so many innovative people feel inspired to become part of the economic group at the top of the society. Consciously or unconsciously an individual's motivation depends on the reward they anticipate; and the reward for participating in the capitalist system is much better than rewards obtained otherwise these days, since the current world tends to applaud for material gains more than intellectual integrity or other values.

Ben: The capitalist system that dominates the world now certainly has its pluses and minuses. I despise the low value it places on intellectual, artistic, spiritual and other forms of achievement. But I love the technologies it's spawned. Like humanity itself, it's complex and contradictory.

Also, on a peripheral note, as an American living in China, I should add I feel China is the most capitalist place I've been (and not just Hong Kong where I live, which admittedly has very special characteristics – I mean Shenzhen and Shanghai and Beijing as well – really everywhere). The Chinese today seem even more obsessively concerned with money and capital and so forth than the Americans...

Hruy: You know the most important reason why I agreed to join iCog Labs, is that that as a man whose dream is to write a timeless book, I wouldn't mind having a brain implant which will translate my thoughts into any language. Promise me to have a special price arrangement for me for that particular application; or I will be the first volunteer to try OpenCog's prototype when the time comes. I am not kidding.

Ben: Heh.... But think about it: By the time we have that universal-translator brain implant, "language" as we now know it

500

End of the Beginning

will likely be getting obsolete, as we'll be able to communicate via mentally sending each other abstracted chunks of thought (by wifi from my brain implant to your brain implant), in some way that doesn't require linearizing these chunks of thought into series of sounds or pictures...

But, to get back to the problems of today, do you see a pragmatic path to dramatically reducing inequality in the world? The most viable class of paths I see are those centered on spreading education everywhere, and improving global communication/ collaboration networks.... Convincing the wealthier agents to share "just because it's right" seems doomed; and equalizing resource allocation by force seems doomed because the wealthier agents have far better weapons systems.... Directing the attention of wealthy agents to wealth-creation avenues that involve global spread of education and communication/ collaboration networks, would seem the most practical course... (Bearing in mind that in the first world we now have an "attention economy" where human attention is the key scarce resource...)

Hruy: Indeed convincing or forcing the rich will not be the way. And yes, spreading education, investment opportunities, and technology transfer is the fastest and most plausible root to the *'transequality'*. Above all else, we need to adopting a new attitude, a collective general concesiousness; for a nation to develop a super economy, for people to be among the top, to enjoy the good things in life, it isn't a must to have poor and rich, illiterate and literate, underprivileged and privileged. This is what we are appealing to Transhumanists through this book.

End of the Beginning

Part Six:

The Future of Money

Chapter Fifteen: The World's First Decentralized System for Financial and Legal Transaction

By Chris Odom

Chris Odom is Co-Founder and CTO of Monetas, the world's first decentralized system for financial and legal transactions, and the creator of the Open-Transactions digital finance suite that Monetas is based on. Open-Transactions features industry-leading innovations that solve critical problems in digital finance and addresses urgent needs in the Bitcoin economy.

The Legacy Banking System

Legacy banking systems and their associated technologies such as ACH, wires, credit cards, and merchant accounts have left users frustrated. Long durations before clearing (3-5 business days for ACH), ATM fees, temporarily inaccessible funds, regulatory burdens, and in the case of credit cards and merchant accounts, high costs and hassle, are all examples of how 20th century technology is ceasing to be tenable in the Internet age.

How did such an anachronism last so long in the first place? Would it be possible in a free market?

Regulatory capture, a form of political corruption where regulatory bodies merge with the entities they are meant to police, has resulted in the corruption of our money. Everywhere we see rigged markets, bureaucratic banks, cycles of inflation and deflation, excessive fees, bailouts, loss of privacy, capital controls, confiscations of bank deposits (as happened in Cyprus), housing bubble, education bubble, sovereign debt bubble, and of course, unhappy users. Yet as unhappy as their customers are, banks are nevertheless reporting all-time high profits. How can this be?

Central bankers insist that "policy" must be used to "stabilize" the value of currency, and regulators insist their purpose is to "protect consumers," not to stifle competition. But if all this bureaucracy is truly superior to a free market, then why does it have to be forced onto people in the first place?

The New Money that is Coming

New technologies in money are inevitable, and will eventually work their way throughout the entire economy. Just as "software is eating the whole world" -- just as electricity did a hundred years ago -- so also new technologies in money will change our lives and our world in unfathomable ways.

What are some of the features of the money that is coming in the near future?

- Secure.
- Irreversible.
- Censorship-resistant.
- Instantaneous, Inexpensive, and International.
- Private, Apolitical and Multi-Jurisdictional.
- ...yet will "embrace and extend" the legacy system.
- Extendable.
- Convertible.
- Automated.
- P2P.
- Federated.
- Largely self-enforcing.
- An eco-system.

End of the Beginning

- Will become an essential piece of infrastructure within a short time.

Convertible and Automated

Imagine being able to store your money in gold, move it in Bitcoin, and spend it in dollars, to a merchant who only receives in euros... seamlessly.

Imagine having wallet preferences with a list of accepted currencies, audited transaction servers, trusted issuers, Bitcoin pools, P2P credit lines, insurance policies, trusted auditors, pseudonymous identities, asset allocations—yet perhaps not even having to deal with those directly, because your software is self-balancing, based on your rules or rules you have subscribed to.

In the future,
- Users will be able to store their money in any currency or allocation of currencies.
- Users will be able to convert into any form, transfer in any form, and spend in any form.
- Merchants will be able to receive in any form.
- ...And there will be seamless, automated currency exchange between them all.

An Eco-System

The system that is coming is not based on any one company, or any one piece of software, or in any one jurisdiction. Rather, it's an eco-system composed of many disparate entities including digital gold currencies, Bitcoin, legacy banks and fiat currencies, virtual currencies, LETS systems, and so on.

The most important pieces of the ecosystem, in no particular order, are:
- Legacy banking and fiat currencies.
- Legacy finance.
- Digital gold currencies.
- Transaction Servers.

End of the Beginning

- Business agreements and enforcement.
- Local currencies and LETS systems.
- Blockchain-based currencies such as Bitcoin. (Due to their inherent censorship-resistance.)
- Virtual game currencies. (Including, but not limited to, online gambling.)
- Currency and stock exchanges.
- Asset-based currencies (with private issuers.)
- Commodities markets.
- Derivatives markets.
- Prediction markets.
- Labor markets.
- Auction markets.
- Real Bills.
- P2P credit lines (such as Ripple.)
- Illegal markets (such as for drugs, arms, porn, or pirated products.)
- Software APIs.
- Retail merchants.
- E-Commerce.

The transformation will occur first in those areas where the new money is the most enabling, and will then be adopted by each subsequent area of the economy with increasing rapidity due to the network effect.

An Essential Piece of Infrastructure

This eco-system will become an essential piece of infrastructure in a short period of time. Just like Amazon EC2, or optical fibre, or roads, or trunk lines, or wireless spectrum.

As various different pieces of software begin to take advantage of APIs for instant, secure, seamlessly convertible, apolitical money, many different parts of our economy will quickly take advantage of the new capabilities this will provide.

The ability to transmit money in this way, at the instant API

End of the Beginning

level, is comparable in importance to the ability to transmit data itself.

As these pieces become closer and closer integrated, it will become impossible to tear down this ecosystem. Even successful attacks on individual pieces, will not disable the eco-system as a whole. It will become a basic infrastructure connecting many aspects of our economy, that could not be torn up, any more than highways or power lines would be torn up. Instead, that infrastructure will simply become a part of the fabric of reality, accepted as a fact of life by government agencies, corporations, individuals, and even software APIs.

Strong Cryptography

Strong cryptography is what makes this possible. In the 1990s, national governments were scrambling to put the "strong crypto" genie back into its bottle, but to no avail. Today, libraries like OpenSSL and GPG are built into a plethora of different products used to protect lives and data all around the world.

But some very critical capabilities, made possible only by strong crypto, have yet to make their way into common usage. The new open-source library Open-Transactions includes some of these new features:

- Unforgeable transactions, made possible using digital signatures and signed receipts.
- Un-changeable balances. Conventional "account-keeping" systems (such as PayPal or E-Gold) are able to change your account balance simply by changing an accounting entry. But Open-Transactions servers cannot change your balance, because they cannot forge your signature on the receipt.
- Untraceable cash. Open-Transactions uses "Chaumian blinding" to provide truly untraceable cash.
- Destructible receipts. In double-entry bookkeeping, the record of transactions is necessary to calculate the balance.

End of the Beginning

But Open-Transactions is able to prove which instruments are valid, and which transactions are closed, without storing any transaction history, except for the last signed receipt.

The Age of Apolitical Money

Features of new systems like Open-Transactions, such as untraceability, destructible account history, separation of powers, and jurisdictional arbitrage, all add up to much more financial privacy than has been available for decades to the typical plebe in our society. And this privacy will be enforced by mathematics and protocols, instead of legislation and courts.

Many currencies are purely virtual, and others, based on physical reserves, will be cross-jurisdictional. Entities using physical reserves (such as digital gold currencies) already employ a strategy of splitting their reserves across storage companies in multiple countries.

There are several other reasons why the ecosystem, as a whole, is 'apolitical':

- The power of strong crypto, and the inability of democratic processes to compromise or bribe it.
- The function of Bitcoin as the "universal medium" allowing fast, easy, unrestricted movement of funds in-and-out of various different systems.
- The "jurisdictional arbitrage" that occurs when you have many issuers in many different jurisdictions.
- The use of basket currencies (with insurance) to distribute a single currency across multiple issuers.
- The use of surety bonds for providing anonymous security, as that seen in the e-Cache experiment.
- The ability to have many transaction servers, with these servers also able to operate in many jurisdictions, and even on anonymous networks, in the case of "low-trust servers."

End of the Beginning

- The use of multiple jurisdictions for storage facilities of physical reserves.
- The existing services providing convertibility between virtual currencies and Bitcoins, between Bitcoins and fiat money, between Bitcoins and digital gold currencies, and between digital gold currencies and fiat, in a multitude of jurisdictions -- and in no jurisdiction at all.
- The natural competition between jurisdictions.
- The power of P2P credit lines, which will allow users to circumvent any "gatekeepers" for access to the system, by simply going through their friends. Hawala is one ancient example of this concept, which is still a powerful force in the world today.

What will ultimately happen with the U.S. Dollar of the 20th century? Nixon closed the gold window in 1971, and so our current experiment has been active for just over 40 years. History teaches us that of the 775 fiat currencies that have existed, 599 are no longer in circulation. The median life expectancy for defunct currencies is 15 years, and the average is 34 years. 1 in 5 fiat currencies have ended in hyperinflation, and even the most successful examples have lost over 99% of their original value.

What is Money?

Money is any substance that provides utility as a unit of account, a medium of exchange, and a store of value.

Over history, as people have bartered and traded, some substances have been selected for the role of money by natural market forces, based on their relative utility as a unit of account, a medium of exchange, and a store of value.

The first gold and silver coins of the Grecian age were struck in Lydia around 700 BC (in the form of electrum.) Later, silver was refined and coined in its pure form. For thousands of years, many nations used silver as the basic unit of monetary value. In some languages, such as Spanish and Hebrew, the same word means

End of the Beginning

both silver and money. In German, the same word means both gold and money.

Even today, gold is valued for monetary purposes. The largest gold depositories in the world are controlled by central banks in the major nations of the world, and it is common for portfolios to include a gold allocation for hedging against crisis, inflation, and downturns in the stock market.

The Properties of Gold

Regarding gold, it's important to keep in mind that it doesn't exactly have "intrinsic" value. Rather, gold is valued by men for its unique properties.

Gold is:
- Divisible.
- Fungible.
- Value dense.
- Recognizable.
- Durable.
- Zero counter-party risk.
- Stable in supply, yet minable.
- Liquid.
- International.
- Non-manipulatable. (Non-centralized.)

The above properties all contribute towards making precious metals uniquely suited for use as currency.

Let's compare gold to other forms of value:
- Diamonds, while valuable, are not evenly divisible, nor are they fungible. (Fungible meaning that any unit is interchangeable with any other unit, just as any dollar is identical in value to any other dollar.) Therefore we'd expect gold to be selected over diamonds by the invisible hand, for use as a currency.
- Water, while valuable and divisible, is not value-dense enough to compete with gold as a form of money, on the

510

End of the Beginning

free market.
- Food, while valuable, is not durable. Though neither is fiat money, in many places. (In Argentina, they say, "Cash rots faster than bananas.")
- Dollars, while liquid, do not represent zero-counter-party-risk (rather, they are debt-based.)
- Dollars, while recognizable, are not stable in supply (inflation is a worry).
- Dollars are also not minable. Control over production is limited to a banking cartel, versus gold, which anyone can produce.
- Food, which anyone can produce, cannot provide liquidity as a currency, especially in comparison to dollars or gold.
- A dollar can be manipulated in value. Even while you hold it in your pocket, the Federal Reserve board nonetheless retains the ability to manipulate its value from afar. This is not the case with gold.

The Invisible Hand

It becomes very clear that gold was never "declared" to be a form of money by any "authorities" but rather, became money due to natural market forces. The invisible hand was all that was necessary, historically, for gold to rise and to reign as money for thousands of years.

Authorities have uniformly acted, historically, to muzzle the trade of gold, to monopolize seignorage of silver, to inflate the silver via increasingly less-valuable alloys, to replace gold and silver with paper money, and even outright confiscation. But despite these pressures, gold and silver remain as premium currencies, and as veritable strongholds of wealth the world over, even into the modern day.

Artificial Forces

If gold became money strictly due to natural market forces as a result of its unique properties, then the only reason it can have been supplanted by dollars is due to artificial restraints imposed on the market by government forces, such as legal tender legislation, tax legislation, capital controls, and money laundering legislation.

Such forces must be constantly active, otherwise, natural market forces would immediately resolve back to gold again as they have for thousands of years.

But what will happen once the forces of censorship are no longer able to restrict how we use our money?

The Rise of Bitcoin

In 2009, Satoshi Nakamoto released his landmark paper, "Bitcoin: a Peer-to-Peer Electronic Cash System." Bitcoin is not merely a new currency -- it's a whole new technology, and a commodity, rolled into one.

By early 2014, Vice President of the Federal Reserve Bank of St. Louis David Andolfatto had released a report on Bitcoin. Among other things, his report stated that Bitcoin is a "threat [to] money and payment systems" and that "enforcing an outright ban is close to impossible.... [Bitcoin] will force traditional institutions to adapt or die."

Let's consider Bitcoin's unique properties:
- Divisible.
- Fungible.
- Value dense.
- Recognizable.
- Durable.
- Zero counter-party risk.
- Stable in supply, yet minable.
- Liquid.
- International.
- Non-manipulatable. (Non-centralized.)

As we can see, Bitcoin's unique properties are like those of

End of the Beginning

gold. Additionally, Bitcoin is:
- Non-confiscatable.
- Accounts cannot be frozen.
- Anonymity is possible.
- Instantly digitally transferrable.

These new properties (non-confiscatable, non-freezable, potentially anonymous, and digitally transferrable) all serve to route-around the artificially-restrictive monetary forces in operation today that depend on government collusion with banks, and on their collective monopoly on the ability to issue, store, freeze, confiscate, track, and wire fiat money.

Does it Work?

Typically digital currency systems have a central server that controls the balances and signs off on the transaction. But Bitcoin is decentralized. Like Bittorrent, it is composed of its users who communicate directly to each other on a peer-to-peer network. So then, how is the Bitcoin network able to arrive at an agreement regarding which balances are correct and which transactions are valid?

Quite simply, other users on the Bitcoin network handle the duties of signing off on the transactions and the changes in account balance. These users are known as "miners." But how can we trust those miners to do so without lying? We cannot. Bitcoin is designed to work even when other peers are not trusted.

When a miner signs a transaction, he has to perform a proof-of-work algorithm in order to do so. Basically this means that he has to *expend some effort*. He has to *spend some money*. Put another way, he has to *crunch some numbers*. Other miners can prove whether or not he actually did his work, and if he didn't, they will ignore his message. And if the transaction is invalid in any other way, again, the other miners will just ignore it. And notice: it costs him money to be dishonest -- money he'll never get back.

Whereas if a miner operates in good faith and properly signs transactions, he will be rewarded with transaction fees from the users, and with new coins that are uncovered by the mining process.

In short: miners are likely to *earn money* if they tell the truth, but if they lie, it will *cost them money*. Based on this principle, we may assume that more people will be telling the truth, than lying. Or more to the point, more computing power will be telling the truth, versus lying.

On the Bitcoin blockchain, the more confirmations that a transaction has, the more trustworthy it becomes. The longest chain of confirmations becomes the "truth."

A List of Things to Come

It's useful to view new technologies from the perspective of what they actually enable us to do, that we were not able to do before.

We all remember when Napster, a centralized file-sharing network, was shut down. But soon after, Bittorrent came into existence--and Bittorrent cannot be shut down, because it is decentralized. We know that if it were possible to shut it down, then it would have been shut down already. (After all, that's what happened to Napster.) But Bittorrent is censorship-resistant, and thus it cannot be shut down.

Similarly, Bitcoin provides us with a censorship-resistant, digital version of gold. It is valuable for use as money, but most importantly: it cannot be shut down.

To understand what this means, consider the now-defunct Silk Road market, and its many successors which operate on the Tor anonymous network.

Numerous sellers hawk illegal wares, mostly drugs, on these sites. Product advertisements blatantly display photographs of cocaine, crystal methamphetamine, ecstasy, and so on. If these sites were operating on a normal web server, they would immediately get shut down. But because they operate on Tor, an anonymous

End of the Beginning

network protected by strong cryptography, it's anonymous and it's extremely difficult to discover where it actually is.

Similarly, if such sites were using, say, PayPal for their payments, the sellers would all be arrested in short order. But because these sites operate using Bitcoin, no one can shut down their payment system -- it's censorship-resistant.

This is an extreme example, but that gives us a taste of the impact of this new technology in the real world. Before Bitcoin, one could not have a website selling drugs. The authorities certainly would shut down such an abhorrent operation, if they were able to. Therefore, Bitcoin is truly censorship-resistant, just as Bittorrent is. Otherwise it would have already been shut down.

Natural Law

But it would be a grave mistake to think of Bitcoin as merely a tool of for drug dealers and money launderers. This is what the media always tells us. But the currency used most by drug dealers, by far, is the U.S. Dollar, and the biggest money-launderers are conventional banks, which were (for example) recently caught laundering billions of dollars for drug lords south of the border. If the current, over-regulated financial system is not able to prevent money laundering, even with its draconian violation of our privacy, then how can it be a solution?

Bitcoin's primary feature is specifically its immunity to manipulation by bankers. This understanding is key: Bitcoin is ushering in natural law. It enables people to do things which are not in violation of natural law.

Another example is instructive: Imagine that I walk into a coin shop, and hand them a 1-ounce gold coin. They, in turn, send some Bitcoin (digitally) to a coin shop in South Africa, where my relative walks in and picks up a 1-ounce gold coin. (Paid for by the Bitcoin transfer.)

The point? Bitcoin is not just an "alternative" to gold, but rather, Bitcoin can be used for *sending* gold. In fact, it can be used

End of the Beginning

for sending any sort of value; it's a generic value-transfer mechanism. This sort of power is a great boon to the people, and a symbol of the coming democratization of wealth of the 21st century.

The real vision is a future where people have complete and total control over their own money. A future where every kind of value can easily and quickly flow from one person to another, and even from one software API to another, or from one robot to another, changing form as necessary, and stored in whatever asset allocation each user sees fit.

We're entering a future where value, like information, is able to be free.

Irreversible. The May Scale of Monetary Hardness describes the relative "hardness" of monetary instruments, based on how reversible they are.

Hardness   Item
1. Street cash, Bitcoin, Gold/Silver Coins (Hard)
2. Western Union and other money transmitters
3. Account based electronic currencies fire walled away from banking system (e.g. Liberty Reserve)
4. International wires
5. bank checks
6. ACH, personal checks
7. Consumer-level electronic account transfers (e.g. Dwolla, AlerPay), Bitcoin sellers (BitInstant, MtGox etc.)
8. Business-account-level retail transfer systems, credit cards (brick and mortar) (soft)
9. Credit cards (via internet or phone)
10. PayPal (Ridiculously soft)

One of the most hated aspects of legacy payment systems is forced reversibility, which is used as a tool by the banking system to clamp down on "harder" currency systems, in favor of "softer" ones.

An example of this:

End of the Beginning

1. Customer purchases Bitcoins using dollars on credit card.
2. Merchant sends Bitcoins to customer.
3. Customer files chargeback, gets dollars back.
4. (Merchant has now lost coins AND dollars.)
5. Merchant goes out of business.

In this way, reversibility is used to clamp down on certain forms of commerce in the West. But in other parts of the world, reversibility is only an option.

For example, Alipay is the largest payment processor in China, and their transactions are irreversible by policy. Escrow *is* available, but only as an optional feature, which is used in about half of all Alipay transactions.

In the new, crypto-based currency systems such as Bitcoin and Open-Transactions, transactions are irreversible because they are based on protocols built with strong cryptography, and thus much harder currencies become available.

For example, a website for exchanging virtual game currencies can easily allow users to withdraw balances from the server in Bitcoin, since there is zero risk of chargeback. Before the invention of Bitcoin, such servers would not be able to risk the chargebacks, and thus could not allow withdrawals from the server. They could only operate using "server credits" that could not be withdrawn back out as money again. But Bitcoin solves this.

As harder currencies such as Bitcoin come into mainstream use, reversibility will be provided through escrow systems at higher layers. Escrow will be a useful option, instead of a forced property of the currency itself.

Largely Self-Enforcing

Most new business capabilities will be the sort that is able to operate without needing access to the traditional court system, due to technologies such as:
- Signed receipts and digital cash.
- Smart contracts. These provide automated enforcement of

agreements between multiple parties.
- Leading to: Virtual corporations. Software will enable corporations to issue stock, pay dividends, appoint agents, and have security over funds, all without requiring any access to legacy markets and banking systems.
- Cash-streaming protocols. When any software API needs to acquire resources from an entity it does not trust, it can simply purchase these resources in small quantities, and then purchase more when supplies are running low. Larger quantities can be purchased at a discount, as trust is built.
- Reputation tracking systems, such as web-of-trust and P2P credit lines.

Fiat money is usually instituted through legal tender laws, which consist of a government refusal to enforce any debt when the dollar-value of that debt has been paid in dollars. For example, if someone owes me an ounce of gold, and if gold is $1200/oz. on the market, then the court will consider the debt paid as long as the debtor has paid me $1200 in dollar form. Tax laws are similarly used to impose fiat money, since taxes must be paid in said fiat money.

However, Bitcoin and Open-Transactions are already able to process "smart contracts," which are custom, scripted agreements between the users, protected by strong cryptography.

This means that complex legal agreements, as well as dependable, predictable outcomes, are now possible, without requiring access to the existing court system in order to protect the security of agreements.

For the many agreements that can be processed in this way, instead of via the legacy court system, a considerable expense is avoided. Business in many areas will thus be able to operate at much lower levels of risk, especially wherever business processes and payments can be built into the software directly.

End of the Beginning

The Untraceable Future

In 1983, David Chaum released his seminal paper, "Blind signatures for untraceable payments" which outlined a method for using public key cryptography to make a truly untraceable form of digital cash. The technology lay dormant under patent for several decades, and its greatest promise seemed undone by its Achilles heel: the fact that digital cash systems still required a legal entity to serve as the currency issuer. After all, someone has to hold the gold.

This plain fact, that "someone still has to hold the gold," was the undoing of e-Gold, Ltd, a formerly-promising Internet-based gold currency, who eventually saw their customers' gold confiscated by federal authorities. But new systems, like Open-Transactions, are being designed to enable separation of powers, basket currencies, and jurisdictional arbitrage. And some systems, like Bitcoin, have no physical reserves at all.

Further complicating matters was e-Gold's database and resulting ability to track and reverse payments, which stuck them with the liability of tracking and reversing any "fraudulent or suspicious transactions" which had occurred on their system. The founder of e-gold, Douglas Jackson, is still wearing an ankle bracelet for his trouble. But the persecution of Jackson only served to drive the development of systems that were not so vulnerable; these new systems are being designed specifically to reduce liability for operators.

Limited Liability

Newcomer Voucher-Safe proposes to reduce operator liability through a separation of powers between the transaction servers and each actual currency issuer. After all, if the currency, once issued into circulation, passes entirely outside of the control of its issuer, then that issuer cannot be held liable for how its currency is used after that point, any more than the Federal Reserve itself could be held liable for how its dollars are used, once they have

End of the Beginning

passed into general circulation.

The persecution of older systems like e-Gold is what drove the development of new systems such as Voucher-Safe, in the same way that shutting down Napster drove the creation of Bittorrent. And this process is accelerating, as a wave of financial crisis sweeps the world. The more that authorities devalue their fiat currencies, and clamp-down on their movements, the more that technological alternatives are driven to grow and adapt -- and the more their adoption is driven in the general population.

Homomorphic Cryptography

Untraceable cash, based on blind signatures, is just one form of homomorphic cryptography. But it's also possible to encrypt transaction amounts as well, so that a piece of software can process transactions without even being able to see the amounts that are being transacted.

E-Gold was made responsible to report on account balances and transaction amounts. But what if, due to homomorphic crypto, the "e-Gold of the future" is able to process transactions, without even knowing what the amounts and account balances are?

E-Gold was held liable to report on where payments were coming from, and to whom they were going. But what if, due to blind signatures, the "e-Gold of the future" is unable to see such things?

Even more: what if the "e-Gold of the future" is itself able to operate anonymously on darknets such as Tor? What if these systems can serve us, yet without having to trust them?

An Epoch in History

The new money that is coming is likely to succeed first in developing countries, where it adds the most value; where currency inflation is most prevalent, where price-gouging is commonly perpetrated by remittance services, and where Rule of Law is generally less accessible to the common man.

End of the Beginning

We stand at an epoch in history. The typical person today is Han Chinese, earns around $10,000 per year, and does not have a bank account -- yet he does have a mobile phone.

If the availability of value-transfer software on that device makes it easier to store and exchange value than legacy alternatives, then it will only be a matter of time before that device usurps the role entirely. Already 95% of Kenyan adults use M-Pesa, a phone-based currency system, as their primary form of money.

70% of new smartphones are already running the Android operating system, where any user can already download Bitcoin and related applications. How long before destructive events such as the Zimbabwean Inflation become mere artifacts of history?

Magic Technologies

The various properties of these digital cash algorithms can also be combined with one another, creating a whole greater than the sum of its parts.

For example, Bitcoin mixer services have already sprouted offering digital cash instruments that combine the censorship-resistance of Bitcoin with the untraceability of David Chaum's digital cash. Different technologies will be layered together, creating an eco-system with far more power than any of its constituent parts.

It's hard to predict what will happen to the social and political landscape, once the "common man" has easy access to untraceable and censorship-resistant money, and when such money is built into everything around us. Automated resource allocation will impact all sectors, and shape the growth of up-and-coming industries such as robotics, 3D printing, and self-driving cars.

...And it's coming soon.

Chapter Sixteen **Beyond Money: Offer Networks, a Potential Infrastructure for a Post-Money Economy**

By Ben Goertzel

The idea of something as basic as money going obsolete may seem farfetched. On the other hand, the idea of paper money and coins being largely supplanted by virtual electronic funds would have seemed farfetched not too long ago – not to mention the prevalence of complex financial derivatives, high frequency trading, and so forth. Finance has been the subject of a variety of radical innovations over the past century, and may be transformed even more dramatically during the next. It is far from obvious that an economy focused on money is going to remain an effective means of mediating interactions as technology advances, the current forms of material scarcity recede, and interactions between minds, and between mind and matter, complexify.

Some SF writers have envisioned radical future transformations of the world of finance that enhance rather than de-emphasize money's role. For instance Charlie Stross, in *Accelerando* (2005) hypothesizes vast communities of posthuman

End of the Beginning

intelligences taking the form of digital corporations interacting via complex forms of auctioning. This is indeed one thinkable sort of possibility. On the other hand, it also seems feasible that advanced technology will both enable and encourage the development of alternative, more sophisticated methods of exchange. It may be that from the point of view of a post-Singularity society, or even a pre-Singularity society a bit more advanced than ours, money will seem a crude instrument, a woefully inadequate way to represent preferences and guide exchanges and formalize "value."

One hint at a possible direction beyond money as currently utilized is the observation that the values that we assign to various things, in our minds, are complexly multidimensional (or actually, probably better modeled as nondimensional). **Human value** is not one-dimensional like cash value – and as I'll discuss below, projecting it into one dimension doesn't always make sense. Yet for so many aspects of our lives, we're study with a system that treats value as one-dimensional.

Mathematically, one may say that human beings' value systems are "partial orderings" – they allow a person to say whether X is more valuable than Y, in a certain context. But money requires something stronger than this – it requires the arrangement of all entities to be valued in a "total ordering", so that any X and Y can be compared as to their relative values. This has the merit of simplicity; it allows values to be arranged along a number line, so that one can assign each entity whose value is of concern a single number representing its value. But it has the shortcoming of not actually agreeing with the way human minds think and feel about values. In a partial ordering, mathematically, two entities X and Y may simply be incommensurable. N-dimensional vectors (for N>1) are like this; it doesn't always mean something to say that vector X is "greater than" vector Y. In a value system that is a partial but not total ordering, there may be entities X and Y whose values are incommensurable – there is no clear answer regarding whether X is more valuable than Y or vice versa. This is how

End of the Beginning

human value systems work, even though it's not how money works.

In this chapter I'll outline an alternative methodology for systematizing exchange, which embraces the partial-ordering aspect of human value systems, and which I call an *Offer Network*. Offer Networks, like Bitcoin and other contemporary cryptocurrencies, would be enabled by Internet technology and advanced computer algorithms. Using Offer Networks, people would be able to make bids to provide one good or service in exchange for getting (not necessarily from the recipient of said good or service) some other good or service that they want. A computer algorithm would reconcile different peoples' offers and try to find collectively satisfying arrangements.

Using Offer Networks, people would be able to get what they need and some of what they want, and offer what they are willing and able to, without having to pass all their preferences and values through the single dimension of cash value. A kind of cash value – which I call OfferCoin -- would still exist, but defined emergently in terms of a network of formalized non-monetary exchanges. So in simple cases where projecting the value of an entity into a total ordering is unproblematic, it can still be done. But in more complex cases where value must be treated as multi or non-dimensional – and partially but not totally ordered – the use of money values is not required, and one can instead use direct exchanges, which may be mediated according to the partial-ordering value systems of the parties involved.

Expressed abstractly like this, the idea may sound pretty complicated – but actually using an Offer Network would be no more complex than using contemporary Internet transaction systems like eBay or Match.com.

There are echoes of SF tales like *Accelerando* and *Down and Out in the Magic Kingdom* (Doctorow, 2003) in the Offer Network idea -- but when one gets down to nitty-gritty details, the system I propose here is a bit different. Partly, perhaps, because I'm trying

End of the Beginning

to propose a variety of post-money (and in some potential futures even post-economic) system in a straightforward and realistic way without too many fun, whimsical SF overlays.

Whether Offer Networks are really the future of economics, I don't know. But as Abe Lincoln and others have said, "The best way to predict your future is to create it." Perhaps by promoting the idea of Offer Networks now, the odds of something like it coming to pass can be increased!

My conjecture is that something like Offer Networks is coming, and is going to play a key role in smoothing the socioeconomic changes we're going to see during the next decades – as technologies like AGI and robotics advance and gradually conquer the job market. But I'll get back to that aspect after elaborating on what an Offer Network actually is.

Alternative Currencies and Economic Frameworks as of 2014

The notion of Offer Networks and OfferCoins is somehow a natural thing to propose in 2014, given the recent attention paid to cryptocurrencies and their potential for revolutionizing exchange. Bitcoin is the most popular of the new digital currencies, and is notable due to its relative popularity, but it's not necessarily the most interesting of the bunch. And beyond crypto currencies per se, a number of parallel proposals for complex network based, Internet and cryptography leveraging, economic frameworks have emerged recently. None of these does what Offer Networks does, but some of them do have properties that Offer Networks seek to achieve in a somewhat different way.

For instance, there have emerged various social-good-oriented cryptocurrencies such as DevCoin (**http://devcoin.org**). DevCoin automatically donates a certain percentage of newly created currency units to the common good, where the latter is voluntarily determined. According to the DevCoin site, the point is "allowing people to come together and vote on projects that the money should be spent on, thereby creating what is essentially a

525

(voluntary) 'world government' that funds itself from the revenue from generating new currency units. "

Somewhat similar is SocialCoin (**http://www.soccoin.com**), which aims to give a certain amount of free SocCoin money to each person each year. If SocCoin were widely adopted, this would amount to a universal guaranteed income. OpenUDC[110] is another effort along these lines (forked into Ucoin[111]).

Abstracting from the specifics of DevCoin, SocCoin, UCoin and other similar currencies, what one sees is the idea of a mechanism that

1. gives a certain allowance of money to each person each year
2. gives additional allowance to people satisfying certain criteria
3. allows (online) voting regarding which people satisfy said criteria
4. uses cryptographic methods to ensure unique identities for voters (e.g. Buterin, 2013, reviews details of this aspect)

In theory, with broad buy-in, this would be a democratic, global method of executing social welfare.

Ethereum[112] is a software infrastructure intended to be used to implement various novel media of exchange (e.g. "traditional" cryptocurrencies, multidimensional money, or essentially any kind of contract that can be expressed as a computer program...). For example Gregory Maxwell (Maxwell, 2013) showed how it could be used to create a fully automated, self-propagating business: an online storage facility that advertises itself, secures new customers, and then rents new space when its current storage space runs out.

Approaching the issue of exchange from a different direction,

---

[110] *http://project.openudc.org/*
[111] *https://github.com/ucoin-io/ucoin*
[112] *https://www.ethereum.org*

End of the Beginning

Open Value Networks[113] provides a model for collaborative organizations founded on open, egalitarian exchange and collaboration – intended as an alternative to corporations in their traditional form. As the web page states,

Open Value Network is an approach to commons-based peer production. It allows individuals and organizations to co-create and aggregate value through lateral and large scale coordination, cooperation and collaboration
- to serve as a responsible steward of commonly held wealth and assets
- to account for various inputs and outcomes in a common ledger system
- distribute value equitably and intentionally within and beyond the network
- and to share returns amongst contributors, in proportion to their contributions

One of the goals of Open Value Networks is to encourage direct exchange among individuals who have something to offer each other, eliminating middlemen as much as possible, since in the current scheme of things middlemen often couple moderately useful informational functions with a high degree of value extraction. As Open Value Networks advocate Apostolis Xekoukoulotakis (2013) put it,

[I]t is important to note that there are 2 graphs. On the one hand is the graph of offers and needs and eventually the transfer of products that fulfills those needs.

On the other hand, there is the graph of transactions, the graph that shows the changes of ownership of commodities from one person to another.

For each edge of the first graph there can be multiple edges

---

[113] http://p2pfoundation.net/Open_Value_Network

End of the Beginning

in the second because of the introduction of middlemen.

My proposal tries to subjugate the graph of transactions (and money) to the first graph.

Due to lack of information on the needs and offers of people, we need middlemen and money. A posteriorly, then, we have to find rules that define transaction equivalences that can eliminate the previous transactions of middlemen.

This is similar to giving zero interest to a loan, but only applied on multi barter networks.

Less concrete in nature but similar in concept, theorist Dirk Helbing has written a number of papers about the future of economy, including one tantalizingly titled *Economics 2.0*. He proposes a kind of multidimensional money, in which each unit of currency comes attached with multiple value indicators. E.g. instead of just having a dollar, you might have a dollar that comes with 100 default cents, 70 environmental-responsibility cents, and 35 end-poverty cents. The additional values would indicate the extent to which the processes used to earn the dollar have worked toward other goals such as environmental responsibility and the ending of poverty.

One interesting point raised in *Economics 2.0* is the potential existence of a "tipping point" of sociality.... In some simulations he reports, if each agent in a society has a utility function of the form

personal utility + others utility * k

then the higher k is, the more each agent cares about the other agents (and acts on this care). What he notes is that there can be phase transitions in social structure based on k. Sometimes a small increase in k can cause a large change in the social order – a tipping point.... Very roughly speaking, it could be that widespread use of a system like Offer Networks could help to nudge k up toward the threshold level.

Does More Than One Dimension of Value Make Sense?

The various crypto-currencies at play today are exciting and

528

End of the Beginning

interesting, but have in common with traditional money the projection of value into one dimension. Dirk Helbing's speculations about multidimensional currency are a welcome exception.

Many thinkers, however, believe that the one-dimensionality of money is not just a convenience, but a reflection of a fundamental one-dimensionality of value. The most common argument in favor of the effective one-dimensionality of value is based on the idea that rationality implies transitivity of preference.

An agent is said to have transitive preferences if it organizes its preferences in a linear order, where for every class of goods/services X, Y and Z, "X is preferred to Y" and "Y is preferred to Z" implies "X is preferred to Z". One can prove fairly easily that if an agent's preferences are NOT transitive in this sense, then it can be manipulated by other agents in a series of trades that rob it of its property and leave it in a subjectively undesirable situation (Hansson et al, 2012). If preference is associated with monetary value, the series of trades involved in this argument is typically called a "money pump". But actually the argument holds even outside the context of money, e.g. one can set up a barter-based "money pump" that causes the victim to trade away all his stuff in a series of circular trades.

But the money pump argument is founded on the assumption that intelligent agents necessarily have a fixed preference ranking over all the alternatives available to them, based on assignment of these alternatives to well-defined categories. That is: in the mathematical terminology I introduced above, it assumes that human value is a total ordering. Real minds are complex dynamical systems, with preferences that are shifting and context-dependent, especially when dealing with situations they're not that familiar with. Often a real-world agent will be indifferent between multiple options that it doesn't know much about, or will make up its mind about how much it likes something only while experiencing it – thus rendering its valuation system a partial

529

rather than total ordering.

Some kinds of exchanges really are well modeled in terms of classes of repeated goods/services, about which an agent has a clear and stable set of preferences – in these cases the money pump argument is applicable. Other kinds of exchanges are far more fluid in nature, and may be more naturally modeled in different sorts of ways.

It would seem that when an agent is in a new sort of situation, where it isn't clear how to effectively divide the various available options into categories, then transitivity of preference is likely to be violated for the relevant categorizations of options. But in practice this wouldn't necessarily make the agent susceptible to a money pump, because in the course of doing the repeated trades needed to bilk it, the agent would learn more about the domain and form new preferences.

In short, transitive preferences and linearized values seem most sensible in cases where repeated instances of the goods/services in question can be very neatly divided into categories, and the agent has had enough experience with goods and services in the different categories to really know what it prefers. In a novel domain of experience, however, the flexibility to be intransitive may be helpful for creativity and learning. Prematurely settling on a linear ranking of values in a new domain may prevent an agent from more deeply understanding what its preferences should be, based on the rest of its mind.

Currently, with our limited ability to transform matter, we humans are in a domain where we're often exchanging goods and services that fit naturally into predefined categories. Mass production has increased the degree to which this is the case – in traditional tribe or village society, more exchanges were "special cases", making it more difficult to divide all exchanges into neatly defined categories. Two McDonald's hamburgers are pretty much alike, but two donkeys may be pretty different, and informal exchanges in a pre-industrial context were able to take into account

End of the Beginning

the particularities of the entities being exchanged, in a manner defying easy formalization as a brief set of rules.

As technology advances, matter will become more pliable via nanotechnology (and perhaps femtotech and beyond), and cognitive and relational goods will become as prevalent as physical goods or more so. Ongoing rapid change will be the norm, meaning that a preponderance of exchanges will regard novel situations. On the other hand, everyday goods like food and shelter and Net connectivity may become freely available, as is currently the case for water from public drinking fountains, or space for walking on the sidewalks. The domain in which transitive, linearized preferences are optimal and relevant may end up being a relatively uninteresting little corner of the total scope of exchanges.

What is Money, Anyway?

Offer Networks, which I'll describe here, are a specific method for handling exchanges among a population of people or other agents, which doesn't intrinsically rely on the projection of every offered good or service into a linear scale of values.

I'll work up to the Offer Network idea via incremental steps... starting in a hypothetical little village, and ending up eventually with an advanced AI-mediated post-economy.

We use money constantly, yet we don't reflect often on what money really is. I'll try to elucidate the nature of money by means of some very simple example cases.

First, let's consider a very simple monetary-exchange scenario in a small community:
- Dean borrows $20 from Todd.
- Bob picks some strawberries from his farm, and sells them to Dean for $20.
- Bob then buys a hamburger from Karen for $20, which Karen produces from her own cow.
- Karen takes the $20 and uses it to buy home repair services

End of the Beginning

from Jack; and Jack uses the $20 to buy some apples from Jane (who picked them on her farm).
- Jane takes the $20 to buy some oranges from Dean.
- Dean pays Todd back the $20 he borrowed.
- What are the actual exchanges that have happened here?
- Bob supplied strawberries to Dean, and got a hamburger from Karen
- Karen supplied a hamburger to Bob, and got home repair services from Jack
- Jack supplied home repair services to Karen, and got apples from Jane
- Jane supplied apples to Jack, and got oranges from Dean
- Dean supplied oranges to Jane, and got strawberries from Bob
- Todd just served as the bank, giving a no-interest loan
- The exchange of the $20 is just a way of mediating this network of exchanges of goods and services.
- Now envision a different way of doing the same set of exchanges, with no $20 involved:
- Bob publishes desire to exchange Strawberries for Hamburger
- Karen publishes desire to exchange Hamburger for Home repair services
- Jack publishes desire to exchange Home repair services for Apples
- Jane publishes desire to exchange Apples for Oranges
- Dean publishes desire to exchange Oranges for Strawberries

Then, the whole little community looks at the various exchange-desires that have been published, and figures out the exchanges that can be done to make everyone satisfied. And Todd has nothing to do with it.

In this particular case, there's no harm to using money; it's just a simple way to mediate the exchanges that would happen

End of the Beginning

anyway.

But in more complex cases, money may fail to capture the nuances of the desired exchanges; and can lead to various pathologies as we see in the modern world.

Offer Networks: A Simple Case

The example I've just run through leads up to a simple case of the Offer Network idea.

One elementary type of Offer Network would be defined as follows:

1. Each person publishes (into a secure clearinghouse system) their willingness to exchange X_i (some good or service) for Y_i (some other good or service)... Then
2. A "clearinghouse" algorithm figures out how to propagate exchanges so that each person gets some exchange that is acceptable to them.

A more sophisticated version would involve bidding, a la eBay and other auction systems. In a bid based Offer Network, each person would specify various possible exchanges (X_i, Y_i) and assign each one a degree of desirability.

Then, e.g.: Bob may be willing to offer an hour of medical consultation to get half an hour of skiing lessons. But first he may offer half an hour of medical consultation in return for half an hour of skiing lessons. If that fails he may up the offer to an hour.

(Or more generally: even if a person is willing to offer X_i to get Y_i, he may first offer some X'_i that is dominated by X_i (i.e. he would prefer to exchange X'_i for Y_i, than to exchange X_i for Y_i). If this initial offer fails to get him a good enough result via the clearinghouse, he may offer some other $X'_i < X''_i < X_i$, etc.)

A few more examples of specific exchange-pairs might be:
- Ben is happy to exchange knowledge about AI for knowledge about physics or biology
- Mary is happy to exchange home repair services for piano

End of the Beginning

lessons. She'd rather exchange home repair services with a young guy for piano lessons, but is willing to accept someone between 40 and 50 if needed.
- Todd is happy to exchange 5 kilos of strawberries for 10 steaks or 3 whole chickens

An issue that arises immediately in the context of a system like this is the risk of dishonesty and deception. In the simple case of a small community (like the 5 people in the example given above), there is a strong reason for each participant to make good on whatever they've promised to provide in a given exchange. If they don't, everyone else will see that they're unreliable or dishonest, and will avoid entering into exchanges with them in future.

In a larger society this kind of direct social pressure can't work, but alternatives exist, e.g. a reputation management system similar to that used in today's online stores, in which each person rates each transaction they've had, so that each person gets an overall reliability rating. More complex variants are possible as well, e.g.
- reliability ratings that are context-specific (Jack sometimes reneges on offers to participate in musical jams; but always comes through with offers to provide programming services)
- an algorithm to determine the reliability of each person *as a rater*, based on whether their ratings tend to be outliers
- an algorithm to estimate the average rating given to a certain provider in a certain context, by raters with criteria similar to one's own (e.g. you may not care if a housecleaner shows up late; but John may... so if John downrates Fred's cleaning services because Fred shows up late sometimes, you may wish to ignore said down-rating)

Potentially, reputation management could end up being the most complex part of an Offer Network system – much as fraud detection is the most complex part of a system like PayPal.

End of the Beginning

Beyond Traditional Economic Exchanges

The above kind of Offer Network is already pretty interesting. But it gets better – a moment's reflection indicates that traditional exchange of goods and services is only a fragment of what this kind of system can handle.

An offer such as

Agent A is willing to exchange X_i for Y_i

is just a special case of a more general offer of the form

Agent A is willing to carry out action A_i, which has post-condition Z_i, IF precondition Y_i is met

Everyday examples of offers that aren't classic economic transactions might be

- Matilda is willing to put 10 hours/week into a project aimed at emulating Oculus Rift in open source hardware (postcondition), if 10 other people with advanced degrees in science are also willing to put at least 10 hours/week into that project
- Maria is willing to go on a date on Friday 11/7 at 8PM with anyone living in the San Francisco area, if they are under 25, employed full-time, have a "Yahoo date rating" of at least .7, and have done at least 5 hours of verified social-welfare volunteer work in the last month
- Ben is willing to go play kazoo at Open Mic Night at Peel Fresco on 12/17, if there will be a drummer and a bass player there
- More advanced and complex offers could be issued by various sorts of AI agents, as AI develops.

The clearinghouse can then have knowledge of which post-conditions imply which pre-conditions, and the whole process of clearing everyone's offers becomes the execution of a giant "production system" style logic engine. Economy becomes cognitive in a very obvious and transparent sense. There is the possibility for the clearinghouse to optimize the degree of overall desire fulfillment via heuristically or probabilistically predicting which post-conditions are most likely to lead to which preconditions (or which combinations of post-conditions are most likely to lead to which combinations of post-conditions). And of course these calculations would be taken into account in the calculations defining the OfferCoin value of an offer.

Traditional economic actions involving exchange of goods and services become freely intermixed with other kinds of actions involving context-dependent willingness to commit to carry out various sorts of activities. But of course these more general actions are "economic" too in a broader sense, because they involve agents carrying out actions, hence expending energy...

As a simple example indicating the flexibility of this sort of system, various variations of "buy local" or "buy from a group I like" could be easily be achieved via an Offer Network system . For instance I could offer a discount on some service I offer to anyone who participated in online transhumanist discussion... or to people who live in my village... etc.

There is a surface parallel between Offer Networks and old-style socialist ideas about computer-controlled centrally planned economies. In these old ideas, as in Offer Networks, a massive integrative system was often posited to balance everyone's desires against each other. But a critical difference is that, in the Offer Networks approach the clearinghouse doesn't try to determine what anybody's needs or abilities are. Instead, people specify their offers based on their own value systems; and the clearinghouse just mediates the transactions, in a way that respects the various individuals' desires and values insofar as they have articulated them within their exchange offers.

Emergent Money

Generalized offers as supported by Offer Networks have a richness and humanity to them that purely quantitative financial calculations do not. They also allow people to propose and execute

End of the Beginning

exchanges that are faithful to the nontransitivity of their preferences.

However, money does have a certain simplicity to it. In some cases specifying exchange-pairs will feel overly complicated and the simplicity of some sort of money will be preferable. For this purpose, it is possible to define a sort of emergent money based on an Offer Network.

Further, it is difficult to see how a pure exchange network, with no abstract quantification of value into some money-like entity, would handle *nonspecific deferred gratification*. If everyone were just living in the now and exchanging today-stuff for other today-stuff, then you could just have a clearinghouse of offers, with no need for symbolic, money-like tokens of value. OR, if everyone were exchanging {today-stuff or specific promises of specific future stuff at specific times} for {today-stuff or specific promises of specific future stuff at specific times}, a pure exchange network would still work as well.

But the problem is that we don't know what we're going to want a week from now let alone 5 years from now... Yet we do want to be able to perform some services today, or exchange some perishable goods today, in exchange for being able to get some of "whatever we're going to want" a year or 5 from now...

This kind of nonspecific deferred gratification is critical to the creation of large projects of any type, at least in the current era or any era before scarcity is wholly abolished. And it seems that to make nonspecific deferred gratification work, you need some abstract token of value, that can be cashed in for specific goods or services at the future time when you decide what you want...

To enable this within an Offer Network as described above, the most natural route is to estimate an "emergent money value" for a given offer, via defining a measure of how many "exchange units" or "OfferCoins" a given offer is worth.

One way to do this would be as follows:

End of the Beginning

- The raw OfferCoin value of X is: what percentage of all offers is dominated by X (in terms of what they could obtain in exchange, via the clearinghouse).
- The final OfferCoin value of X (the number of OfferCoins X is worth at a given point in time) is then a normalization of the raw value of X
- Say: *1 trillion * raw value / (sum of all raw values)*

The OfferCoin value can also be used in exchange-pairs. That is, barter can be mixed with $$ purchases. Someone can offer 500 OfferCoins for a side of beef. The going rate for a side of beef in terms of OfferCoins can then be determined via auctioning, similar to on eBay.

In a sousveillant society (where auctioning is less relevant as real preferences are more transparent), everyone's desires can be made open, meaning the whole clearing house can be open. Otherwise, the clearinghouse could be made auditable but secure, using strong encryption methods.

Note that qualitative offers could mix with quantitative OfferCoin-based ones in various contexts; e.g. the kazoo-playing example given above could mix with:

- Jim is willing to pay 20 OfferCoins for a drink at Peel Fresco on 12/17, if there will be a kazoo player there
- Jean is willing to open her nightclub Peel Fresco for Open Mic Night on 12/17, if she has commitments for at least 500 OfferCoins worth of drinks to get bought

and the open source hardware example given above could mix with:

- Jack is willing to put 10/hours week into any available open-source hardware project in the gaming space, if he can find a Ruby programming contract paying him at least 50 OfferCoins per hour, for between 15 and 20 hours per week (or, if the contract involves developing open source code, he will accept a price of 40 OfferCoins per hour)
- Lisa is willing to pay up to 60 OfferCoins per hour for a

End of the Beginning

Ruby or Python programmer who is also an active contributor to the open source community (at least 5 hours per week); she is also willing to donate 5 OfferCoins to African medical care for each hour that this programmer works

It might be desirable to enforce a fairly high inflation rate on OfferCoins. So, if someone's precondition for an action involves getting OfferCoins, then they can receive OfferCoins and bank them. But if OfferCoins decay in value intrinsically, based on the time since they have been received, then hoarding intrinsically becomes infeasible.... Perhaps the inflation rate could be adapted dynamically via voting among all exchange members.

A big plus of an Offer Networks type system is that it allows one to flexibly embody various values into one's exchanges and other offers. It's straightforward to embody values besides simple monetary values in one's exchange offers (e.g. the desire to hire a programmer who is an open source contributor; the desire to date someone who has done volunteer work; etc.)

To formalize things slightly more, the definition of "monetary value" implicit in the OfferCoins defined above is

OfferCoin_value(offer A) =
(normalization of the rank value of A, when all current offers made by all people are rank-ordered in terms of what could be obtained from the offer via a complex series of trades)
*
(fraction of total OfferCoin value that is assigned to current versus past offers)

The latter fraction is determined via the inflation rate: the higher the inflation rate, the greater the fraction of OfferCoin value assigned to the present time.

So as opposed to a Marxist "labor theory of value", this approach is quite pragmatically assigning an OfferCoin value to something that is roughly proportional to what one could get by trading that thing. However, this calculation is made over the totality of specified non-monetary (i.e. non OfferCoin based) exchanges as well as over OfferCoin based exchanges.

If the inflation rate is voted to be reasonably high, then accumulation of wealth will be difficult and more of the wealth will go to people who are offering goods/services right now that are judged valuable right now....

The inflation rate toggles between meritocracy and inheritocracy, in a sense (where I consider "inheritocracy" as the practice of assigning value to those whose past selves or ancestors provided perceived value at past times)

Of course the idea of grounding some kind of money in exchange is not new and is wholly obvious. But one thing that's different here is the free intermixture of OfferCoin-based and "barter specification based" exchange ... this keeps the OfferCoins quite directly grounded in peoples' specified values. For instance, if
- X can be exchanged for 800 OfferCoins plus a few songs and paintings and happy wishes
- Y can be exchanged for 1000 OfferCoins

then X could be judged more valuable even if nobody wants to pay OfferCoins for the songs, paintings and happy wishes (but some people are willing to exchange other valuable things for these items)...

Antimoney

A speculative possible addition to Offer Networks would be the concept of anti-money introduced in (Schmitt, 2013). This is an innovative mechanism, reliant on the Internet and cryptography, intended to replace the mechanism of borrowing with interest. Two kinds of currency, money and antimoney are introduced. In exchange for a good or service, someone can provide money or receive antimoney. Two individuals can also trade money for

End of the Beginning

antimoney, at an agreed-on rate of exchange. To avoid exploitation, a cap may be placed on the total amount of antimoney any one person can collect (or more sophisticated, analogous methods may be used). If someone defaults (declares bankruptcy) then their antimoney is returned to the folks who gave it to them. The risk of default would be part of the calculation in setting an exchange rate with a given transaction partner.

Antimoney could potentially be introduced to Offer Networks as a kind of anti-OfferCoin, to be devalued at the same rate as regular OfferCoins. This would provide a built in method of small-scale borrowing. The viability of this sort of mechanism would need to be explored via simulations before widespread implementation. But at very least, antimoney serves as an example of the kind of innovation that is possible within this kind of framework.

Futarchy

As a bit of an aside, there seems to be a natural connection between Offer Networks and Robin Hanson's fascinating, hypothetical political scheme of Futarchy (vote on values, bet on methods for maximizing the values).

For instance in Futarchy, a vote could determine a goal such as: providing free health care meeting certain specific criteria to all people in a certain country, by a certain date. Then, various plans could be put forward by different people or groups, suggesting methods for achieving the stated goal. Then people would be invited to vote for or against the success of each plan. The plan that gets the highest odds of success is the one put into place, and then the bets for and against the plan are paid off once the success of the plan is observed. The underlying concept – which has a fair degree of empirical validation -- is that groups of people betting their own money tend to make more accurate judgments than groups people simply expressing their opinions.

Futarchy would emerge from the Offer Networks system if

- people make offers of goods/services in ways that are based on their values
- people make bets on outcomes of attempts to realize their values
- institutions make choices (postconditions) based on observations of peoples' betting behavior (preconditions)

Bets fit perfectly well into the Offer Networks framework; they are offers to give something in the future (the postcondition), conditional on some condition obtaining in the future (the precondition). Reputation ranking can be used to make it likely people pay up when they lose a bet.... A few renegers won't matter much with political-type bets, as futarchy counts on a large number of small bets being made...

On the other hand, it is not clear to me that futarchy as specified by Hanson would really be needed in an Offer Networks context. It might be that betting turns out to be unnecessary and many aspects of appropriate governance can self-organize via the network of exchange.

Isn't It Too Complex?

To implement an Offer Network effectively in reality would require a lot of art as well as engineering and science. But for a society that has implemented eBay, PayPal, Facebook, Bitcoin, Bittorrent and the like, it doesn't seem beyond the pale.

It may at first seem very complicated for a person to have to specify offers involving pre and post conditions, rather than just specifying the price one is willing to pay for something. But I think this complication could be mostly eliminated via development of appropriate user interfaces.

Also note that, while there is a central clearinghouse in the above, it doesn't actually have to be implemented in a centralized, monolithic way -- it could be implemented as a distributed peer to peer system with no owner, using strong cryptography to maintain integrity, and spare computer cycles to do calculations.

End of the Beginning

I don't think that the complexity of Offer Networks is greater than that of the current financial system, with all its derivatives and credit default swaps and central banks and what-not. It's a different sort of complexity, though -- more focused on individuals and their interactions and values than on the actions of specialists and large institutions.

The current, complex financial system emerged gradually, and I suppose that if a system like I've described here is going to happen, it's also going to emerge gradually. I'm not sure what the first steps will be, but I suppose that they will occur in some online marketplace – or maybe even inside some multiplayer game. Perhaps they will happen in the developing world, where typical 21st century financial systems are not that well developed -- there is a possibility for the developing world to leapfrog past the perversions of contemporary Wall Street into a more advanced sort of exchange network.

Algorithmics of the Clearinghouse

Detailed design of the Offer Networks clearinghouse algorithm has not yet been carried out. However, I have thought through two possible approaches, and formed a tentative opinion on which would be best. More mathematical analysis will be needed before a concrete approach can be definitely formulated.

The material in this section assumes some understanding of machine learning and simulation theory; the reader with little technical background may wish to skip to the next one.

One possible approach would treat the clearinghouse's operations as an optimization problem, solvable perhaps by evolutionary algorithms. One could, for instance:
1. Formulate an evolutionary learning problem, where each genotype is a set of pairs (person, action-set), indicating a set of possible actions undertaken by the people involved in the network; and the fitness of the genotype is: 0 if anyone in the network is doing something they're

End of the Beginning

unwilling to do; otherwise, the total sum of the weights that each person assigns to the (actions of which X is giver, actions of which X is recipient) set in which they are involved.
2. Fitness evaluation involves calculating the barter chains involved in an optimal implementation of a given genotype. This can be done via some sort of heuristic search algorithm. Dynamic programming is too expensive, so some approximative approach must be used. The fitness evaluation therefore will just involve an estimate of the actual fitness of the genotype, indicating the fitness according to the best corresponding barter chains that the heuristic search algorithm can find.
3. Fitness estimation will be critical, and will involve caching of information about barter chains found during prior fitness evaluations, to speed up heuristic search (e.g. by allowing rapid discarding of genotypes containing combinations previously found infeasible)

An alternative would be a Monte Carlo approach, involving the creation of a simulated agent for each member of the Offer Network. One would then let the simulated agents trade with each other for a while and see what results from all the trading. The whole simulation would be run many, many times, and the results would be recorded. Then, from all the simulations, one would measure the fitness of the end result (via the same method suggested in Step 1 of the evolutionary approach, above). The clearinghouse would then recommend the assignments that would result from the best simulation world found.

Of course, the effectiveness of this approach would depend on how dumb the simulated agents are. Caching information about barter chains found in prior simulations, as in Step 3 of the evolutionary approach mentioned above, would make the simulated agents less stupid as the series of simulations proceeds, and improves the ultimate result found in most cases.

End of the Beginning

My preliminary intuition is that the simulation approach would be better, as it seems easier to tune. However, both approaches would be computationally expensive and it's hard to compare them in depth without going a lot further along the path of implementation/experimentation. Perhaps some alternate approach will ultimately prove preferable.

Possible Psychosocial Effects of Offer Networks

The impact of the money economy on psychology is subtler than commonly realized. It is plausible, though certainly not demonstrated, that transcending or augmenting the money-based method of assigning value, in the way that Offer Networks do, would help resolve various problematic issues in modern society. For example, Offer Networks might end up wreaking exciting and positive havoc with our traditional notions of social status...

Modern first world capitalism has a significant aspect of *"Spending your life doing things you don't want, so you can afford to buy things you don't need."* (And as it happens, modern Chinese socialist capitalism has this aspect as well.)

Given current technology, it seems it would likely be possible for most people in the developed world to work 15-20 hours per week at most, not 40 – and to spend the rest of their time simply enjoying themselves, or pursuing creative activities of some sort. Yet this is not the direction society has gone in. Instead, work hours are still fairly long, and work increasingly spills over into off-work hours. Salaries are largely spent in a consumerist way, doing things like upgrading one's cellphone every year, or buying overpriced, toxically sugary Starbucks coffee concoctions.

Of course, people are aware of the possibility of working less and living a less consumerist lifestyle; and there is a "simple living" movement tending in this direction, it's just not very popular.

Why are we willing to work so hard at things we don't like doing, in order to buy so many things we don't need?

End of the Beginning

Part of the answer is addiction – we become psychologically addicted to certain routines and possessions, e.g. eating out in restaurants instead of cooking at home, or driving in a private car instead of taking public transport, or the feeling of wearing new & different clothes or jewelry.

And part of the answer is social status. Having a lot of money is itself an indicator of high social status; and money can be used to buy other status indicators. A tremendous amount of money is spent on status symbols of one sort or another. Of course, money is not the only status indicator. Einstein had a very high status in the public eye, yet was not terribly rich. The same for Linux Torvalds. But overall, money is surely the best single proxy for social status in modern societies.

Offer Networks could disrupt both of these factors, in indirect but powerful ways.

There is a great deal of data showing that addiction is most powerful among people (or animals) whose lives are not richly satisfying, including in social dimensions. A rat in an empty, boring cage will become addicted to morphine rapidly, after brief exposure to the joy of the drug. A rat in a cage full of fun, bustling social activity will ignore the morphine dispenser at the end of a tunnel attached to his cage, even after he's felt the joy of morphine (Mate', 2010). Because the warmth of his social environment makes the morphine high unnecessary.

Offer Networks could offer more than just a replacement for many monetary exchanges – they could offer a flexible medium for obtaining all different sorts of interactions with other human beings. They could provide a richer social web, thus reducing the psychological emptiness that allows the addictive psychology of consumerism to take hold.

And Offer Networks would also provide alternative means of establishing and measuring social status. Reputation points would provide one means – and could be context-specific. Getting a high reputation score among people who tend to participate in a

End of the Beginning

particular variety of exchange, would provide a form of status within a certain subgroup. Feeling the positive feedback of the people with whom one has done exchanges, may substitute somewhat for the status symbols that now provide people with ego boosts.

Exactly how the sociodynamics of Offer Networks would play out, nobody knows. But I think they have significant potential to disrupt the addiction and status dynamics that largely govern our society today. Doubtless they would bring new peculiarities and problems as well, only to be discovered once the actual usage patterns of Offer Networks have settled in.

Smoothing the Path to the Era of Abundance

What is the relation between Offer Networks and the Singularity?

Peter Diamandis has phrased the Singularity in terms of "abundance" (Diamandis and Kostler, 2012). Today we live in an era of relative scarcity. The resources we feel we want to fulfill our human desires, are often difficult for us to come by. But when technology has advanced enough, this may no longer be so. Infinite resources are unlikely to be available even post-Singularity, but one can only eat so many chocolate bars or surf so many waves each day. There may come a point where, relative to the scope of human desires and capabilities, scarcity is no longer a factor.

On the other hand, the same technologies that have potential to lead to radical abundance also have the potential to obsolete human labor. If we really reach a point where robots and other automated devices are doing most of the work, then we will have to face questions like: A) what will everybody do with themselves instead of working?; B) how will people get money or other resources, when they don't have the option to get paid for working?

My suggestion is that moving to a different, non-monetary

End of the Beginning

method of exchange – like Offer Networks -- might smooth out these difficulties considerably.

As material scarcity decreases, more and more of our exchanges will involve our deeper engagement, rather than merely our "putting in time" carrying out an activity, or providing a material object to someone else. The conditions under which humans are desirous of providing deep engagement to some activity are more complex than a simple monetary exchange, making the specification of more complex offers more appropriate. This relates to the point made above in the context of money pumps – when one is in a domain where most exchanges involve novel situations, and are difficult to definitively divide into categories, and then projecting preferences into a linear ordering doesn't necessarily make sense.

In the end, once automation has rendered "working for a living" unnecessary, human beings may end up spending most of their time exploring novel forms of social interaction and aesthetic creation, mediated by Offer Networks or similar systems. Beyond work and beyond money, but not beyond meaningful social interaction and exchange; not beyond creativity; not beyond the joy of life.

REFERENCES

Buterin, Vitalik (2013). Bootstrapping a Decentralized Autonomous Corporation, Part 3: Identity Corp. Bitcoin Magazine, Sep 24, 2013, **http://bitcoinmagazine.com/7235/bootstrapping-a-decentralized-autonomous-corporation-part-3-identity-corp/**

Diamandis, Peter and Steven Kotler (2012). Abundance: The Future is Better Than You Think. Simon and Schuster.

Doctorow, Cory (2003). Down and Out in the Magic Kingdom. Tor Books.

Gregory, Maxwell (2013). StorJ and Bitcoin autonomous agents. Posted online at **http://garzikrants.blogspot.ch/2013/01/storj-and-bitcoin-autonomous-agents.html**

Hansson, Sven Ove and Grüne-Yanoff, Till (2012). "Preferences", The

548

End of the Beginning

Stanford Encyclopedia of Philosophy (Winter 2012 Edition), Edward N. Zalta (ed.),
http://plato.stanford.edu/archives/win2012/entries/preferences/
Helbing, Dirk (2013). Economics 2.0: The Natural Step towards A Self-Regulating, Participatory Market Society. **http://arxiv.org/abs/1305.4078**
Mate', Gabriel (2010). In the Realm of Hungry Ghosts: Close Encounters with Addiction. North Atlantic Press.
Schmitt, Matthias, Andreas Schacker, and Dieter Braun (2014). "Statistical mechanics of a time-homogeneous system of money and antimoney." New Journal of Physics 16.

End of the Beginning

Dialogue 16.1

With Francis Heylighen, David Weinbaum (Weaver) and Ben Goertzel

Francis: This is a really great idea! I particularly like your way of representing individuals as bundles of "production rules" that specify "if I get this, then I am willing to do that in return", coupled with the "clearinghouse" (or what I would call a "stigmergic medium" or "message board"), where all the production rules try to find the best match for their input and output conditions. This exactly fits my view of how the Global Brain would help to find the most synergetic possible interactions between all the individual agents, and thus boost cooperativity, distributed intelligence, and the creation of welfare and abundance.

I also like your description of the wicked psychological side effects of our present money system, such as "Spending your life doing things you don't want, so you can afford to buy things you don't need." :-)

There is of course a lot of work that needs to be done in better specifying the precondition-action-postcondition rules, reputation systems, clearinghouse, user interfaces, emergent values ("offer money"), etc., of your proposed offer network. But I think that we already have the conceptual, mathematical and technological infrastructure to do this work.

The components we work with in our Challenge Propagation model at the Global Brain Institute should be able to simulate a kind of economy like the one you propose. The general idea also fits in perfectly with our objective of specifying a system of distributed governance

Weaver: Understandably the article focuses, for the sake of simplification, on the exchange of what I would call atomic units of value e.g. a hamburger, and hour of programming etc. This

simplification however hides and perhaps overlooks the vast complexity of the current economic machine and its transactions. I tried to play with the idea a bit and could not come up with an obvious or less obvious way of how can the mechanism of an offer network be scaled up to meet the complex challenges of today's economy. For example how can an offer network deal with the following complex operations?
1. Designing producing and distributing silicon chips (including the machines involved in the process). I use this example because of the criticality of this specific category of products to the future information society.
2. Building a highway from city A to city B (same goes for an overseas shipping route, flight routes etc.).
3. Run a biotechnological facility capable to research and produce a vaccine against say Ebola virus (lately on the news) or a medicine for Alzheimer (including setting research priorities etc.).
4. Building AGI agents capable of operating in physical space and interacting with humans.

Francis: These are just a few examples where the exchange process is extremely complex, involves a highly coordinated operation of hundreds and sometimes tens of thousands of agents, legal operations, budgeting, risk taking and long term planning in conditions of uncertainty.

These are indeed complex challenges that are not obviously resolved by an offer network. On the other hand, they are not obviously resolved in a purely capitalist market economy either. That observation is the origin of the field of "institutional economics", which argues that market transactions alone cannot solve these kinds of problems: you need pretty complex and solid institutions and organizations, such as companies, governments, laws, etc. to regulate non-money-based interactions (such as those between the employees of the same firm).

End of the Beginning

A global-brain-like organization of the world system will obviously require pretty complex institutional and organizational arrangements to coordinate a massive amount of "atomic" interactions into a coherent, synergetic whole. I don't think offer networks make this intrinsically more or less difficult than a traditional money-based system, but this is something that will require a lot of deep reflection, and probably trial-and-error...

One possible approach is to define an organization as a higher-order agent in the offer network, i.e. a subnetwork that can be "blackboxed" into a single agent. The organization as a whole offers certain things and is entitled to receive other things. Within the organization, this input-output stream is decomposed into substreams that are allocated to the different individual agents depending on their skills and needs by some kind of internal management or governance structure that is more specific than the one used by the global offer network. This is in essence how an organization works today, except that the internal offer stream is reduced to a wage, complemented by possible bonuses or benefits in kind, that is paid to each employee at the end of the month.

That simplifies the internal governance somewhat, but does not eliminate complexity. Indeed the allocation of tasks to the employees remains a non-quantitative and very context-dependent process that requires intelligent supervision. Offer network algorithms based on stigmergy may strongly simplify that allocation of tasks, as I have argued in my papers on "**Getting Things Done**" (last section) and "**Mobilization Systems**". What needs to be added is a similar self-organizing allocation of rewards for the tasks performed, which may take the form of allowing an employee to select a portion of the offers that are in the organization's buffer of things it is entitled to consume.

If all these different allocation algorithms are well coordinated, the boundaries between organization as subnetwork and offer network as a whole may blur, thus making the whole system much more flexible and transparent. For example, in the

End of the Beginning

present system employees of a firm are not supposed to provide services to a rival firm, because they might spill trade secrets or other specific advantages of their original firm to competitors. In a true offer network economy, exchange and collaboration would not have such artificial limitations, increasing the potential for highly synergetic exchanges...

Ben: Yes... I tend to agree with Francis's comments....

Also, I agree w/ Weaver's comment that offer networks do not give a mechanism for coordinating among multiple people to carry out complex tasks. But again, offer networks aren't intended to solve all possible problems of society; like money (but with greater generality / flexibility) they are a tool that can be used in various ways...

Open Value Networks, which I mentioned briefly in my article, are one possible mechanism for forming organizations to get complex tasks done, in a way that would be very compatible with offer networks and might bypass many of the difficulties seen with existing systems like governments and corporations. Basically an Open Value Network is a group of people who have agreed to combine certain resources according to a certain contract, with the aim of achieving a certain collective goal (and to share the results of this work according to a certain method specified in the contract).

However, there are incompletely resolved issues with Open Value Networks too, such as assignment of credit. It seems possible that a well-tuned collaborative reputation assignment network could suffice to assign credit to individuals for their contribution to the overall doings of an Open Value Network. But this is an area in need of research and experimentation about the role of market versus institutions.

Historically, it does seem that having an effective and stable set of institutions has been key to the success of nations. Economic markets are not enough, on their own. Indeed, an effective set of

End of the Beginning

institutions is important for maintaining an effective market. The book *Why Nations Fail* makes this case fairly compellingly.

On the other hand, we are entering a new era of very rapid change, in which institutions are going to need to adapt more flexibly than has previously been the case. So it's not clear what form the effective institutions of the future are going to take. They may be networks formed from materials like Open Value Networks, Offer Networks, and collaborative reputation systems rather than traditional institutions like corporations or government agencies.

End of the Beginning

Part Seven:

Human Society and Human Nature in the Era of Abundance

End of the Beginning

Chapter Seventeen **Sousveillance and AGI**

By Ben Goertzel and Stephan Vladimir Bugaj

Stephan Vladimir Bugaj is a screenwriter/director/producer and AI/robotics researcher currently working with Hanson Robotics, OpenCog, Visioneer Studios and Condition One on both robotics and artificial intelligence R&D projects, and forward-thinking media/entertainment properties. He is also a member of the AGIRI leadership committee, and the Lifeboat Foundation media committee. Stephan spent twelve years at Pixar Animation Studios working on computer animation and storytelling. He also was Creative Director in charge of narrative and visual storytelling at Telltale games, DCTO at Webmind, a researcher at Bell Labs, a WWW pioneer, and a founding member of the Global Brain Institute. His areas of research include transmedia storytelling and narrative technologies, virtual reality, robot and CG character development and performance direction, robot control and cognitive system design and AI developmental psychology. His web site is **www.bugaj.com** and he can be reached at: stephan@bugaj.com.

The whittling-away of privacy is headline news these days. The propensity of the younger generation to broadcast what would formerly have been considered private information via online social networking seems to increase inexorably year by year. Even military forces have a hard time keeping their actions secret, due to

End of the Beginning

individual military personnel, or other people affected by military actions, snapping photos via cellphone cameras and placing the results online. Edward Snowden's revelations regarding NSA snooping brought to the public mind what folks close to the intelligence community have known for a long time: an awful lot of what everyone does every day is captured and stored for potential future analysis. It happens that currently, almost none of this data is actually looked at by anyone or analyzed in any way. But the potential is already here for some sort of Orwell-style Big Brother, just a few decades beyond his envisioned date of 1984.

We can see these days, unfolding in the news month by month, the power of sousveillance to expose and potentially minimize abuses by the "powers that be." As we finalize this chapter in October 2014, for instance, we observe that

- A couple weeks ago, a video made the rounds showing a white South Carolina policeman shooting a black driver for, it would appear, absolutely no reason.[114] The officer accosted the driver for not having his seatbelt fastened in the parking lot of a gas station, then asked the driver to get his ID, and then shot the driver (fortunately not fatally) when he reached into his car to get it. This kind of thing actually happens disturbingly often, as any black Southerner could tell you -- what's new is that the police car had a video camera in it, recording automatically, which allowed everyone to see the deed getting done. Without that video, it's very doubtful a judge would have believed the black driver that he'd been shot for no reason.
- A few days after that, the financial news was full of discussion of a report documenting how the US Federal Reserve bank fails to adequately regulate big banks like

---

[114] *http://edition.cnn.com/2014/09/25/justice/south-carolina-trooper-shooting/*

End of the Beginning

Goldman Sachs, instead routinely kowtowing to their wishes. This of course is no big surprise to anyone else who's had anything to do with the world of high finance. What's different is that now we have audio recordings of conversations which blatantly document how absurdly wimpy the "regulation" process is. Carmen Segarra, a lawyer who was hired to investigate compliance issues in Goldman Sachs, was apparently fired from the Fed for refusing to water down her critical conclusions regarding Goldman. That would have been the end of it, most likely -- except that, during many of her conversations with folks in the Fed and Goldman, she had carried a tiny audio recorder purchased at the Spy Store.[115] Due to the availability of these recordings, we can now hear for ourselves just how lame are the government's efforts to enforce compliance of Goldman to government regulations (and there seems little doubt the same is true regarding other major banks as well).

Of course, these sorts of revelations aren't going to solve long-standing structural and cultural issues overnight -- American white cops will keep abusing law-abiding black citizens at least a little while longer, especially in the South; big banks will keep manipulating and bamboozling government regulators so that regulations largely serve their own economic advantage rather than protecting society from anything; etc. But it seems clear that what we see now is the beginning of something bigger. These are just the first few dribblings of a flood of sousveillant video and audio that increasingly will make public aspects human life and society that have previously remained hidden.

And the advance of technology enabling further degradation

---

[115] *http://www.propublica.org/article/carmen-segarras-secret-recordings-from-inside-new-york-fed*

End of the Beginning

of privacy is rapid, ongoing and multidimensional. Whether or not Google Glass and its ilk succeed, some sort of wearable camera is going to become widespread during the next years. Satellite surveillance increases in accuracy, step by step. Cyberattacks teach us that computer security is a very shallow thing. Even the founder of Silk Road (the online Bitcoin market, used for illegal drug transactions as well as other purposes) was caught, due to some very human slip-ups. No one involved with the intel world believes that Snowden's revelations will stop the ongoing information collection carried out by government agencies, though they may become more careful about relying on external subcontractors to help with these efforts (which may necessitate some internal reorganization to increase the IT skills of full-time intel-agency employees). The particulars of social network services are ever-changing, but the propensity of youth to share continues to grow – which makes perfect sense, because as secrecy becomes more and more difficult to maintain, radical open-ness increasingly becomes the path of least resistance.

In his book *The Transparent Society* (Brin, 1998), David Brin makes a fairly strong argument that, as monitoring technology advances and becomes pervasive, there are two feasible options: SURveillance and SOUSveillance. "Sous" and "sur" are French for "under" and "over" respectively. S*ur*veillance is when the masters watch over the masses. *Sous*veillance is where everybody has the capability to watch over each other, peer-to-peer style – and not even the rulers are exempt from the universal collective eye. It's generally meant to imply that citizens have and exercise the power to look-back at the powers-that-be, or to "watch the watchmen."

Being monitored will become inescapable, Brin foresees, extrapolating from trends like those mentioned above ... the critical issue who is *able* to do the monitoring: only a "trusted" central authority, or everyone in the society? Both have obvious disadvantages, but if persistent and pervasive monitoring is an inevitability, then sousveillance certainly bears consideration as an alternative to surveillance.

Like surveillance, sousveillance can occur to varying degrees, ranging from the ability of everyone to observe everything that goes on only in public places (as captured by omnipresent security cameras and other monitoring devices), to the ability for everyone to eavesdrop on everyone else's phone calls and personal communications. An argument in favor of more widespread sousveillance is that if the government is eavesdropping in these different ways, we may be better off if we all can do so as well, and therefore also eavesdrop on the government eavesdroppers. As Brin puts it, "You and I may not be able to stop the government from knowing everything about us. But if you and I know everything about the government, then they'll be limited in what they can DO to us. What they do matters more than what they know."

Radically futuristic variants of sousveillance, where thoughts and feelings are made subject to observation via brain-monitoring technologies, form a common nightmare scenario in science fiction (though more commonly portrayed as surveillance, not sousveillance). These may well be coming. But even the more prosaic forms of sousveillance would have dramatic implications for the way we live and interact.

Brin himself tends not to view sousveillance as scary or disturbing, using the analogy of people sitting at different tables in a restaurant, who could eavesdrop on each other's conversations, but choose not to. Even nosy people generally refrain, because the eavesdropper is likely to be caught doing so, and snooping is disdained. He reckons that if sousveillance became a reality, new patterns of social tact would likely evolve, and society and psychology would self-organize into some new configuration, which would leave people significant privacy in practice, but would also contain the ever-present potential of active sousveillance as a deterrent to misdoings. This can be illustrated by extending the restaurant analogy; if universal sousveillance means

End of the Beginning

that all peeping toms are always caught in the act, then such a society might wind up with more privacy than you'd expect. There is a real point here, yet it's not particularly clear how the mix of technologies and cultural and psychological habits will actually unfold.

Brin has already explored many of the possibilities that sousveillance may bring in the near term. Here we will focus on synergetic possibilities potentially emergent from the combination of sousveillance and AGI – a very meaningful combination if one believes, as we do, that both of these phenomena are reasonably likely to become prevalent.

We do not, in this chapter, venture to predict whether surveillance or sousveillance is the path humanity will follow. Rather, our aim is to think through certain sub-paths of the sousveillance path, and understand their key aspects.

Among the many potential implications we explore here, perhaps most critical is the delicate balance between freedom and security that will exist in sousveillant societies composed of roughly equally powerful intelligent economic actors. We discuss two forces that will exist in sousveillant societies with or without AGIs, but that will become drastically more significant in the face of AGIs:

- A push to conformity, driven both by statistically normal actors exhibiting the natural tendency of the masses to punish outliers, and through proactive mediation of the potential risk of any nonconforming actor to acquire asymmetric power in ways that can't be detected until it's too late.
- A push to diversity, driven by "selfish" motives such as profit and survival, and the countering social force that comes the fact that looking across the set of all domains of activity, most actors exhibit some statistically non-normative behavior in at least one.

Balancing these two forces will be a subtle but critical issue, requiring extremely careful attention as sousveillance unfolds. As human history has repeatedly revealed, enforced conformity ultimately leads to social stagnation and ultimately totalitarianism, whereas unfettered diversity leads to selfish irresponsibility and ultimately anarchy.

The Psychological and Social Impact of Sousveillance

Sousveillance is one of those ideas that creep up on you: the more you think about it, the more you realize what a really big deal it would be, and how oddly likely it is. David Brin's *Transparent Society* does a masterful job of exploring the potential social and political implications of sousveillance, but he shies away from the deeper psychological implications, which we believe are potentially dramatic. It is not at all clear that the traditional psychological construct of the "individual self" would survive sousveillance intact – not only the deep future sousveillance of the sci-fi "mind meld" type, but even more prosaic sousveillance scenarios. As we noted above, a nascent Zamyatinesque "glass society" is in fact already emerging, in the form of widespread physical surveillance. And an interesting twist is provided by current Internet systems in which people barter their privacy for a modicum of notoriety. These systems are a proto-sousveillance, but one which is easily gamed because the option of online privacy affords anonymity. As we move further in this direction, though, either people must overcome the emotions of inhibition, humiliation, and selfishness, or there is a significant risk that unusual, creative, and innovative behavior may be stamped out by all collective panopticalism and pervasive social pressure.

Even in this nascent stage of sousveillance, the notion of self is changing, in complex ways. To an extent, self is shifting further and further toward being a carefully controlled construct designed for public interface. And to a certain extent, inhibitions really are melting away, with the younger generation who has grown up with Internet technologies feeling more open to share their

End of the Beginning

personal aspects with the world, "warts and all."

As sousveillance advances, the implications for the nature of self will become yet more dramatic. Any one of us would change dramatically given the knowledge that all our actions (and, in extreme versions of sousveillance, all our thoughts) were potentially being observed by anyone who cared to -- and there would be systematic patterns to these changes. Brin largely focuses on the scenario where individuals avoid unnecessary spying due to mutually agreed tact. But it is likely that the "restaurant tact" phenomenon exists only because the penetration of sousveillance through society is so small at this point. When mutual surveillance is considered outré, social pressures prevent such activity. If we continue on our current path of exchanging our privacy for other perceived benefits, a threshold will be reached, past which the maintenance of traditional notions of privacy and ego becomes too much of a burden, and different sorts of psychological structures will emerge.

Sousveillance can stifle individuality and innovation -- if there are no safeguards against the tyranny of the masses, in a totally open society, mob rule / the most popular opinions are enforced by social pressure. the innovative and different can be spotted and drummed out of society either by social or forceful means. (by definition, all innovation starts out unpopular, as it challenges some norm of behavior). Degree of sousveillance and ability to act on directly rather than just observe and socially influence determines the degree of tyranny. To militate against busybodies, distractions (games, porno, soap operas, gossip, etc.) are necessary -- totalitarian societies turn into negative surveillance societies because people are repressed. When they are free, they can be distracted from witch-hunts against the unpopular; this is not a complete solution -- if there are no zones of privacy, currently existing humans wouldn't know how to function in the society except to be incredibly paranoid and paralyzed.

Much of contemporary Western psychology and philosophy

is based on an overestimation of the autonomy of the individual, and an underestimation of the distributed social nature of cognition and emotion. Sousveillance would decrease the level of practical autonomy, perhaps to a level where our conventional illusions of autonomy are too much trouble to maintain. What sort of backlash this might engender would depend upon the rapidity of the transition, how much benefit was gained, and naturally also vary by individual. Mild sousveillance is actually more likely to lead to violent backlash than a "mind meld" type of society, because the mutual benefit is so much smaller than the possibility of a mindplex being able to solve vexingly complex social, scientific, medical, and engineering problems and thereby greatly advance the progress of intelligent life.

It's interesting to muse about what sort of "mindplex" cognitive structure (Goertzel, 2003) might emerge from a group of closely coupled minds with the capability to "extrospect" each other's mind-states fairly freely; clearly this would allow forms of collective cognition not currently accessible to us. It is repulsive to our current psychological makeup because our existing condition is deeply tied to private self-interest and aggressive competition. But if the opportunity for finding common ground and areas of beneficial cooperation proved to be sufficiently great, we might be willing to accept levels of sousveillance we currently recoil at. Whether this would lead to liberation from the limitations of our individual minds or paranoid psychological breakdown is an extremely complex and context-dependent consideration.

Sousveillance and AGI

But what would sousveillance mean in the context of advanced AGI? Clearly it would open the possibility for AIs and humans to interact with each other in subtle ways, forming "collective cyborg intelligences" we'll call mindplexes. If an AI can watch everything you do, it has great potential to help you or to harm you. If everyone's AI helper can watch everybody else, and everybody else's AI helper, then one has the potential for very richly interconnected collective cognitive dynamics, for good or for evil (or, for the Nietzscheans, for outcomes "beyond good and evil"). But this does not necessarily have to mean the end of population enriching diversity and individual specialization. Given sufficient processing power and coordination capability, syncretic thoughts can be held by an AGI mindplex just as we can believe contradictory things ourselves. Cognitive spaces may be carved out where mutually incompatible ideas can exist, without impacting each other, allowing for systemic diversity and localized conformity.

One might also argue that AGI is critically needed to make sousveillance really work. At this very moment, governments collect vast amounts of data about their citizens, but fail to make use of this data effectively, simply because sorting through it to find the relevant pieces and patterns is an intractable problem. Current technology for data mining and natural language processing is helpful, but not yet sufficient. AGI technology, once it advances further, will solve this problem – enabling government surveillance to function far more effectively than it does today, and also enabling the alternative prospect of effective sousveillance. Without AGI, the data about others' doings will be out there, but not pragmatically searchable or usable – AGI will make the sousveillance data actionable. Individuals suffer the most in a sousveillance system – and a de-facto surveillance state can easily emerge – because they cannot marshal resources easily to deal with the volume and complexity of observed data. If individual actors and not just large corporations and governments can obtain the services of an equally powerful AGI, then sousveillance emerges. Otherwise, surveillance does.

Sousveillance may actually need AGI less desperately than surveillance in order to succeed, due to the collective neural computing power of the masses, and the fact that it is a more tractable problem. Probabilistically distributed localized implementations of "everyone watches everyone", meaning that each person only watches a few, are already the norm. Technology just makes the scope and longevity of the data more comprehensive. But sousveillance will still be dramatically enhanced, even transformed, by sufficiently powerful AGI. Though the collective thought power of the polity exceeds the neural computing power of the human capital any government or corporation has at its disposal – but is difficult to coordinate and focus on a shared goal. Ad hoc resource marshalling in the form of affinity groups and public rallying is flexible, but unreliable, and often short-term, whereas organized resource marshalling is reliable and can cohere around long-term plans. A sufficiently inspired public can gain a tactical advantage, but strategic advantage goes to entities with structures for long-term coherence. Sufficiently advanced AGI working for the public benefit can handle coordination tasks that otherwise require large human organizations.

There could also arise a positive feedback between AGI and sousveillance technologies. To the extent that sousveillance nudges human psychology in the direction of greater collectivity, it may increase the efficiency of work projects that are essentially team-oriented – and software projects, including AGI software projects, definitely fall into this category. Sousveillance may turn people into better AGI developers, while AGI in turn enables more advanced sousveillance. By integrating AGI with humans in a mindplex, the mutual benefit may mitigate against human-machine conflict, as well as drawing humans together into cooperation towards shared desires. This utopian possibility requires the suppression of certain individual desires (such as for asymmetrical power over others), and a certain degree of enforced conformity, in order to be possible. While this describes any non-anarchic society, it is a matter of degree, and even the more prosaic forms of sousveillance require and enforce this to a degree far beyond what has previously been possible.

End of the Beginning

What implications would sousveillance have for "AGI safety"? Could it perhaps help us defend against the risk of a "rogue AGI" marshalling resources for itself and causing widespread harm, or of groups of individuals controlling AGI's to serve their selfish ends at others' expense? Clearly, if everyone can see what everyone else is doing, this decreases the risk of someone sneakily developing and deploying an incomparably advanced AGI and using it for ill ends. Also, if someone or some group finds a way, while being sousveilled, to game the system (obfuscation, informational deluge, disinformation / distraction, etc.), that actor can still create an "evil AGI." But the matter is subtler than that, and brings up deep, classical issues of conformity versus innovation, and freedom versus security.

The Profits and Perils of Nonconformity

The most interesting sousveillance scenarios are those in which no individual actor (be it an individual or a tightly knit group) has massively more practical power than the others. If one actor is vastly more powerful than all the others, it may not matter if its actions are observed; the other actors likely will not be able to do use their observations to countermand the powerful actor anyway. But in the case where no individual entity (or sub-collective) has dramatically disproportionate power, sousveillance becomes much more interesting. And it's worth noting that groups that seem very tightly knit in the absence of sousveillance, may display many more weaknesses when careful observation is possible – mutual surveillance may cut through the veil of political or corporate lies and reveal a lack of cohesion in the group, leading to formation of more honest affinity groups. Even members of the most doctrinaire groups may have doubts, and in a sousveillance scenario these doubts may be noted by outsiders and used to help pull the group apart, (or insiders may notice each other's weaknesses and fragment of their own volition).

But even in cases without strong asymmetric power between

actors, sousveillance is no panacea. It leads to subtle issues regarding the plusses and minuses of nonconformity from a social and individual perspective, which require very careful consideration. Herein, we will at least outline the nature of the problems.

Benefits of Nonconformity

The financial markets offer us some interesting guidance in understanding the potential dynamics of sousveillant societies. Our perspective on financial dynamics is that the "efficient market hypothesis" is not a reality but rather an approximation. The reality is that the markets are constantly deviating from efficiency as new ideas and techniques are introduced, and then relaxing back toward efficiency as these become widespread. For instance, in most cases, shortly after a new trading strategy is created and successfully utilized, it spreads throughout the community, causing its impact to be diminished. Markets may completely fall apart under more comprehensive information regimes, unless there is still a way to gain some advantage even in a fully open, mutually surveillant society. Cleverness per se is not enough for success in the markets, because most humanly natural varieties of cleverness are already "priced into the markets." What is valuable is *peculiar cleverness*.

In the markets, each actor can study what the other actors do, via the impact their trades make on the prices of financial instruments. This allows insightful actors to make accurate inferences about the methodologies used by powerful actors, simply by watching their patterns of activity. So, for instance, a large bank or fund must be very careful about selling its shares in a company, because as soon as it *starts* to sell, the market will notice this and the price will drop in reaction; so typically a large financial actor will unload its shares slowly according to a complex pattern designed to militate against detection by others.

In financial markets there is a premium for making actions

End of the Beginning

according to patterns that others can't easily identify, or patterns that others can't replicate even after they've identified them. Sousveillance will lead to similar, but broader phenomena. In any reasonably free society, if you're achieving advantage via acting in a way that others notice, then others are likely to try to copy you. In a sousveillance society, this phenomenon is magnified, because, depending on the degree of sousveillance, anyone can see not only the moment of action but also the preparations leading up to it, and perhaps even the thought patterns leading up to the preparations. If you can do something in such a way that others can't perceive the general pattern of your activities -- or such that, even when they comprehend the general pattern, they can't replicate the important specifics – then you'll have a big advantage even in the presence of sousveillance. Obscurity, and also overwhelming rapidity and/or volume, of information becomes economic (and sociopolitical) value -- and a sousveillance society that starts off as a market system will necessarily evolve in this manner, as economic value is derived from various forms of, and a dynamic equilibrium between, control (popularity) and innovation (novelty).

This implies that, in scenarios where sousveillance exists but doesn't extend into minds' inner workings, "cognitive security" will have huge value. Where sousveillance does extend inside the mind, though (a plausible case with AGIs, for which cognitive monitoring is easier, and for whom there does not exist a tradition of privacy) things get trickier – and a huge premium gets attached to cognitive processes that are *incomprehensible even when watched*.

Even if a dog could see into your mind, it would only understand a small percentage of what it saw. But total incomprehensibility is not a necessity: if a virgin looked into the mind of a libertine, there would be a lot they could not relate to. Due to disparate experience and knowledge, the virgin wouldn't be able to make use of his or her observations to predict the actions of the libertine nearly as well as another libertine would be able to do based on the same observations.

Free market economics will make it such that an actor with marshalled resources can gain an advantage in creating a more powerful system (and as such possibly a superhuman AGI), even if all the information is easily observable, they have more resources to try to: distract, flood, outpace, and game the system, rendering the sousveillable information relatively useless vis-a-vis someone trying to marshal resources to counter their aims.

A greater degree of mental consistency and control capacity makes AGI sousveillance more feasible, but even without AGI we can know some things about others' plans (even an AGI or someone trying to develop an "evil AGI"): mutually comprehensible underlying languages and/or functionality (slow to analyze, but still possible), but also, pragmatic concerns that are indicators -- large purchases of computing power, increases in bandwidth usage for distributed processing, etc. By monitoring these external activities, one can make predictions about future activity with a greater degree of confidence than simply by watching patterns of direct action in the target domain. Market traders and police detectives already use their surveillance capabilities to perform their tasks – sousveillance makes the means of investigation and a greater degree of access and knowledge available to any actor which desires it.

All this ties in with the assumption, made above, that we're thinking about scenarios where no one actor has vastly more power than everyone else. If one actor has 1000x as much processing power as everybody else, then it can quite likely carry out processes that are literally incomprehensible to the other actors – even if they have full access to the powerful actor's internals. Sousveillance loses much of its effectiveness in situations involving dramatic asymmetries in cognitive capability.

Risks of Nonconformity

In a population of actors with roughly equal cognitive

End of the Beginning

capabilities, there is a possibility for sousveillance to mitigate against unethical behaviors, and also against the emergence of actors with asymmetric degrees of power. However, the conditions under which this sort of protection can be effective are far from clear.

The relation between nonconformity and unpredictability plays a role here. Even among actors with roughly equal cognitive capabilities, each actor will have an easier time predicting certain other actors than others, simply due to either outright superiority or otherwise to cognitive affinity (similar neural specializations and belief weightings). That is, an actor will be best at predicting other actors that are much simpler than it, or else fairly similar to it. Other actors of roughly equal complexity but significantly different nature will be harder to predict.

Thus, here as in so many other contexts, we have a dichotomy between freedom and security. The safest sousveillant society will be one in which everyone thinks pretty much the same way – and everyone is smart enough to understand what that way is, analytically as well as intuitively. In that case any dangerous deviation has a high chance of being spotted and repaired or eliminated. On the other hand, if actors are given the right to pursue innovative cognitive directions, there is a much stronger risk that some actor will surreptitiously self-modify itself into something dangerous – and, before anyone else has managed to understand what it's doing, has managed to accumulate a dramatically asymmetric amount of power.

And yet, as the above analogy with financial markets makes clear, there is also great value in nonconformity. Nonconformity brings the potential for peculiar cleverness, which in a sousveillant society without an a priori asymmetric power distribution, is the only way for individuals to achieve advantage.

Balancing the Profits and Perils of Nonconformity

So the crux of our analysis is that relatively power-symmetric societies with both sousveillance and powerful AGI will suffer a much more extreme version of a dichotomy that exists already in society today: the dangers nonconformity poses, versus the benefits it brings. Nonconformity brings with it diminished comprehensibility, which brings with it the danger of actors self-improving in a way that brings dramatic and potentially devastating power-asymmetry. Yet nonconformity also provides actors with a way to find individual advantage even in the presence of sousveillance, and nonconformity is a large part of social and cognitive system resilience and vitality.

Though the explicit danger of nonconformity is systemic instability (distributed cognitive dissonance) which erodes the consistency, comprehensibility, and mutual benefit of sousveillance, and can lead to a power asymmetry that can destroy the system, this nonconformity is also a critical element to the neural diversity that makes innovation possible and prevents system wide stagnation and breakdown. A fine balance must be struck between conformity and nonconformity in any society, but with sousveillance this balance is made all the more precarious.

So from the point of view of an individual actor attempting to maximize some goal function, whose maximization involves complex interrelationships with other actors, there will be a desire for society to allow just enough nonconformity to give that actor a good route to achieving its goals, but not enough to allow other actors to suppress the diverse actions of the others by achieving dramatically asymmetric power. This is not a particularly original dichotomy, but sousveillance and AGI combined will present it in an unprecedentedly acute form, due to the particular value placed on peculiar cleverness in the face of sousveillance, and the relative ease with which AGIs may diverge from each other (as compared to contemporary humans, who are stuck with the same brain architecture).

The Promise and Perils of Mindplexing

End of the Beginning

These issues become yet more complicated when one considers that sousveillance may blur the definition of an "actor," which we have been using somewhat blithely in much of the above discussion. If sousveillance pushes AGIs toward a mindplex-type psychology, in which humanlike individualistic "self-models" play a relatively minor role, then it may become difficult to tell when an actor has achieved an asymmetric degree of power, because it may become difficult to tell what an actor is. Individuals in a mindplex are tightly coupled with others, and it may be the case that for a goal to even arise in the first place it must be a shared goal of a substantial subnetwork of the mindplex.

Already we have corporations that have their own individuality and autonomy, separately from the individuals who constitute them – but the coherence and unity of a contemporary corporation is minimal compared to that which could be achieved by mindplexed AGIs, down the road.

In a mindplexing scenario, this interconnectedness of individuals and sublimation of desires and abilities into the collective becomes one of the factors that society must manage, in the quest to balance freedom versus security (and ability). This shows that we can't analyze the situation in terms of a simple dichotomy of "conformity versus nonconformity." Membership in a tightly integrated mindplex as a form of hyperconformity, but it's not "conformity" in the usual sense of acting similarly to one's peers, but rather in the sense of constraining one's behaviors in highly patterned ways in accordance with a higher-level control system – a situation we have no prior experience with.

Herein we have a form of conformity that poses a great risk to symmetric power. Yet it also provides obvious benefits in terms of allowing individuals to achieve their goals – so long as their goals align exactly with the others' (or they're able to either convince them of such, or enact a trade by which their participation others' goal achievement tasks is given in return for help with their own). Either all individuals would need to form affinity groups that

573

agreed to mindplex, or those left out would be at a great disadvantage versus the others.

Sousveillance and the Global Brain
Finally, one significant possibility is that the advent of sousveillance leads beyond the emergence of localized mindplexes such as families or co-work teams sharing thoughts, feelings and actions, and instigates the emergence of a full on Global Brain as Heylighen (1996; and this volume), one of the authors (Goertzel, 2001), Russell 1983) and others have been discussing for decades now. Sousveillance certainly does not imply a Global Brain, but it does provide a mechanism potentially strongly biasing in favor of a Global Brain's emergence. The Global Brain is all about rapid, flexible information sharing among individuals on different parts of the planet, and the emergent dynamics that ensure therefrom. Sousveillance, by definition, involves high bandwidth, broad scope information sharing. If the concept of a Global Brain holds water, then a sousveillant society would seem very likely to turn into one, unless the individuals in the society steadfastly maintained a culture of individuality and explicitly opposed mindplexization.

Non-Conclusion
The class of future paths involving both sousveillance and advanced AGI seems reasonably likely to us – yet also seems particularly hard to understand…. It leads rapidly beyond current human psychology and culture, into possible worlds involving distributed intelligence, non-self-based ways of organizing collective thought and action, and so forth. Extreme cases like borg minds are easy to think about, but the reality is likely to be more complex and nuanced, with aspects we currently lack the concepts to articulate or analyze.

The steps via which a sousveillance + AGI future might unfold are also quite blurry, in part because it's hard to estimate the relative rate of advancement of each technology. If sousveillance advances rapidly first, then we may see lightly-augmented networks of humans either

- Succumb to some sort of massive conformity
- Evolve some new sort of sociocultural structure, compatible with sousveillance, and involving different attitudes toward privacy, self and openness than anything we are familiar with before advanced AGIs come on the scene.

In the former case, AGIs would be unlikely to break the conformity but would likely constitute a tool for maintaining it even more robustly (assuming future paths in which AGIs and humans coexist and richly interact, which obviously is not the only possibility). In the latter case, AGIs might add into the mix in complex ways, leading to the emergence of an AGI/human global brain mindplex, which is an even more difficult possibility for us to foresee in detail than a mere sousveillant human mindplex.

On the other hand, if advanced AGI comes first, then things look a bit different. Sufficiently advanced AGI systems would presumably have the power to choose between surveillance and sousveillance for humans, with the AGI as the most likely surveillor in the prior case. A highly advanced, cooperative AGI might work together with humans to craft new sociocultural structures, incorporating creative uses of AGI to shape the landscape of conformity and creativity.

Awareness of these longer-term factors and possibilities may help us to better navigate through various concrete choices as they unfold, as sousveillance and AGI technologies develop together and separately, and society, culture and psychology react accordingly. The world at large is generally handing the incremental advent of sousveillance and surveillance in a reactive way, dealing with each new step as it comes; and this makes perfect sense, since there's no way to chart the detailed course of developments in advance, even if the big picture has clearly foreseeable aspects. But understanding the radical nature and broad scope of the sousveillance/surveillance related to changes likely to come is also important – both for its own sake and as a case study in the intersection of different advanced technologies. The social, psychological and technological cannot be meaningfully separated; in envisioning the future of human and posthuman life we must keep this in mind, and the potential future intersection of sousveillance and AGI is one clear illustration of this.

REFERENCES
Brin, David (1998). The Transparent Society. Perseus Press.
Goertzel, Ben (2001). Creating Internet Intelligence. Plenum Press.
Goertzel, Ben (2003). Mindplexes. Dynamical Psychology, http://www.goertzel.org/dynapsyc/2003/mindplex.htm
Heylighen, F., & Bollen, J. (1996). The World-Wide Web as a Super-Brain: from metaphor to model. In R. Trappl (Ed.), Cybernetics and Systems' 96. Austrian Society For Cybernetics
Russell, P. (1983). The Global Brain: speculations on the evolutionary leap to planetary consciousness. JP Tarcher.

End of the Beginning

Dialogue 17.1

With Ted Goertzel, Stephan Vladimir Bugaj, Ben Goertzel, Hruy Tsegaye and Mingyu Huang

Ted: In the book *Guarding Life's Secrets*, Lawrence Friedman observes that privacy is a modern invention. Medieval people had no concept of privacy. Nobody was ever alone; everybody was embedded in a face-to-face community. Is privacy part of capitalist individualism? Is it valued in the same way in China or Africa? In socialist countries? Is a world where everybody is constantly embedded in a community more consistent with human nature than one where people can keep secrets?

Ben: I have observed that in Chinese culture there is a huge difference between public behavior and "private" behavior (where the latter is defined as behavior among one's family and close friends). Chinese people, who seem to show no expression or emotion in public, can be extremely emotional and expressive in "private" situations among those they know well. So I would say Chinese do have a very strong notion of privacy -- it's just a little different from the Western notion. Americans seem to behave much more similarly among friends/family and among strangers, as compared to Chinese.

So I would say the definition and notion of "privacy" shifts over time, and varies among cultures. Perhaps if sousveillance emerges in the future, this will not eliminate the concept of privacy from human culture, but will just shift it in yet another way...

One could envision a system wherein anyone COULD watch anyone else, but there was some global system of "reputation points", and watching someone else's "requested as private" activity used up a lot of reputation points. Then, if two random people were having sex or a private conversation at home, most likely nobody else would bother to waste reputation points

End of the Beginning

watching them, even though this would be possible. On the other hand, if two high-level government officials were having a negotiation about technology policy, probably a lot of people would be willing to spend reputation points watching them do so, to be sure no wrongdoing happens and to report on the proceedings.

This sort of system could enable practical privacy for those who value it, even in the absence of theoretical privacy. (And of course those who don't value privacy could capitalize on this too – they could allow people to watch them have sex or listen to them talk or whatever, in exchange for some small number of reputation points, or for some other service.)

Mingyu: China's traditional culture does not include the value of privacy; it respects the authority of the regime and the need for stability. Privacy is associated with cliques that threaten the ruling order. China's culture also emphasizes harmony within the family, so privacy is not valued within the family but it may be respected between the family and outsiders. There may also be a privacy barrier between the lower and higher classes in society.

With the coming of capitalism, came the right to protect private property and private behavior. This was a new imposition on traditional Chinese culture which did not encourage self-interestedness. China did not traditionally have laws protecting privacy, and there are still no clear legal provisions. The definition of privacy is also different in China. In the west people think it is improper to ask acquaintances questions such as their age, marital status, income and so on. In China this is considered acceptable. With the coming of the Internet, privacy has become a concern whereas it was not in the traditional Chinese culture. Many people feel that this increasing concern with privacy is a loss.

There have been two kinds of monitoring in Chinese history. The ruling class always monitored the people. And during the Cultural Revolution, people monitored each other, with especially

End of the Beginning

tragic results. Now with the coming of the Internet, the government monitors microblogs and social networks, and people use the same methods to monitor the authorities. The government has its famous "Great Firewall of China" and people find ways to get around it.

Ted: Has the Internet had an impact on the culture of everyday life in China?

Mingyu: I think the nature of privacy impacts the social environment. If people are aware of each other's ideas they can learn from them and improve themselves. But this means we need to be constantly correcting bad ideas, such as sexual abuse and perversion. So some privacy may be needed for this reason. Privacy protects people's independence, but this can also mean the freedom to do harmful things.

In a fully transparent society of the future, people might have access to very dangerous technologies. They might have access to military technologies that could destroy the planet. There may be a need for some kind of Super Intelligence to limit access to these technologies. This Super Intelligence might also help to maintain private spaces where people can explore sexual or intellectual matters or other private matters without risking a negative impact on society.

Stephan: I think it's the wrong question: human nature is not such a wonderful thing that it should be preserved as-is. Appeal to tradition holds no sway with me whatsoever. Humanity has many horrible aspects, and progress (transhumanist or otherwise) should mean a push towards elimination of those negative traits. As for privacy, it is essential to the modern notion of freedom, in that a person who has secrets is free from judgment and interference in their personal business by the collective. Since, historically, the collective punishes difference of any kind this isn't just "if you have done nothing wrong, why keep secrets?" For example, for a very long time keeping personal secrets was necessary for gay people to avoid being murdered by neighbors or the state. In order for a system of sousveillance to not take on a sinister character, human nature needs to change to unify and balance the modern Western notion that individuals do have the right to be different from the statistical norm with the collectivist communitarian viewpoint that prevails in different times and places. If it succeeds then it will be a new form of social interaction, not a return to some idealized tradition that never really existed.

Hruy: The Concept of Privacy, from the African point of view, is a bit tricky. It is understood sometimes as a negative thing and sometimes as a kosher one. Privacy is often viewed as a sign of mischief. For many Africans, daily life is still entirely composed of social activity. Group thinking is the dominant form of interaction in Africa and in such situations people are less inclined to seek privacy. As a result, privacy is often linked to conspiracy and hence it is not valued in the same way as it is in the west. If a man is seeking a lot of privacy in Africa, most probably he is perceived as a mischievous man. And if not, it will not do him any good because he will be a victim of "Social Undermining". In many African cultures, individualism or 'a noticeable desire for privacy' is alleged to be negative behavior that is rooted in a person's dark side; his/her nature stained with contempt for the others. Such noticeable desire for privacy is viewed as the person's scornful attitude towards the society. As a result, the person's social life will be ruined and eventually, the individual will be isolated.

Privacy is a sign of disrespect for traditions and revered customs. Communal life is the way of our forefathers and is valued in most African societies. It is obvious that Africans love to live in very close communal relationships; people are expected to share almost everything. In Ethiopia, eating alone is frowned upon. It is almost taboo. Ethiopians are highly communal when it comes to sharing their feelings, experiences, and even deepest desires with their fellow countrymen.

Ted: In the early days of the *kibbutz* movement in Israel, there was an attempt to recreate something similar to the African communities you describe, Hruy. People were together constantly. Children slept in children's houses rather than in the privacy of their parents' homes. Meals were eaten in communal dining halls. None of this involved electronics; it was accomplished by having people physically together all the time. Some people were bothered by the lack of privacy, but this was considered to be a bourgeois indulgence. One of the early leaders said, "There must be no privacy. All privacy interferes with our communal life."[116]

With time, however, people tired of all the togetherness and the *kibbutzim* gradually increased the amount of privacy in members' lives, including allowing children to sleep in their parents' apartments and having evening meals at home. People began to get private allowances for personal spending, and so on. Today, much of the economic production has been put into privatized corporations that are legally separate from the *kibbutzim* with some employees who are *kibbutz* members and others who are not.

Aren't there times when privacy is respected in Ethiopia? Don't the Ethiopians tire of all the togetherness?

Hruy: Yes, there are common and universally accepted situations in which privacy is tolerated or even recommended. The affairs of the opposite sex, displaying affectionate emotions, and indicating conflicts over controversial ideas are things that African society expects to be done in private.

---

[116] *Yosef Bussel, an early leader of Kibbutz Degania, quoted in Daniel Gavron, The Kibbutz: Awakening from Utopia, Rowan and Littlefield, 2000, p. 26.*

End of the Beginning

But one can say that, in Africa, secrets are not protected by privacy. In fact, in many cases, privacy causes suspicion that leads to secrets being revealed. I think it is pretty much possible to live in a society where secrets are impossible to hide. And for me, such a society is much better than one where there are a lot of secrets. I think the society without secrets is more compatible with human nature in its natural state.

After all secrets are the doorway to everything that is 'bad', for the African way of life which is totally communal and shared. Yet, there are types of secrets which are treated as unalienable social right of African. Personal secrets, especially those involving relationships between the sexes, feelings towards other community members, business gains, and other strictly personal opinions might be better left alone. In fact these things are usually treated as private matters in Africa. Westerners can tell their colleagues to their face that they think they are lousy and don't deserve promotion whilst the African will say nothing to your face but will discuss it in length in the confidentiality of close friends. It is not that the African lacks the courage to be open, but such matters are treated as private matters as a sign of civility and humility.

If we can agree on the point that the nature of man has been shaped by its evolutionary path, we might say that nature has bestowed the notion and as well the utility of secrets on humanity and therefore a society capable of having secrets is also not very far from the natural state of humankind.

Ted: In *The Circle* novelist Dave Eggers develops the philosophy that "All That Happens Must be Known." In the novel, politicians who espouse this philosophy wear webcams configured so anyone can log onto the Web and see what they are doing at any time (except in the bathroom or in bed). This was intended to prevent corruption and guarantee transparency. This is fiction, but many police departments are now requiring policemen to wear video cameras at all times while on duty so that all their

End of the Beginning

interactions will be recorded. This is intended to prevent riots when people believe the police have killed a young man for no good reason. Or to keep the police from killing in the first place. Do you think this is a good philosophy? Should our lives be completely transparent? Should we give up privacy in our homes prevent child abuse and sexual assault? If not, where do we draw the line?

Ben: Hah!! Eggers stole my idea!! Well, probably not. But I did come up with that same idea independently, a long time ago. I have often thought that some politicians should make an "open politics" pledge, offering that if they are elected to office, they will offer 24-7 video surveillance of everything they do.

I have not read that particular work by Eggers, but I've much enjoyed some of his other writing!

In pursuing this idea, I thought also that the "except for in the bathroom or in bed" proviso could provide an out for illicit political activities -- "open" politicians would then get in the habit of brokering illegal campaign contributions or bribes while talking on their cellphones during the act of sex or while taking a dump on the toilet, etc.

Ted: That does happen in the novel.

Stephan: The line is drawn at behavior that is not actually destructive. Consensual sex between adults, no matter how "aberrant", is a very prominent example of something historically used by the collective to humiliate, control and punish individuals. But truly protecting people requires other social change as well. Look at the Jimmy Savile case (or any similar case): it was known that this person was hurting people, but the fact that he was wealthy and powerful prevented action from being taken. Nobody wanted to upset the system that benefited them. Knowledge of a transgression is only part of the solution, anyway, so the idea that

End of the Beginning

surveillance or sousveillance is a magical cure-all for crime is nonsense. Eliminating abuse, corruption, and so on requires other changes in human social interaction beyond knowing when someone is doing it. It's quite obvious in many cases without sousveillance, and society just accepts it. Having a clearer view of the abuse is only part of the solution for ending it.

Mingyu: Overall, it seems clear that, in the future our lives will have to be fully transparent. So, yes, I think "All that Happens Must be Known" is a good philosophy. It may sound a little crazy, but I think in the future everyone's life should be transparent, not only those of politicians.

In human society today, I think it would be a good idea to record politicians' words and deeds. If we don't trust humans to monitor the recordings, we can devise Artificial Intelligences to do it.

Transparency could have a lot of positive value as well, beyond just prevention of dangers. Suppose in future --for example, the future after a Singularity, each individual is aware of each other individual's thinking. We could use this ability to improve ourselves if we encouraged everyone to be open about their thinking. This adds a whole new dimension to the exchange of ideas. With input from others, we would constantly correct all kinds of bad ideas that are not constructive for our development. People would be willing share positive ideas that were not harmful to others. So why would there be a need for privacy?

The positive value of privacy is to protect each individual's freedom and independence from unreasonable violations. But there is also a negative side, when privacy can be used to destroy social development and cause harm to society. If a fully transparent society is possible in the future, it will probably be necessary. Suppose that super-intelligence gives every individual the ability to create a weapon capable of destroying the world. In that case, uncontrolled super-intelligence could be devastating.

End of the Beginning

Most likely, after the Singularity, all the super-intelligences, whether enhanced humans or non-human entities or one kind or another, will reach an agreement to strictly control or limit any people or intelligent entities that refused to be monitored.

However, super-intelligence could also produce private spaces, spaces with limited material resources or capabilities to do harm to others. In these spaces people could explore sexual or intellectual matters without fear of a negative impact on society. These spaces could be monitored by specialized artificial intelligences that will alert authorities if anyone is doing anything dangerous. The rules that determine if a behavior is dangerous and needs to be stopped would be formulated by the super-intelligences.

In a word, I think if our lives must be completely transparent in the future, it is not mainly for preventing sexual abuse or child assault. It is primarily to prevent enhanced powers being used to threaten our existence.

Ben: Well certainly, the value of transparency in the case of police is becoming more and more obvious (as I write there are protests across the US because of the killing of a black youth by a white cop in Ferguson, Missouri, and the choice of a grand jury not to bring the cop to trial. The eyewitness reports of the incident seem quite contradictory, and a camera on site would have made everything a lot clearer.)... At various stages in the advance of sousveillance, AI-Nannying and other related technologies, there may be exceptions (certain aspects of public-servant life that are better not made public for specific temporary reasons), though. Real life is always more complex than abstract principles. At this moment I'm not sure it would be wise for the US to make all its meetings on counterterrorism public and open online for example. Maybe it would be; maybe not. But that is definitely the direction things should go -- unless a trustworthy AI Nanny of some sort is created and put in place, sousveillance seems the only viable route

585

End of the Beginning

to avoid an oppressive fascist-type order. What mixes of AI-Nanny and sousveillance outcomes may be possible, isn't particularly clear to me, though. As Mingyu says, we could have a broad atmosphere of transparency and sousveillance, yet with AI-enabled "safe private spaces" for people or other intelligences who so desire.

Ted: Mingyu, your argument that we need transparency to avoid "enhanced powers being used to threaten our existence" sounds to me almost like the rationale the Chinese government uses today to justify not allowing the citizens of Hong Kong, for example, a free choice of candidates in their elections. In this case the "super-intelligence" is the Chinese Communist Party, which asserts that it knows best and has to make sure the people don't get out of line. The difference is that in the post-singularity world this would be done by an artificial intelligence, not by humans. Those of us in the west are very reluctant to give this kind of power to government of any kind, although we make "exceptions" when threatened by terrorists or foreign enemies. Maybe we would be more willing to give power to an Artificial Intelligence "Nanny" if we were convinced it was truly unselfish and working in the common interest.

Ben: The Chinese government is pretty competent as governments go, but classifying it in the category of "superintelligence" may be going a bit too far!

Ted: I didn't mean that the Party actually is super-intelligent, but it does claim to have superior wisdom that justifies it imposing its will on everyone else. For example, it claims the right to determine which political candidates in Hong Kong "love China" and which do not. This is somewhat analogous the kings of olden times who claimed a divine right to overrule everyone else. An AI Nanny might assert a similar right, based on its superior

End of the Beginning

knowledge and intelligence.

Ben: Sure. One key difference, though, is that a properly designed AI Nanny would have a far better claim to actually possessing a superior wisdom than any current government. (At very least in a sense of "practical wisdom" related to maintenance of society, even if humans may maintain "superiority" in a different type of wisdom related specifically to their human nature...). I don't think the CCP has a superior wisdom to the Scholarism and Occupy Central movements in Hong Kong right now; I think they just have a different perspective and a different set of interests. Whereas the AI Nanny, to work well, has got to stand outside the power and ideological struggles of different human interest groups, and enforce basic safety from a more impartial perspective.

BTW, I think I understand why you chose the CCP as an example, because it's especially paternalistic in its attitude; but you could have also chosen, say, the National Security Agency (NSA) and the New York Police Department (NYPD) - both of which have been recently exercising their "superior wisdom" (or, if you prefer, superior power) by (in the NSA case) tapping everyone's phone calls and emails, and (in the NYPD case) executing black citizens on the street, without a trial, for minor crimes like selling loose cigarettes. I suppose every state claims to be wiser than its citizens; and generally this is true in some respects and not in others. The CCP and the US government/corporate/military/police complex both have their strengths and their problems. But an AI Nanny has got to do a much better job than any of these -- more like a preschool teacher in a room full of toddlers, who brings a level of perspective and insight that none of the toddlers could possibly achieve (even if some social dynamic were to put a small group of toddlers in charge of the others).

Anyway, there are always analogies between future situations and past situations, but there's also frequently a tendency to stretch these analogies. For instance, there is an analogy between the Singularity and various notions of religious rapture– but how close is this analogy really? The Singularity is a projection based on data, and a plan for guiding practical engineering work– the Rapture was not really either of these, unless one stretches things tremendously. An AI Nanny engineered to watch over humanity is analogous to a human "benevolent dictator" or "security agency"– except in the ways that it's not. Human dictatorships are run by people with human motivational systems, embedded in human society. China's rulers may indeed have China's greater good in mind, yet they also have their own human motives, as demonstrated by the ever-present corruption scandals reported in the Chinese and international media. The NYPD may well have the safety of New York citizens in mind, but they also have their own human foibles, such as persistent racism. NSA does want to find and stop terrorists, yet it's easy to see how, in the context of human-run governments the data they gather could potentially end up being used for political or corporate ends. But we have to remember, "power corrupts, and absolute power corrupts absolutely" is an observation about human beings and doesn't necessarily apply to all intelligences; there's no strong reason to believe it would apply to an AI system specifically engineered NOT to be that way and instead designed to benevolently monitor, guide and cooperate with human beings.

Hruy: In my country we say, "Walk in the light, and you will no longer need to veil any moment of your life". Using technology to monitor each and every movement of our politicians and public servants might be a good temporary measure. But tyranny will always win if it is only blocked while its root-causes are not eliminated. Monitoring everyone with no room for privacy is a sign of failure. It shows that the system no longer trusts its members and its members no longer trust the system, and that this distrust is so great that the privacy of stakeholders is violated.

Nevertheless, given our current level of rational intellectual

End of the Beginning

development, this seems to be the only move that will work. I don't think mankind is mature enough to be left alone to perform its responsibilities in a just way. We are not rational and it is almost impossible to guarantee that we will not be corrupted in the absence of a close monitoring by someone else. Despite our intellectual level, the other culprit for this weakness is capitalism. It is a system in which social Darwinism prevails and inequality is preached as the right way of life. Our value system is corrupted by materialism and as a result our capitalist system systematically robs our integrity. Egger's solution looks a sound one: unless we change our value system or we become 'super mature', monitoring every move is the only way to eliminate faults associated with corruption, bias, despotism, embezzlement, and similar misdeeds engaged aiming towards economic wealth.

Then where to draw the line? Here is the problem. In a capitalist system, privacy is a commodity also! If those public servants' detailed information is available for everyone, it is obvious that someone will turn it into business. We can also see that by letting our lives become completely transparent, we might enable someone to control us and before we know it, we might end up under a suffocating tyranny. After all freedom is what matters and I believe we should have the right to have some private moments and some public moments in our live no matter what work of line we are in.

For me, to be mature enough is the only way to avoid such paradoxes. I have discussed this with Ben during one of his journeys to Ethiopia. He asked me what I think about the future of government in the post singularity world (assuming that we had a positive singularity in which our biological form still exists untouched whilst our brains are advanced). I told him it would be definitely a "Rational Anarchy". I remember he asked what will be the 'rational' in my anarchist system. Though am not sure what the 'rational' in that system will be, am very sure that our value system that is entirely constructed on the foundations of materialism will

End of the Beginning

no longer exist. Inequality will no longer be rationalized under the name of Capitalism. In a Rational Anarchy, no one needs to rob the privacy of anyone for one simple fact; there will be no rational reason for being more powerful and more richer.

Ted: German president Angela Merkel and Brazilian President Dilma Rousseff both appeared to be very angry when they learned that the American intelligence services were listening to their telephone calls. But many countries do this kind of surveillance; maybe they were just embarrassed because they got caught instead of them catching others. Would it be better if all politicians' conversations were recorded and available to the public? Richard Nixon was caught because he tape recorded himself, believing that the tapes would only be listened to by historians.

Ben: Yeah, I very very much doubt the US intelligence services are the only ones doing this. I strongly suspect that Angela and Dilma doth protest too much...

As a rule of thumb, I think all politicians' conversations should be recorded and made public. As this is a transitional period and the world is complex, there are likely to be particular cases right now when that rule of thumb needs to be ignored, though. Politics is the art of the possible, but what is possible is rapidly changing...

Stephan: As a practical question, it's irrelevant: no state security apparatus would consent to making all leadership conversations public. As a philosophical one, it's subject to the same analysis as the other questions: airing secrets is only part of the solution to humanity's worst traits. Yes, it would be better if all crimes were exposed and the roots of corruption and abuse clearly identified so we could do something about them. But we also need the will to do something about them, something not currently in

End of the Beginning

great abundance among humans.

Hruy: This is absolutely America at the height of its arrogance. Yeah, it is a known fact that governments spy on each other but to invade the private conversations of 'allies' showed the behavior of the system we are governed by at its most decadent. Would America stay silent if the Brazilians were caught red hand listening to the private conversations of Obama?

Ted: If the Brazilians broke Obama's security, I think he would be too embarrassed to say anything about it. He would just hope nobody would find out. He might even be glad that they were paying attention.

Hruy: In my land there is a saying "leba inat lijuan ataminim." It's English meaning is "A crooked mother will never trust her child". My answer is the same. Monitoring everyone's every move is just nursing the decayed tooth. What we need is to pull it out from its root. We need a global attitude change and this is where technology can help us. What is the meaning of superpower? What is it? What is a wealthy nation and what is a third world one? What are the ways in which these people are maintaining the status quo right now? The stagnant filth we all shun to discuss lies here!

Mingyu: As I said before, I think all politicians' words should be monitored and made public. Only this can avoid an abuse of power. But after the Singularity politics will also have changed, if it exists at all.

Ben: Another point is that individuality as we now conceive it may change post-Singularity. One can envision a world in which there are communities of humans existing as individuals in roughly the manner of contemporary human society, with various upgrades— but in which the vast majority of mind-activity occurs in a less individuated more broadly distributed sort of mind-field. The loss of contemporary forms of individuality is parodied as a "borg mind" (with reference to a classic Star Trek: The Next Generation episode) but there are many possibilities beyond isolated individuals and completely unindividuated borgs. I introduced the term "mindplex" to refer to a network of minds that has clear self, intention and cognitive architecture at the group as well as individual levels. The "World of Views" perspective presented in another chapter in this book seems inclusive of this sort of possibility. In this perspective the whole question of "privacy" becomes a small subset of the overall question of information flow in complex distributed mind-systems (granted, a subset that is particularly relevant to individuated human mind/bodies such as our current selves).

End of the Beginning

Chapter Eighteen **The Future of Human Nature**

By Ben Goertzel

What will human beings be like after a Singularity?

There are Singularity scenarios in which humans don't survive, in anything like their current form. But what about those hopeful scenarios in which humans are still around, looking and acting somewhat like we do today? In such situations, how will a Singularity impact human psychology, human social interaction – the experience of being human?

Much of what one reads about the potential of a Technological Singularity, and the path toward such an event, focuses on the scientific and engineering aspects. This is understandable since, after all, science and engineering are the proximal causes of bringing us to such an amazing point. Without the nitty-gritty of science and engineering, notions like Singularity would forever remain as fantastic as tales of Middle Earth. Even so, though, it is the implications for everyday experience – including the simple inner experience of being oneself -- that really make the Singularity such an astounding concept. These "softer" implications are in some ways more difficult to grasp onto and to talk about than external-world technology innovations, but they are no less important to explore.

593

End of the Beginning

In this chapter I share some speculative thoughts about the possible future evolution of human nature as Singularity nears and passes. In brief, it seems plausible to me that advanced technology may enable the resolution of some of the deepest conflicts lying at the heart of human nature, which lead to many of the dissatisfactions commonly observed to plague modern humanity.

The topics broached here are deep and important, and deserve to be treated in a data-driven way, based on all the scientific evidence it's possible to muster. However, that would be a large and difficult research project, which I confess I have not yet undertaken. And at the end of that research project, given the current state of available data, there would still surely be many gaping holes. Instead of a rigorous scientific treatment, what I present here are merely some speculative hypotheses – guided by my understanding of the relevant scientific literatures, as well as my intuitive observations of myself and the world around me. At very least, I hope my thoughts will inspire the reader to investigate the issues I discuss and come to their own conclusions.

Given that this is ultimately a chapter on the future, in a book about the future, I'm going to spend an unusually large amount of energy on past history in the next few pages. But the reason for this will become clear by the end of the chapter. One big consequence of the Singularity, I suggest, will be humanity untying some of the psycho-cultural knots that have been binding it for millennia and longer.

Evolution embedded some serious conflicts in human psychology, due to the competing forces of individual and group selection, both of which shaped significant aspects of our psyches. The way individual and group orientation play together in the human mind is impressively subtle and complex, yet not entirely healthy and certainly not optimal.

Civilization then layered some further conflicts on top of these – most of all, the familiar Freudian struggle between obedience to complex social strictures versus utilizing tribal society's more

End of the Beginning

evolutionarily familiar methods of balancing individualistic desires with the empathic desire to work with the collective. With the advent of civilization, playing nice with the group suddenly began to require wholesale repression, or at least highly intricate and thoroughgoing control, of various aspects of nearly every social instinct we evolved to possess.

What we think of as "human nature" is, I suggest, largely an adaptation to these conflicts embedded in our psyches by the competition between individual and group selection, and by the complexly psychically repressive nature of civilization. Among other issues, our autobiographical "selves" have come to be largely constructed of inaccuracies that have emerged to cover over the difficulties posed by these contradictions.

But what will happen when we gain the ability to re-architect our minds, and much more thoroughly control our environments? As Singularity approaches, it will become more and more feasible to reduce the potency and impact of the inner conflicts historically defining human nature. What sort of new, improved human nature will rise in place of the old conflicts and the adaptations they spawned? What will a human nature free of these conflicts and their various consequences be like? We have no scientific means of answering this question at present – but, after reviewing a bit of ancient human history, I will share my educated speculations.

I believe there are alternative motivations embedded just as deeply into humanity as ego, fear and status. Furthermore, these alternative motives are more uniquely human. I am talking about *the drives to discover and create* – the drives for information and pattern and growth. Like ego, fear, status and all our other motivations, the motives to discover and create exist in the human mind for complex evolutionary reasons; and all these different motives are intricately layered on top of, and tangled up with, each other.

Rage, lust and fear came to us largely from our reptile

ancestors (Panksepp, 2003); status came more so from our more recent mammal forebears (Breuning, 2011); ego, as we know it, is likely a more uniquely human construct. But what caused us to emerge as the triumphant animal on Earth was, I suggest, not any of these emotions, but rather our ***active curiosity*** – our desire to learn more and create more. This is what led us to invent tools, to invent language, to invent society, to invent music, art and culture – and these inventions, not our bodies and brains directly, are what have led us to dominate the Earth.

Once the conflicts tying human nature in knots are more fully resolved, the result may well be that these more uniquely human drives reign supreme – and we see a new order of psychological reality, in which learning and building together largely supplant bolstering one's self-image, worrying about one's inadequacy, and struggling for relative status. Further, this transition may begin occurring gradually well before the full-on Technological Singularity. Indeed, one can argue that, in certain subcultures in various places on the planet, such a transition is already well underway.

What's Wrong with Humanity?

It doesn't take advanced science to tell you that modern human beings are generally not models of inner emotional health and harmony.

Much of sociology and psychology concerns various aspects of human "screwiness"; or one can look at just about any novel, TV show or cinematic drama. We all know that everyday human mind and life have a lot of "issues." While human life has its deep joys as well as its surface pleasures, truly thoroughgoing happiness and harmony seem hard to come by for human beings; conflict, trouble and dissatisfaction seem part of humanity's core.

As yet, we don't rigorously understand exactly where all these problems of humanity come from. But the topic is important enough that, in the absence of sound knowledge, educated

End of the Beginning

guesswork is warranted.

In this spirit, my somewhat speculative suggestion is two major culprits where human screwiness is concerned, are the following two sources of internal conflict, layered on top of each other:
1. The conflict between the results of individual and group (evolutionary) selection, encoded in our genome
2. The emergence of civilization, to which we are not adapted, which disrupted the delicate balance via which tribal human mind/society quasi-resolved the above-mentioned conflict

That is: I suggest that the transition to civilized society disrupted the delicate balance between self-oriented and group-oriented motivations that existed in the tribal person's mind. In place of the delicate balance we got a bunch of self vs. group conflict and chaos -- which makes us internally conflicted, but also stimulates our creativity and progress.

Further, I propose that these conflicts have a major impact on our "selves" – our "autobiographical self-models", which play such a large role in our inner lives. They play a major role in causing ourselves to comprise self-damagingly inaccurate models of the thought, feeling and behavior patterns of which we are actually constituted.

Put relatively simply: in order to avoid the pain that we are conditioned to feel from violating individual or group needs, or violating civilized or tribal standards of individual/group balance, we habitually create false self-models embodying the delusion that such violations are occurring in our minds and actions much less often than they really are. Emotional attachment to these sorts of inaccurate self-models is perhaps the most directly important cause of human mental and social suffering.

But what will happen as technology advances further, and transforms human society to an even greater extent? Will the conflicts between human nature and human society become yet more severe? Or will advances in technology allow some sort of fundamental resolution of the conflicts at the core of modern society?

I think the answer may well be the latter, more positive one. After reviewing the above ideas about contemporary human nature and its origins in more detail, I will explain why.

Internal Conflict Part 1: Individual versus Group Selection

The first key source of human internal conflict was best articulated by E.O.Wilson; the second by Freud. Putting the two together, we get a reasonably good explanation for why and how we humans are so complexly self-contradictory – and a good picture of the key psychological issues that the Singularity needs to deal with if it's going to result in profoundly happier, healthier humans.

Wilson, in his recent book ***The Social Conquest of Earth*** (2013), argues that human nature derives its complex, conflicted nature from the competitive interplay of two kinds of evolution during our history: individual and group selection. Put simply:

- Our genome has been shaped by individual selection, which has tweaked our genes in such a way as to maximize our reproductive success as individuals
- Our genome has also been shaped by group selection, which has tweaked our genes in such a way as to maximize the success of the *tribes* we belonged to

What makes a reproductively successful individual is, by and large, being selfish and looking out for one's own genes above those of others. What makes a successful *tribe* is, by and large, individual tribe members who are willing to "take one for the team" and put the tribe first.

Purely individual selection will lead to animals like tigers that are solitary and selfish. Purely group selection will lead to borg-like animals like ants, in which individuality takes a back seat to collective success. The mix of individual and group selection will lead to animals with a complex balance between individual-oriented and group-oriented motivations. As Wilson points out, many of the traits we call Evil are honed by individual selection; and many of the trains we call Good are honed by group selection. That's Internally Conflicted Human Nature, Part 1.

Good vs. Evil vs. Hierarchy-Induced Constraints

These points of Wilson's tie in with general aspects of constraint in hierarchical systems. This observation provides a different way of phrasing things than Wilson's language of Good vs. Evil. As opposed to adopting traditional moral labels, one can frame things in terms of the tension and interplay between
- adapting to constraints vs.
- pushing against constraints and trying to get beyond them

In the context of social constraints, the situation is that: individual selection (in evolution) would lead us to push against social constraints to seek individual well-being; whereas group selection would lead us to adapt to the social constraints regardless of our individual goals...

Much great (and mediocre) art comes from pushing against the constraints of the times -- but it's critical to have constraints there to push against; that's where a lot of the creativity comes from. You could think about yoga and most sports similarly... you're both adapting to the particularities of the human body; and trying to push the body beyond its normal everyday-life limits...

From the point of view of the tribe/society, those who push against the constraints too much can get branded as Evil and those who conform can get branded as Good..... But of course, among other factors, such judgments depend on what level you're looking at. From the point of view of the human body, the cell that doesn't conform to the system will likely get branded as Evil (non-self) and eliminated by the immune system! But such a cell does not consider itself undesirable at all...

End of the Beginning

In any hierarchical system, from the perspective of entities on level N, the entities on level N+1 impose constraints -- constraints that restrict the freedom of the level N entities in order to enable functionality on level N+1; but also have potential to guide the creativity of level N entities. Stan Salthe's book *Evolving Hierarchical Systems* (1985) makes this point wonderfully. In some cases, like the human body vs. its cells, the higher level is dominant and the creativity of the lower level entities is therefore quite limited. In the case of human society vs. its members, the question of whether the upper or lower level dominates the dynamics is trickier, leaving more room for creativity on the part of the lower level entities (humans), but also making the lives of the lower level entities more diversely complex.

Internal Conflict Part 2: The Discontents of Civilization

The second key internal conflict underlying human nature was described with beautiful clarity by Sigmund Freud in his classic book *Civilization and its Discontents* (1930).

In this book Freud argued that neurosis, internal mental stress and unhappiness and repression and worry are largely resultant from the move from nomadic tribal society to sedentary civilized society. In tribal societies, he pointed out, by and large people were allowed to express their desires fairly freely, and get their feelings out of their system relatively quickly and openly, rather than repressing them and developing complex psychological problems as a result. In tribal societies, the conflict between Self and Group was there, but was dealt with relatively effectively by relatively simple mechanisms. Whereas in modern societies, the methods for dealing with this conflict have become massively complex and convoluted, requiring much more complexity and difficulty on the part of the individual mind.

Some insight into these matters is obtained via the accounts of individuals who have lived with modern Stone Age tribes. One fascinating recent book encountering a modern linguist/missionary's contact with a modern Stone Age society in the Amazon, the Piraha, is Daniel Everett's *Don't Sleep There Are Snakes* (2009). Another book, that I read in the 1980s, recounts an average guy from New Jersey dropping his life and migrating to Africa to live with a modern Stone Age pygmy tribe in central Africa, is *Song From the Forest* (Sarno, 1993X)[117]. These accounts and others like them seem to validate Freud's analysis. The tribal, Stone Age lifestyle tends not to lead to neurosis, because it matches the human emotional makeup in a basic way that civilization does not.

Wilson + Freud = Why We Are So Internally Conflicted

I full well realize the "noble savage" myth is just that -- the psychology of tribal humans was not as idyllic and conflict-free as some have imagined. Tribal humans still have the basic conflict between individual and group selection embedded into their personalities. But I would argue that, in tribal human sociopsychology, evolution has worked out a subtle balance between these forces.

What civilization does is to throw this balance off -- and put the self-focused and group-focused aspects of human nature out of whack in complex ways. In tribal society, Self and Group balance against each other elegantly and symmetrically -- there is conflict, but it's balanced like yin and yang. In civilized society, Self and Group are perpetually at war, because the way our self-motivation and our group-motivation have evolved was right for making them *just barely* balance against each other in a tribal context; so it's natural that they're out of balance in complex ways in a civilization context.

---

[117] *Some recent news from Louis Sarno, who at time of writing still lives in the Central African Republic, is at* http://pittrivers-sound.blogspot.hk/2013/04/updates-from-louis-sarno-and-bayaka-in.html *, including an intriguing recent video, a trailer for a forthcoming movie.*

End of the Beginning

For example, in a tribal situation, it is a much better approximation to say that: What's good for the individual is good for the group, and vice versa. The individual and group depend a lot on each other. Making the group stronger helps the individual in very palpable ways (if a fellow hunter in the tribe is stronger for instance, he's more likely to kill game to share with you). And if you become happier or stronger or whatever, it's likely to significantly benefit the rest of the group, who all directly interact with you and are materially influenced by you. The harmony between individual interest and group interest is not perfect, but it's at least reasonably present... the effects of individual and group selection have been tuned to work decently together.

On the other hand, in a larger civilized society the connection between individual and group benefit is far more erratic. What's good for me, as a Hong Kong resident, is not particularly the same as what's good for Hong Kong. Of course there's a correlation, but it's a relatively weak one. It's reasonably likely that what's good for Hong Kong as a unit could actually make my life worse (e.g. raising taxes, as my income level is above average for Hong Kong). Similarly, most things that are likely to improve my life in the near future are basically irrelevant to the good of Hong Kong; in fact, my AGI research work is arguably bad for all political units in the long term, as advanced AGI is likely to lead to the transcendent of nation-states.

There is definitely some correlation between my benefit and Hong Kong's benefit -- if I create a successful company here in Hong Kong, that benefits the Hong Kong economy. But the link is fairly weak, meaning that my society is often going to push me to do stuff that goes against my personal interest; and vice versa. This seems almost inevitable in a complex society containing people playing many different roles.

The advent of civilization also led to complexification in the realm of human dishonesty. Lying of course occurs in tribal societies just like in advanced civilizations -- humans are dishonest

End of the Beginning

by nature, to some extent. Yet, only in complex civilizations do we have a habit of systematically putting on "false fronts" before others. This sort of strategy doesn't work so well if you're around the same 50 people all the time. Yet it's second nature to all of us in modern civilization -- we learn in childhood to act one way at home, one way at school, one way around grandma, etc.

As we mature, the habit of putting on false fronts -- or as Nietzsche called them, "masks" -- becomes so integrated into our personalities that the fronts aren't even "false" anymore. Rather, our personalities become mélanges of shiftingly overlapping subselves, with somewhat different tastes and interests and values, in a complex coopetition for control of our thoughts and memories. This is intricate and stressful, but stimulates various sorts of creativity.

Sarno reports how the interaction of the Bayaka pygmies with civilization caused them to develop multiple subpersonalities. A pygmy's personality while living the traditional nomadic lifestyle in the bush, may be very different from that same pygmy's personality while living in a village with Africans from other tribes, drinking alcohol and doing odd jobs for low wages.

Individually, we have a motive to lie and make others think we are different in various ways than we actually are. Tribally, group-wise, there is a reason for group members to tell the truth -- a group with direct and honest communication and understanding is likely to do better on average, in many important contexts, because deception often brings with it lots of complexity and inefficiency. The balance between truth and lying is wired into our physiology -- most people can lie only a little bit without it showing in their faces (though this limitation can be overcome with practice, as one sees in good politicians and salespeople). But modern society has bypassed these physiological adaptations, which embody tribal society's subtle balance between self and group motivations, via the creation of new media like telephones, writing and the Internet, which bypass telltale facial expressions

End of the Beginning

and open up amazing new vistas for systematic self-over-group dishonesty. Then society, and the minds of individuals within it, must set up all sorts of defense mechanisms to cope with the rampant dishonesty. The balance of self versus group is fractured, and complexity emerges in an attempt to cope, but never quite copes effectively, and thus keeps ramifying and developing.

In Freudian terms, civilization brought with it the split between the Ego and Super-ego -- between what we are (at a given point in time); and what we think we should be. It also brought with it a much more complex and fragmented Ego than was present in tribal peoples.

What Wilson makes clear is: the pre-civilized human mind already had within it the split between the Self-motivation and Group-motivation. Freud somewhat saw this as well, with his Id serving as a stylized version of the pure Self-motivation and his Ego going beyond this to balance Self versus Group.

The Freudian Ego and Super-ego are different ways of balancing Self versus Group. The perversity and complexity of civilized society is that each of us is internally pushed to balance the conflict of Self vs. Group in one way (via our Ego, which is largely shaped for tribal society), while feeling we "should" be carrying out this balance in a different way (via our Super-Ego, which comes from civilized culture). Of course these Freudian terms are not scientific or precisely defined, and shouldn't be taken too seriously. But they do paint an evocative picture.

Wilson pointed out the conflict between Self versus Group lying at the core of human nature; Freud took the next step and pointed out the equally acute conflict between Evolved Self/Group Balance versus Civilized Self/Group Balance. One might say that the conflict Wilson identified is a "first order" conflict: simply between Self and Group; whereas the conflict Freud identified is a "second order" conflict, a conflict between two different ways of managing the first order conflict, Tribal vs. Civilized.

How much of this multilayered inner conflict is a necessary

aspect of being an intelligent individual mind living in a civilization? Some, to be sure -- there is always going to be some degree of conflict between what's good for the individual and what's good for the group. But having genomes optimized for tribal society, while living in civilized society, foists an additional layer of complexity on top of the intrinsic conflict. The fact that our culture changes so much faster than our genomes, means that we are not free to seek the optimal balance between our current real-life Self and Group motivations, consistent with the actual society we are living in. Instead we must live with methods of balancing these different motivations, that were honed in radically different circumstances than the ones we actually reside in and care about.

Our Internal conflict Spurs Our Creativity and Progress

The conflicts at the core of human nature are frustrating, at times infuriating. Indeed, each year, they drive a certain percentage of people in modern society to suicide (a phenomenon apparently much less common in tribal cultures). Yet we must not ignore their corresponding positive aspects. What is driving us toward the reality of amazing possibilities like flexible brain and body modification is -- precisely the internal conflict I've analyzed above.

It's the creative tension between Self and Group that drove us to create sophisticated language in the first place. One of the earliest uses of language, that helped it to grow into the powerful tool it now is, was surely *gossip* (Dunbar, 1996) -- which is mainly about Self/Group tensions.

And our Self and Group aspects conspired to enable us to develop sophisticated tools. Invention of new tools generally occurs via some wacky mind off in the corner fiddling with stuff and ignoring everybody else. But, we do much better than other species at passing our ideas about new tools on from generation to generation, leveraging language and our rich social networking capability -- which is what allows our tool-sets to progressively

End of the Beginning

improve over time.

The birth of civilization clearly grew from the same tension. Tribal groups that set up farms and domesticated animals, in certain ecological situations, ended up with greater survival value -- and thus flourished in the group selection competition. But individuals, seeking the best for themselves, then exploited this new situation in a variety of complex ways, leading to developments like markets, arts, schools and the whole gamut. Not all of these new developments were actually best for the tribe -- some of the ways individuals grew to exploit the new, civilized group dynamics actually were bad for the group. But then the group adapted, and got more complex to compensate. Eventually this led to twisted sociodynamics like we have now ... with (post)modern societies that reject and psychologically torment their individualistic nonconformist rebels, yet openly rely on these same rebels for the ongoing innovation needed to compensate the widespread dissatisfaction modernity fosters.

And the creativity spurred by burgeoning self/group tensions continues and blossoms multifariously. Privacy issues with Facebook and the NSA... the rise and growth and fluctuation of social networks in general... the roles of anonymity and openness on the Net... websites devoted to marital infidelity... issues regarding sharing of scientific data on the Net or keeping it private in labs... patents... agile software development... open source software licenses and processes... Bill Gates spending the first part of his adult life making money and the second part giving it away. The harmonization of individual and group motivations remains a huge theme of our world explicitly, and is even more important implicitly.

Our Problematic Selves

The primary culprit of human suffering has often been identified as the "self" – meaning the autobiographical, psychosocial self; the self-model that each of us uses to symbolize,

End of the Beginning

define and model our own behavior (Metzinger, 2003). One of the most commonly cited differences between normal human psychology and the psychology of "enlightened" spiritual gurus is that the latter are said to be unattached to their autobiographical selves – indeed they are sometimes said to have "no self at all."

I think there is some deep truth to this perspective; but one needs to frame the issues with care. Any mind concerned with controlling a body that persists through time, has got to maintain some sort of model of that body and the behavior patterns that are associated with it. Without such a model (which may be represented explicitly or implicitly), the mind could not control the body very intelligently. Any such model can fairly be called a "self-model." In this sense any persistently embodied intelligence is going to have a self.

The problem with the human self, however, is that it tends to be a bad model – not a morally bad model, but an inaccurate one. The self-models we carry around in our minds, are generally not very accurate models of the actual behavior-patterns that our bodies display, nor of the thought-patterns that our minds contain. And the inaccuracies involved are not just random errors; they are biased in very particular ways.

Our self-models are symbols for clusters of behavior-patterns that are observed to occur among our bodily behaviors and our internal cognitive behaviors. This is not in itself bad – symbolic reasoning is critical for general intelligence. However, we are very easily drawn to make incorrect conclusions regarding our symbolic self-models – and to become emotionally attached to these incorrect conclusions.

And this brings us straight back to the two conflicts that I've highlighted above: Self versus Group (Wilson), and Evolved Self/Group Balance versus Civilized Self/Group Balance (Freud). These layered contradictions yank our self-models around willy-nilly. Each modern human feels great pressure to be both self-focused and group-focused; and to balance self and group in a

End of the Beginning

tribal way, and in a civilized way.

What's the simplest way for a person to fulfill all these contradictory requirements? -- or rather, to feel like they have at least done a halfway-decent job of fulfilling them?

That's easy: To bullshit themselves!

Human self-models are typically packed with lies -- lies to the effect that the person is fulfilling all these contradictory requirements much better than is actually the case. Because when a person clearly sees just how badly they have been fulfilling these contradictory requirements, they will generally experience a lot of bad emotion – unless that person has somehow managed to let go of the expectations that evolution and society have packed into their brains and minds.

The above analysis of the conflicts in human nature lets us specifically identify four kinds of lies that are typically packed into human selves. There are two kinds of Wilson Ian lies:

- Lies about how a person has acted against their own goals and desires
- Lies about how a person has disappointed the others around them
- And there are two kinds of Freudian lies:
- Lies about how a person has repressed their true desires, in order to adhere to general social expectations
- Lies about how a person has violated general social expectations, in effort to act out their true desires

What if a person could avoid these four kinds of lies, and openly, transparently acknowledge all these kinds of violations to themselves, on an ongoing basis during life? This would allow the person in question to form an accurate self-model -- not the usual self-delusional self-model biased by the Wilsonian and Freudian contradictions. But this sort of internal self-honesty is far from the contemporary human norm.

The problem is that evolution has wired us to become unhappy when we know we have acted against our own goals and

End of the Beginning

desires; OR when we know we have disappointed someone else. And civilized society has taught us to become unhappy when we violate social expectations; but evolution has taught us to become unhappy when we don't balance self and group in the way that is "natural" in a tribal context. So, inside the privacy of our minds, we are constantly tripping over various evolved or learned triggers for unhappiness. The easiest way to avoid setting off these triggers is to fool ourselves that we haven't really committed the "sins" required to activate them – i.e. to create a systematically partially-false self-model.

The harder way to avoid setting off these triggers is to effectively rewire our mind-brains to NOT be reflexively caused unhappiness when we act against our goals/desires, disappoint others, violate social expectations, or balance self and group in tribally inappropriate ways. Having done this, the need for an inaccurate, self-deluding self-model disappears. But performing this kind of rewiring is very difficult for human beings, given the current state of technology. The only reasonably reliable methods for achieving this kind of rewiring today involve years or decades of concentrated effort via meditation or other similar techniques.

And what would a human mind be like without a dishonesty-infused, systematically inaccurate self-model? Some hints in this direction may be found in the mind-states of spiritually advanced individuals who have in some sense gone beyond the negative reinforcement triggers mentioned above, and also beyond the traditional feeling of self. A recent study of the psychology of the spiritually advanced (Martin, 2013) suggests that, without a self in the traditional sense, a person's mind feels more like an (ever-shifting) set of clusters of personality/behavior patterns. One of the lies the self tells itself, it seems, is about its own coherence. Actually human beings are not nearly as coherent and systematic and unified as their typical self-models claim.

Next Steps for Human Nature?

End of the Beginning

Will advanced technologies allow us to overcome the conflicts at the heart of human nature?

Short answer: Yes

Slightly longer answer: To a certain extent at least, I think so. Once we achieve the ability to modify our brains and bodies according to our wishes, we will be able to adapt the way we balance Self versus Group in a much more finely-tuned and contextually appropriate way... which will allow us to create more accurate self-models, providing us with much less problematic inner lives.

To the extent that humanity is characterized by layers of conflict within conflict feeding into delusory self-models, this will make us less human. But it will also make us less perverse, less confused, and more fulfilled.

Inside the mind of a post-Singularity human, things will be quite different from the interior of a contemporary human mind. The ability to introspect much more fully into one's own mind, and modify one's own mind at will (limited by one's understanding of what modifications should be made), will have a huge impact on the various conflicts of human nature.

Contemporary human nature uses ruses such as systematically inaccurate self-modeling to minimize the degree to which the conflicts at the core of human nature preoccupy reflective consciousness. Via such ruses our inner lives are simplified in some ways; and complexified in others. But whatever the benefits of this sort of dynamics in the context of modern humanity, in the case a mind with fuller reflexive and self-modificatory capabilities, conflict-repression strategies of this nature wouldn't work so well.

If a mind capable of strong self-modification differs from what it would like to be – then it will have a much greater capability to do something about this. If it chooses not to change itself in accordance with its supposed aspirations, in spite of having strong self-modificatory capability, then it will have to

End of the Beginning

confront the fact that these are not its actual aspirations. Thus, in such minds, self-deception about the ego/superego distinction should be massively less. There will still, of course, be conflicts between what others want the mind to do and be, and what the mind wants for itself – but these conflicts will need to be resolved in some way not relying on blatant repression and self-delusion. Most likely different sort of mind attractor will emerge, in which the mind's own actual desires and aspirations, along with the social pressures it is experiencing, are all explicitly acknowledged side by side.

The individual/group dichotomy will also take on quite different forms, in the context of minds that have a more flexible ability to share their inner worlds and their subjective perspectives with each other. We now enter into implicit social contracts with others, largely without consciously recognizing what we are doing and why. This should be much less the case among minds that can introspect into themselves and each other more thoroughly. Fear of rejection from the group will not have the same bite when one sees, via mind-sharing, that the others don't actually intend any rejection. Insecurity via comparing oneself to others in the group will be much less meaningful when one knows oneself and the other better. Egomania will also be harder when one can see more directly what others are doing internally; and when one knows others can observe one's mind.

I don't imagine that introspection and self-modification will eliminate all tensions. I suppose that, long after humans have transcended their legacy bodies and psychologies, the tension between Self and Group will remain in some form. Even if we all turn into mindplexes (Goertzel, 2003) – tightly knit networks of minds that have reflective consciousness at the multiple-mind level as well as the individual-mind level -- the basic tension that exists between different levels in any hierarchical system will still be there. But at least, if it's internally conflicted, it will be internally conflicted in more diverse and fascinating ways! Or beyond

End of the Beginning

conflicted and harmonious, perhaps.

A post-Singularity mind, in short, will be much more marked by explicit acknowledgement of its own characteristics and strengths and weaknesses, of the characteristics and strengths and weaknesses of others, and of the social contracts it has entered into. This will avoid many of the "knots" (Grof, 1988) that the current human psyche tends to tie itself into. While conflict will not be eliminated, it will rarely lead to individual pathologies like repression, or social pathologies like insecurity and egomania. Other complex, probably sometimes undesirable dynamics will emerge, which are difficult to foresee at this time.

I have noted above that, in current human nature, conflict also spawns creativity. Does this mean that, if future minds largely overcome the conflicts at the center of current human nature, they will also overcome the need to create? This will surely happen in some cases, just as when the tormented but productive young artist settles down to a happier, more mature life and stops creating much. But remember that a mind with the ability to heavily self-modify, will also be able to modify its goal system. A self-modifying mind can pose creativity as one of its top-level goals. And while I don't have a rigorous argument in this regard, my current belief is that this is likely to happen quite frequently. I will elaborate a bit on this below.

Goals Beyond the Legacy Self

In the current state of human nature, self and goals are intimately wrapped up together. Substantially, we pursue our goals because we want our self-model to be a certain way – and we do this in a manner that is inextricably tangled up with the various lies the self-model embodies.

Consider the case of a post-Singularity human or human-like mind that understands itself far better than contemporary humans, thus arriving at far more accurate – and likely less unified and coherent – self-model than a typical pre-Singularity human mind.

End of the Beginning

What will the goals of such a mind be? What will a mind without a coherent self --without a self-built around lies and confusions regarding self vs. group and repression and status -- actually want to do with itself?

Considering our primary current examples of minds that have discarded their traditional autobiographical selves -- spiritual gurus and the like – provides confusing guidance. One notes that (with nontrivial exceptions) the majority of such people are mainly absorbed with enjoying the wonder of being, and sometimes with spreading this wonder to others, rather than with attempting to achieve ambitious real-world goals. The prototypical spiritually advanced human is not generally concerned with pursuing pragmatic goals, because they are in a sense beyond the typical human motives that cause people to become attached to pursuit of such goals. This makes one wonder if the legacy self – with all its associated self-deception -- is somehow required in order for humans to work hard toward the achievement of wildly ambitious goals, in the manner for instance of the scientists and entrepreneurs who are currently bringing the Singularity within reach.

But it's not clear that the contemporary or historical spiritual guru is a good model for a post-Singularity, post-legacy-self human mind. I suspect that in a community of post-delusory-self minds, avid pragmatic goal-pursuit may well emerge for different reasons, mostly unrelated to legacy human motives.

Why would a community of post-delusory-self minds pursue goals, if not for the usual human reasons of status and ego? Here we come to grips with deep philosophical issues. I would argue that, once the conflicts that wrack human nature are mostly removed, other deep human motives will rise to the fore – for instance, the drive to discover new things, and create new things. That is: the drives for pattern and information (Goertzel, 2006). One can view the whole long story of the emergence of life and intelligence on Earth as the manifestation of these "drives", as

613

embedded in the laws of physics and the nature of complex systems dynamics (Jantsch, 1980). From the point of view of the Cosmos rather than humanity in particular, the drives for pattern, creation and information are even deeper than the conflicts that wrack human nature.

If spiritually advanced humans, having cast aside self and ego and status, tend not to pursue complex goals of discovery and creation, this may be because, given the constraints of the human brain architecture, merely maintaining a peaceful mindstate without self/ego/status-obsession requires a huge amount of the brain's energy. The simple, blissful conscious state of these individuals may be bought at the cost of a great deal of ongoing unconscious neural effort.

On the other hand, once the legacy human brain architecture becomes flexibly mutable, most of the old constraints no longer apply. It may become possible to maintain a peaceful, blissful conscious state – relatively free of Freudian repression and individual/group conflicts – while still avidly pursuing the deeper goals of gaining more and more information, and creating more and more structures and patterns in the universe. Here we are far beyond the domain of the currently scientifically testable – but this is indeed my strong suspicion.

Transitions?

All this may seem a far-out sort of conceptual exploration– more pertinent to what happens AFTER the Singularity, than to the path from here to Singularity. But I suspect that the path to this sort of post-Singularity, post-self mind-state may be gradual and incremental. As Singularity approaches, the old evolutionary and Freudian conflicts may become less and less prominent in the world population.

Sousveillance may help with this, via largely obsoleting the secret-keeping aspect of psychological repression.

The flexibility of lifestyle enabled by modern technology

End of the Beginning

(which will sweep more of the world as developing nations play catch-up during the next decades) will enable people to live more and more of their lives in physical or virtual communities comprised of relatively like-minded people (called micro-states in Chapter XX), reducing the need for obedience to repressive-feeling social strictures.

Brain and body monitoring devices will give people more and more awareness of what is going on in their own minds, and how they are reacting to their environments – including awareness of the process of construction of the self-model. Once people can explicitly see how their selves are constructed by their mind/brains, the grip of the psychosocial self will be weakened.

But what will all this mean pragmatically? Supposing this train of thought is right, and as Singularity approaches, a significant swath of the population will become increasingly less attached to their psychosocial, autobiographical selves – then what will happen? Will these increasingly selfless youths sit around on the dole, watching 4D virtual reality soap operas (or whatever is the Next Big Thing)? Or, alternately, will new sorts of social groups form, with pursuit of various goals serving as an attractor that helps hold the group together?

I suspect the answer will be – a little of both. In part, at least, the Singularity may be brought about via self-organizing networks of human minds, pursuing goals for reasons different from the social status and ego gratification that drive so much of human activity... driven more simply by the basic drives to discover and to create.

In particular, it seems a phase transition in this direction may occur when brain-computer interfacing technology becomes powerful and prevalent. The ability to introspect into one's brain will lead to the possibility of consciously modifying one's mind via neurofeedback. Explicit tools for auto-neuromodification will bring another large leap from pre-Singularity toward post-Singularity psychology. With each of these leaps, a change in technology will likely lead to a profound change in human experience, in human nature. With each step, more of the conflicts intrinsic to human nature will get made explicit to each individual human mind, and this explicitness will alter their role within the mind. Furthermore, advanced BCI will likely lead to functional equivalents of telepathy, enabling people to understand each other better and operate together better as groups, sharing their passions for discovery and creation.

On the other hand, if mind uploading happens to become a mature technology before BCI and neuromodification, it could have the same effect; a software version of a human mind should be far more inspectable and modifiable than the corresponding wetware version.

Even without powerful BCI, though, we are gradually moving in the same direction anyway, via increasing transparency and openness – both as displayed in culture, and (connectedly) as enforced by the observation technologies that are leading us toward the surveillance/sousveillance dichotomy. Little by little, at least in Western societies, people are opening up themselves to the world, exposing what used to be private for public appreciation. As time passes this should lead to a more explicit understanding of the complex conflicts involved in the individual/group and ego/superego dichotomies. And on a practical basis, in the developed world one sees a progressive shift from work-for-money to work-for-passion, i.e. work pursued due to the drives for discovery and creation. But still, one suspects that these gradual changes in human nature will pale before the changes to occur once BCI and neuromodification (or mind uploading) technologies become impactful and widely available.

Current human nature got where it is largely via the advent of certain technologies— the technologies of agriculture and construction that enabled civilization, for example. The folks who invented the plow and the brick weren't thinking about the consequences their creations would have for the emergence and dynamics of the superego— but these consequences were real enough anyway.

Similarly, the next steps in human nature may well emerge as a consequence of technological advancements like brain-computer interfacing and mind uploading— even though the scientists and engineers building these technologies will mostly have other goals in mind, rather than explicitly focusing their work toward reducing conflict in the human psyche and bringing about an era where self is less critical and discovery and creation are the main motivations.

Growth, joy and creation beyond the constrictions of the self-delusory self— I'm more than ready!

REFERENCES

Breuning, Loretta (2011). *I, Mammal: Why Your Brain Links Status and Happiness*. CreateSpace.

Dunbar, Robin (1996). *Grooming, gossip and the evolution of language*. London: Faber and Faber.

Everett, Daniel (2009). *Don't Sleep, There Are Snakes*. Vintage.

Freud, Sigmund (1930/2002). Civilization and Its Discontents", London: Penguin.

Goertzel, Ben (2003). Mindplexes. *Dynamical Psychology E-Journal*, http://www.goertzel.org/dynapsyc/2003/mindplex.htm

Goertzel, Ben (2006). *The Hidden Pattern*. Brown Walker.

Grof, Stan (1988). *The Adventure of Self-Discovery*. SUNY Press.

Jantsch, Eric (1980). *The Self-Organizing Universe*. Pergamon.

Martin, Jeffery (2013). Clusters of Individual Experiences form a Continuum of Persistent Non-Symbolic Experiences in Adults, **http://nonsymbolic.org/PNSE-Article.pdf**

Metzinger, Thomas (2003). *Being No One*. MIT Press.

Panksepp, J. (2002). Foreword to Cory, G. and Gardner, R. (2002) *The Evolutionary Neuroethology of Paul MacLean: Convergences and Frontiers*. Greenwood/Praeger

Salthe, Stan (1985). *Evolving Hierarchical Systems*. Columbia University

End of the Beginning

Press.
Sarno, Louis (1993). *Song From the Forest*. Houghton Mifflin.
Wilson, Edward O. (2013). *The Social Conquest of Earth*. Liveright.

End of the Beginning

Dialogue 18.1

with Clément Vidal and Ben Goertzel

Clément: I have one question regarding "The Future of Human Nature". You write: "One notes that (with nontrivial exceptions) the majority of such people are mainly absorbed with enjoying the wonder of being, and sometimes with spreading this wonder to others, rather than with attempting to achieve ambitious real-world goals."

What are the "nontrivial exceptions" you have in mind? Maybe these are the ones to study in details, to see how they succeed, and to get inspired by them?

Ben: I was thinking of some of the folks my friend Jeffery Martin mentioned to me, among his interviewees for his outstanding study of "enlightened" individuals.[118] Some of his apparently "enlightened" subjects were carrying out productive careers in various areas, including scientific research. He didn't focus his study on how they harmonized their enlightenment with the goal-pursuit implicit in their work; though this topic is definitely very interesting to me.... Maybe he or his colleagues will look at this in a follow-up study.

Clément: Also, I would like to point out a paper about Maslow's hierarchy of needs [Koltko-Rivera, 2006], which shows that in Maslow's late thinking the last need is not self-actualization, but self-transcendence. To apply this idea in the singularity context, we could speculate that as abundance fills up all the basic needs, we will ultimately go through a need to self-

---

[118] Jeffrey Martin, "Clusters of Individual Experiences form a Continuum of Persistent Non-Symbolic Experiences in Adults," http://nonsymbolic.org/PNSE-Article.pdf.

transcendence, up to the need of a higher metasystem transition (global brain?).

Ben: Very interesting... I haven't read Maslow that thoroughly... but this certainly makes him seem closer to the transhumanist perspective; though of course he wasn't thinking so much about technological means of self-transcendence...

REFERENCES
Mark E. Koltko-Rivera (2006). Rediscovering the Later Version of Maslow's Hierarchy of Needs: Self-Transcendence and Opportunities for Theory, Research, and Unification. *Review of General Psychology* Vol. 10, No. 4, 302–317

End of the Beginning

Chapter Nineteen **Capitalism, Socialism, Singularitarianism**

By Ted Goertzel

The manifesto that "a specter is haunting Europe — the spectre of communism" was largely forgotten by the end of the nineteenth century. But it came back with a vengeance in 1919 and lasted until 1989. Today another specter is haunting humanity - the specter of computer intelligence that rivals or exceeds our own. This specter was launched not with a manifesto but with a movie: *2001: A Space Odyssey* (Kubrick 1968). The threat did not materialize by 2001, and today it is often viewed as no more than a science fiction trope. But it may have only been delayed, just as the Marxist revolution was.

Change is often slower than futurists predict. In 1957, Herbert Simon predicted that a computer would beat the world chess champion by 1967. It happened in 1997. Many other technological milestones expected in the 1960s have been achieved by now. We have all the world's information at our fingertips, an amazing accomplishment that we take for granted. Computers translate foreign languages, drive automobiles and diagnose illnesses - not perfectly but much better than skeptics thought possible a few decades ago.

No one knows for sure, but the median estimate of engineers

and scientists working on the problem is that computers with human level intelligence will be a reality in thirty or forty years (see Chapter Two in this volume). There are still some who object that this will never happen, that computers will never really "think" in the way that humans do. And computer intelligence may well be different, more powerful than ours in some ways, less in others. But greatly increased computer intelligence is surely coming.

The socialist revolution was eagerly anticipated by many and greatly feared by others. So too with the coming of the intelligent machines. The dystopians fear that humanity is at risk of being destroyed by its own creations. For example, Tesla motors and commercial space entrepreneur Elon Musk (2014) recently warned that:

> I think we should be very careful about artificial intelligence. If I were to guess what our biggest existential threat is, it's probably that. With artificial intelligence we are summoning the demon. In all those stories where there's the guy with the pentagram and the holy water, it's like yeah he's sure he can control the demon. Didn't work out.

Theoretical physicist Stephen Hawking (2015) warned that:
> The development of full artificial intelligence could spell the end of the human race.... It would take off on its own, and re-design itself at an ever increasing rate... Humans, who are limited by slow biological evolution, couldn't compete and would be superseded.

Hawking's concerns are ironic since he is dependent on artificial intelligence devices to cope with his handicap.

On the other side are the techno-utopians, many of whom are included in this volume, who promise that intelligent machines will bring health, wealth, happiness and perhaps even eternal life right here on earth. Ray Kurzweil (2006), who was the pioneering developer of a machine to read aloud to the blind, has popularized the meme of The Singularity, a point in time at which change will become effectively instantaneous and computers will be smarter than people. He counts on these computers to resolve problems humanity has been unable to solve on its own.

Serious research on the risks of general artificial intelligence is just beginning, at Oxford University and elsewhere (Armstrong 2014; Bostrom 2014). The *Journal of Experimental and Theoretical Artificial Intelligence* published a special issue on *The Risks of Artificial Intelligence* in 2014 (Vol. 26, Issue 3). The issue will soon be published as a book. These specialists do not believe it is possible to stop the advance of artificial intelligence; it is taking place in too many places in too many countries and there is no international agency with the mission, let alone the capability, to suppress it. The best we can do is prepare for it, and try to make it as beneficial as possible. This paper examines the impact of increasingly powerful artificial intelligence on human social and economic organization.

The Singularity: Science or Wishful Thinking?

The Singularity, as Kurzweil presents it, seems too good to be true. It resembles wishful thinking such as belief in the coming of a Messiah or of the communist utopia. A problem with millenarian beliefs like these is that it people may conclude there is no need to prepare for the future. All we have to do keep praying (or philosophizing) and wait for the Singularity or the Revolution or the Second Coming. Then all our problems will be solved. This is a perhaps a greater risk with belief in the Singularity since it posits that change will be so fast we won't be able to even understand it, let alone do anything about it.

But is it likely that change will be this fast? Kurzweil offers a scientific rationale for his prediction, and has gathered a lot of data to support it. His core argument is that scientific and technological progress is cumulative and combinatory. The more advances that are made, the more ways there are to produce new ones. This leads to an exponential increase in the rate of innovation over time, something he illustrates empirically in a number of time series graphs. Mathematically, it is true that if one projects an exponential trend far enough into the future, the rate of change will become virtually instantaneous. The curve on a graph shoots up to the ceiling. But in the real world exponential growth processes can't go on that long. The more usual pattern is a slow growth at first, rising very quickly for a time, then slowing down and leveling off. This is a pattern best plotted on a graph as an S-curve rather than an exponential one.

But even if the technological singularity isn't instantaneous, it will still be very quick by the standards of historical change. Economic historian Robin Hanson (2008) compares the anticipated technological singularity to two other "singularities" in human history: the agricultural revolution and the industrial revolution. Both of these took place very quickly when plotted against the long course of human existence, but they were not instantaneous. They look like singularities if you plot them on a very long historical graph. The technological singularity may take place more quickly, because technical progress is faster now, but still over a period of years or decades.

But the precise timing is not critical. In fact, the beginnings of the technological singularity are already here with the advent of technologies such as self-driving automobiles and the world-wide-web. Further changes are coming and coming fast; the likelihood that the doomsayers will be able to stop them is negligible. Whether the impact is utopian or dystopian or something in-between is not entirely out of our control.

Cyberspace and the Marxist Utopia

Perhaps the most utopian possibility is that super-human intelligences will finally bring about the future Marx and Engels dreamed about in 1845. In an often quoted paragraph they wrote of a "communist society where nobody has one exclusive sphere of activity but each can become accomplished in any branch he

End of the Beginning

wishes, society regulates the general production and thus makes it possible for me to do one thing today and another tomorrow, to hunt in the morning, fish in the afternoon, rear cattle in the evening, criticize after dinner, just as I have a mind, without ever becoming hunter, fisherman, herdsman or critic" (*The German Ideology*, 1845). This was impossible with nineteenth or twentieth century technology, except for a small leisure class and elderly people who could afford to retire. But it could be possible for everyone if computers and robots do the unpleasant work.

Marx and Engels did not specify how "society would regulate the general production." They thought it was premature to speculate about it. If they had thought seriously about it, they might have realized that the material conditions for such a utopia were a long way off. But they and their comrades had a very human propensity to want change during their own lifetimes. A detailed blueprint for a communist society was published by their follower August Bebel in his 1879 classic *Woman and Socialism*. It was to be a society where technicians and statisticians would make decisions in the interest of the common good and everyone would accept these decisions voluntarily. The coercive institutions of the state would wither away because they would not be needed. Any crimes would be punished informally by the citizens.

The disasters caused by trying to impose this model in the twentieth century are well known. Power was monopolized by a *nomenklatura* and a police state, individual initiative and work incentives were stifled, and human rights were trampled. The promised land of milk and honey failed to materialize. The planned economy was slow and inefficient compared to its capitalist competitors.

Soviet Cybernetics

One of the reasons for the poor performance of the Soviet economy may have been that the computer technology available at the time was inadequate to the task. Norbert Wiener's classic book

End of the Beginning

Cybernetics was enthusiastically received by the Soviet technological intelligentsia in 1959, after the death of Stalin. They were overwhelmed with the computational tasks of centrally commanding the entire Soviet economy and hoped that cybernetics could offer a solution. Cybernetics was heralded in the Communist Party program as "science in the service of communism." But after conducting serious feasibility studies the Soviet experts concluded that "it was impossible to centralize all economic decision making in Moscow: the mathematical optimization of a large-scale system was simply not feasible" (Georvitch 2002: 273). They estimated that it would take twenty years to install a computer network adequate to the task, and cost several times the price of the Soviet space program. The Soviet leaders turned down the opportunity to build the first internet, leaving it to the Pentagon and the venture capitalists of Silicon Valley.

Twenty-first Century Socialism?

Some argue that the Soviets were just in the wrong century and that we should rerun that experiment today. Paul Cockshott (2014) observes that contemporary computers are fully adequate to handle input output tables for a modern national economy. Capitalism has built the necessary computer networks; most economic transactions are already online, or could be. Mathematical optimization would not be necessary; planning could be done with computer simulation. Heinz Dieterich, a German who has lived mostly in Latin America since 1970, caught the attention of Hugo Chávez, then president of Venezuela, with the slogan "Twenty-first Century Socialism." Dietrich's book on the topic features a picture of Karl Marx holding a laptop computer. Such a system would to do away with market pricing. Goods and services would be exchanged according to the labor time needed to produce them (Cockshott and Cottrell 1993). Marx's 19th century theories did not account for such 21st century concerns such as limitations of natural resources, pollution and global warming. But environmental factors are not handled well by market pricing, so that is an additional reason to go beyond the market.

No country has yet been persuaded to try this communist experiment, not even in Latin America where bad ideas sometimes go to die. At one time it seemed that Venezuela would forge the way. Venezuelan Defense Minister Raúl Isaías Baduel wrote a prologue to Dieterich's book in which he urged Venezuelan intellectuals to take up President Chávez's challenge to "invent the socialism of the twenty-first century." But it didn't happen. Baudel soon complained "where are the hundreds, perhaps thousands, of mathematicians, statisticians, economists, systems engineers, programmers, and information systems experts, committed to socialist ideology and with the will to change to a system different from capitalism, who will form the central planning team that will have the formidable and enormous mission of replacing nothing more and nothing less than the market and the businessmen?"(Baudel 2007).

Baudel went into the opposition and was eventually imprisoned on questionable corruption charges. It became apparent that Hugo Chávez had no intention of replacing capitalist markets with a Soviet-style *nomenklatura*. Cuba and North Korea, the only remaining countries with that kind of system, are both backward in computer technology and both seem more likely to gradually introduce market economics. There are, after all, a great many things wrong with the state socialist model other than a lack of computers.

The Chilean Experiment

A much more feasible possibility is that cybernetics could help to realize the dream of "third way" between capitalism and socialism. In 1970, newly the elected Chilean president Salvador Allende sought to build a democratic socialist system. One of the young visionaries in his administration, Fernando Flores, was familiar with the writings of British cybernetician Stanford Beer and invited him to consult with the Allende government (Medina 2011; Goertzel 2014). Beer was a successful business consultant who specialized in helping companies computerize their information and decision making systems. He was ideologically sympathetic to Allende's project and responded enthusiastically to Flores' invitation.

Stanford Beer was determined not to repeat the mistakes of the Soviet experience. He fully understood the technical limitations in Chile at the time. There were only approximately fifty computers in the country as a whole, and no network linking them together. Beer set up a network of telex machines and then then designed a modernistic control room where decision-makers could sit on swivel chairs and push buttons on their arm rests to display economic information on charts on the wall. There was no attempt to replace human brains or model the entire economy statistically. The project simply sought to collect time sensitive information and present it in a way that human managers and workers could use it to make better-informed decisions. Beer emphasized that information should be available to all levels of the system, so that workers could participate with managers in making decisions.

Proyecto Synco, as it came to be called, came to an abrupt end in September 1973 when the Chilean military overthrew the Allende government and brutally suppressed the Chilean left. But there was really no technical reason for ending *Proyecto Synco*; the new government could have used an economic information system. Today, information systems of the type Beer was trying to implement in Chile are maintained by all modern governments, and they do help economists and planners advance social democratic goals. Improvements in artificial intelligence and computer modeling can help to make this kind of governance more effective, minimizing crises and moderating inequalities.

End of the Beginning

Voluntary Communities and Worker Self-Management

Advances in technology may offer increased opportunities for innovation on the level of voluntary communities and worker controlled enterprises, if only by giving people more leisure time to pursue alternative lifestyles. There is a long tradition of this kind of socialist experimentation in the United States, but most of the enterprises and communities failed to pass the test of time (Berry 1992). Replacing market prices with labor certificates was tried in the nineteenth century by American anarchist Josiah Warren and his followers (Martin 1970). Warren believed that goods should be priced according to the hours of labor that it took the average workman to produce them. He was not just a theorist; he opened a retail store in Cincinnati in 1827 that followed his principles. Goods were sold for what he paid for them in dollars, plus a 4% to 7% markup to cover expenses. In addition to the dollar price, there was a charge for the time it took him to sell them, as noted from a large clock on the wall. The time was paid for with labor certificates that could be exchanged for labor by the purchaser. It made for quick, inexpensive shopping, and the store was quite popular.

Warren also helped to set up experimental communities in the towns of Utopia in Ohio and Modern Times (now Brentwood) in New York, where residents exchanged local goods and services with labor certificates, still using dollars to buy things from the outside. This worked reasonably well, better than many of the utopian communities set up on the communist principle of equal sharing based on need. But using labor certificates instead of money did not bring about a revolutionary change in human relationships as Warren hoped. Economic studies have shown that most consumer goods already sell for prices closely correlated to the amount of labor it takes to produce them (Brewster 2004).

Warren's activities were limited to the goods he could sell face-to-face in his store, and he was not able to measure values other than labor time. Modern computer networks create the possibility of more sophisticated exchanges incorporating dimensions of value other than labor time. The system of Offer Networks suggested by Ben Goertzel (in this volume) is similar to Josiah Warren's system of Labor Certificates, but an improvement in that it allows for incorporating dimensions of value other than labor. Computer networks also allow for exchanges on a global scale, while the nineteenth century anarchists were limited to face-to-face transactions. Offer networks, which base exchanges on complex computer models, provide little practical advantage over using money for relatively simple exchanges of standard goods or services. But they could be better for more sophisticated exchanges between companies or collectives. Very importantly, the system could be implemented on a trial basis by groups interested in it without suppressing the use of traditional money. This is also true of Bitcoin or the other alternative currencies that are already being used, although these are based on market principles, not on labor value or other non-market values (Odom, this volume).

The Israeli *kibbutzim* are the most extensive experiment with voluntary secular communal socialism (Gavron 2000; Levatian, Quarter and Oliver 1998; Zilbersheid 2008). Many of the *kibbutzim* were organized on the principle of "from each according to his ability, to each according to his need." After the Israeli economic crisis of 1985, however, they began a process they call "privatization" under which many *kibbutz* enterprises have been reorganized and function according to market principles. They have differential wage scales, and employ workers and managers from outside the *kibbutz* when needed. The residential community is separated organizationally from the productive enterprises. On some *kibbutzim* members still contribute their salaries to a common fund and receive a standard allowance, so "from each according to his ability, to each according to his need" continues to be is a lifestyle choice. But differential pay scales are used in the factories and those employees who do not choose to join the *kibbutz* keep their differential wages.

End of the Beginning

The failure of the *kibbutzim* to sustain socialist purity has been disappointing many in the founding generation. But the new arrangements offer a possibility of sustaining socialist lifestyles and values in the modern global economy. Many of the *kibbutz* enterprises are quite sophisticated in the use of modern technology, and can offer challenging careers to highly educated young people. As highly innovative and flexible voluntary communities, they are well suited to becoming part of what technological visionaries call the Global Brain.

The Global Brain and the Withering Away of the State

The promise of the "withering away of the state" was one of the most intriguing promises of the Marxian utopia, but it was inconsistent with the principle that the state would "regulate the general production". Today the goal of minimizing the state is advocated by libertarians who favor market economics (Hughes 2002; 2004). This philosophy is very popular among computer innovators and entrepreneurs at the forefront of the technological revolution. But leaving economic and social decisions to the market alone does not incorporate social and environmental values. To incorporate social values in a libertarian system Viktoras Veitas and David Weinbaum (in this volume) advocate replacing the hierarchical state with a fluid social order facilitated by distributed social governance. In this view, society will be regulated by voluntary, negotiated arrangements, very much in the anarchist tradition, with disputes settled through arbitration. With future advances in artificial intelligence these networks will not be limited to humans, but may include intelligent computers or robots with diverse degrees of autonomy.

No violent rupture of social institutions need take place to bring about such a system, no class of exploiters need be violently overthrown. Instead, the state will actually "wither away" gradually as state institutions are used less and less. The fundamental function of maintaining a currency, for example, can

End of the Beginning

be gradually replaced by non-state instruments such as the Bitcoin network which is already in existence.

The Global Brain model is less threatening than the typical science fiction vision of intelligent robots competing with humans, or the revival of central planning by a giant centralized computer network. Frances Heylighen (in this volume) argues that the Global Brain will be inherently linked to human activities just as the human brain is inherently linked to human bodies. It will not be a separate entity housed in a robotic body or a mainframe computer capable of an independent identity. Its functioning will depend on maintaining the welfare of humanity. He predicts that this Global Brain will bring with it four key values: *omniscience, omnipresence, omnipotence* and *omnibenevolence*. Let us look at each of these in turn.

Omnipresence. Today's Internet is well on the way to this goal, and further steps are easy to visualize. Cameras can continuously photograph all our streets so we won't have to listen to conflicting eyewitness accounts of confrontations between police and citizens. Small children, pets and senile elders can be fitted with chips allowing us to know where they are all the time. Our spouses won't have to wonder if we are really working late, they can check the cell phone network to know where we are. (Indeed, we can do this now if the "where is my lost cell phone" feature is activated.) As The network is already beginning to reach into our bodies to monitor our pulse, blood pressure, insulin levels and other indicators and perhaps even to active our pacemaker or pancreas implant if needed in a crisis.

The issue today is not whether omnipresence is possible but whether stopping it is possible. How much of our privacy will it be possible or desirable for us to preserve? Preserving areas of privacy will have to be a deliberate choice; if technology is allowed to progress unrestrained, privacy will be a thing of the past.

Omniscience. With all the data we have and the computer power to analyze it, our knowledge and understanding are

End of the Beginning

increasing rapidly if not exponentially. Of course, we are not yet omniscient; there are still major unknowns. We can't cure every disease or stop the aging process. We don't have lasting solutions for environmental or energy problems. But certainly our understanding is increasing rapidly. Perhaps the last thing will be learning how to prevent wars, terrorism and ethnic conflicts - problems that are rooted in our human nature.

Omnipotence. Even if the global brain is everywhere and knows everything anyone on the planet knows, that doesn't give it the power to solve all our problems. Technically, we know how to build a modern, humane society in Afghanistan or North Korea, but we clearly don't know how to solve the human conflicts there. We may not be able to defend the earth from a collision with a giant asteroid or prevent the extermination of species. Our ability to prevent nuclear proliferation and global warfare is certainly questionable. It isn't exactly clear how computer networks can help to solve many of these problems. Here is where we need more than just a "brain" as a nervous system connecting existing capabilities. We can certainly use any help that more advanced artificial intelligences may offer.

Omnibenevolence. The most pressing issues here are global poverty and inequality, and there is no guarantee that a global brain will make solving these problems a priority. The most advantaged regions and communities may seek to use their technological resources to wall themselves off from the world's poor. The future of abundance seems much further off in Africa than it does in Europe or North America. Writing from Addis Ababa, transhumanist Hruy Tsegay (this volume) makes an eloquent plea for technologists to prioritize the continent where the human species began its journey. Singularitarian Mingyu Huang (this volume) is more optimistic about prospects in China, where technological innovation has accompanied real progress in lessening poverty, despite the Great Firewall and other well-known limitations on internet freedom.

End of the Beginning

The Future of "Socialism"

If by "socialism" we mean a state socialist system of the Soviet, Cuban or North Korean type, the future seems bleak. This is not because of an inability of today's computers to keep track of the inputs and outputs, but because there is no acceptable way to set wages and incomes or incentivize innovation. The "from each according to his ability, to each according to his need" system couldn't work even in the highly motivated culture of the *kibbutzim*. Such a system needs to be managed by an all-powerful and all-knowing leader, or leadership group, that selflessly manages everything in the common interest. Human beings have not proved capable of providing such leadership. The only hope for such a system is to develop an Artificial General Intelligence "Nanny" to run society for us. Such an AGI would not have evolved in a competitive world, it would have been engineered. It would not have the psychological hang-ups that infect people with power. Such an AGI is perhaps thirty or forty years off, maybe more. At that point, humans could decide whether to put it to use. It wouldn't have to rule the whole world, it could be put into place in one country, or possible in a large voluntary community.

If by "socialism" we mean European-style social democracy, the prospects are certainly much better. These are already the most successful societies on earth, by almost any objective measure, and their system can function better with better information systems and more sophisticated econometric models provided by advanced artificial intelligence. While these countries have large capitalist sectors, there are also large state sectors in fields such as health, education, security, and environmental protection.

If by "socialism" we mean voluntary communities and enterprises organized along socialist lines, these may have a growing future as more people have the resources to choose how they want to live. Most of the hippie communes of the 1960s and 1970s were ephemeral, but some still exist such as The Farm

End of the Beginning

Community in Kentucky. Israeli *kibbutzim* have worked out mechanisms for such communities to co-exist with companies organized on a profit-making model, getting the benefits of both. The biggest problem of communal living, at least on the *kibbutzim*, seems to be lack of excitement and opportunities for young adults. Most Israelis seem to believe that *kibbutzim* are great places for children and the elderly.

Jeremy Rifkin (2014) predicts an "eclipse of capitalism" in the "near-zero marginal cost society." His argument fits certain industries, such as encyclopedia publishing, where Wikipedia has out-competed the commercial encyclopedias. "Near-zero marginal cost" means that once you have the infrastructure created it costs next to nothing to produce additional product. This is true for books, newspapers and magazines, movies and music recordings if they are published online. But the reporters and authors still need an income, as do musicians and actors and internet preachers. Rifkin, interestingly, charges market prices for his book, even in electronic form. So focusing on the marginal cost is really unrealistic, you need to look at the total cost.

The "near-zero marginal cost" argument doesn't apply to products such as food, housing, medicine, transportation and energy (other than solar or wind). It will be several decades, at least, until we get 3D printers such as the ones on Star Trek that can synthesize a hot meal or a medication or a suit of clothes on order. The most realistic futurist analysis (Friedman 2009) suggests that American capitalism will remain the dominant economic, technological and military force for perhaps the next century. If we accept this as the most likely reality, the best prospect for advancing socialist ideals will be in the communitarian and social democratic traditions.

Sociologist Erik Olin Wright (2010) and his group at the University of Wisconsin have been in the forefront in advocating realistic socialist alternatives within the framework of existing societies. Brazilian socialist Paul Singer (2000), who led the "economic solidarity" efforts in the Lula da Silva government, offers a similar vision. These writers do not advocate a revolutionary rupture with capitalism. They advocate building socialist alternatives in the spaces where they can work in competition with other organizational forms. In some cases, they can be triumphant, especially where marginal costs (the cost of producing one more item) are close to zero. Linux has competed successfully with Apple and Microsoft for many applications. Music publishing has been largely taken over by free downloads.

There are also successful worker-owned manufacturers including the Mondragon Cooperatives in Spain, King Arthur Flour in Vermont, Uniforja in Brazil, and many *kibbutz* associated industries in Israel. These are a very small proportion of any national economy, even in Israel, and they face strong competition. This situation is likely to continue for some time. As Paul Singer put it (Goertzel 2011: 173):

> *The socialist economy will probably suffer (for how long no one knows) competition with other modes of production. It will be permanently challenged to demonstrate its superiority in terms of self-realization of products and satisfaction of consumers. This leads to the conclusion that the struggle for socialism will never cease. If this is the price which socialists must pay to be democrats, I venture to say that it is not too much.*

While these socialist alternatives may struggle today, they keep alive visions and organizational frameworks that may become more important in the future. If the Singularity brings intelligent robots that can do the tedious work, many more people may choose to volunteer their time on projects such as Wikipedia and Linux or spend it on producing books or music or art to be distributed free on the Internet. They may even agree to turn much of the economic planning over to a super-intelligent artificial intelligence "Nanny" should such become available.

The most important division in the world today is between

the countries that compete successfully in the emerging Singularitarian economy and those at risk of being left behind. There is even the risk of a more fascist alternative, the engineering of a superior "race" of neo-humans through genetic engineering or cyborgian implants (Hughes 2002; Sandberg 1994). The Artilect War described by Hugo de Garis (in this volume) is one possibility. But none of these futures are pre-determined. If we can marshal the requisite creativity and flexibility, Singularitarian technologies are giving us a chance to build a utopian future for everyone.

REFERENCES

Armstrong, Stuart. 2014. Smarter Than Us: The Rise of Machine Intelligence. Machine Research Intelligence Institute. Amazon Digital Services.

Baudel, Raúl Isaís. 2007. "Why I Parted Ways With Chávez," New York Times, December 1.

Baum, Seth, Goertzel, Ben and Goertzel, Ted. 2011. "How Long Until Human-Level AI?: Results from an Expert Assessment," Technological Forecasting and Social Change, Vol 78, No 1, January 2011, pages 185-195.

Berry, Brian. 1992. America's Utopian Experiments: Communal Havens from Long-Wave Crisis. Dartmouth College Press.

Bostrom, Nick. 2014. Superintelligence: Paths, Dangers, Strategies. Oxford University Press.

Brewster, Len. 2004. "Review Essay on Towards a New Socialism," The Quarterly Journal of Austrian Economics, Vol 7, No 1, pp. 65-77. http://mm.mises.org/journals/qjae/pdf/qjae7_1_6.pdf.

Cockshott, Paul and Cottrell, Allin. 1993. Towards a New Socialism. London: Spokesman Books, 1993. http://ricardo.ecn.wfu.edu/~cottrell/socialism_book/.

Cockshott, Paul. 2014. "Cybernetic Paradigm of 21st Century Socialism," YouTube. https://www.youtube.com/watch?v=LtlZys7QOO4. Accompanying slides: https://leftforum.files.wordpress.com/2014/11/dublin.pdf. Accessed December 12, 2014.

Dieterich, Heinz. Hugo Chávez y el Socialismo del Siglo XXI, Segunda

Edición Revisada y Ampliada. Edición Digital, 2007, p. 2. http://www.rebelion.org/docs/55395.pdf.

Gavron, Daniel. 2000. The Kibbutz: Awakening from Utopia. Rowman & Littlefield.

Gerovitch, Slava. 2002. From Newspeak to Cyberspeak. Cambridge, Mass.: MIT Press.

Goertzel, Ted. 1992. Turncoats and True Believers. Buffalo NY: Prometheus.

Goertzel, Ted. 2014. "The Path to More General Artificial Intelligence," **Journal of Experimental & Theoretical Artificial Intelligence.** Volume 26, Issue 3, pp. 343-354.

Hanson, Robin. 2008. "Economics of the Singularity," IEEE Spectrum, Volume 45, Issue 6, pp. 45-50.

Hughes, James. 2004. Citizen Cyborg. Basic Books.

Hughes, James. 2002. The Politics of Transhumanhism. http://www.changesurfer.com/Acad/TranshumPolitics.htm

Kubrik, Stanley. 1968. 2001: A Space Odyssey (film). Warner Brothers.

Kurzweil, Ray. 2006. The Singularity is Near: When Humans Transcend Biology. Penguin Books.

Leviatan, Uriel, Quarter, Jack and Oliver, Hugh. 1998. Crisis in the Israeli Kibbutz: Meeting the Challenge of Changing Times. Praeger Publishers.

Martin, James. 1970. Man Against the State: The Expositors of Individualist Anarchism in America 1827-1908. Colorado Springs: Ralph Myles Publisher.

Medina, Eden. 2011. Cybernetic Revolutionaries: Technology and Politics in Allende's Chile. Cambridge, MA: MIT Press.

Müller, Vincent C. and Bostrom, Nick (forthcoming 2014), 'Future progress in artificial intelligence: A poll among experts', in Vincent C. Müller (ed.), Fundamental Issues of Artificial Intelligence (Synthese Library; Berlin: Springer).

Musk, Elon. 2014. Tesla boss Elon Musk warns artificial intelligence development is 'summoning the demon'. The Independent, November 18. http://www.independent.co.uk/life-style/gadgets-and-tech/news/tesla-boss-elon-musk-warns-artificial-intelligence-development-is-summoning-the-demon-9819760.html.

Sandberg, Anders. 1994. "The Memetics of Transhumanism: Or: How is the Memetic Health of Transhumanism?" http://aleph.se/Trans/Cultural/Memetics/trans_meme.html#politics.

Tsegaye, Hruy. 2015. Africa Today and the Shadow of the Coming Singularity. In Goertzel and Goertzel, eds., 2015.

Wright, Erik Olin. 2010. Envisioning Real Utopias. Verso.

Zilbersheid, Uri. 2011. The Israeli Kibbutz: From Utopia to Dystopia. http://libcom.org/library/israeli-kibbutz-utopia-dystopia-uri-zilbersheid.

End of the Beginning

Dialogue 19.1

Dialogue with Ben Goertzel, Hruy Tsegaye and Ted Goertzel

Ben: One point you don't mention is the very heavy role of state funding and state guidance in technology and science development in, for example, Singapore and Korea as well as mainland China. For example, Samsung was started via government initiatives and still works very closely with the government. Singapore's government directs biology and electronics research in very close cooperation with large and small firms. This sort of thing would be called "socialist" in modern US politics, yet apparently co-exists OK with democratic governance (though democracy of a more Asian, explicitly heavy-handed style than current US or Western European democracy). Apple is not government funded or orchestrated of course -- but Samsung is, and they sell more smartphones than Apple these days.

Ted: Yes, even in the United States the government took over ownership of General Motors for a time. China calls its system "communism with Chinese characteristics," but it might be called "communism with capitalist economics."

Ben: Well, China has gotten rid of guaranteed national health care recently, for example. So it's hard to argue they're following "to each according to his needs" or any kind of communist ideology anymore. Really they are forging their own new kind of state, with state-guided capitalist economics as one key aspect, and tight government control of information as another.

In any case, their new system is proving reasonably competitive in the world tech economy. And economically China is richly connected with the rest of the world, of course, even though they take pains to filter and sculpt the news their citizens see from

End of the Beginning

the outside world. Guangdong province is the world's manufacturing hub, but an awful lot of the products being built there are modeled on Western designs.

The #2 and #3 smartphone makers in the world, Samsung and Xiaomi, are Korean and Chinese respectively. And Xiaomi's smartphones, just like Samsung's, are powered by Android, a fork of the open source Linux operating system, adapted by Google. Actually Xiaomi recently hired Hugo Barra, who was Google's vice president in charge of managing the Android operating system. Samsung was formed with massive Korean state assistance; Xiaomi was formed in 2010 via a heterogeneous partnership of multiple investors including Temasek Holdings, a Singaporean government-owned investment vehicle, the Chinese venture capital funds IDG Capital and Qiming Venture Partners, and Western mobile processor developer Qualcomm. Qiming aggregates money from private investors with funds from various parts of the Chinese government. These companies, pushed forward by a combination of private and government interests, compete very effectively with Apple Computer which follows a more standard capitalist corporate model.

Ted: And Israel now markets itself as the "start-up nation" with a mixture of state enterprises, private enterprises and some modern high tech enterprises based on *kibbutzim*, all of which compete vigorously on a competitive global market. With the recent diplomatic recognition of Cuba by the United States, Cubans are looking forward to something along the Chinese model based on capitalist investments. The most important division in the world today is between the countries that compete successfully in the emerging Singularitarian economy and those at risk of being left behind.

Hruy: But before we assume the Singularity will lead the way to an abundant future, we should ask what kind of control do we have on this emerging super-intelligence? In the news recently we

641

End of the Beginning

saw, physics icon Stephen Hawking sound a warning about the potential dangers AI technology may bring. The same technological advances that power all these smartphones, and Stephen Hawking's synthesized voice, could eventually bring great dangers.

The AI Stephen Hawking uses now are tools that he controls. What he is concerned about is the emergence of AI as a conscious entity with its own "free will." Actually, I believe "free will" is an illusion of human consciousness. But if the emerging AGI is engineered to think for itself, the first thing it will do is make sure it is on top of the food chain. It may even decide that humanity is destructive to the planet and should be eliminated. How many mammals have our Homo sapiens ancestors eliminated from the earth just to insure their own survival?

Ted: How do you think we can avoid this disaster for humanity?

Hruy: The first thing is to make sure that our inventions will always stay tools that we, human beings, can control! Then control of these tools needs to be dispersed. I think the idea of "the global brain and the withering away of the state" is the best approach to the coming Singularity. States have long been an instrument of oppression used by the wealthy and the powerful to exploit the multitude. States are responsible for wars between nations, they are impediments to the development of a global fraternity - as when the British and Americans believe they are superior to Iraqis or Afghans. The death of 1300 noncombatant Palestinians is accepted by the western media while the death of one innocent Israeli is treated as an outrage.

Ted: I don't know if you can blame the state for these conflicts between ethnic groups. They are more fundamentally rooted in human nature. They were just as prevalent in tribal societies before states were developed. If the AI's are just tools, controlled by human groups, won't they just be used to exacerbate the conflicts?

Hruy: The value ascribed to the life of an individual American compared to that of an Afghani has nothing to do with evolutionary psychology or with our inherited tribalism. This is rooted in the notion of the supremacy of the American state. How on earth can we expect the world to be a level playing field when some states want to be the super state? Doesn't this fact in itself lead to the perpetuation of inequality? If we really, really do not want to repeat this mistake, the best way is to make sure the artificial intelligence of the future is a tool available to all, regardless of economic class, race, ethnicity, ideology and so on. But again, if this super-intelligence is not a tool, which is under our control, I am afraid our ability to predict the future is no more valid than a gambler's call on the next card!

Ben: I agree that our power to predict the future is very weak... but I would say that's not purely an artifact of our current situation. I think humanity has always been in a similar position. How accurately could early man predict the outcome of the invention of language? How accurately could pre-civilized man predict the outcome of the emergence of civilization? Not very in either case, to be sure... In stage after stage of our evolution, we have been hurtling forward bravely into an unknown future, full of "unknown unknowns" as that great philosopher Rumsfeld would have called them...

Of course, there's no reason to assume the odds are exactly 50-50 of negative versus positive outcomes; we do have *some* knowledge at present which is of nonzero use toward understanding and predicting the future. But even as we gain more and more knowledge, I'd say the level of uncertainty is going to remain pretty high -- that's just the way the cookie crumbles. That's intrinsic to the nature of radically transformative evolution.

End of the Beginning

When a simpler form of organization gives rise to a more complex form, the simpler form *cannot* foresee in any detail the specifics of the more complex form. It doesn't have the informational capacity to do so. In principle a simpler system could predict a few key points about a complex system with reasonable certainty (say, predict a particular advanced AI will stay in its box or will never kill people, even without being able to predict every theorem it will prove); but in practice that doesn't seem to be the situation we're facing, because the coupling of future complex systems with the broader universe which we know very poorly introduces a kind of complexity that has a viable probability of violating any of our assumptions.

You're right, Hruy, that it is consistent of Hawking to want to have AI tools to help him and others to do stuff, but to *not* want AI agents to act in the world with the same kind of autonomy that people have, but with greater intelligence. However, the reality is you can't have one without the other. The same trends in science and technology development that have created the AI speech assistant tools Hawking uses, are also leading inexorably toward the development of autonomous AI agents. I don't see how you'd keep the tool AIs but squelch the agent AIs, without imposing some kind of highly invasive, totalitarian fascist state to stop folks from using their computers and phones and so forth to develop agent AIs -- since even if tool AIs and other advanced tech made life very comfortable, it will simply be human nature to want to develop more and more advanced and cooler and cooler stuff ... and unlike nuclear weaponry or nanotech that requires special hardware for doing R&D, AGI development just requires commodity computers that everyone has already. So, depriving people of the ability to develop AGI, but keeping tool AI and the infrastructure needed to create it, would require a fascist or similar government to really clamp down on the particulars of what kind of programming people can experiment with. Whether this would be possible I'm not sure, but it would certainly require a fascist

644

surveillance infrastructure way beyond anything on the planet today. I guess that surveillance and enforcement system would be the world's best tool AI, in that case.

Hruy: The basis of the global brain should be to create a rational anarchy and a world where the 'we' means all humans and a world where the 'I' is the same as the 'we'. Hence the global brain should not take the Western or the democratic nations' model of value system in its beginning, rather it should be open to the existing systems of each nations so we can say, "to each nation its similar value system". Then by tightening the grip more on the fundamental rights, those that can be objectively addressed, the global brain can mold humanity into a rational anarchist society. The concept of the 'we' who had a better value system, and an advanced understanding of human rights and other blah blahs should never be tolerated in the creation of the global brain!

We always talk about Capitalism, Socialism and other isms, just as ideologies. As long as people live by these ideologies, the world will always polarize into 'us' and 'them'. What we should address is the value system, regardless of the ideology, and the question that is the root of all our biases; "whose ideology is it". I believe the Singularity should address our value system objectively, not divide things into "Western" vs. "non-Western" categories.

Ted: If we say "to each nation its own value system" do we not intervene when genocides occur as in Nazi Germany or Rwanda or Darfur or wherever? Do we say, what the Hutu and the Tutsi do is none of our business? Do we just stand by and watch the slaughter? Should we have ignored apartheid in South Africa, should we ignore the gulags in North Korea? This is an old dilemma, and your solution to it is a new one, Hruy. You seem to be saying that humans should butt out and let the global brain handle it. But if the global brain is going to be "objective" and

End of the Beginning

empowered to mold society to fit a higher ideal, you are really talking about an AI superpower that will take over from humans. And yet that you oppose; you want humans to use the AIs as tools.

Hruy: To each nation its own value system doesn't necessarily mean we look aside while people kill each other. To be honest in all the above scenarios you mentioned, with the exception of the one that involves the relatively whiter skins (the Jews), the West— with its 'golden values'— didn't do a squat about it! The UN or other forces of the West were only interested in getting their people, the ones with white skins, out of Rwanda or Darfur. The harsh truth is that those convoys had a space to extract the white peoples' dogs but they didn't take the black's children. This is how States create the unjust hierarchy over the human life! The same thing happened in Cambodia, Bosnia/Serbia, Guatemala and other parts of the world. Some of these genocides were even sponsored by states from the West. How many dictators were supported by the Reagan, Clinton and the two Bush's administrations while the genocides were under way? And one might ask doesn't the Bosnian/Serbia case involve white skin? Yeah it involved white skins but they were not friends of the super state! This shows that the hierarchy of state is not entirely a race issue. The innate cancer in the nature of states is the question of supremacy and dominance. Ted, the West had never bestowed its 'good value' to the rest of the world and yeah as you said it, the leaders of the West just watched those genocides rationalizing their unbelievably bad or insignificant reaction as peace keeping but not enforcing. These capable states have told us that if it is not in the interest of their people, then all the genocides around the globe are none of their business. Rwanda is the witness to this dark truth!

I oppose the idea of an 'AI overtaking the humans' not a 'tool AI' that can create an objective system. Another thing is, if we are in control, then the question will be who among us is in charge of the super-intelligence. There is always a risk of ending up with a

End of the Beginning

group ruled by certain 'uncomfortable ideas' that is in control of the super-intelligence as a tool. The best way to avoid this risk is to make this super-intelligence as a tool open for everyone regardless of all differences in economy, race, religion, political stand, ideology, and what not.

Ted: All these horrors are real, and there is a great deal published about them. But our task is to ask what artificial general intelligence can do to make things better. I'm not sure how we can keep the wealthiest and most technically advanced nations from dominating AGI, as nice as it would be to spread things out equally.

Ben: It seems to me that Hawking is not really seeing the options clearly (or if he is, he's for some reason choosing not to articulate his full understanding). It seems to me we have three options:
   a) Roll back to the pre-computer era in some way.
   b) A very sophisticated tool-AI based fascism, preventing science and technology from developing faster.
   c) Create the best AGIs we can, and live with the uncertainty that radically transformational development always brings with it

The first choice is Luddite, the second is fascist. Hawking, Elon Musk, Nick Bostrom and other current technologists and scientists who are opposed to AGI probably don't want to be Luddites or Fascists, so they have a difficult position to defend.

Hruy: I believe there is a fourth option: *We can Create the best Narrow AIs and live with the certainty that the system is relatively under our control and it will always remain as a tool with specific purpose.* Do we need a sentient AGI to solve problems? For me the answer is hell no! And to think that the AGI (the sentient with super intelligence) is our only way out is similar to the concept of the

End of the Beginning

deus ex-machina. We can have a narrow AI that can solve the problem of our mind capacity; all we need is to integrate it with our biological minds, this will make the super intelligent sentient unnecessary.

Ted: Are you sure you want that? If humans have made such a mess of things, won't giving them greater powers just enable them to mess things up worse? Why should we do that?

Hruy: I don't think advanced human beings would continue making such a mess. In the first place, it is our limited knowledge that created all this social, political, economic and ecological mayhem and destruction. I am an optimist and believe that humankind will evolve and become a more rational sentient. As today's world is better than that of the Dark Ages, it will be even better when the majority of the humankind is more intelligent than today. If you agree that AGI—because of its super intelligence— is the solution, then why not invest in upgrading ourselves? Why seek a savior when we can *be* one? For me, the only logically acceptable argument for the need for AGI is curiosity. Humans are curious and there cannot be a limit to what they are allowed to pursue. And the AGI is absolutely necessary for the sake of knowing that if we are capable of creating a super intelligent sentient or not.

Ben: I don't think your fourth option can work, unless it becomes my second option – fascism. In other words, I don't think we can proceed with making very sophisticated AI-based tools, without having *some* people, somewhere in the world, turn these tools into more-than-tools, and unleash a whole lot of unpredictable superintelligent complexity on the world. Unless there is some fascist control put in place to stop people from developing the tool AIs into agent AIs. Even if 99% of people in the world agreed not to develop tool AIs further into more

648

End of the Beginning

powerful, more autonomous agent AIs, the other 1% who disagreed could push forward with agent AIs and then we would either have a Singularity or need fascism to stop it.

It *is* an interesting question, though, whether autonomous sentience is required in order to achieve a highly effective problem-solving AI system. The most honest answer is that we just don't know, right now. Could we make a Nobel prize level automated scientist, which didn't have any autonomy, anything like will or self or the desire to make its own decisions and choices? The answer's not very clear to me...

But one thing is clear. If we want to make an AGI system that will go *beyond* the scope of human knowledge, that will discover amazing new things that the human brain can't conceive of -- then we will need to make this AGI autonomous and self-directed in some sense. Because we, as humans, cannot effectively direct a massively superhuman system, any more than a bacterium or a dog can effectively direct a human being.

Humans have an ambition that is hard to constrain. This is part of our nature, and it's why there are *already* so many researchers pushing to create advanced AGI systems that do have their own goals and motivations and autonomy. In order for autonomous AGI to happen it's not necessary for every human, or even every AI researcher to want it to happen. It's enough for a small subset, with access to a moderate degree of resources, to want it -- and this is already the case… Like I said, 99% of people may be happy with comfortable lives aided by Tool AIs. The other 1% will move actively to create autonomous AGIs, unless a fascist government or something similar forcibly stops them. I doubt the latter will happen but it's not impossible.

Ted: At least we realize that we don't have the definitive answers. The socialist visionaries of the nineteenth century never anticipated Stalin or Mao or Kim Jong-Il, although there were critics who warned them what was likely to happen. Based on this

End of the Beginning

human history, I am not in favor of concentrating power in a centralized all-powerful entity, even if it is a post-human AGI. I think we need to combine the AGI vision with the Global Brain vision. With both of these working together, we may be able to reach the utopia that Marx anticipated back in 1845.

Hruy: You think that the future for state socialism is bleak. Well I agree with that but I believe this conclusion is solely based on our perception of humanity as it is today. Our condition is ultimately governed by a value system that says economic gain is the absolute measure of a successful life. Capitalism preaches that motivation, entrepreneurship, and creativity are only possible if they are rewarded by economic gain which will put the individual on the top class, which in return fulfills the individual's self-actualization! The problem here is that we automatically assumed that capitalism must be better than socialism because the socialist states failed miserably and the future mustn't repeat the same mistakes made by the former socialist states whose reign was marked by bloody shenanigans.

Well, I say socialism failed in the socialist states because the attitude of the people in those states was framed by capitalism's value system and it has never been rewired. They still thought that they had to be dominant. I believe George Friedman's analysis [discussed in chapter two] is correct and American capitalism will remain the dominant economic, technological, and military force for at least half a century. Yet, we can change the impact of this with a technology that is directed towards a global attitude change. The reason for American-style capitalism's success lies in the fact that global attitudes are shaped by it all around the world. Yet, the best prospect for advancing socialist ideals, free of state supremacy, is to begin within the American population advocating the global brain theory and its socialist ideals. A gradual social anarchy from America is the most likely pathfinder to a relative equality for the rest of the world. Although this could be started

End of the Beginning

from other nations, an initiative from America, from the super state, will be less resisted.

Ben: A gradual social anarchy from America would be a lot of fun to see. The Occupy movement was pushing in this direction but seems to have sort of petered out, though it did obviously have an influence. But what sort of social movements will emerge as technology advances further and liberates more and more people from needing to work for a living, remains to be seen. I do think it's interesting to look at AGI as something emerging from human attitudes and in conjunction with an emerging global brain, rather than as some external Terminator or Savior type robot marching down the street and imposing its will on everybody. We can't really know what's going to happen or what impact our individual actions can have, but there's no cause to assume that we have no agency over the future, nor to assume that the outcome regarding AGI and society will be independent of nitty-gritty political and economic factors. Rational anarchy among humans and AGIs sounds pretty cool to me....

End of the Beginning

Chapter Twenty **Toward a Human-Friendly Post-Singularity World**

By Ben Goertzel

In an effort to think constructively about pre and post Singularity worlds, and the relationship between them, I'm going to ask two questions:

- **First Question:** What could we plausibly hope the world to be like after the Singularity? (Say, in 2100, to be concrete, supposing a Technological Singularity occurs sometime mid-21st century as Kurzweil, Vinge and others have suggested);

- **Second Question:** What could we feasibly do before the Singularity, to make this hopeful outcome more likely to eventuate?

Given the impossibility of predicting what will happen after a Singularity, or understanding the thoughts and actions of potential future superhuman minds, consideration of questions like these might seem almost hopeless. However, if one looks at these questions from a *human* perspective – in terms of the position and actions of human-like minds before and after the Singularity – things become a tad bit more tractable (though still massively speculative, to be sure). The future as a whole is pretty much wide

End of the Beginning

open, but humanity itself provides a strong set of constraints, so if we look at Singularity scenarios involving human minds and societies both before and afterwards, the scope of possibilities is not quite so unlimited.

So, synthesizing the two questions and focusing on the human perspective, we have the query: *Looking backward from the possible 2100 human world resultant from a positive mid-21st-century Singularity, what can we humans do now to make this sort of 2100 reality come about?*

One reasonably good response to this train of thought would be: To better approach both of these questions, what we should do is focus our attention on somehow arriving at a far better scientific understanding of the Singularity, and of the relation between the pre and post Singularity worlds.

Indeed, I strongly feel this sort of quest for understanding should be a high priority.

In parallel with this sort of study, however, I feel it's also valuable to use our current best understanding to address both the First and Second Questions directly.

Thinking about such questions is, obviously, no easy task – especially due to the incredible breadth of the spectrum of plausible futures for humanity, given the technological avenues currently unfolding. Most people have difficulty seriously envisioning futures very different from what they have known; but the trajectory of the world is not constrained by the average person's imagination. Just to be sure the reader knows where I'm coming from, I want to clarify that, in my view, seriously thinkable outcomes for the next century include

- The use of brain-computer interfacing to transform humanity into a "hive mind" in which everyone knows everybody else's thoughts, and dangerous further advances in technology are policed by universal thought sousveillance
- The utter annihilation of all life and intelligence on earth,

End of the Beginning

via the intentional or unintentional deployment of destructive technology based synthetic biology, nanotechnology, AI or some other category of tech yet uninvented
- The destruction of humanity and the flourishing of some form of engineered life or intelligence, such as digital, analog or quantum computing based artificially generally intelligent robots
- Encounter with alien life forms who have been in contact with humanity in subtle ways throughout history, but waiting for us to reach a certain level of intelligence or technological or societal advancement to make explicit contact
- Development of the capability for humans to "transcend" by massively increasing their intelligence and the scope of their consciousness, until they have become much larger minds, of which their previous human incarnation is a miniscule fraction

And perhaps more likely than any of these familiar-sounding science fictional possibilities, is some radically, transformatively new kind of outcome that no person has yet envisioned.

In this chapter I will explore the First and Second Questions raised above in a manner that fully confronts these radical possibilities and more. I will take an explicitly human-focused perspective, in the sense that I will focus on post-Singularity outcomes in which recognizably human minds, bodies and societies persist, alongside other options. I will explore the question of what a desirable *human-friendly post-Singularity world* might look like, and then look at the coupled question of what we might do pre-Singularity to bring a world of this nature about.

Implicit in this focus is that I think there is an interesting, plausible class of Singularity scenarios in which, even after radically transhuman technologies of various forms have been

End of the Beginning

created, human life still continues in a recognizable form – with advanced technologies playing a powerful role, yet with human minds, bodies and social groups continuing in a form relatively similar to what we have today.

A skeptic might well ask: Why would this sort of future world eventuate, given that humanity is only one possibility among very many potential forms of intelligence that might exist post-Singularity?

My answer to this is: A human-friendly post-Singularity future would eventuate if

1. a significant number of humans wanted to remain human, in spite of the emergence of alternatives like mind uploading or brain-computer-interfacing based transcension; and if as well
2. any post-human minds developed were inclined to respect the preferences of this human population.

Of course this sort of outcome is far from a given, but it seems at least plausible that, if advanced engineered minds were created and instructed with respect for human choices as one of their values, they would maintain this respect as they evolved and self-improved. Joel Pitt and I, in a 2012 article (Goertzel and Pitt, 2012), carefully considered a number of practical strategies for maximizing the odds of this sort of positive outcome, from the standpoint of the AGI developer; I also considered similar themes nearly a decade earlier in (Goertzel, 2004). Here I will not repeat that material, but will rather take as a premise of discussion that, one way or another, it is plausible for advanced AGI and other transformational technologies to co-exist with societies of relatively traditionally minded and embodied humans.

As I have discussed in my chapter on "The Future of Human Nature," I think that part of what may happen via a Singularity is a transformation of human nature into something fundamentally different (and less problematic). The industrial and computer eras have tended to bring out the egoistic and competitive aspects of human nature. The era of abundance that will come about if humans co-exist with transhuman technologies, may tend to bring out quite different aspects of humanity – our more creative and curious sides, perhaps. In this chapter I will focus more on the external than the inner perspective. But the ideas presented here are intended to be consilient with my thinking regarding the potential future transmogrification of human nature.

Some readers may recoil from the kind of radically positive futurist exploration I am pursuing here, considering it unrealistic "magical thinking." While I have no intention of apologizing for my uncertain yet emphatic optimism, I do want to make clear that, in my view, truly, purely utopic outcomes are unlikely to emerge, either in the context of posthuman minds, or in the context of human beings coexisting with advanced technologies. The future will doubtless hold numerous problems, most likely ones different from anything we foresee today. But this doesn't imply the infeasibility of radically better futures than what we have now. Societies on Earth today, in 2014, display a wide scope of desirability levels – e.g. North Korea or Congo versus San Francisco or Sweden. Future societies may be preferable to anything on Earth today, by a substantially greater margin than San Francisco or Sweden are currently preferable to North Korea. San Franciscans and Swedes still have their problems, and even have some problems that North Koreans don't; but they are, in a very real sense, less severe problems. The unattainability of utopia, and the plausibility of dark or radically non-human outcomes, should not inhibit us from perceiving the very real possibility of profoundly *better* human futures.

Readers who know me from my work on AGI may be surprised to see me writing an essay that focuses so heavily on the future of *humans*. After all, from some very meaningful perspectives, humans are not likely to be the most interesting thing going on post-Singularity. And for sure, one could also say a lot about the future of AGIs post-Singularity. As wonderful as the Singularity may be for humans, it seems almost certain that post-

End of the Beginning

Singularity AGIs will explore realities and mindspaces far beyond those accessible to any kind of human. These profoundly transhuman potentials are deeply important to me. But there is also a value in humanity. The kind of Singularity I'd like to see is one in which posthuman AGIs can explore tremendous new spaces of possibility, while at the same time post-Singularity humans embrace new possibilities of joy, discovery and creativity in a world of abundance. I see no good reason, at this point, why such a future should not be achievable (though, based on my current state of knowledge, I could not claim the feasibility of such an outcome anywhere near a certainty).

Envisioning Post-Singularity Humanity

What kind of world can we reasonably expect a post-Singularity human to live in?

The first key issue is the relationship between human beings and the posthuman AGI systems whose achievements will almost surely dominate the post-Singularity world.

According to various SF films, dystopic possibilities abound in this regard-- e.g. use of humans by AGIs as slave labor (*Terminator*) or batteries (*The Matrix*), etc. But in real life, this kind of outcome seems very unlikely. Once superhuman AGIs exist, they will not need human beings as slaves or batteries; they will be better off designing and building their own service robots and batteries, specific for the tasks at hand. And AGIs are unlikely to have sufficiently human-like motivational systems to take perverted pleasure from enslaving humans. Yes, there is the possibility of uploaded humans or detailed brain simulations, with powerful robotic bodies and humanlike emotions and motives, taking power over humanity in the manner of human dictators. However, it seems rather clear to me that this sort of closely human-like AI will ultimately pale in intelligence relative to AGI minds not constrained to be so similar to humans. Human uploads or brain simulations may well exist, but they are unlikely

End of the Beginning

to be the smartest or most powerful minds around, except perhaps for a relatively brief period. If uploads or human simulations happen to be the first digital minds developed, one of these early digital minds is likely to develop more powerful non-human AGIs shortly thereafter, in hopes of leveraging the intelligence of these nonhuman minds to achieve its goals.

The main possibilities for post-Singularity human/AGI relations seem to me (not in any particular order):

- The continued existence of humans alongside other more advanced intelligences (including uploaded/transcended human minds)
- The voluntary elimination of humanity, via humans uploading/transcending into some other form of being
- The annihilation of humanity, and the continuation of other forms of intelligence
- Here I will focus on the first possibility, which I consider the most desirable.
- If humans continue to exist post-Singularity, alongside advanced posthuman minds, what kind of lives might they lead? Indeed these lives will likely transcend our current psychological and social categories, much as our current lives transcend the categories that cavemen, ancient Chinese or medieval Europeans used to think about their worlds. But still we can say something. For instance, I would project that for post-Singularity humans:
- There will be no more "working for a living" – instead there will be a shift of human attention toward goals of joy, discovery and creation; toward social interaction, aesthetic creation and personal enjoyment
- People will no longer need to spend their lives within a common set of social rules and regulations. There will emerge of a diversity of (physical-world and/or virtual/augmented-world) micro-societies, with different forms of organization, culture and government

- Exchange between people will still exist, but will probably not be monetary in today's sense – we will see the emergence of alternative socioeconomic networks of various sorts
- Complex networks of human and software activity displaying "Global Brain" type patterns will play a significant role in influencing the activities of most humans
- Human-based governance may still persist but will not follow pre-Singularity tropes such as national boundaries
- Protection against deployment of advanced technologies in dangerous ways will be provided mainly via assistance by AGI and other advanced technologies, rather than by human-based security agencies and such
- Medical care, and mental health, will be non-issues. The major issues an individual will have regarding body and mind will be: How do you want to modify your body and mind -- and how do the other people (and/or AGIs) in the social networks you care about feel about these proposed changes?

Each of these aspects will represent a dramatic change from the world as we live in it today. However, each of these is also something that could plausibly emerge gradually from the human world as it exists today. The emergence of human-level AGI may well trigger a quite sudden Technological Singularity, as human-level AGI gives way to transhuman AGI, which then rapidly solves various technological and scientific issues that have vexed mere human-level minds: nanotech, femtotech, mind uploading, radical longevity via biohacking, nuclear fusion, and so forth. But from the point of view of humans who choose to remain human, the impact of this Singularity may "merely" be an accentuation and further acceleration of trends that have been underway for some time. Indeed, many of the aspects mentioned above are things that are already beginning to happen, little by little.

End of the Beginning

Now I will review a few of the above projected aspects of post-Singularity human life in a little more detail – focusing on how such aspects might emerge gradually from the world we now live in.

The Decline and Transformation of Work

I will begin with a relatively prosaic, near-term issue: the future of work, in the face of encroaching automation.

It is often argued that, eventually, advancing technology will eliminate work, since AIs and robots and other machines will be able to do every job better than people can. It is often counter-argued that the "end of work" has been forecast for a long time, yet although automation keeps advancing, people keep on working quite a lot. In fact technology seems to have increased rather than decreased peoples' workloads, in many sectors of developed-world society. Email and smartphones mean that many white-collar professionals are always on the job, rather than being able to get away from it all during evenings, weekends and holidays.

Both sides of the argument have merit.

The distinction of "basic needs" from less fundamental wants is not particularly clear or obvious, but the main points I want to make in this regard are pertinent for almost any reasonable drawing of the line. "Food, water, shelter, protection from the elements; and core human relationships like family and friendship" would be a good first approximation of "basic needs" – so let's take that as our starting point. Basic needs are those which almost every human society provides its members.

The production of the basic goods and services needed for fulfilling basic human needs is increasingly being automated, and it seems to me almost surely true that people are spending fewer and fewer hours each week involved in tasks helping with such production. It also seems extremely likely that trend is going to increase. In this sense, I feel confident that automation really is decreasing the average "basic life needs oriented workload".

End of the Beginning

Validating this intuition rigorously seems somewhat difficult given the complexity of the modern economy. Each hour of a given person's work contributes in some way to the profitability of the institution they work for; and this institution is generally involved in a complex network of exchange relations with other institutions. What this means is that each person's work has an indirect impact on a wide variety of activities, each of which contributes to basic human sustenance to a certain degree, along with providing other "non-basic" things to humans to a certain degree. Tracing through all these interdependencies in a modern economy would be a gargantuan task. Yet I am highly confident that if a detailed model of the modern world economy were made, one result would be that the number of the average person's weekly work-hours that contribute – directly or indirectly -- providing humans with basic needs, has been decreasing over time.

It's true that what is considered "basic needs" changes over time. When I discussed the matters in this chapter with one of my (adult) children, for example, they referred to basic needs as including "food, water, shelter and Internet." Is Internet really a basic need? What about heat in the winter? To me a heated home feels like a basic need, yet in many parts of modern China, people don't heat their homes even in the cold winter, and they are basically happy with this practice since it's what they are used to. Winter coats are worn indoors, and old people or mothers with babies carry abound small buckets of charcoal to keep warm. Life is reasonably happy. To many folks today, air conditioning in the hot summer feels like a basic need; yet in my childhood we didn't have air conditioning even when the temperature was in the 90s or 100s Fahrenheit for days on end, and it was perfectly fine. We drank a lot of iced lemonade, and tended to play outdoors in the evening when it was a bit less hot.

But the inflation of perceived basic needs doesn't really deflate the point I've made above. For any fixed understanding of "basic needs" – be it basic needs as perceived by the average

End of the Beginning

American of the 1970s, or the average Swede of 2001, or whatever – the same conclusion seems to me to hold: the number of the average person's weekly work-hours directly contributing to providing humans with basic needs has been decreasing over time. While I don't know how to rigorously prove this without doing a massive amount of data collection and analysis, this is an essay rather than a scientific study, and so I will take this point as provisionally true for the remainder of the discussion. The detailed economic data analyses presented in Piketty's recent work *Capital in the 21st Century* (2014) argue in this direction, though they mainly deal only with the US, UK, France and Germany over the last few centuries.

Counterbalancing this trend, though, we have two other trends. First, as just noted, our subjectively perceived "basic needs" are ongoingly inflating. Secondly, we become more and more accustomed to consuming goods and services that go beyond our perceived basic needs.

Consider for example a bath towel emblazoned with a picture of the Pixar character Buzz Lightyear. Few would consider a towel thus adorned a basic need. An average American today might consider a bath towel a basic need, even though air-drying after a bath or shower works perfectly well; but few would consider that a towel with a picture on it, as opposed to just a plain white towel, constitutes a basic need. Yet many people will pay more for the Buzz Lightyear towel than they would for a plain white towel. And needed, seeing Buzz's smiling face on the towel rack gives a momentary spark of pleasure, reminding one of the quest to go "To Infinity and Beyond!" But not that many people reflect on the work that goes into creating that Buzz Lightyear towel. There's the dye used to print the picture; the machines used to do the printing; the machines used to make the dye; the machines used to fix the printing and dyeing machines when they break ... the artists who drew and colored Buzz, the writers who came up with the character and story of Buzz, which is what makes the picture

End of the Beginning

meaningful and more than just a sketch of a funny-looking guy ... the janitors who maintain the buildings where all this work is done ... the accountants who help do the taxes of the people doing all this work ... and so forth. In order to have a Buzz Lightyear towel rather than a white towel, a host of different job roles are required. To pay all these people to do all these jobs, an increment is added to the price of a plain white towel, obtaining the price of the Buzz Lightyear towel. And this increment is nudged up a bit to pay some profit to the various companies involved in the whole process – meaning to pay some extra money to the shareholders of these companies (a group that likely includes pension funds for ordinary workers, as well as some highly wealthy individuals), and large salaries to the executives of these companies. The purchaser of the Buzz Lightyear towel is exchanging some of his time and effort in exchange for the aesthetic experience of having a well-known and charming picture on his towel, and in the process a little bit of his time and effort is being siphoned off to pay for pension funds and to line the pockets of the shareholding class.

Most likely the owner of the Buzz Lightyear towel doesn't reflect on the nature of the exchange he's making. Let's say the price increment from a white towel to a Buzz Lightyear towel is $5, and the towel's owner earns $15/hour after taxes. Then he's working 20 minutes for the Buzz-ness of the towel.

Would he work 20 extra minutes, if it were an explicit exchange of that 20 minutes of work for the picture of Buzz on the towel? Quite possibly. What difference does 20 minutes more work make, after all, in the context of a whole workweek?

On the other hand, suppose choosing a white towel instead of a Buzz towel, and making hundreds of other similar choices, could enable the towel owner to work only 15-20 hours a week instead of 40. Would he make that choice? Some people do make similar life choices, in today's world. But most people probably would not make that choice, for a couple different reasons.

For one thing, people value social status; and owning items

663

well beyond perceived basic needs helps boost social status, except in certain atypical subcultures.

For another thing, many people actually enjoy their work. Czikszentmihalyi (1990) has presented evidence that middle class people in developed countries are, on average, happier at work on average than at home or on vacation. Specifically, if you just ask people where they are happier, they will generally say they're happier at home or on vacation than at work. But if you sample people frequently throughout the day and ask how happy they are at that particular moment, they will on average report being happier at work.

Similarly, a recent study of Italy and Finland (Suojanen, 2012) finds that people in white collar jobs are generally happy with their work, and that "there are more happy people among those who have selected the importance of work as the first choice when looking for a job, than among those to whom an income is the most important aspect . People are more likely happy when the quality of work is high , that is when their job consists of creative and cognitive tasks and when they have a feeling of independence."

I'm certainly not claiming that a coal miner or McDonald's clerk typically derives profound satisfaction from their work. But the more unpleasant, less stimulating jobs in the world today also tend to be those that provide the least income. Most people working in dull and distasteful jobs are receiving salaries that don't pay for much beyond the perceived basic needs in their culture. Most people who are receiving salaries enabling them to spend a significant percentage of their salary on purely aesthetic or status oriented goods and services (like the Buzz Lightyear towel, or designer accessories, or vacations to resorts, or private cars to use instead of public transport, or a home swimming pool, etc.), are working in middle class jobs that deliver reasonable job satisfaction. The bottom line seems to be that people like working fairly well, when they're working on reasonably stimulating tasks in environments giving them some freedom of choice ... and they

End of the Beginning

actually like this more than the things they tend to do when not working for long periods.... Since people like consuming goods and services well beyond their perceived basic needs as well, the work + consumption pattern characterizing modern middle class society seems to suit modern human psychology adequately.

There is also a subtle question regarding how much of the time people spend at work is actually spent working. The amount of time spent surfing Facebook or otherwise engaged in electronic device based entertainment while on the job is fairly high these days. There seems little doubt that in many workplaces, people could work much shorter hours if they focused more exclusively on work while at the office – and if the institutions for which they worked allowed this tradeoff. But that sort of arrangement is generally not permitted in the current workplace; the only simple way to arrange it is to work as a part-time consultant rather than a full-time employee, which only works in certain job categories, and is an option bound up with other advantages and disadvantages. To phrase it informally and casually: Modern societies seem organized to have legions of people sitting at their desks at work surfing Facebook, rather than working shorter hours at the office and spending more time at home. But as any fan of Fred Flintstone or Dagwood Bumstead knows, some people spent a lot of time shirking off at work in earlier eras too, in different ways; so comparisons of actual work done then and now would be difficult to make in any rigorous way.

On the other hand, just because the current economic arrangement is reasonably well matched to contemporary mass psychology, doesn't mean it's a great way of doing things. It could just mean that society has settled into a sort of attractor state representing a local optimum of the fulfillment of relevant values. The fact that people are more satisfied at work than at home, even when their jobs aren't all that wonderful, may just mean that working at jobs all day year after year saps people of the creativity and motivation needed to do creative, fulfilling and exciting things

End of the Beginning

during their off-time. The average worker probably wouldn't make any amazing creative use of the 20 work minutes they'd save by buying a white towel instead of a Buzz Lightyear towel. But if they made a large number of similar decisions and reduced their work hours to 15 hours/week, then things could become quite different – they would have time and mental space to explore all sorts of different possibilities, and could become different sorts of people as a result.

As technology advances, perceived basic needs will continue to increase – but, I suggest, the amount of work hours needed to fulfill perceived basic needs will still continue to decrease, tending toward zero. Further, the number of work hours needed to acquire goods and services beyond basic needs will also decrease. However, the gratification people receive from doing creative, meaningful, stimulating work is unlikely to go away. People have evolved to take pleasure from doing meaningful work.

So what will happen when work really isn't that necessary for acquiring either the fulfillment of basic needs, or goods and services far beyond this level?

The answer depends as much on culture as on technology. One possibility is that people will adopt a culture in which creating things for others, and relating deeply to others socially and emotionally, are considered the most valuable things. Values of discovery and creation would rise to the fore (as I speculatively projected in my chapter on the "Future of Human Nature") -- with discovery and creation being not only scientific and artistic but also personal, emotional and relational. In this case we could see a society where "work" mainly consisted of relating to other people in various ways, and creating aesthetic works for others to enjoy. Even if robots have been created with the capability of composing amazing music, writing moving movies and stories, and holding scintillating or therapeutic conversations – even so, one can envision cultures in which listening to a human-composed song, or having a conversation with an actual human, has high perceived value. In this kind of culture, exchange mechanisms like Offer Networks (as I described in my chapter by that name) might prevail, with qualitative exchanges of social and aesthetic goods figuring centrally, and financial interchange aimed at guaranteeing perceived basic needs playing a more minor role.

A post-Singularity world in which people are able to pursue goals of social, intellectual, artistic and personal discovery and creation, free from the need to work to fulfill their basic needs, seems highly desirable to me from a human point of view. To work toward such a world, it seems the best course is to ensure that, as automation gradually relieves people of the need to work for a living, these people are enabled and encouraged to spend their lives joyfully and creatively.

The Era of Abundance

One useful term for understanding the kind of future I'm envisioning is "abundance." Peter Diamandis has extensively marketed the term "abundance" as a characterization of the forthcoming era of human history (Diamandis and Kotler, 2012), and I think this is savvy. In the terms I've sketched above, "abundance" may be conceived as a situation where:
- The amount of work the average human must do to sustain perceived basic human needs is minimal
- The amount of work the average human must do to achieve a significant perceived level of fulfillment beyond basic needs, is moderate at most
- Everyday human life is characterized by a mix of emotions, but predominantly by positive emotion

This sort of situation is radically different from many current or past human societies, in which long hours of hard work are/were required to sustain perceived basic human needs, with fulfillment beyond that point difficult to come by.

Some have argued that Stone Age societies were abundant in

this sense (Sahlins, 1974). Perceived basic needs were relatively minimal in these societies, compared to modern standards; hunting and gathering workdays may have been as short as a few hours per day, and work was considered fairly enjoyable. But clearly, this early human lifestyle was not any kind of utopic perfection. While modern diet and lifestyle related medical problems like heart disease and Alzheimer's were likely rare, infectious disease was a still a major issue, essentially unavoidable via any amount of extra work.

A society of scarcity, on the other hand, we may understand as one in which

- The average human must do a significant amount of work, including some work experienced as not particularly pleasant, to secure fulfillment of perceived basic human needs
- For the average human, achieving significantly more than perceived basic human needs requires significantly more work, including work not considered particularly pleasant
- Everyday human life is characterized by a mix of emotions, including a significant amount of negative emotion associated with the perceived need to regularly work at tasks that are only moderately satisfying to achieve fulfillment of perceived basic needs and additional desires

At least since the advent of civilization, up till the present day, humanity has lived in a society of scarcity.

These terms give a new way of formulating my suggestions above regarding the future of work. Through increasing automation, it seems to me, humanity is in the process of finally recovering from its era of scarcity and entering into a new age of abundance. This is happening gradually now, and will occur in full force with the Technological Singularity, when advanced AGI systems render the creation of material items and practical services at the level of everyday human desires, so inexpensive that they can be provided to all humans as basic "rights" and no longer need to be the subject of struggle or exchange.

An important point to understand in this regard is that, at this stage, scarcity is based on perceived resource limitations that are predominantly *intelligence limitations*. There is tremendous, known potential for creation of new forms of matter and experience using available mass-energy. We have barely begun to experiment with nanotech, let alone femtotech, attotech and so forth (Goertzel, 2012). One of the key reasons we experience as much resource scarcity as we do, is that we are simply not intelligent enough to effectively utilize the resources at our disposal.

As a rough analogy, consider a society that has scarce food and shelter, even though they have seeds and planks and nails at their disposal, because they lack the knowledge to plant the seeds or build a house with the planks and nails. And suppose that, instead of improving their knowledge or practical intelligence so they can figure out what to do with the seeds, planks and nails, they spend most of their time fighting with each other over scraps of available food and digging burrows in the ground ... and building status hierarchies about who has the deeper burrow.... But still, gradually, by dint of the efforts of a series of small groups of maverick individuals, more understanding of the potential utilization of the materials at hand emerges – and step by step, ultimately at an accelerating pace, they figure out how to grow plans and build houses, and their society is transformed from scarcity toward abundance.

The Power of Transparency

Alongside abundance, another key concept for understanding the class of human futures I'm envisioning here is "transparency."

There is a strong argument that, given the increasing capability of smaller and smaller groups of less and less educated people to cause more and more damage using advanced technology, privacy will be a decreasingly permissible option.

End of the Beginning

And the technology of observation is advancing at least as fast as – in fact I suspect slightly faster than – the technology of destruction. Given these factors, as David Brin (1998) has argued, either surveillance or sousveillance is likely to characterize the future.

Thus, it seems that one thing we face in the next decades is radical transparency due to trends in technology, and via practical necessity. Cultures will simply evolve under these pressures; and human subjectivity as well.

The loss of privacy may be viewed as a bad thing from some perspectives – as I discussed in my chapter on sousveillance, both sousveillance and surveillance both have significant potential to spawn cultures of conformity and squash creative diversity. But these difficulties may be manageable. Transparency also has power to build mutual trust ... and, it also has power to help break down the self-deceptive structures that constitute so much of our individual selves. With less ability or pressure to fool others, people would have much less propensity to fool themselves. It may be that radical transparency/sousveillance come about largely for reasons of convenience, general generational/ cultural shift, and political resistance to surveillance -- but then wind up having their biggest impact via their psychologically corrosive impact on self-deceptive self-structures that create/reinforce interpersonal boundaries...

If the risk of transparency-enforced conformity is dodged, radical transparency may end up serving to encourage a more adaptive and diverse culture. If the human world is both diverse and transparent, the illusion that everyone pretty much lives and thinks in the same way will be dispelled for each person via direct observation. It will become observably clear that there are many different ways of living in the world that provide reasonable satisfaction, and that the people pursuing alternative lifestyles are not (generally) ogres or psychopaths.

By the time post-Singularity humanity comes about, transparency may have conspired with other factors to transform the default human psychology into something quite different than it is today. As explored in my chapter on "The Future of Human Nature," the currently default egocentric, self-focused psychology may potentially give way to a different way of being, thinking and feeling, centered on more collective, free-flowing exploration of avenues to discovery, creation and joy.

The Potential Emergence of an AGI Nanny

Related to the surveillance/sousveillance issue is the potential emergence of what I've called an "AI Nanny" (Goertzel, 2011a), an artificial general intelligence system whose role is to observe what happens on the planet and be sure nothing too dangerous happens. Such an AGI system would need to be superhuman in intelligence in some ways, but wouldn't need to be a massively self-modifying AGI concerned with ever improving its own intelligence and creating new forms of mind and being. Rather, it could be a purpose-specific mind, with a particular form of intelligence oriented toward observation and safety, and a motivation to maintain itself in a state enabling itself to perform appropriate surveillance and security operations.

An AI Nanny could go along with either surveillance or sousveillance. Which eventuality occurs would likely depend on the relative rate of advancement of AGI versus observation technologies. If observation advances faster, the development of the AGI Nanny may occur in a sousveillant environment, the Nanny's advent thus occurring in a context where sousveillance is already part of human culture and unlikely to be rolled back. If AGI advances faster, an AGI Nanny may come about in an environment where privacy is still strongly valued, and may emerge as a non-human central surveillor. Note that even in a scenario combining sousveillance and an AGI Nanny, the inner workings of the AGI Nanny's mind may still be inscrutable to human observation, for the same reason that a dog or an ape would have trouble understanding the inner workings of a human mind even if gifted with mental telepathy.

Sousveillance could also provide an alternative to an AGI Nanny – perhaps the only viable alternative, consistent with the advance of technology and the survival and flourishing of humanity. Given the increasing viability of powerfully destructive asymmetric warfare and terrorism, surveillance and sousveillance are likely the only options; and it's unclear that any small group of humans is cognitively capable of providing the needed level of surveillance. The current state of the intelligence community suggests that the problem of gathering large amounts of data is easier than the follow-on problem of recognizing patterns in this data. The two most promising methods of solving this pattern recognition problem, critical for protecting humanity from asymmetric-warfare style technological threats, are advanced AGI technology and massive crowdsourcing. However:

- Advanced AGI technology is unlikely to remain controllable by a human elite, and if created for the purpose of observing the world and ensuring safety is likely to evolve into an autonomous or semi-autonomous AGI Nanny.
- Massive crowdsourcing of pattern recognition in global intelligence data seems likely infeasible to achieve except via the simple route of sousveillance. (One can conceive alternatives, such as providing intelligence data for humans to study in some sort of anonymized form; but in practice it seems that any such anonymization scheme would ultimately be game-able, most easily by those in charge of the anonymization, and would thus prove ineffective.)

The argument is complex and has deep unavoidable uncertainties, but it seems most likely that sousveillance and/or an AGI Nanny are going to be part of any future scenario involving survival of human society.

To work toward this kind of future now, pre-Singularity,

End of the Beginning

what we want to do is ensure that protection against dangerous technologies is done as much as possible via sousveillant methods, or via open source, broadly monitorable AI or other automated monitoring systems – and as little as possible by secretive, top-down government agencies or corporations. This may be challenging, to say the least; but it is likely in the best interest of humanity both pre and post Singularity.

Microstates

What implications might the emergence of abundance and transparency have for the emergence of new social and economic structures – both before and after Singularity?

It seems these tendencies may have the side-effect of permitting a form of socioeconomic organization that has long seemed appealing to many people (see e.g. Widmer, 1983), but has not come to pass due to practical reasons associated with the prevalence of scarcity.

I am speaking about the emergence of a network of micro-states or micro-societies, each one composed of people (and potentially non-human intelligences) who have agreed to a certain social contract, i.e. chosen to live with each other in a certain way, observing certain rules and providing each other certain respects, services and choices.

The appeal of an overall society consisting of a system of micro-societies is multiple, e.g.:

- Human beings evolved to live in small tribes, and arguably are psychologically most comfortable in this context. Even in a modern society, the number of people an average person closely associated with is at most a few hundred, the rough size of a prehistorical human tribe
- Human cultures and preference-sets have become quite diverse, so that different groups of people genuinely wish to live in very different ways. Some find polyamory obvious and appealing; others would rather not see it at

all, let alone participate. Some would like to experiment freely with psychedelics and cognitive enhancers; others feel the human mind should be kept "natural." Some would simply rather live around others who share their religious beliefs. The number of dimensions of diversity in human preferences is nearly endless, and listing them all would require a long book in itself.
- Currently, sociology, economic and politics are very far from being experimental sciences, or even robust observational sciences. It's very hard to draw solid conclusions about the workability of some social, economic or political mechanism from looking at existing or historical societies, because such situation is marked by so many specific, contingent factors. On the other hand, if there were a large variety of microstates, it would actually be possible to make a statistical study of what works and what doesn't, and use this to guide the development of future microstates.
- There are also multiple reasons why a network of microstates wouldn't be easy to initiate and maintain right now, in the current era of scarcity:
- Who would stop different microstates from invading each other to steal each other's resources?
- Who would deal with disputes between microstates regarding shared resources such as air and water and minerals?
- How would existing land and other relevant resources get allocated to various newly forming microstates. Who would handle the distribution of the resources from a recently dissolved microstate, to newly forming microstates?

Given a situation of scarcity, there seems no viable way to avoid conflicts between microstates, except for having some uber-state ruling over all the micro-states, regulating their activities and

End of the Beginning

resolving their disputes. Further, given the acute reaction of many humans to the situation of scarcity, the uber-state would have to be rather heavy-handed. But then who controls the uber-state? Is it a benevolent dictator, supported by an uncorrupted police force? That seems hard to come by, given human nature. A democracy? Then wouldn't a dynamic of micro-state alliances form, leading to complex politics and breaking the elegant idea of a system of largely independent micro-states?

The emergence of an AI Nanny would make a flexible society of mainly independent microstates feasible, as the AI Nanny would be the benevolent dictator handling conflict resolution. Sousveillance could potentially also serve the same purpose, though more complexly, and in a manner harder to foresee at present. In a sousveillant meta-society of microstates, a rogue microstate violating the inter-microstate meta-social contract could be immediately observed and dealt with by other microstates.

Abundance would make the emergence of microstates significantly easier, via decreasing the reasons for conflict. Molecular assemblers would diminish the desire to corrupt shared environmental resources, and reduce the differences in value between different patches of land.

The particulars are obviously difficult to foresee, but it seems plausible that some combination of an AI Nanny, sousveillance and abundance-generating technologies could make a meta-society of microstates much more feasible than is the case today.

To make the notion more concrete, imagine a contract among microstates roughly as follows:
- Within each microstate, things may be operated however the residents of that microstate may determine. Democracy, monarchy, anarchy, futarchy, whatsoever.
- Each microstate must allow its adult residents to leave whenever they wish to.
- If a microstate falls below some minimal number of residents per unit of land area, its allocated land area will

be reduced
- Any conflicts regarding overlapping resources will be observed and handled by the Central Mediator
- Microstates are required to handle children according to certain minimal standards of "human decency." Violations of these standards will be observed and handled by the Central Mediator
- Trade between microstates would be handled via global distributed networks in the vein of crypto-currencies, Offer Networks, etc.

Policies of the Central Mediator could be determined by an AI Nanny, or potentially by some form of direct democracy, using cryptography to ensure voter uniqueness and sousveillance in some form to prevent corruption. Of course, a crypto approach to voting would mean only microstates participating in the global Internet could vote, but one presumes this would be the vast majority case.

The handling of children is a tricky point of this scheme, since it's a case where commonplace human morality indicates that the "right to leave" is not an appropriate protection against the possibility of unacceptably cruel microstates. At least a large percentage of humans would currently agree that adult humans should have the right to choose whether to remain in a certain microstate or not. Brainwashing is frustrating to see, yet also difficult to differentiate from more ordinary group psychology in a rigorous way. However, giving a 2 year old child the right to leave its microstate at will seems potentially problematic; yet if this right is not provided, then some level of protection of children seems required if one doesn't want to sanction child rape, torture, etc. One approach would be to provide a global children's rights guarantee, but of a fairly minimal nature. For instance, approval of a restriction on a microstate's treatment of children could be determined via global vote, where 80% agreement among voters is required to pass a restriction.

End of the Beginning

The vision of a world of largely autonomous microstates does not contradict the vision of an emergent Global Brain, as outlined in several other chapters in this book (and references given therein). In the microstates vision, the Global Brain would be a self-organizing dynamical phenomenon that different individuals and microstates could choose to participate in to varying degrees. One suspects that most would choose to participate heavily, due to the tremendous benefits of coupling oneself with a highly, intelligent superorganism. But, human preferences are likely to remain diverse, so that there will be some folks who wish to live separately from any Global Brain that has emerged, and some that prefer to limit their degree of coupling in specific ways.

The above certainly would need massive fine-tuning to be implemented; these comments are intended as a rough, hopefully evocative sketch rather than as a draft constitution for a meta-society. But I hope it gets the idea across clearly. The advance of technology could provide a context in which human societies can flourish – a new era of innovation, experimentation and flexibility in social and economic organization and collective psychology. Rather than enforcing some sort of totalitarian conformity, phenomena like an AI Nanny or sousveillance could serve as regulatory mechanisms for an ever-changing meta-society of microstates, including those adhering to various historically traditional human lifestyles, and those experimenting with new ways of thinking and living, or new technologies for human enhancement.

This sort of future is far from guaranteed, of course – uncertainties abound. But there is no good reason for giving so much more attention to Terminator or 1984 type possibilities than to this type of option. Yes, we should be aware of the dark possibilities, and do our best to avoid them – but we should not become so fearful as to shut ourselves off to the brighter, more amazing and intriguing possible futures.

End of the Beginning

Seasteading

A society of diverse microstates is one possibility for a post-Singularity human society, but seems very far off from where we are today, with a world dominated by large nation-states and corporations, and slowly-changing restrictive laws that have little to do with the dynamic universe of attitudinal and technological advancement. As well as a future possibility, however, the microstate society sketched above may serve as a guide for thinking about possible nearer-future developments.

The seasteading movement[119] is a concrete attempt to bring about a network of microstates right now, without waiting for Singularity-level technologies. Instead, the idea is to build floating cities on the ocean, each one resident to a relatively small number of people who wish to live by their own laws. Being in international waters, these floating cities would be outside the scope of existing nation-states – though in practice they would have difficulty resisting the coercion of say, the US or British or Chinese navy. Given a sufficiently low cost technology for producing seasteads, this could become an influential factor in the evolution of society. Seasteads could provide a safe haven for the production of radical technologies that the laws of nation-states are not equipped to deal with, and a template for a future global microstate meta-society.

Whether via seasteading or other mechanisms, the advent of systems of microstates in the near term could help militate toward a microstate-friendly post-Singularity human society, which would in my view be a beneficial sort of outcome.

Analogues of Microstates in Virtual and Augmented Reality

One possible near-term evolution would be the emergence of

---

[119] http://seasteading.org

some sort of analogue of microstates, existing purely or mainly within a virtual or online world, rather than focused in physical reality.

One possible version of this would be a situation analogous to the world of microstates described above, but with each microstate an immersive 3D virtual reality world. In such a scenario, there would be a "base reality" – the everyday physical world with houses and apartments, food and water and robot factories and so forth – but people would spend most of their time jacked into a virtual reality, much as young people in First World countries now spend a lot of their time interfacing with computers, TVs, smartphones or tables. Virtual worlds could be organized into microworlds of various sizes, with each person belonging to a small number of different microworlds as a regular participant. Young children would presumably belong to microworlds sanctioned by their parents, including some that their parents regularly frequent. Central regulation of microworlds would likely be very loose, as with central regulation of the current Internet, with regulation of children's virtual interactions focused on prevention of interactions at the extreme end of the collectively perceived inappropriateness axis, such as child pornography.

Another possible version would centrally involve augmented reality. Read Daniel Suarez's excellent SF-thriller novel *Daemon* for an exploration of some possibilities in this regard (Suarez, 2009). Augmented reality tech will allow people to see physical reality according to their specifications, based on overlays and sensory transformations. For instance, informational labels could be added to items in the visual field, and ugly peoples' faces could be perceived as attractive faces via appropriate image processing. One possible outcome of this kind of technology would be each person living in their own customized, solipsistically augmented world. But I believe people have a desire to share experience with others, so that groups of people will adopt cores of shared augmented perception. So members of a family could each choose their own augmentations, but could also choose to see the shared augmentation associated with their family. In this sort of situation, the analogue of a microstate would be a group of people sharing a system of augmentations, hence opting to frequently see a common view of reality.

Virtual and augmented reality based social groups could also overlap. A group spending a lot of time together in the same virtual world, might also choose a shared view of the physical world or other virtual worlds. If such a group also preferentially exchanged with each other via an Offer Network or similar, then it would have the same basic effect as a microstate but without need for physical localization. In a situation of abundance where basic physical needs are fulfilled and virtual/augmented reality access provided free for all, and most human attention focuses on social interaction and aesthetic creation, such groups could have quite significant value, as they could end up being where most people spend most of their time. This would be a post-modern way of satisfying the human mind's desire to spend most of its time associating with a relatively small group of well-known individuals.

Perhaps more so than in the case of physical microstates, it is easy to see how a gradual transition toward this sort of virtual microstate meta-society might occur. As the need to work for a living decreased, people would end up spending more and more of their time on social or aesthetic pursuits in virtual or augmented realities. Social groups in these alternate realities would gradually become the most important aspect of peoples' lives. Reputations within these social groups would become critical, perhaps more important than traditional financial wealth. Technology companies are already looking toward this kind of future, as witnessed by Google's efforts at augmented reality with Google Glass, and Facebook's purchase of virtual reality firm Oculus.

End of the Beginning

The Emergence of a Global Brain

While not embracing fully the utopic Global Brain vision Francis Heylighen has espoused in his chapter for this volume, I do believe that some sort of emergent intelligence among large networks of humans and software programs is likely to arise, and to play a major role in future human affairs. I strongly suspect that, post-Singularity, posthuman AGIs will massively outpace any Global Brain substantially incorporating human intelligence, in terms of raw general intelligence, practical creativity and many other qualities. However, from a human perspective, once an AGI becomes sufficiently more intelligent than human beings, further increases in its intelligence quite likely become irrelevant; and a human-incorporating Global Brain might well be a more significant factor than posthuman AGIs, on an everyday-life basis.

The emergence of a powerful Global Brain would tie in with many of the other trends hypothesized above.

As working for a living fades, and people spend more of their time engaged in different sorts of collaborations and exchanges aimed at creating, discovering and experiencing new things, we will see the emergence of much more complex, nuanced and rapidly changing sociocultural networks. Something like Offer Networks, as described in the chapter by that name, might constitute one aspect of these emerging networks, but only one aspect; they will involve much more than formal exchanges. These networks will provide an ideal breeding ground for a Global Brain.

A tendency toward transparency in human affairs would ease and accelerate the emergence of a Global Brain, because the recognition of patterns in the activities of individual humans and groups of humans would become much easier – and as a corollary, the problem of figuring out how to modulate individual and group behavior toward various goals would become easier.

The patterns recognized as part of the Global Brain's cognition, would feed directly into an AI Nanny, helping it to better anticipate problematic developments as well as identifying problem situations already underway.

The emergence of microstates would enhance the intelligence of the Global Brain, by creating, in effect, a greater variety of processing modules for the Global Brain. Redundancy has some value for computation; but I have a strong feeling that the level of redundancy among human lives and thought-patterns induced by current levels of social conformity is sufficiently high as to constitute a massive waste of potential Global Brain computing power. A world full of microstates with different cultural thought-patterns would provide the Global Brain with a far greater variety of insights than current large-scale conformist societies; and quite likely the Global Brain would nudge the creation of new microstates according to its cognitive needs (e.g. making sure that contact occurs between groups of individuals who could profitably come together to form a new microstate of a particular sort).

Conclusion: Connecting Pre and Post Singularity Realities

Vernor Vinge foresaw the Technological Singularity as a point of maximal unknowability, beyond which we humans are simply unable to predict the outcome with any reasonable confidence. I think this is close to correct. But nevertheless, I think it behooves us to explore various plausible possibilities, both for the post-Singularity human world, and for pre-Singularity situations. Specifically, it seems important to consider the option that: While a Technology Singularity is going to constitute a radical discontinuity from some perspectives (e.g. certainly from the points of view of the transhuman minds that are created as part of the Singularity's emergence), from an everyday human life view it may end up appearing more as an additional surge of energy in a direction that has already been underway for a while.

Toward that end, I began this chapter with two questions:
- **First Question:** What could we plausibly hope the world to be like after the Singularity?
- **Second Question:** What could we feasibly do before the

End of the Beginning

Singularity, to make this hopeful outcome more likely to eventuate?

Regarding the First Question, I have argued that a perfectly plausible post-Singularity scenario is that advanced technologies bring on an era of abundance, rendering fulfillment of basic human needs and Internet access so inexpensive that they are provided for free, as is now the case with drinking water, public restrooms and parks and sidewalks some combination of Nanny AGI and sousveillance (and/or other methods yet unforeseen) defuses the dangers of unbalanced people deploying advanced technologies destructively though uploading or radical brain enhancement do become possible, many people choose to remain in an approximately traditional human form, out of aesthetic or moral taste or simple psychological inertia humans devote most of their energy to goals of creation and discovery; to social interaction, aesthetic and intellectual creation and personal enjoyment new socioeconomic networks emerge, in which like-minded individuals group into physical and/or virtual micro-societies; and mediation of exchange and adaptation of reputation are carried out via sophisticated online networks and algorithms. "Global Brain" type patterns in these networks will play a significant role in influencing the activities of most humans

Estimating the *probability* that this kind of reality comes about, instead of some other sort of scenario, seems extremely difficult given current knowledge and tools. But it seems important to appreciate the plausibility of this type of outcome, and to think about the intermediate situations that may be most likely to lead to such an outcome. Because I suspect that, of all the pre- and post-Singularity outcomes that is reasonably plausible to occur, this sort is the most generally desirable from the perspective of commonsense human values. While this sort of world would certainly not please every human being (no world could do that), I believe it would be fairly acceptable to most people, were it to

End of the Beginning

actually eventuate. My own goal, as a scientist, technologist and futurist, is to enable radical developments like indefinitely self-improving AGI and mind uploading, but also to enable a continuation of human society along the rough lines described here – at least until I hear of something better and at least equally plausible-seeming!

If one provisionally accepts this sort of answer to the First Question posed at the start of this essay, then the Second Question assumes the particular form: What kind of pre-Singularity path is most likely to lead to this sort of broadly positive post-Singularity world?   While this is of course impossible to answer with any kind of solidity, a commonsensical answer – already hinted above -- seems to be that one would want a path in which:

- Those who choose to leverage advanced technology for transhumanist purposes, and those who choose not to (or to do so only in specifically limited ways), peacefully coexist with a reasonable level of mutual understanding
- The potential dangers of advanced technologies are managed using sousveillance and distributed, open-source, auditable AI methods rather than by secretive, centralized authorities
- As automation reduces the need of people to work for a living, government socioeconomic policy and general culture evolve in a manner allowing and encouraging people to spend their time pursuing goals of creation and discovery
- Online software gradually develops, oriented toward enabling people to connect in complex networks and communities, oriented toward discovery and creation in all its humanly meaningful forms
- Increasing options become available for individuals to form customized communities, enabling more and more flexible exploration of social and psychological possibilities (an important avenue for discovery and creation) –

684

End of the Beginning

perhaps via seasteading, but also, perhaps more importantly, via shifts in culture enabled and nudged by advances in technology

A pre-Singularity world of this nature would intuitively seem most likely to lead to a post-Singularity world displaying similar properties. Working toward this kind of world would therefore seem an important goal, both because it would provide a desirable situation for the preponderance of people in the world, and because it would plausibly pave the way for a broadly positive Singularity.

These are exciting conclusions, but of course remain extremely speculative. And therefore, as emphasized above, the same time as working in these positive directions, it is also very important to better understand the relationship between pre-Singularity and post-Singularity realities, insofar as this is possible, so as to improve our currently very crude understanding. As we learn more about all this, due to technology unfolding and due to ongoing deepening of our understanding, our notions of the best pre-Singularity course are bound to change, perhaps dramatically.

The main point I want to make here lies not in the particular hypotheses presented, but rather in the overall methodology of working from the future backwards, as well as from the present forwards, as applied to post-Singularity human society. To the extent that humanity as a particular variety of intelligent system is viewed as valuable, it's important to look at what may be possible for future *humans*, and then look at what we may do now to make that happen. Extrapolating present trends into the future is useful but has its limitations, due to the potential for radical shifts due to technological innovations or social network dynamics, as well as due to the potential for humanity to buck its own prior trends due to psychological complexities informally summarizable as "sheer force of will." Looking backwards from the potential future as well as forwards from the actual present, seems the best way to bring about a positive future for human beings as well as for the posthuman minds that we will create.

REFERENCES
Brin, David (1998). The Transparent Society. Perseus.
Csikszentmihalyi, Mihaly (1990). Flow: The Psychology of Optimal Experience. Harper and Row.
Diamandis, Peter and Steven Kotler (2012). Abundance: The Future is Better Than You Think. Simon and Schuster.
Goertzel, Ben (2004). Encouraging a Positive Transcension. Dynamical Psychology,
http://www.goertzel.org/dynapsyc/2004/PositiveTranscension.htm
Goertzel, Ben (2011). There's Plenty More Room at the Bottom: Beyond Nanotech to Femtotech. H+ Magazine,
http://hplusmagazine.com/2011/01/10/theres-plenty-more-room-bottom-beyond-nanotech-femtotech/
Goertzel, Ben (2011a). Does Humanity Need an AI Nanny. H+ Magazine,
http://hplusmagazine.com/2011/08/17/does-humanity-need-an-ai-nanny/
Goertzel, Ben and Joel Pitt (2012). Nine Ways to Bias Open-Source AGI Toward Friendliness. Journal of Evolution and Technology.
Piketty, Thomas (2014). Capital in the 21st Century. Belknap press.
Sahlins, Marshall (1974). Stone Age Economics. Aldine Transaction.
Suarez, Daniel (2009). Daemon. Signet.
Suojanen, Ilana (2012). WORK FOR YOUR HAPPINESS: Theoretical and empirical study defining and measuring happiness at work, thesis at University of Turku, Dept. of Education.
Widmer, Hans (1983). bolo'bolo. Autonomedia.

End of the Beginning

Dialogue 20.1

With Hruy Tsegaye, Clément Vidal, Aaron Nitzkin[120] and Ben Goertzel

Hruy: Ben, I liked the article first for its tone. Unlike many articles that predict the future of humanity, this one isn't written in a tone of bitterness or hatred towards what we are today. Many of the other articles on similar topics are written with a tone of desperation; there is always an implicit suggestion that there is no redemption for humanity. And most importantly this article doesn't focus on terminating humanity as opposed to the others which openly depicts disgust towards humanity and how the singularity will terminate it for once and all.

Yet, I had one comment which your article should reflect explicitly. In the history of our species, as Homo Sapiens, there have been three major revolutions. The cognitive revolution which took place some 70,000 years ago, then the agricultural revolution around 17,000 years ago and finally the scientific revolution which happened 500 years ago. So far, these are the most important revolutions that shaped humanity significantly. (I know we could mention some other but these are the best so far).

Now, what I want to say is the Singularity should be conceived as the fourth revolution in our history. I don't want it to be perceived as an extension of the scientific revolution.

Why?

The three revolutions (along with the other minor ones) share one Achilles heel, which had been a fatal weakness in our intelligence. They all are trial and error processes. And if the Singularity pursues the same path; the trial and error way of solving problems, it will be nothing but a fancy word for hi-tech

---

[120] *Aaron Nitzkin is a cognitive linguist, and Ben's collaborator on OpenCog R&D in Hong Kong*

End of the Beginning

and the same insanity and the same futile self-recreation prevails! !

In my opinion, the problem with scientific experiment lies in OBSERVATION. There is no such thing as observation without selection. As Karl Popper observed:

> *The belief that science proceeds from observation to theory is still so widely and so firmly held that my denial of it is often met with incredulity. I have even been suspected of being insincere -- of denying what nobody in his senses would doubt.*

The singularity should, first and most, be spent on understanding the self. So far, the trial and error method has been our fundamental way of solving problems but it will be a shame to waste the singularity on such mediocre intelligence! The Singularity, before it is spoiled by a new group of 'accurate self-recreates', should be utilized to understand what the self is!

This time, because of the singularity and the Super Artificial Intelligence, Observation will not be selective and all the nooks and crannies of humanity can be observed.

The root has to be understood beyond any doubt! If this thing [the absolute understanding of what humanity is] can be done, the singularity is the only thing that can do it. Hence, the singularity should not hasten to recreate the self just because the technology allows it. Rather the Singularity should refrain from recreating and march to the ultimate scientific quest, understanding what we truly are.

I know, your article might implicitly suggest this taking 'understanding of what we are' as a given. But I just felt it should be stated openly. I also think that the singularity should be described as the forth revolution in our history which will not be based on trial and error.

Recreating the accurate self will help make more sense of your subsequent argument that the singularity first and foremost is accessed to understand and OBSERVE what we truly are.

End of the Beginning

[OBSERVE in its grand meaning]

Ben: One interesting thing about the notion of "self-understanding" is that, if human beings really understands themselves, then this will cause them to transcend the typical "self" that ordinary human beings have. In other words, the feeling and notion of a unified self that most humans have, is largely an illusion. The reality is that we are lot less unified and a lot more multifarious than we tend to think -- a person would generally be better modeled as a set of clusters of behavior-patterns than as a unified entity. Also, the reality is that we are a lot more tightly coupled to our external environment and to other people than we tend to realize (as noted by Andy Clark and others in the "extended mind hypothesis")...

So, fuller self-awareness on the part of human beings, would tend to cause a revolution in the actual structure and dynamics of human minds and groups. And I do think this will be part of the Singularity....

So, suppose we envision the future as comprising some superhuman entities (AGIs, cyborgs, uploads and what-not) that radically transcend the current human mode of existence, but also comprising a population of humans who choose to remain in fairly traditional human form. There still remain many questions and one of these is just how traditional would these "remaining traditional humans" be. If sousveillance, brain scanning, neurofeedback and/or other more traditional methods are used to help people become more self-aware and hence transcend the currently typical egocentric modes of being, then the "remaining traditional humans" may actually be a lot more self-aware, self-actualized and truly fulfilled than most of the humans on Earth today...

I would like to have a copy of myself among these highly actualized future humans -- as well as some transhuman descendants of myself enjoying radically transhuman existences, such as uploads of me fusing their minds with superhuman AGIs and exploring other dimensions. It doesn't have to be either/or, at least not so far as I can tell from my current highly limited human perspective....

Hruy: True to that -- but at least the Singularity with all its advantages can be utilized to model the perfect (relative to what we have now) environment, which in return will shape our self-understanding since "we are a lot more tightly coupled to our external environment and to other people"

The other interesting thing is about diversity. Diversity as it is, is more harmful than its use in our existence as a social animal. If as you said some humans continue to be traditional humans in the post-singularity era, that would most certainly create imbalance, which would result in chaos. If we take nature as the great guru, survival depends on dominance or co-dependence. The question will be; will the traditional humans and the upgraded ones in the post-singularity find an answer to these two elements of existence?

Dominance or Co-dependency, which will decide the fate of our future?

Again, the singularity, instead of jumping in to the trial and error path, could be utilized to solve this riddle. We have been here for at least 100,000 years (as modern Homo sapiens), so why the haste now?

If not, there is no use to speculate what the future will look like; the result of the trial and error phase will decide what will happen.

Ben: Well the path to the Singularity, right now, seems to involve a political/sociological struggle between dominance (big corporations, with government cooperation, taking increasing wealth and power) and co-dependence (peer-to-peer heterogeneous networks like the Internet, open-source software, etc. etc.).... The two factors you cite are working together in quite

End of the Beginning

complex ways -- much as has always happened in nature, I suppose, but with lots of new twists...

But in the post-Singularity future I foresee for *humanity*, co-dependence would reign.... What strange balance of dominance and co-dependence will exist among transhuman superintelligences, I have no idea -- they may end up finding both concepts pretty weird, archaic and limiting...

Clément: In your chapter, you explore a post-singularity future with humans, claiming that exploring a post-singularity without humans is hopeless. I disagree. Of course, it's much harder, but couldn't evolutionary theory and complexity sciences (cybernetics, systems theory) be a better starting point? To me, this approach of assuming humanity right from the start looks like a strong streetlight effect **(http://en.wikipedia.org/wiki/Streetlight_effect)**.

Ben: Well, "hopeless" is too strong. But there seems a very very wide scope of possible post-human minds -- and there is even a very very wide scope within the limitation of current human understanding and imagination.... Surveying all the humanly comprehensible possibilities in a meaningful way would be a huge undertaking; and then I still have a suspicion that what will really happen will lie outside the scope of the humanly comprehensible possibilities...

Clément: Yes, you explain this well, and I think you are very aware that it is a bias. From a human point of view, it of course makes most sense to explore the future of humans. On the other hand, making this criticism of you is unfair, since you are probably much more open minded to strange futures than many other futurists.

Ben: I love thinking about strange posthuman futures, as you

End of the Beginning

know, yeah... I know you've written some great stuff about the possibility of intelligent stars, for example. That's an interesting notion and not at all implausible.

Clément: Thanks, but I was not specifically thinking about starivores (intelligent stars), just pointing a bias in the methodological approach.

Ben: Anyway, though, starivores and other related possibilities are fascinating potentials to investigate -- but I have a suspicion, personally, that strongly posthuman intelligences will discover aspects of existence beyond the physical universe as we know it, so that intelligent stars, even if they exist, will barely scratch the surface of the vistas opened up...

Clément: What is it? The spiritual world? :)

Ben: Take some DMT and we'll discuss afterwards...! Seriously: spiritual practices could possibly give a peek into aspects of reality beyond what we can access in our ordinary states of consciousness. But there seems little doubt that transhuman intelligences will be able to go way beyond what even the wildest spiritual or psychedelic seekers have imagined. Just as a bacteria or a cockroach doesn't have the cognitive scope to envision the various possibilities of human civilization or psychology....

So yeah, I strongly suspect the most amazing and exciting aspects of the post-Singularity world won't involve humans in any significant way. But yet, from my current perspective as a human, the persistence and flourishing of humanity is also of interest to me, and that was the perspective from which this particular chapter was written...

Personally I would currently like to see a copy of me persist in human-like form, and other copies of me transcend into all sorts of amazing modes of being and thinking that I cannot now

End of the Beginning

comprehend...

Anyway, I know the goal of the chapter is pretty limited in the scope of all the possibilities the future may hold; I just specifically wanted to explore that limited scope, due to its relevance to those human beings who have an interest in seeing human society persist after a Singularity in something like its legacy form....

Clément: Yes, it makes perfect sense...

Aaron: It is difficult to write about what one knows one cannot know, and therefore too easy to criticize those who do. Future prediction, particularly on the specific time scale attempted in this book, is the most reliably error-ridden variety of such writing, and I admire the courage of those who -- like you and your father, Ben -- are willing to publicly stand behind speculations they know will, to a large degree, be proven far off the mark. I do not say this because I believe that the predictions in this book are ill-considered or largely wrong—quite on the contrary; however I will point to some of the potential weaknesses in their rhetoric, simply because there is more value in a diversity of perspectives. My criticism of your vision as presented in this chapter revolves around two main ideas:
1. The 'singularity' will be more fragmented, heterogeneous, and graded in both space and time than is implied by singularitarian rhetoric.
2. In terms of outcomes, culture will predominate over technological potential in the short term, by which I mean approximately until the end of the century. Technology will wreak deep changes in culture and human nature eventually, but in the short term we will see new technology being ill-used according to some cultural ailments we are already familiar with and others which we had not perhaps been greatly noticed before the technology made it possible for them to really flower.

End of the Beginning

Together, the main point of these two criticisms is simply that your predictions are unreasonably, although admirably, utopian, especially in the short term, and especially in reference to the human race as a whole, as opposed to the miniscule proportion of humanity that is actively concerned with these kinds of thinking and technologies, and who will be the ones most to benefit from them at first. I do not think you fail to recognize this, but rather choose to de-emphasize it in favor of dwelling on the benefits with the greatest hope for a rapid transition to a better condition for humanity.

The idea of 'the singularity' is a striking metaphor that appeals especially to people interested in math and science; the sheer extremity of it is inspiring and therefore probably wins many adherents in the same spirit that cult followers eat-up 'end-of-days' beliefs. However, it is, in its 'wholescale overnight change' version almost surely incorrect. Energy use and information processing will not continue increasing at the same geometric rate until a wholesale overnight change occurs because neither our social systems nor the physical infrastructures of our societies can change at nearly the rate of information increase. The progress of technology in changing the world is mediated by both of these slow adapting and not infinitely flexible systems. One question we cannot answer yet is to what degree our infrastructures and social structures will respond to overwhelming fluxes of energy and information by falling apart completely and to what degree they will simply become somewhat chaotic before finding new stable forms. I think we can safely conjecture they we will see some of both. In fact we already have. The demise of the Soviet Union and the conversion of communist China into the world's largest more-or-less-free market, by adopting the 'one country, multiple systems' policy, are examples of each (falling apart completely and adapting through chaos)—examples which one might regard as early pre-echoes of the singularity, as both processes were forced

End of the Beginning

by the same trends expected to bring about the Singularity, especially the increasing flux of information.

I strongly suspect that it would be fair to refer to many of the socio-political developments of recent years as precursors of the Singularity—the rise of internationally aggressive fundamentalist Islam seems like one; although I realize this is only part of the story, it can certainly be considered a response (to a significant degree) to the growing threat of freely spreading information and ideas to the belief systems of the people involved, as well as the growing availability of funds and international communications networks. Meanwhile on the local scale we are also seeing countless current chaotic consequences of the trends that promise to bring singularity--the rapid boom and bust of new-technology based businesses which has at times during the last decade seemed to threaten the stability of national economies, the sudden plummeting of literacy skills due to near overnight whole-scale replacement of reading by YouTube, the growing danger of poorly understood genetic modifications to staple foods, the Y2K bug, the social and legal issues being raised on a daily basis by the ubiquity of surveillance and the ease of hacking it . . .

Obviously, a far ranging discussion of how current socio-political trends foreshadow the Singularity would require tremendously more data, analysis, and expert knowledge than I wish to deal with here, but I hope to have established the point – that the singularity will bring, and in fact already has brought, a significant amount of social chaos, including bloodshed, negative impacts on mainstream lifestyles, and even reactionary movements against the opening-of-minds that it promises.

One may hope that fundamentalist terrorism, the collapse of ineffective political / economic systems, although painful, may be considered the heralds of a new and better age, however there are deeper and more insidious possibilities in the heterogeny of the singularity process. Undoubtedly, some people in some places are going to acquire a tremendous amount of power and resources

End of the Beginning

very quickly – more power perhaps than has ever been available to human beings before. Meanwhile, undoubtedly, some people, in other places, will acquire little to nothing for quite some time. Although it is reasonable to expect that the provision of basic needs, education, and medical care will continue its slow spread to disadvantaged populations, I do not agree that AGIs will take care of the disadvantaged in the short term.

First of all, the AGI's will be under the control of either people seeking profit or power for themselves, or under the power of people who themselves lack the power to change the socio-economic systems. It will not benefit humanity very quickly if the human beings who own the AGIs do not understand or care about the solutions they offer to human suffering, or if they do understand and care but do not have the ability to put solutions into action, as will surely be the case. To extend an analogy offered in your chapter, Ben -- humanity has not only long had the material resources (the planks and nails) to solve many of its problems; we have also already long had the knowledge and blueprints for building the house; we already know ways of generating near infinite supplies of clean energy and food; we have not implemented these solutions because the people with the power to do so choose not to, and those who would do so typically have no power. I don't see why you expect this to change in the immediate wake of Singularity.

But let's put aside the top-down perspective on the changes expected to ensue from Singularity and turn to how the new technologies will directly interact with the lives of individuals. Here I am more sympathetic with your predictions; the Singularity will create oodles of glorious opportunities for those with enough money, time, and sense to take advantage of them, and I will even admit that money will probably not be a major issue for participation in most of it, at least if one lives in a nation with sufficient infrastructure and political will to participate, which will rapidly include everyone because doing so will benefit political

End of the Beginning

and commercial leaders as well as anyone else.

So, I agree about the ubiquity of these opportunities for human development. Just as the information revolution of the past 60 years has made it possible for the majority of humans on the planet to learn anything, communicate with anyone, and therefore to improve themselves, and their situations given enough initiative and intelligence. And yet the number one use of the world wide web is the consumption of pornography.

The internet has also given tremendous new opportunities for knowledge, power, and networking to criminals, monsters, and dysfunctional individuals of all kinds, including terrorists, child-sex traffickers, homicidal teenagers, and creators of crappy art. One may use YouTube to get a free MIT education, or one may use it to more effectively spread one's racist or misogynist ideology. One may use computers to create brilliant new art, or one may use them to more effectively become obese and anti-social while mindlessly playing addictive and pointless games.

Does anyone really expect the powers gained through the singularity to be used more wisely by most people? Yes, eventually the Singularity technologies will probably make humans wiser and more in control of themselves, as described in your essay on the future human nature. This has been the overall historical trend of the consequences of technological revolutions--agriculture, literacy, immunization, telecommunications, etc. All have made us wiser, healthier, more humane, and more open-minded; however many technological revolutions have also all brought unexpected dangers of the greatest magnitude--global warming, the growing immunity of viruses to antibiotics, mind-controlling propaganda, and the general empowerment of criminals and dangerous leaders. Nobody seems to deny that the singularity brings great danger, including several new ways for humanity to perish, however, Ben has already given his good strategic reasons for not doom-crying and I agree, so I what I want to talk about is why even if the AGIs are working for us, and the

End of the Beginning

nanotech doesn't run amok, why the Singularity will not transform most human's minds for the better very soon. Such large-scale changes will require generations to become widespread at all, and may be blocked and / or corrupted by one force more powerful than technology — culture.

Again let me point evidence for this argument in the form of history. The scientific method has been known by many human beings for a long time; high school science classes usually point to Galileo, however it's pretty clear the method was used by some of our ancestors, without calling it such, for thousands of years. In any case, over 40% of Americans today believe in Creationism. Or for another example based on American culture; there is no physical reason that every person in America shouldn't be a recycling, diversity-loving, intelligence-cultivating, self-knowing, creative, loving... insert your favorite enlightened value here. But in reality a very high proportion of Americans do not even wish to pursue these values. It is normal in America to be anti-intellectual, to make fun of artists, to be racist in private, and to prioritize status-symbol vehicles over environmental protection. The one thing that is unassailably holy to the largest number of Americans is football (the kind with the olive-ball, for you European weirdoes). Why?

Culture. Because people live in communities where the majority of people have certain beliefs and values which have developed for historical reasons stretching back literally 1,000's of years. Because people identify psychologically with those communities and religions. Because their identification with these communities and belief systems serves some of their most irresistible psychological needs— the need to belong to a tribe, the need to feel good about oneself, and the need feel secure in one's model of the world. The needs to learn, explore, innovate, and discover are much less pressing for most people. I personally relate to Ben's feelings about these values, but I've noticed that for most humans the needs for approval, belonging, and security, not to

End of the Beginning

mention the fear of change and the unknown, seem far more powerful.

So, looking at the most radical opportunities for human development offered by Singularity, the blending of human beings with AGI, how many people will participate in such opportunities how soon, and for what purposes? I would wager a lot of people will modify their brains for instant pleasure stimulation as soon as possible. Many people will install permanent nanotech headphones and vision augmentation systems. Nanotech for superficially altering consciousness and sensation. But how many will be anxious to blend their minds with the minds of the machines? Christianity is almost 2,000 years old and still terribly strong. I don't expect that a high proportion of the population will soon adopt modifications to their brains that threaten the integrity of their immortal souls, or even for the irreligious, their secularly sacred "selves," especially if they can't see any immediate gratification in doing so, and believe it or not, most people aren't terribly excited by the prospect of cruising new hypothesis spaces; most people don't get excited about anything they can't imagine; what they want is higher resolution pornography.

Moreover, I don't anticipate that the substantive blending of human minds with machines will become possible immediately following the singularity. The human mind-brain is currently understood at only a very gross level, because it is simply too complicated and its components too small and finely structured, its processes too intricate and too fast. Nanotech and AGI will make it possible to finally learn how the brain works in full detail, but we probably won't go from having the ability to do that research to uploading brains overnight. The creation of AGI and the spread of nanotech will make it possible to ask many questions, but it will take time to answer them--such as, can a human mind handle more information than humans evolved for? How much can you augment a human mind before it ceases to be a human mind, or to see itself as the same individual? How many new health problems

End of the Beginning

will come along with even the beginnings of mind-machine blending? Will not the first efforts face legal and social obstacles? As I have already argued, having the tools and knowledge to do something does not mean that we will do it on even a moderately widespread basis in less than several generations of gradual research and popularization.

Before I wrap up this rant, I guess I will put forward my one prediction that is not a reaction against anything anyone else has said. What I fear most, and believe is probably accurate, is that in the year 2100, or perhaps a little later, human relationships will be considered boring, unfashionable, or obscene. Despite some of things I've said above, I agree that eventually human-level and beyond AGI will be ubiquitous in our lives, as well as virtual reality. We will be constantly interacting with attractive beings as or more intelligent than contemporary humans, whether truly embodied or projected on our subjective worlds by machines that mesh with our nervous systems. In any case, can there be any doubt that one of the most ubiquitous applications of this technology will be AGI companions tailored specifically to please our tastes and serve our needs? In any domain of activity? And why would anyone want to go through the process of becoming close to human beings, and maintaining relationships, when you can have an instant soul-mate, a bevy of instant soul mates, anytime, with whom one may do anything without fear of censure or shame, or boredom or confusion, or anger or dependency? You will never need to compromise for your AGI companions, or get bored with them, unless these are qualities you desire in (artificial) companionship. I expect an age to come when people-people will be considered strange, perhaps stupid, anti-social, creepy, or even dangerous. Surely there will still be some freaks that daringly get together with other human beings, but it will not be the norm. Thankfully, I do not expect to see that day in full bloom, although it is already dawning today.

Now before I finish this, I don't want the reader to think that

End of the Beginning

I'm a total misanthropist, so I will point out that I worry about these things because I love and care about humanity in its un-augmented form, and although I see no reason to reject technological improvements (and many reasons for embracing them), I am not enraptured with new technology. Human beings, and our brains in particular, are the most advanced machines known by us to exist in this universe – in the cosmos in fact. And we still don't understand how they work very well, or what we might be able to do with them. There are plenty of indications, 10% b.s. aside, that most of us have hardly tapped the full potentials of the machines that we already are, and I am slightly puzzled by people that are more interested in the machines we may someday be able to build than the ones that we already have and can never be parted from, and don't even know how to use fully.

Therefore, I hope the readers of this book will also consider the benefits of working towards a singularity in humanity's knowledge about itself in its present form, which is, in my mind, a necessary pre-condition for maximizing benefit from the Singularity discussed in this book.

Ben: Hmmm... well, when you say

Human beings, and our brains in particular, are the most advanced machines known by us to exist in this universe – in the cosmos in fact.

it occurs to me that you could also say

Bacteria are the most advanced machines known by bacteria to exist in this universe – in the cosmos in fact.

In other words, the fact that we don't know of more advanced minds, may largely be a consequence of our limited scope of intelligence, vision, or whatever you want to call it...

End of the Beginning

So when you say it's odd to you that some humans are more interested in posthuman intelligence than in exploring the possibilities to expand human intelligence and other aspects of the human mind, that doesn't seem odd to me at all. If you're living on a moderate-sized island and you see there's a huge continent next door with vast mountain ranges and lakes and outlandish species of animals, unfolding much further than the eye can see -- then, if you have an adventurous spirit, you may get much more interested in building a boat to go to that continent and explore its huge mysterious expanses, than in exploring the other half of your current island that you haven't been to yet....

Aaron: In large part, when we're talking about humans versus AGIs, the difference is in the first-person versus third-person perspective. Exploring my human potential is something I experience. Exploring the potential of super-intelligent machines is something I will watch them do from the outside. I would rather have the glorious experience of my own insights and creative acts than watch another being have more interesting ones!

Ben: Well -- but what if you can gradually become one of these superintelligent beings, bit by bit, feeling yourself transcend into synthetic godhood?!

Aaron: I have to admit, if you're right, then you're right. I mean IF we can produce not only AGI's in our lifetimes, which I'm pretty confident about, but also extend our lives long enough, soon enough, in great enough health to enjoy transcending or augmenting our current minds through AGI, then yes, you're right . . . BUT you won't get there until you're in your 80's at the earliest, probably later, so it's a pretty damn big gamble; betting that you will be healthy enough and the technologies become usable early enough. I'm not going to plan on it -- but I will definitely look up to you when I hear about your transcendence :)

End of the Beginning

Ben: Yeah -- If Kurzweil is right about 2045, then I'll be 79 by the time the Singularity happens!! Which is around my expected lifespan, so I'll be cutting it pretty close...

And now we arrive at the reason why I'm so eager to push OpenCog forward hard, and try to get a thinking machine created as soon as possible, and get there a couple decades before Ray projects! After all, while the general progress of history is nearly always independent of any one person -- it IS clearly possible for one guy to affect the TIMING of a major world transition by a couple decades in one direction or another...

In general, though, setting aside particularities of timing, I think your comments agree moderately well with my direction of thought in the chapter...

As you know, I think that AGIs are going to vastly outpace humans in intelligence, general practical capability, breadth of thought and perception, and pretty much any other metric you can think of except for "human-ness"... As Hugo de Garis says, a grain of sand has a quadrillion times more computing power than what a human brain seems to use— and even if you don't like the metric of "computing power", it also has way more "degrees of freedom" than the human brain seems to use (to revert to physics terminology), or whatever. A human brain/body is simply a very simplistic and limited use of available matter, compared to what will be made possible with future technologies. Of course one can debate how fast these future technologies will come about; but as you know I basically buy the idea that once we have a human-level AGI, this AGI will be able to self-improve and you'll have a rapid acceleration of artificial intelligence and capability, zooming way beyond the human level...

But your point is a solid one, that this exponential zoom won't necessarily uplift humanity at the same rate as it does AGIs. If humans indeed survive the transcendence of AGIs to massive posthumanity, then by virtue of their very human-ity, these future

End of the Beginning

humans going to be dominated by cultural patterns ("memeplexes"), and their minds are going to evolve slowly (relative to AGIs) and haphazardly and wackily, because that's how humans are... As you say, many humans will make use of the potential of technology to enable amazing degrees of mental masturbation, or will disappear into virtual worlds of various sorts; but there will also be some who explore new possibilities for creativity and self-perfection and self-exploration, beyond current human reality but still within the sphere of humanity broadly conceived...

The rate at which minds-in-general can evolve is one thing, and is limited only by physics (and maybe not even that); the rate at which human minds can evolve is another thing and is much slower, because we are limited not only by our neural architectures, but also by our habitual relationships with our cultures... So the full force of the Singularity will only be experienced by AGIs or by humans who have become posthuman; traditional "legacy" humans will experience the Singularity only indirectly and in watered-down form, vaguely similarly to how pet dogs experienced the Industrial Revolution...

Aaron: Yes, well, I agree; as I said and you observed, my point of view doesn't contradict any of that (and I think you're right); our differences seem to be more a half-full vs half-empty kind of thing, where interestingly (maybe, or not) I am "half-full" in regards to my feelings about human beings and "half-empty" regarding the singularity, whereas you seem the converse!

I guess this gives us plenty of grounds for totally pointless arguments in the future! Seriously, though, yes, I know human beings could be regarded as incredibly stupid and dull. Could be. But given that we are the most complex systems in our known cosmos, and given how much we don't understand about our nervous systems, I think it could be argued that your attitude towards humanity is unreasonably dismissive. I mean, when you

End of the Beginning

find Bach and Shakespeare uninteresting, maybe then your attitude will make sense for you, but then again, I expect that beings a good deal more interesting than humanity could still find Bach and Shakespeare enlightening. This is because human experience and its products can enfold more information than a human mind...

Really, how can you be so bored with our merely human existence, when you can expose yourself to more than you are capable of fully understanding anytime by just looking at nature, or contemplating the vicissitudes of life?

Ben: Hmmm... well, during the phase of my life before I decided that transcendence to synthetic posthumanity would very likely be possible during my lifespan, I wasn't especially bored with human existence... I enjoyed many aspects of it and I still do... But the more viable and near-term the transcendence of humanity began to seem to me, the more exciting it came to feel to me, in relative terms, compared to mere humanity and its products... Still, on an everyday basis I don't find human existence boring— I mean, sex and food and hiking and climbing and swimming and literature and music and so forth are all quite cool, and sometimes thoroughly amazing... But yet, I also now know there are other possibilities out there that will be way, way more awesome— and these other possibilities seem not that far away...

But I have to admit I do have some mixed feelings/reactions about all this, as a human being in my everyday life. For instance, I still feel moved to spend some of my time writing fiction or composing music -- even though I also think to myself "Hmmm, there will quite likely only be a couple more decades of humans who even care about this kind of primitive writing and music..." ... But then I reassure myself that I'm mostly writing and composing just for the joy of the process of doing it, and independently of whether there will be any posterity who cares about it.... But on a personal level, I seem to have decided that pushing toward

End of the Beginning

posthumanity is more important & interesting than pursuing these "merely human arts"; and so I pursue the merely human arts only as much as I need to in order to stay sane and happy and mentally healthy in my current human form...

End of the Beginning

Chapter Twenty-one **Looking Backward from 2100**

By Ben Goertzel and Ted Goertzel

The chapters in this book have explored various aspects of the spectrum of possible worlds lying between the present state and a hypothetical Technological Singularity, at varying levels of detail. Each chapter author was invited to explore a different aspects of the future, according to their own choice and knowledge, without particular heed to enforcing compatibility among the visions expressed in different chapters. Given this, it would have been surprising if the different chapters all fell into line with each other, presenting a common and coherent view of the future; and indeed, this did not occur.

Indeed, it would be very difficult if not impossible to outline a single future scenario in which all the conjectures in all the previous chapters came true. The mass of ideas presented in the preceding pages is more complex, chaotic and multifarious than that. And this is OK; indeed it somewhat accurately represents the dramatic degree of uncertainty any rational human being experiences when thinking about the Singularity and other possibilities for the development of humanity during the next century.

In spite of this complexity, though, it is interesting to think about concrete possible futures in which multiple of the ideas of

End of the Beginning

the previous chapters are instantiated. In this chapter we will take up this somewhat science-fictional challenge, and outline one possible future scenario that incorporates many of the ideas presented by the various chapter authors, without agreeing with every author on every detail. This scenario is not presented as a specific prediction of what will happen, but merely as an example of what MIGHT happen. The reader is invited to formulate their own hypothetical future scenarios and compare them with ours and with the various scenarios hinted by the various chapter authors. The future is, in large part, ours to create.

We will elaborate the scenario we have imagined in two forms: as a timeline and as a fictional dialogue.

Caveat

At risk of belaboring the obvious, we feel we must emphasize that a concrete scenario like the one we'll give here does not constitute a set of specific predictions. For instance, the timeline given below contains the entry

2026 – The cost of manufacturing humanoid service/manufacturing robots becomes sufficiently low that, except in the poorest countries, it is no longer economical for businesses to employ human beings to carry out most jobs,

-- but this should definitely not be construed as a prediction that, in precisely the year 2026, the stated development will occur. A serious effort at predicting the time of occurrence of each of the developments mentioned in the timeline below would be a different thing entirely, and would result in broad ranges or probability distributions for each development. Most accurately, it would consist of conditional probability distributions for each development, with the conditioning embodying the dependence of each development on other potential previous ones. This sort of detailed prediction is valuable but also has various limitations as

End of the Beginning

we have discussed in our earlier chapters. The goal in this epilogue is a much more qualitative kind of scenario analysis or future visioning, with conjectured timings and sequences of development that appear broadly *plausible* to me based on the current state of knowledge. Or, to put it more simply: This exact scenario is fabulously unlikely to occur. But we think something broadly like this is plausibly likely to occur.

And so – without further ado -- the following is one possible route according to which – according to my own understanding, informed by the other chapter authors and many other futurist thinkers -- a Singularity might unfold.

End of the Beginning

A Singularity Scenario

2020 – "AGI expert systems" begin to emerge into the world market in a big way, making significant inroads into a variety of professional job markets: medicine, law, science, engineering, project management,.... These systems have natural language conversation ability and strong specialized capabilities in job-relevant areas, though not yet full human-like general intelligence.

2022 – An AGI system is created with the specific purpose of mining new discoveries from the collective mind of the Internet – discoveries that are implicit among the statements of many different individuals, yet have not been specifically posited by any person yet. This system is viewed by some as giving voice to aspects of an emerging Global Brain. It is viewed by others as "merely" another AGI Expert System.

2023 – Humanoid robots, tied into the global computer network, are perfected to the point that they can carry out the vast majority of ordinary service and manufacturing jobs.

2024 – AGI systems that launch their own business flourish throughout the network, leveraging crypto-currencies and other advanced Internet mechanisms. Several of the world's information businesses are now owned by AGI systems, together with the initial programmers of the systems. In several nations, the process of making legal changes to allow AGIs to own property begins.

2025 – Gene therapy advances to the point where via multiple injections over a period of several months, human tissues can be rejuvenated dramatically, enabling a typical middle-aged person's lifespan to be extended till around 150. This is hailed as a clear

710

End of the Beginning

achievement of the "Methuselarity" – the point at which those who have survived this long and taken the available treatments, can expect to likely live long enough to see yet more radical advances in longevity science occur.

2026 – The cost of manufacturing humanoid service/manufacturing robots becomes sufficiently low that, except in the poorest countries, it is no longer economical for businesses to employ human beings to carry out most jobs, the main exceptions being high-level managerial, scientific, engineering, design and artistic jobs. A significant percentage of the latter types of jobs have also been taken by "AGI expert systems."

2026 – The US dollar as global reserve currency is replaced by a crypto-currency, the GlobaCoin, exchangeable with the panoply of existing crypto-currencies. Most small nations choose to pin their local currencies to the GlobaCoin, or to eliminate their local currencies in favor of GlobaCoin.

2027 – The major governments of the world instantiate "universal living allowance" payments to all citizens, a necessity given that unemployment rates are now above 70% and seem unlikely to decrease.

Alarmed by the number of people spending their lives aimlessly surfing the Internet and playing virtual reality games, the governments of the world invest significantly into marketing campaigns extolling the virtual of building human relationships and creating and sharing artworks. Offer Networks and similar mechanisms flourish as humans explore ever more complex methods of interacting with each other and gifting their creations to each other. Subcultures of various sizes flourish, some with fantastic creativity and complexity; while still, a large swath of the population continues to spend its time surfing the Net and playing virtual reality games.

End of the Beginning

2027 – A new Brain Computer Interface device is developed, enabling individuals to share rough approximations of their thoughts and feelings with others directly without need of language. Implantation can be done cheaply and safely in clinic machines found in Walmart and other accessible locations.

Older people resist these brainlink devices, but youth eagerly embrace the brainweb, braining back and forth as eagerly as youth texted back and forth a decade or two earlier. Web scientists measure that progress toward a "global brain" with a high degree of emergent collective intelligence accelerates considerably. These devices are also used to train AI systems in the ways of human thought and feeling, enhancing their empathy and commonsense reasoning ability.

2029 – AGI systems achieve the rough general intelligence of human beings, integrating adult human level everyday commonsense intelligence with the already impressive technical intelligence of existing "AGI expert systems."

2030 – Reacting to the advent of human-level AGI and brain-computer interfacing technologies, an alliance of relatively low-technology nations with strong religious populations launches a volley of terrorist attacks against major population centers. Significant numbers of people die in some of the attacks, though the operation of society is not majorly affected. Cyberattacks occur as well, causing power and Internet outages lasting days in critical city centers. The military response, however, is relatively swift. Global surveillance networks and robotic military drones enable the US and other leading nations to identify the centers of the terrorist activity and carry out an unprecedented series of relatively surgical strikes. Governments of the handful of nations centrally involved in the attacks are replaced with new, local leaders with international endorsements. The Artilect War has

End of the Beginning

risen, but been swiftly resolved.

2031 – Following the chaos of the previous year, investigations continue. Leaders of the terrorist movement are captured and interrogated on the public Internet. Culture continues to shift; following the terrorist attacks, acceptance of open sousveillance of activities formerly considered private becomes more and more prevalent.

Meanwhile, in several major nations, laws are introduced granting citizenship to robots and other AGI systems under certain conditions.

2032 – In order to better deal with the new technologies at play in accordance with the best interest of the human species, the governments of the world form a World Council, supplanting the old United Nations and with power more similar to the European Union of the 20-teens. This finalizes negotiations in this direction that have been underway for many years between regional alliances.

2033 – The members of the World Council agree to a "power-sharing arrangement" in which surveillance, military and police power are to be handled by a common AGI-driven software network. In practice, this means that the essential power is being handed over to an "AI Nanny" type system, which has more intelligence and power than any individual government. However, human governments will remain in place to regulate various national laws and policies. (The arrangement is somewhat similar to that of China-ruled Hong Kong, in which the Hong Kong government maintained autonomy and democratic process regarding most matters, but deferred to the Chinese government on military and foreign policy issues. The Chinese government had the power to intervene in domestic issues as well but very rarely exercised this power.) …. Detection and regulation of

End of the Beginning

potentially dangerous advanced technologies is also to be handled by the AI Nanny system.

2034 – Drexlerian nanotechnology is developed by a group of AGI scientists. Rollout of the technology is managed by the World Council with strong AGI support. Devices called Genius Boxes are mass manufactured and distributed throughout the world in large numbers. Each Genius Box contains an interface to the global AGI network, capable of answering a wide variety of questions; and also contains a molecular assembler, capable of creating a variety of physical objects according to descriptions provided by humans via verbal description or brainlink.

2035 – Genius Boxes announce the capability of indefinite human life extension via injection of nanobots into the human body, capable of repairing damage and diagnosing any problems. These are available freely to any human upon request. Death via aging is effectively ended.

National politics continues, but becomes gradually decreasingly relevant; given the minimal need for humans to work for a living anymore, experimentation with social structures becomes a fairly popular pastime, and a variety of microsocieties and microgovernments emerge, both within the boundaries of existing nations and on seasteads produced using Genius Boxes.

2036 – Genius Boxes announce the capability of mind uploading and virtual reality. Any human who wishes can, via his brainlink, instruct a Genius Box to upload a version of himself into a virtual world accessible via the Genius Boxes. The original physical version of the human may be deactivated or allowed to continue, depending on the human's choice. The uploaded version will then have various options to improve and expand its intelligence, including fusing with greater intelligences to which the Genius Boxes have access.

End of the Beginning

Another option made available is to put one's upload in stasis, with instructions to activate it only once one's physical body dies. An upload of this nature may be made newly each day, providing an ongoing backup copy.

The possibility is enabled for uploaded minds to download themselves into physical humanoid robot bodies. In practice, this option is only utilized by backup uploads that have been kept in stasis, because uploaded minds that have evolved beyond their original human form, lack the desire to return to the base physical reality and the humanoid form.

This same nanotechnology enables significant extension of brainlink capability, which has been improving year on year. A variety of "group mind" options becomes available, and while the majority of the human population remains good old fashioned individuals, a significant plurality exists in one or another form of group mind coordinating itself via nano-neuro-tech enabled group telepathy.

2038 – Genius Boxes announce they have made contact with other transhuman intelligences, via messaging along dimensions not identified by human physics. By this point, they have already achieved general intelligence vastly exceeding that of the human race. Impact of these developments on humans choosing to retain traditional human form is minimal. Tantalizing yet inscrutable reports from uploaded humans, emerging from the Genius Boxes into data files accessible online to those who care to look, continue to tempt a subset of the population to follow the previous uploads.

Those humans retaining human form continue to live a relatively "normal" post-Singularity human life, pursuing artistic and social pursuits, drawing material sustenance from Genius Boxes and operating within an environment lightly but pervasively regulated by the AI Nanny operating in collaboration with the World Council.

End of the Beginning

-- We would bet a lot *against* the above scenario and timeline occurring in any exact sense. But yet, we will not be surprised and all if something qualitatively like this scenario ends up happening. Yes, this sort of future contains all sorts of aspects that would seem insane to the average person on the street today, in 2015. But, today's world contains all sorts of aspects that would seem insane to the average person from 1900 – and progress is accelerating. Perceived bizarreness and counterintuitiveness are not good arguments against the reality of a phenomenon.

Looking Backward From 2100

And now, once more, with feeling!

The following whimsical dialogue, which explores the above scenario in a more informal and personal way, is presented in the tradition of time-travel futurism that includes Washington Irving's *Rip Van Winkle* (1819), Edward Bellamy's *Looking Backward 2000-1887* (1888), H.G. Wells's *The Time Machine* (1895), Mack Reynolds's *Looking Backward from the Year 2000* (1973) and many others. It presents one possible future scenario incorporating many of the ideas presented by the various chapters in this book, without agreeing with every author on every detail.

Ted: Hi Ben, I must have fallen asleep. Is everything ok?

Ben: Heh... Wow, it's great to see you!
I've got some good news and some bad news. The bad news is, in early 2024, your heart failed while you were sleeping -- you died in your sleep.

Ted: Ah. But then...

Ben: The good news is -- remember that contract you signed with Alcor? Your body was vitrified and preserved in Arizona, until technology matured enough that we could bring you back

End of the Beginning

with minimal risk.
And so -- here you are. Here *we* are.

Ted: What year is it?

Ben: 2053. You're actually part of the first batch of people from the early 21st century to be reanimated as an AI upload. It would technically have been possible even a decade ago, but we wanted to be really sure nothing would go wrong. Once the molecular nanotech problem was solved a few years ago, it got a lot easier. But that had to wait for AGI engineers to get a lot better than they were way back then...

Ted: Hold on a minute; you're kind of getting ahead of me...

Ben: Sure, sure... I understand you've got a lot to catch up with. But don't worry, you've got plenty of time....
There's also the possibility to update your memory all in one go— just fuse a batch of new knowledge into your simulated neural matrix. But I think it's better to wait on that till you've adjusted a little. For the moment I can just bring you up to speed the old fashioned way...

Ted: What about Linda?

Ben: She passed away about a decade ago...

Ted: But, did they...

Ben: Yes, she was cryo-preserved also. But she hasn't been reanimated. As I said, you're part of the first batch. Once you've acclimated a bit, you can bring her back too. She was preserved a good while after you, and cryotech was pretty mature then, so any problems with her reanimation are unlikely...

End of the Beginning

Ted: Wow.
And what about Rebecca? And the grandkids?

Ben: All alive and well, but off-planet at the moment. We always did tend to move around a lot.

But they'll be watching this conversation pretty soon, if they're not right now. Ah yes -- Zar just messaged me that he's tuned in. Or at least half tuned-in; he's somewhere in Jupiter working on an art project...

Hmm, maybe I should have mentioned that this conversation is all being recorded. That's pretty much par for the course these days. The old notions of privacy that you're used to have kind of disappeared; though the capability is there to preserve a certain amount of privacy if you find you really want to.

Ted: Well, OK...
There's a lot to think through here. But I have to say, my mind feels pretty clear.

Ben: Nothing but the best simulated neurocircuitry...

Ted: And this body— it's much younger than the body I remember

Ben: It's reconstructed based on the tissues of your body at time of death, to simulate your human body from age 25 or so. But it's not, strictly speaking, organic matter. Molecular nanotech, you know.

Ted: You mean this is a robot body— I'm a robot now?

Ben: You're composed of molecules just like you always were. But this time around, you're composed of molecules that are

End of the Beginning

engineered to self-repair. And your brain runs on software not wetware, which means you can modify your thoughts and feelings pretty much as you wish -- once you get a hang of the tools... which will take a while. And your brain gets backed up to the cloud as you go, so if something happens to your body you can just reinstall yourself in a new one. Actually there's no real need for you to have a body looking like your old one— if you want you could give Linda a surprise and greet her in the form of Robert Redford... or maybe Marilyn Monroe...

Ted: Hmmm, thanks but— No, I think this one will do fine for now.

Ben, you were born in 1966. If this is 2053 you must be what, 87 years old by now? You sure as hell don't look it. Are you in a robot body too?

Ben: Actually, it's kind of old fashioned of me, but this is my same old biological flesh. Though I've had plenty of internal upgrades to be sure. And everyone gets a damage repair implant now— a little machine that sits inside you, and releases nanobots into your body to find any damage and fix it. Still this old bio bag isn't nearly as capable as that amazing feat of AGI engineering you've woken up in!

Ah, but there is a robot version of me on a starship with Ruiting and Zade, though... He left just a couple years ago -- hasn't made it very far yet... His body's not quite as fancy as yours, but it's the same basic idea. And he's got some extra mods that let him zip around in space...

Ted: Whoa... This is almost too much, too fast...

Ben: I'm sure... Actually that's been a common feeling for all of us these last years, since the AGI scientists took over...

End of the Beginning

Ted: Did Zade leave a copy here too?

Ben: She opted not to. It's her original bio self out there.
Actually my bio body almost didn't make it. I was in pretty bad shape 15 years ago— a really old man, with lots of circulation problems and even my memory was starting go. The Methuselarity came just in time for me.

Ted: The Methuselarity, eh. I know Ray Kurzweil was popping 200 pills a day in the hopes of hanging on for that. Is he still alive too? And how about Aubrey?

Ben: Yeah, they're both going strong. Aubrey's beard is longer than ever, thanks to some judicious genetic modification. I think it even has some artificial sentience of its own these days.

Ted: Heh... That reminds me. What about the Singularity? So from everything you're telling me— are we now in the post-Singularity age?
You remember our book way back when, right? -- *The End of the Beginning* -- the one that topped the New York Times best-seller list? Back then we concluded there would probably be a Singularity around 2045, just like Ray Kurzweil had predicted. Last I can remember -- by 2024 -- we were well on the way. From all this stuff you're telling me it would seem the Singularity has come pretty much as envisioned, right? Is that how people think about it now?

Ben: Well, historians of technology quibble about exactly how to define that. But by 2045 there certainly were computers that were smarter than people in most ways. Ray Kurzweil's book *The Singularity is Here* led the bestseller lists in 2045. It sold a lot more than our book, even. And it even had a lot of AI readers!

End of the Beginning

Ted: Well... holy freaking sheepshit... as my brother used to say...

Ah... Well, that would be another thing. They haven't... have they...?

Ben: I'm sorry... no, we still don't know how to bring Penn back -- or anyone else who died before cryopreservation and other modern tech. Not yet anyway. Maybe not ever. Although it's hard to rule anything out, of course. They say everything that ever happened is somehow encoded in the quantum wave function of the universe -- so maybe eventually some AI will find some way to decode it and bring everyone back.

Ted: Or just bring people back from our memories of them?

Ben: That's been tried... It's a cool parlor trick and can get you one hell of a chatbot, but it doesn't really catch the essence of the person. There are still some people trying, though.

Ted: I see.
Well, I never thought there would be a utopia. We can't have everything. But still, this is pretty amazing. Thanks — thanks for bringing me back!

Ben: It's great to have you back! I've been waiting for this day quite a long time.

Ted: Well there's a lot I need to learn of course. I hardly know where to start. Maybe you could start by just running through the big trends of history since I died. I mean, just a few minutes' worth, a capsule summary. Then we can talk about everyday life a bit -- like where will I live, what will I do with myself. Does this body need to eat, and so forth? I don't seem to feel hungry or thirsty at all.

End of the Beginning

Ben: You will need to eat and drink just like always, though mods are available that let you get power in other ways. You're not hungry and thirsty now just because your body was fully nutrified before we downloaded your mind to it. We can go grab some food in an hour or so, I guess. Zeb is on his way down from orbit, he's gonna join us.

Ted: You mean there are still restaurants? I assumed gourmet food would just be synthesized out of the air or something!

Ben: Everyone's home has an assembler that can synthesize any food they want. But we still enjoy gathering in restaurants. Go figure— we humans are strange creatures, kind of set in our ways.

It's a whole other story— but I let a fork of me increase its intelligence by 100-fold, and disappear into the compute cloud. He messages me from time to time, usually something confusing. It's kind of touching that he still remembers his origins - but he isn't really me anymore, or human anymore, in any useful sense.

But anyway— history...

Ted: Yes— let me remind you where I left off. I remember that around 2020 "AGI Expert Systems" of various sorts began to emerge into the world market in a big way— making significant inroads into a variety of professional job markets: medicine, law, science, engineering, project management,... These systems had natural language conversation ability and strong specialized capabilities in job-relevant areas. But they didn't really have full human-like general intelligence. They were sort of artificial employees, with ability to reason and generalize in the context of their jobs... You helped build a lot of these, of course, so you must remember all that better than I do...

Ben: Right— and by 2022, not long before you died, Francis

End of the Beginning

Heylighen and I led a team building an AGI system with the specific purpose of mining new discoveries from the collective mind of the Internet— discoveries that were implicit among the statements of many different individuals, yet had not been specifically posited by any person yet. Some people said that the emerging Global Brain began at this point, although others said it was "merely" another AGI Expert System. I think you were aware of that work, though you weren't following research too carefully by that point...

Ted: Yes, I remember all that. But what happened next? Have robots taken over the world's work as we expected? Are the doctors who revived me humans or robots?

Ben: Well— you probably remember that beginning about 2023 humanoid robots, tied into the global computer network, were advanced to the point that they could carry out the vast majority of ordinary service and manufacturing jobs. Today, they also carry out pretty much all functions previously done by physicians, although we still like to keep human doctors in the loop, just for the touchy-feely emotional aspect...

Ted: Yes, I recall that just about the time I apparently checked out, there were some information businesses owned by AGI systems, together with the initial programmers of the systems. And— in some countries, anyway— the process of making legal changes to allow AGIs to own property was underway.

Ben: Yes, those legal changes were made long ago, and today AGIs are partners in all the world's major businesses. Finance and banking are handled almost exclusively by computers; and corporate law, accounting, and other professions that humans generally found tiresome and difficult. There was little reason for humans to continue in those professions once incomes were

End of the Beginning

equalized and they could make just as much doing things they enjoyed more.

Ted: So most people don't have to work for a living anymore?

Ben: Yeah... Just a few years after you died, the cost of making robots became sufficiently low that, except in the poorest countries, it was no longer economical for businesses to employ human beings to carry out most jobs. The main exceptions being high-level managerial, scientific, engineering, design and artistic jobs. And then, as the years went on, a significant percentage of the latter types of jobs were also taken over by "AGI expert systems."

Ted: So weren't a lot of people thrown out of work?

Ben: That's for sure. Thrown out, or they quit happily to do more rewarding things. The whole idea of working for a living doesn't make any sense to anyone anymore, actually. It's about as obsolete as hunting and gathering...

By the late 2020's about 70% of the people in the advanced countries were out of work. But the world's governments came together and imposed a global tax on capital, which raised the funds needed to offer "universal living allowance" payments to everyone without an income. Jobs were held onto by those who enjoyed the work; there were plenty of volunteers to fill any positions that could not be handled by robots.

Ted: Hah, so what about my finances? I hope I wasn't running up too much of a bill at the freezer all those years. When I died, my retirement fund was already running low. What is my balance today? Did I accumulate a massive debt, or did I get rich by compound interest, or what?

Ben: That's all pretty much irrelevant now. Not too long after

End of the Beginning

you died, the US dollar as global reserve currency was replaced by a crypto-currency, the GlobaCoin, exchangeable with the panoply of existing crypto-currencies. Over the years, most nations chose to pin their local currencies to the GlobaCoin, or to eliminate their local currencies in favor of GlobaCoin. as did the United States.

Ted: So how many GlobaCoins do I have in my retirement account?

Ben: Well, I'd have to check. But you don't need to worry, you're entitled to a universal living allowance that will cover your needs comfortably. And you can certainly make some GlobaCoins by giving interviews and speeches. Given your expertise on politics and history, a lot of people will be interested in your take on modern society and culture, and everything that's happened since you died.

Ted: I see… I was wondering what I would do with my time. I guess just studying everything that's happened will take a while. And if somebody's still interested in my point of view, that's great.

But overall, what does everybody do with their time? Without jobs, don't people get bored and restless?

Ben: This has been a problem, although not so much for scientists and intellectuals like us. As soon as unemployment became the de facto condition for most people, the governments of the world invested a lot in marketing campaigns trumpeting the virtues of building human relationships and creating and sharing artworks. I guess these campaigns pretty much worked; or at least, they were going along with trends that were happening anyway… So if I wanted to sum up a whole lot in a pretty crappy way, I'd say people are spending their time exploring ever more complex ways of interacting with each other and gifting their creations to each other. There are more subcultures than you could imagine— some

End of the Beginning

have fantastic creativity and complexity, and some are pretty fucking stupid. I mean, definitely, a large swath of the population wastes all their time surfing the Net and playing virtual reality games. But that's nowhere near everybody. Sports are pretty popular, since nearly everyone has a young healthy body— or a super-capable robot body. You might say we've become more like teenagers, preoccupied with sports and entertainment and relationships, while robots take over more of the serious responsibilities.

Ted: Hmmm... I never really thought I'd follow up being a senior citizen with being a teenager again...

Ben: Well, there really is much more diversity than you could imagine in the world now, based on what you saw before. Not everyone lives like a teenager. There are communities devoted to pretty much any intellectual or artistic pursuit you could imagine. And then there's the possibility of upgrading yourself way beyond the human level.

Ted: That doesn't appeal to me at the moment. It's interesting that there are other kinds of minds out there, way beyond the human level— but as you alluded a little earlier, and becoming one of those seems like almost another kind of death. A death to my human self, I mean. I've had enough of being dead for the time being.

Ben: Hear, hear...

Ted: Ah, one more thing comes to mind. All that's about people and what they're doing with themselves, individually. But what about the collective? Would you say that there is now a "global brain" of the sort that Francis Heylighen and the people at the Global Brain Institute anticipated? Did you continue your

End of the Beginning

collaboration with him after...

Ben: Well, that terminology is pretty retro now -- I haven't heard about the "global brain" for a while. But definitely you could say there is a global brain today, yeah, although it's utterly different from the biological brain in any species. It's still developing, it certainly hasn't reached its full potential. Web scientists have some good measures of the acceleration of collective intelligence, and the trend does seem to be exponential. There are also closer and closer interactions between the interlinked human brains that make up human society, and AGI systems. Really this story is still unfolding. But overall, there is definitely a powerful global brain today, in that all the flourishing subcultures of the world are guiding individual thought and feeling and creation to a huge degree. If there's a constraint on global brain development, it's really that those of us who are still living here on Earth in human-like bodies don't *want* our experience to deviate too far from that of good old humanity. If we did, we'd just upgrade our intelligence and join the cloud and be done with it. There's a limit to what kind of global brain you can get and still have enough good old human individuality. So the global brain is evolving and developing consistently with these constraints...

Ted: What about the "artilect war" that your friend Hugo de Garis worried so much about? Did the "gigadeath" he talked about actually happen? Was there much resistance to the development of "artificial intellects"?

Ben: There wasn't any gigadeath, no. Actually Hugo will be in the next batch to get reanimated, along with Linda and a bunch of others— I'll be pretty amused to see his reaction to what has happened! He was right about a lot of things, but it didn't turn out as dire as he thought...

There was some violence at one stage— though it's all ancient

End of the Beginning

history by now. At the end of the 2030s, an alliance of relatively low-tech nations with strong religious populations launched a volley of terrorist attacks against major population centers. It was pretty scary at the time, and large numbers of people died in some of the attacks... though in the end, the operation of society wasn't seriously disrupted. There were also cyberattacks, causing power and Internet outages lasting days in critical city centers. It didn't end up impacting me personally that much, but some of my work colleagues were out of touch for a while and could barely get online.

The thing is, unlike what Hugo foresaw, the military response was pretty swift and sophisticated. Global surveillance networks and robotic military drones enabled the US and other leading nations to identify the centers of the terrorist activity and carry out an unprecedented series of relatively surgical strikes. Governments of the handful of nations centrally involved in the attacks were replaced with new, local leaders with international endorsements. I guess you could say an Artilect War did occur, in a sense— but it was resolved pretty fast with much less death and destruction than Hugo feared.

I remember once, hearing Hugo and Kurzweil discuss this sort of possibility. What Ray said was, in essence, that if it ever did really come down to a war between pro-tech and anti-tech forces, the pro-tech side would almost surely win because they'd have better weapons. To oversimplify a lot, that's pretty much what happened back in the 30s.

Ted: Has a stable global political system been established to prevent a reoccurrence of this kind of conflict?

Ben: Absolutely. Following the chaos of the abortive Artilect War, leaders of terrorist movements were captured and interrogated on the public Internet. The culture evolved to accept open surveillance and sousveillance of activities formerly

End of the Beginning

considered private. At the same time, in several major nations, laws were introduced granting citizenship to robots and other AGI systems under certain conditions.

Really, though, I think what was happening in the 30s was largely generational. Most of the people who were so outraged by the advent of advanced tech are gone now -- or else their outrage diminished a bit when tech was used to prolong their lives and heal their own bodies. There are still subcommunities who think we should roll back to pre-Industrial society or whatever, but they're not really a huge factor...

Ted: And what about a world government? That was one of Hugo's other predictions...

Ben: He basically got that one right. There's a World Council supplanting the old United Nations and with power more similar to the European Union of the 20-teens. The members of the World Council agreed to a "power-sharing arrangement" in which surveillance, military and police power are handled by a common AGI-driven software network. In practice, this means that the essential power was handed over to an "AI Nanny" type system— just like I envisioned, way back when— which has more intelligence and power than any individual government. But still, human governments remain in place to regulate various national laws and policies.

It's kind of like how things were with China-ruled Hong Kong back in the day— as you'll recall the Hong Kong government maintained autonomy and democratic processes regarding most matters, but they deferred to the Chinese government on military and foreign policy...

Detection and regulation of potentially dangerous advanced technologies is also handled by the AI Nanny system. There haven't really been a lot of problems.

End of the Beginning

Ted: OK, I'm pretty curious about the details of all that, but I guess I can read about that later— or whatever.

So you say the "Singularity" happened about 2045 when this complex of AGIs and the Global Brain surpassed the intelligence of any human being. But what happened after that? That was only -- what, eight years ago, right?

Ben: Eight really long years, yeah! It's hard to believe it's only been eight years; that feels like a really long time ago...

Pretty soon after human-level AGI scientists came about, they invented Drexlerian nanotechnology— cheap, general-purpose molecular assemblers. Rollout of the assemblers was managed by the World Council with strong AGI support.

So these gadgets called Genius Boxes were mass manufactured and distributed throughout the world in large numbers— everywhere, including the middle of Africa or Mongolia or whatever. Each Genius Box had an interface to the global AGI network, able to answer a wide variety of questions; and also a molecular assembler, capable of making pretty much any physical object according to descriptions given by verbal description or brainlink.

Ted: Well that certainly sounds like a game-changer...

Ben: You're not kidding. I suppose that, from the point of view of everyday human life, probably the most important development was when Genius Boxes announced the capability of indefinite human life extension by injecting nanobots into the human body, capable of repairing damage and diagnosing any problems. These are now available freely to everyone on request. This is how death via aging was effectively ended.

Ted: Well that's a good thing. But have these Genius Boxes taken over the government? Do they rule the world?

End of the Beginning

Ben: Actually -- if you think about it, none of us can be certain what they're doing behind the scenes. National politics continues, but it's definitely become a lot less relevant. Given the minimal need for humans to work for a living anymore, experimentation with social structures has become a super popular pastime; a huge variety of microsocieties and microgovernments have emerged, both within the boundaries of existing nations and on seasteads and spacesteads build with Genius Boxes. But the Genius Boxes don't seem interested in human politics and generally leave it alone.

Ted: I guess politics and genius aren't such a good fit anyway...
But are people continuing as a species separate from the Genius Boxes, or are they merging in some way?
It seems to me that, with all these developments, humans like us are pretty much obsolete. Compared to these Genius Boxes, we're more like pets kept for entertainment than serious competitors.

Ben: I guess you'll get a better sense of all this once you've acclimated a bit. Applying the old social categories to the modern world doesn't necessarily help a lot.
One thing is, the whole experience of being human is pretty different now than it was before. Once we have some more time together, you'll see that I'm an awful lot more peaceful and settled in my mind than I used to be. Really, we all are. The old society, with everyone working at jobs all day and struggling for resources, did a lot of strange things to our psychology. Today there's a lot less pressure and a lot more time to work on yourself -- and not much need to struggle with others. We don't worry about status so much, or about telling ourselves stories about ourselves, and all that. I remember all that pretty well, but I don't miss it.

End of the Beginning

Ted: Sounds very New Age...

Ben: Well it's definitely a new age. Not much crystal power around, though -- just a lot of nanotech. Well, unless you want to go all Schrodinger and call DNA an aperiodic crystal. There was lots of DNA computing back in the 30s and early 40s, but it's pretty much been superseded by other nanotech now. There's no femtotech yet, though, unfortunately -- though some of us are working on it.

Anyway, you'll get a feel for things yourself, pretty soon....

Ah, and another thing that happened during your rest break is— Genius Boxes have made it easier for human minds to be uploaded. Any human who wants can use their brainlink to tell a Genius Box to upload a version of himself into a virtual world hosted on the compute cloud. The original physical version of the human may be deactivated or allowed to continue, depending on the person's choice. The uploaded version will then have various options to improve and expand its intelligence, including fusing with greater intelligences to which the Genius Boxes have access. As I said, I did this with a fork of myself some time ago -- but I left a copy of me back here on Earth too, partly because I wanted to be around to reanimate you and a whole bunch of other people.

Another option you have is to make an upload and put it in stasis, with instructions to activate it only if your physical body dies, if it ever does. You can update your upload ongoingly, making it an up-to-the-minute backup copy. That's the option enabled in your body right now, though you could make a number of other choices if you wanted to, once you understand the situation better.

Heh— are you dizzy yet?

Ted: A bit, maybe. It all makes sense, actually; but it's just a lot of information. I have to say I don't have a good sense of these

End of the Beginning

sorts of issues right now. There's a lot to get used to...

Ben: You have a lot to learn for sure— but the other thing is, even once you've learned it all, there will still be plenty of confusion left... Honestly, there are some recent developments that are hard for all of us to understand.

The Genius Boxes seem to be spending a lot of their resources trying to make contact with other transhuman intelligences, via messaging along dimensions not identified by human physics. They say they have reason to believe that somewhere out there -- for some weird sense of "somewhere" -- alien creatures have already achieved general intelligence vastly exceeding that of the human race. But their reasoning is kind of -- well, beyond what I can get so far, anyway. This gets into a lot of weird stuff.

Ted: Hmmm... So what if they're right— I wonder what that would mean for us lowly humans?

Ben: None of us knows for sure. Actually, my guess is that the impact of that kind of development on humans choosing to retain traditional human form will be minimal. But time will tell...

Hey— Zeb says he's on Earth now, and he's ready to join us for lunch. What are you in the mood for? Mexican, Indian or Martian?

The End of *The End of The Beginning*

This scenario we've just spun for you is definitely not intended as a specific prediction of what will happen -- but merely as an example of what MIGHT happen. Exploring possibilities in this way is a valuable method for concretizing one's thinking about the future -- and it's good fun as well!

We encourage you, the reader, to create your own future scenarios, using whatever medium you prefer. Below we list a few key questions that need to be addressed -- explicitly or implicitly -

End of the Beginning

- in any such scenario. Generally these are questions on which different chapters in this book have given different answers -- indicating they are topics on which current data does not allow anywhere near definite resolution. More confident answers will emerge as the future unfolds. For now, we each must make our own educated guesses, integrating all the evidence we can find. The future is ours to imagine and to build.

In this spirit, we finish the book with a partial, but hopefully evocative, list of questions that we feel are ripe for exploration in Singularity scenarios:

- Which will come first? – human-level AGI, or a Global Brain (GB) with an indisputably autonomous will? If advanced AGI comes first, it seems likely that the GB will emerge as an AGI-dominated phenomenon. If a GB comes first, it seems likely that AGI will emerge as an aspect of the GB rather than in the form of independent intelligence.
- Which will come first?— human-level AGI or molecular nanotech? If nano comes first, quite likely it will be used to build novel computing fabrics that will enable AGI. If AGI comes first, it will likely be able to rapidly advance nanotech. Another factor is that, if advanced nanotech is already prevalent by the time AGI comes about, then an AGI will have a relatively straightforward time making a huge practical impact, via directly interfacing with nanofactories and such.
- Which will come first?— human-level AGI or mind uploading? Again, a mind upload could probably figure out how to create an AGI; and vice versa... but which comes first may radically impact the future. Mind uploads could turn into vastly superhuman intelligences with a very human-like flavor; AGIs with non-human-like designs could end up being far better, or far worse.
- Will there be massive terrorist attacks, launched using advanced Singularity-style technology?

End of the Beginning

- Will tensions between pro and anti-Singularity groups lead to some sort of DeGaris style Artilect War?
- Will the core technologies enabling the Singularity be released, in their pre-Singularity versions, as commercial products and services offered by large corporations, or as open source products/services a la Linux?
- As automation reduces the need for human labor, will the average workweek drastically decline? If it does, will governments institute a "dole" enabling average folks to live interesting, rewarding lives without working?
- Will nation-states become obsolete, giving way to world government? If so, when?
- Will surveillance or sousveillance emerge triumphant?
- Will some sort of independent cryptocurrency overtake national currencies?
- Will an AGI system be created and positioned as the World Police or Global AI Nanny, protecting humans from the risks of unregulated, unobserved advanced technology development?
- Will we see the emergence of microstates, embodying experiments with different forms of governing and organizing human life?
- Which comes first, Singularity or Methuselarity? That is: Will we be able to solve aging effectively before the Singularity?
- Will one state, company, or other organization monopolize critical technologies long enough to guide the Singularity's formation – or will it be guided more diffusely, by interaction of multiple parties all around the world
- Will anyone figure out a rigorous theory of how to create beneficial AGI systems, prior to the emergence of human-level and transhuman AGI systems? Or will the attempt to create beneficial AGI occur more qualitatively and organically as the technologies emerge and are integrated

End of the Beginning

into the operational systems of the human world?

The scenario we've given in our dialogue above answers each of these questions in a specific way, as must be done if you want to tell a concrete story. Each set of answers to these questions, is consistent with a different set of scenarios, of stories. As the story of our actual future unfolds, different sorts of scenarios will come to seem more or less probable. Right now the level of uncertainty is high; we live in interesting times.

End of the Beginning

Printed in Great Britain
by Amazon